Railroads 2018

Transportation Research Record: Journal of the Transportation Research Board Volume 2672, Issue 10

ISBN: 978-1-5108-8179-2

Printed from e-media with permission by:

Curran Associates, Inc.
57 Morehouse Lane
Red Hook, NY 12571

Some format issues inherent in the e-media version may also appear in this print version.

Published by Sage Publications USA

ISBN (Print) 978-1-5108-8179-2
ISBN 2018 Printed Set (All Issues) 978-1-5108-7735-1

Printed by Curran Associates, Inc. (2019)

Additional copies of this publication are available from:

Curran Associates, Inc.
57 Morehouse Lane
Red Hook, NY 12571 USA
Phone: 845-758-0400
Fax: 845-758-2633
Email: curran@proceedings.com
Web: www.proceedings.com

Volume 2672 Number 10 December 2018

TRR
JOURNAL OF THE TRANSPORTATION RESEARCH BOARD

Contents

Railroads

Articles

Transportation Officials, Washington, D.C.
K. Jane Williams, *Acting Administrator, Federal Transit Administration, U.S. Department of Transportation*
Paul Zukunft *(Admiral, U.S. Coast Guard), Commandant, U.S. Coast Guard, Washington, D.C.*

---

* Membership as of February 2019.

TRB

TRANSPORTATION RESEARCH BOARD

CORE SPONSORS

Transportation Departments of the 50 States and the District of Columbia

Federal Government

U.S. Department of Transportation
- Federal Aviation Administration (FAA)
- Federal Highway Administration
- Federal Motor Carrier Safety Administration (FMCSA)
- Federal Railroad Administration (FRA)
- Federal Transit Administration
- National Highway Traffic Safety Administration (NHTSA)
- Office of the Assistant Secretary of Transportation for Research and Technology (OST-R)

U.S. Air Force Civil Engineer Center (AFCEC)
U.S. Coast Guard (USCG)
U.S. Department of Energy
U.S. Department of the Interior (DOI)
U.S. Environmental Protection Agency (EPA)

Nongovernmental Organizations

American Public Transportation Association (APTA)
Association of American Railroads (AAR)
California Air Resources Board

The National Academies of
SCIENCES · ENGINEERING · MEDICINE

The **National Academy of Sciences** was established in 1863 by an Act of Congress, signed by President Lincoln, as a private, nongovernmental institution to advise the nation on issues related to science and technology. Members are elected by their peers for outstanding contributions to research. Dr. Marcia McNutt is president.

The **National Academy of Engineering** was established in 1964 under the charter of the National Academy of Sciences to bring the practices of engineering to advising the nation. Members are elected by their peers for extraordinary contributions to engineering. Dr. C. D. Mote, Jr., is president.

The **National Academy of Medicine** (formerly the Institute of Medicine) was established in 1970 under the charter of the National Academy of Sciences to advise the nation on medical and health issues. Members are elected by their peers for distinguished contributions to medicine and health. Dr. Victor J. Dzau is president.

The three Academies work together as the National Academies of Sciences, Engineering, and Medicine to provide independent, objective analysis and advice to the nation and conduct other activities to solve complex problems and inform public policy decisions. The Academies also encourage education and research, recognize outstanding contributions to knowledge, and increase public understanding in matters of science, engineering, and medicine.

Learn more about the National Academies of Sciences, Engineering, and Medicine at **www.national-academies.org.**

The **Transportation Research Board** is one of seven major programs of the National Academies of Sciences, Engineering, and Medicine. The mission of the Transportation Research Board is to increase the benefits that transportation contributes to society by providing leadership in transportation innovation and progress through research and information exchange, conducted within a setting that is objective, interdisciplinary, and multimodal. The Board's varied activities annually engage about 7,000 engineers, scientists, and other transportation researchers and practitioners from the public and private sectors and academia, all of whom contribute their expertise in the public interest. The program is supported by state transportation departments, federal agencies including the component administrations of the U.S. Department of Transportation, and other organizations and individuals interested in the development of transportation.

Learn more about the Transportation Research Board at **www.TRB.org**.

TRR
JOURNAL OF THE TRANSPORTATION RESEARCH BOARD

Article

Transportation Research Record
2018, Vol. 2672(10) 1–11

DOI: 10.1177/0361198118792756
journals.sagepub.com/home/trr
SAGE

Evaluating the Impact of Highway-Railway Grade Crossings on Travel Time Reliability on a Highway Network Level

Zifeng Wu[1], Laurence R. Rilett[2,3], and Yifeng Chen[4]

Abstract

Highway-rail grade crossings (HRGCs) have a range of safety and operational impacts on highway traffic networks. This paper illustrates a methodology for evaluating travel-time reliability for the routes and networks affected by trains traveling through HRGCs. A sub-area network including three HRGCs is used as the study network, and a simulation model calibrated to local traffic conditions and signal preemption strategies using field data is used as the platform to generate travel time data for analysis. Time-dependent reliability intervals for route travel time are generated based on route travel-time means and standard deviations. OD level reliability is calculated using a generic reliability engineering approach for parallel and series systems. The route travel time reliability results can be provided as real-time traffic information to assist drivers' route-choice decisions. The OD level reliability is a way to quantify the impact of HRGCs on highway network operation. This effort fills the gap of reliability research for HRGCs on the route and sub-area network level, and contributes to improving the efficiency of decision-making for both traffic engineers and drivers.

A highway railway crossing is the location at which a railway line intersects with a roadway in the form of either an at-grade crossing or a grade-separated crossing. The number of highway–rail grade crossings (HRGCs) grew with the growth in highway miles. In cities and towns with a grid street layout, particularly in the midwest and west, a crossing over the railroad was often provided for every street, resulting in about 10 crossings per mile (*1*). In 2005, there were 248,273 total intersections of vehicular and pedestrian traveled ways with railroads (*1*). That is approximately 2.4 crossings per railroad line mile.

Compared with grade-separated crossings, HRGCs have greater impact on the safety and operation of the roadway system. The general form of the basic accident prediction formula considers factors of the average annual daily traffic (AADT), pavement type, number of lanes on a highway and the average daily train traffic, maximum timetable speed, and number of main tracks (*1*). Over the last decade, studies have been undertaken to investigate the factors that have contributed to HRGC crashes and various safety enhancement techniques (*2–10*). The FRA has analyzed grade crossings located on the federal-aided highway system equipped with flashing lights and automatic gates and has forecasted that annual hours of delay for autos could increase by between 35 and 123 million between 2002 and 2022 (*11*). The impact of HRGCs grew with the growth in vehicular and train traffic. Traditionally, the Nebraska Department of Transportation (NDOT) identifies potential locations for new grade separation structures based on the exposure factor, crash costs, elimination of vehicular delay, and other appropriate factors (*12*). The exposure factor is calculated as the product of the number of vehicles and the number of trains at a given HRGC for a day and used as a quantitative indicator to evaluate the impact of train traffic on the highway network. The Texas Department of Transportation (TxDOT) commissioned research to

[1]Transportation Analyst, AECOM, Pompano Beach, FL
[2]Nebraska Transportation Center, University of Nebraska-Lincoln, Lincoln, NE
[3]Department of Civil Engineering, University of Nebraska-Lincoln, Lincoln, NE
[4]Traffic Engineer, AECOM, Southfield, MI

Corresponding Author:
Address correspondence to Zifeng Wu: lilianwu2015@gmail.com

develop a prioritization system for highway–rail at-grade crossings that addresses warrants to identify low-volume, passive crossings, and a broader priority index that considers more variables than the original index (*13*). Rodriguez and Danaher summarized an evaluation methodology to prioritize improvements on HRGCs for corridor planning projects based on existing conditions (*14*).

From an operational perspective, HRGCs introduce delays and exacerbate congestion on highway traffic if trains travel through during peak hours. Chaudhry and Gates evaluated and compared the adequacy of simulation-based models and deterministic analytical methods in terms of queue spill-back length estimation from the traffic signals in the proximity of railroad crossings (*15*). When an HRGC is located within 200 ft (61 m) of a signalized intersection, the standard practice is to interconnect the railroad signalcontrol equipment and the highway signal control equipment, and use a preemption signal control strategy when a train is approaching (*16*). Cho studied the impacts of an individual HRGC on the signalized intersection in its proximity with different transition preemption strategies (*17*). Chen extended the scope of study to the impacts of multiple HRGCs at a corridor level with regards to safety and corridor delay, with the current and proposed transition preemption strategies for dual tracks (*18–20*).

Reviewing previous literature reveals a lack of research into the impact of an HRGC on the network level. The measurements to quantify the HRGCs' impact on highway operation were limited to overall or average delay and exposure factor. Despite the increasing popularity of reliability in roadway traffic topics, and the common existence of HRGCs in many cities, research has not been conducted on the impact of HRGCs on highway travel-time reliability. This paper studies the impact of HRGCs from a sub-area network level though a well calibrated and validated simulation model for a real-world network in Lincoln, Nebraska. Travel time reliability is used to quantify the HRGC impact. For road users, the travel time reliability of an HRGC-included route and alternative routes can be provided as real-time traffic information to assist drivers' route choice decisions. For example, they might choose to take an alternative route that is not impacted by the train arrival because it has a link that is grade-separated. For the traffic system manager, the network level OD reliability can be used to assess and monitor the operational impacts of train traffic for bi-modal transportation networks and as an additional metric for prioritizing improvement projects. This paper fills the gap of reliability research for HRGCs on the route and sub-area network level, and contributes to improving the efficiency of decision-making for both traffic engineers and drivers.

Bi-Modal Simulation Model

Study Network

The bi-modal transportation network used in this study is bounded by Cornhusker Highway, Holdrege Street, 27th Street, and 48th Street, in Lincoln, Nebraska, as illustrated in Figure 1*a*. This is a 2.4 km by 3.2 km urban transportation network that includes three HRGCs: 33rd Street HRGC, Adams Street HRGC, and 44th Street HRGC, as shown in Figures 1*a* and 1*b*. The railway crosses 27th Street and 48th Street at separated grade, via underpass and overpass, respectively. Corresponding to the map in Figures 1*a* and 1*b* is the VISSIM 5.4 simulation model of the network. The thick blue arrow line represents a 3.2 km section of the BNSF mainline railroad tracks while the dark-gray lines represent the roadway transportation network. The three red circles represent the locations of the HRGCs and the blue rectangles represent grade separated roadway-railway crossings.

The OD matrix input to feed this simulation model is estimated based on empirical intersection turn data (*21*). The simulation model was calibrated to local conditions using traffic volume data measured on the major network arterials by the City of Lincoln (*20, 22*). The 16 numbered points shown in Figure 1*b* indicate the locations where traffic volume data were collected. The dynamic traffic assignment function in VISSIM 5.4 is used for trip distribution and assignment, so the results can better capture the interaction between supply and demand. The simulated travel times are used to estimate the reliability metrics introduced later in this paper.

Railway Traffic Modeling

The three signalized intersections located at points (11), (13), and (16) in Figure 1*b* are located at 560 ft (170 m), 100 ft (30 m), and 110 ft (33 m), respectively, from a HRGC. The traffic signal timings follow standard train preemption logic when a train is present. The goal is to operate the traffic signal to allow vehicles to clear off the HRGC before a train arrives. The train preemption logic of these three intersections were programmed using VISVAP in the VISSIM environment. The empirical train speed and length distributions collected by the Doppler radar located at an upstream location of the 33rd Street HRGC were used to simulate the rail traffic in the simulation model (*23*).

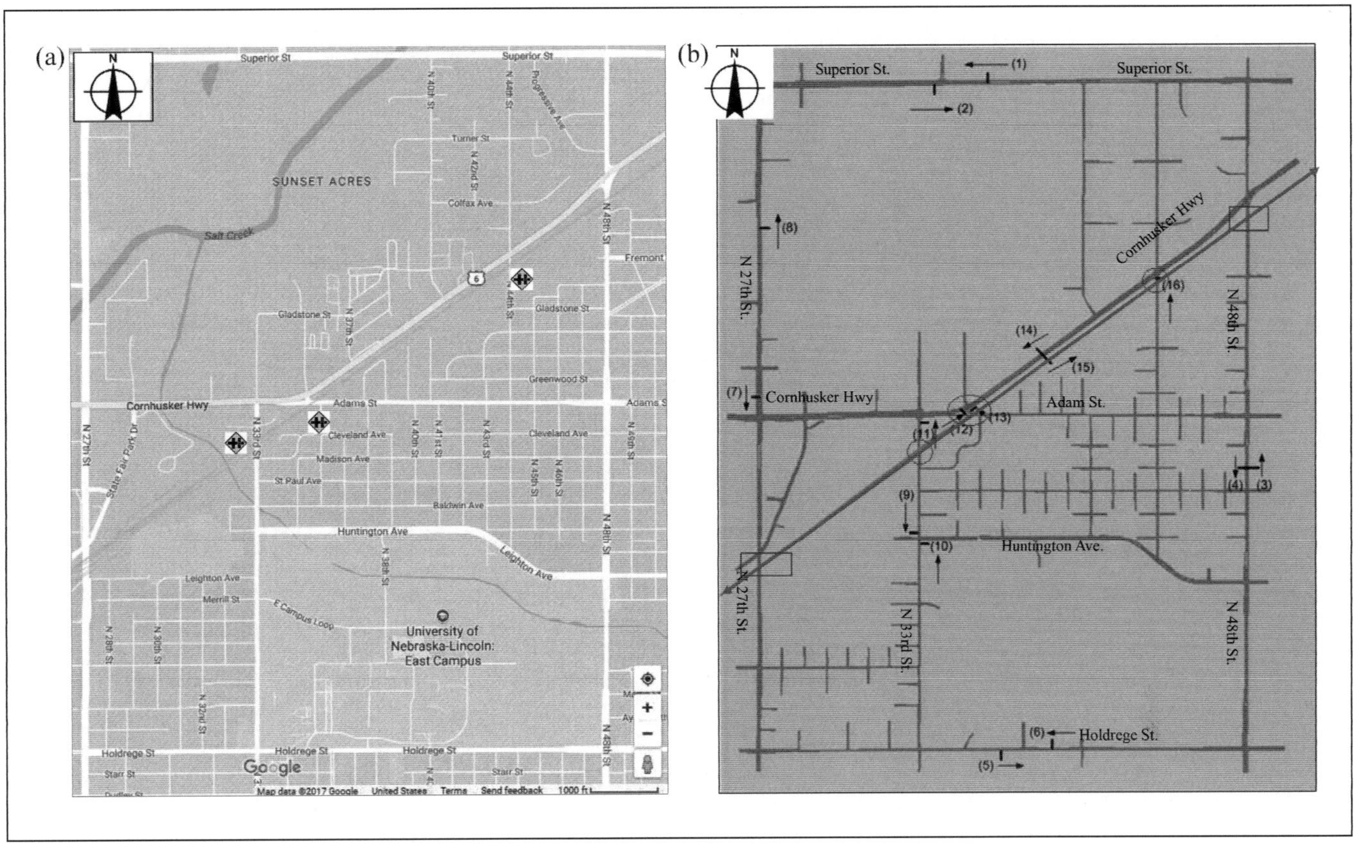

Figure 1. The bi-modal study network.

Impact of Trains on Vehicle Arrival Time Reliability

Train traffic in a bi-modal transportation network generally increases delays to vehicle traffic at HRGCs as the vehicles wait for the train to clear the HRGC. Drivers who use paths that include HRGCs will be particularly affected. The arrival time reliability is used to measure this effect, as shown in Equation 1. If drivers can be informed, in real time, of the reliability intervals of their route travel time when there is a train present in the corridor, they would have a better idea about potential delays and could choose routes with better arrival time reliability. It is hypothesized that this will reduce total network delays and increase overall network performance.

The reliability interval of the arrival time for a driver traveling between node i and node j, is a function of the departure time d_k and the associated travel time reliability interval for the nth time interval:

$$R^{\alpha}_{ijn} = \bar{x}_{ijn} \pm m^{\alpha} \cdot s_{ijn} \; \forall i = 1 \ldots I, \; \forall j = 1 \ldots J, \; \forall n = 1 \ldots N \tag{1a}$$

$$A^{\alpha}_{ijkn} = d_k + R^{\alpha}_{ijn} \; \forall i = 1 \ldots I, \; \forall j = 1 \ldots J, \; \forall n = 1 \ldots N, \; \forall k = 1 \ldots K \tag{1b}$$

where:

R^{α}_{ijn} = the predicted reliability interval of travel time between nodes i and j for a departure time within the nth five-minute interval, at a confidence level of α.

m^{α} = the coverage parameter at a confidence level of α. This parameter is user defined and controls the spread of the interval. In the case study in this paper, it is set to 1 based on a preliminary study (*22*).

$\bar{x}_{ijn}$ = the predicted mean travel time from node i to node j for vehicles departing within the nth five-minute interval.

s_{ijn} = the predicted standard deviation of travel time from i to j for vehicles departing within the nth five-minute interval.

I, J = the number of origins and destinations, respectively.

N = the total number of five-minute intervals within the analysis period.

A^{α}_{ijkn} = the predicted reliability interval of the arrival time for the kth vehicle traveling between node i and node j departing within the nth five-minute interval at the confidence level of α. Note that the kth driver departs from node i at d_k.

d_k = the departure time of the kth driver. This corresponds to the nth five-minute interval.

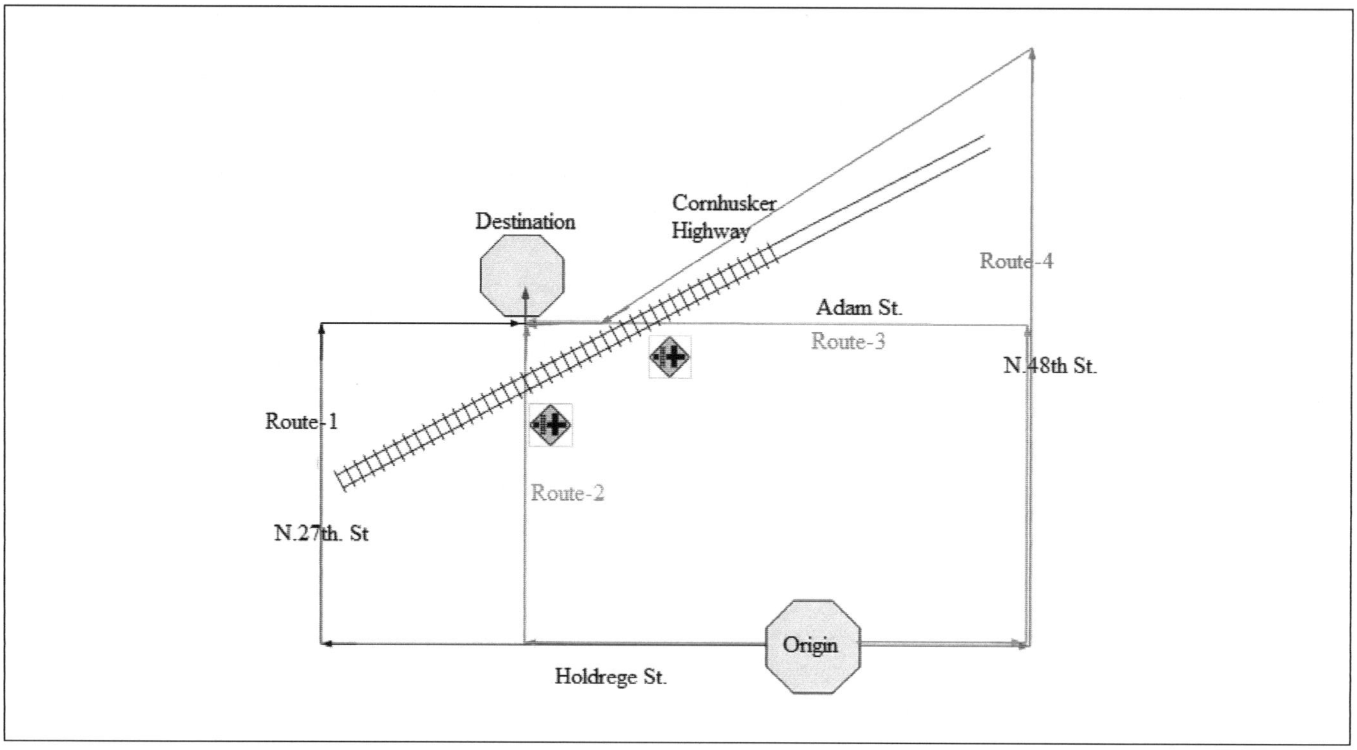

Figure 2. Four routes selected to study the OD pair from campus (origin) to UPS (destination).

K = the number of vehicles traveling from node i to node j.

α = the confidence level, which is the percentage of drivers who depart node i during the nth time interval and can expect to arrive at node j within the predicted reliability interval A^{α}_{ijk}. In the case study in this paper, this is set to 95%.

One key step to implementing the reliability interval (RI) in practice is to choose an appropriate value of m^{α}. The coverage rate of a RI corresponding to a given m^{α} depends on the efficiency of prediction models. The coverage rate is defined as the percentage of the actual arrival times that are within the predicted reliability interval. Therefore, the value of m^{α} needs to be decided by a preliminary study so the predicted reliability interval can provide an acceptable coverage rate for real applications. In this study, the simulation model is used to predict the link travel times for the next time step. The averages and standard deviations of route travel time are estimated by the first-order approximation method using the predicted link travel times (*24*).

Route Set in the Study Network

One OD pair in the study network is selected to demonstrate the application of the arrival travel time reliability. The origin is the University of Nebraska-Lincoln's east campus, and the destination is the UPS customer service center, as shown in Figure 2. Four potential routes are selected, and they are representative of the options available to a driver. The four available routes are shown in different colors in Figure 2. Drivers on Route 2 have the opportunity to change to Route 1 before making their first right-turn. Similarly, drivers on Route 3 have the opportunity to change to Route 4 before making their second left-turn. Therefore, in this analysis, when a train is present in the corridor, Route 1 and Route 4 are considered to be alternative routes for Route 2 and Route 3, respectively. Relevant information for each route is shown in Table 1. The free flow travel time is the travel time at the link speed limit.

Reliability Information for Individual Drivers

Figures 3*a*–3*d* show the time-dependent reliability intervals under the one-train scenario for Routes 1, 2, 3, and 4, respectively. The x-axis is the simulation time in seconds. The y-axis is the route travel time in minutes. The green dotted lines bounded with squares in Figures 3*b* and 3*c* indicate the presence of the train at the two HRGCs on 33rd Street (Route 2) and Adams Street (Route 3). Because Routes 1 and 4 have grade-separated HRGCs and are not affected by the presence of a train, there are no green lines shown in Figures 3*a* and 3*d*. Each reliability interval of route travel time is calculated

Table 1. Basic Information of the Four Routes to Study

Route No.	Physical length (km)	Speed limit section		Number of signals	Free flow travel time (min)	Presence of HRGC
		Speed limit km/h (mph)	Section length (km)			
Route 1	3.96	56 (35)	1.45	6	5	No
		64 (40)	1.79			
		72 (45)	0.72			
Route 2	2.29	56 (35)	2.29	3	3	Yes
Route 3	4.67	56 (35)	2.98	6	8	Yes
		40 (25)	1.53			
		72 (45)	0.16			
Route 4	6.16	56 (35)	1.26	9	10	No
		40 (25)	2.88			
		72 (45)	2.03			

for the trips that depart within the corresponding five-minute interval shown on the x-axis.

The means and standard deviations of route travel times for both the one-train scenario and the no-train scenario are summarized in Table 2. As shown in Table 1, Route 2 is shorter in physical distance than its alternative Route 1, and Route 3 is shorter than its alternative Route 4. However, Table 2 reveals that the presence of a train on Route 2 and Route 3 generates enough delay to make their alternatives more attractive to drivers. As an example, consider the five-minute interval from 600 to 900 seconds. The average route travel times of Route 1 and Route 2 in the no-train scenario are 6.2 and 4.3 minutes, respectively. In the one-train scenario, the average Route 2 travel time increases to 7.1 minutes, which is 15% longer than Route 1. In addition, the standard deviation of Route 2 in the one-train scenario is 2.6 minutes while that of Route 1 is 1.4 minutes, an increase of 86%.

Route 2 is less impacted than Route 3 for the one-train scenario. The underlined numbers in Table 2 indicate the four five-minute intervals where Route 3 is negatively impacted, and the one five-minute interval where Route 2 is negatively impacted. These results indicate the drivers' difficulty in evaluating the traffic conditions for each route with the presence of a train. Providing time-dependent reliability interval information can assist drivers in route choice decision-making.

Impact of Trains on Route and OD Level Reliability

System managers may be interested in quantifying the impact of train traffic on the route performance and network performance for a given OD pair. This section demonstrates the metrics and techniques that can be used for these applications.

Scenario 1: Impact on Route Travel Time Reliability

The route travel time reliability is measured by the probability of the total areas between the reliability interval bounds of route travel time with a customized evaluation threshold E_r, defined by local traffic agencies. It is assumed in this research that local agencies or system operators want to evaluate the network in terms of the degree to which the network meets drivers' expectations. The evaluation thresholds that represent drivers' expectations need to be defined first. For example, drivers may expect a longer travel time if they choose a route with a longer physical distance and more signals. Therefore, the evaluation threshold, E_r, is defined based on the ideal route travel time, as shown in:

$$E_r = a * t_i \tag{2a}$$

where:

E_r = the evaluation threshold of rth route between two nodes.

a = the inflation factor which is used to reflect driver expectations. It is hypothesized that in large cities, such as Chicago or Los Angeles, this factor will be larger than in smaller cities. In this paper a is set equal to 1.5.

t_i = the ideal travel time of rth route.

The ideal route travel time t_i is based on the free flow travel time t_f and reasonable delay time at signals:

$$t_i = t_f + t_d \tag{2b}$$

It is assumed in this effort that drivers expect to pass 75% of the signals without stopping, and stop once at the remaining 25% of the signals. Each stop at a signal is assumed to result in a delay of 0.5 min, without loss of generality. Note that the assumed values in the following equations can be adjusted with location-specific empirical data:

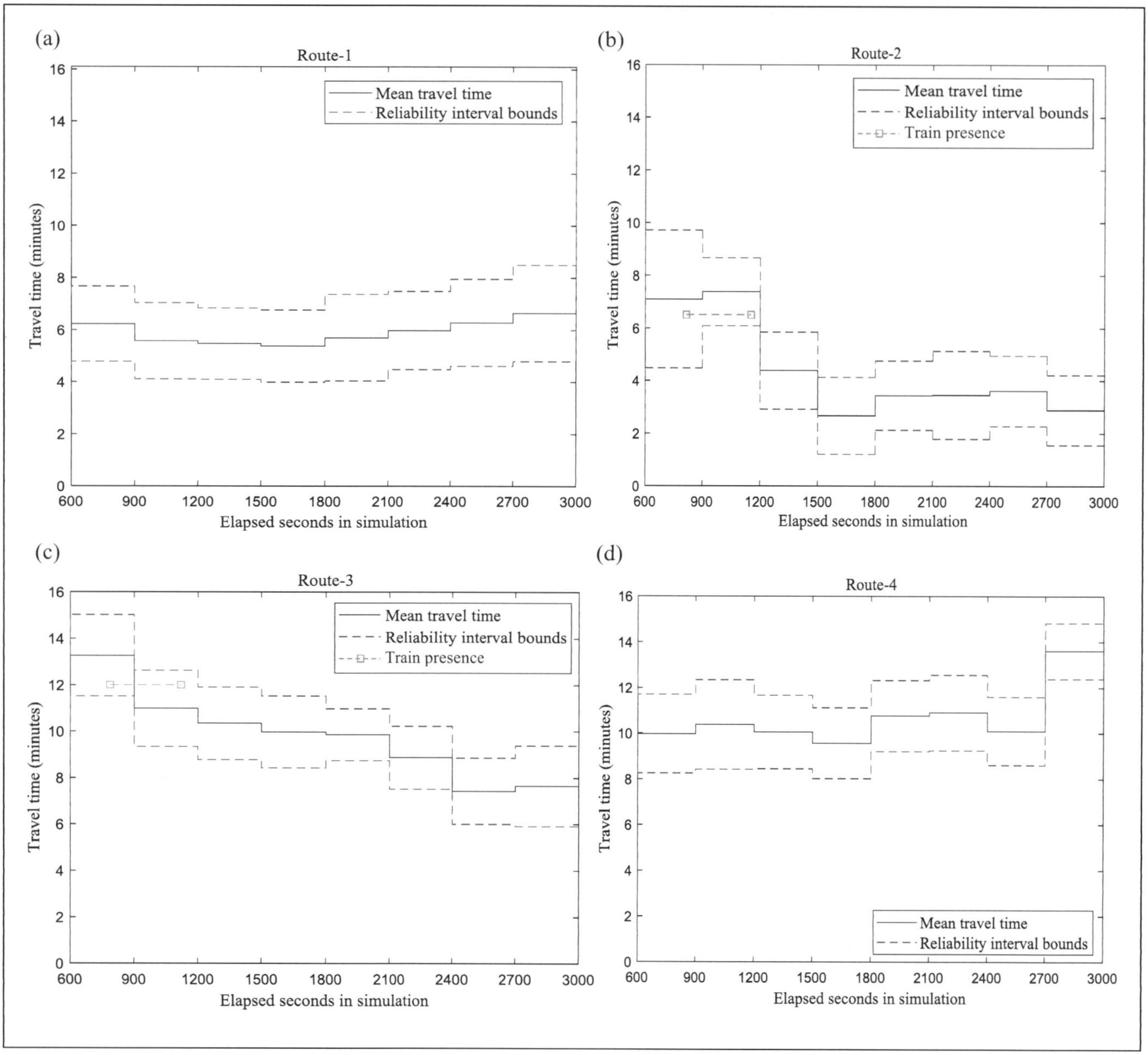

Figure 3. Time-dependent reliability intervals for route travel time under the one-train scenario.

$$t_f = \frac{d_r}{v_r} \times 60 \quad (2c)$$

$$t_d = t_s \cdot n_r \cdot p \quad (2d)$$

where:

t_f = the free flow travel time of rth route (minutes).

d_r = the distance of rth route (miles).

v_r = the speed limit of rth route (miles per hour).

t_d = the estimated delay at traffic signals (minutes).

t_s = the estimated average delay when stopped at a traffic signal (minutes). In this paper, it is set equal to 0.5 minutes.

n_r = the number of traffic signals along the rth route.

p = the expected percentage of traffic signals where a vehicle will stop for a red signal (percent). In this paper p is set to 25%.

Alternatively, traffic agencies may define their acceptable route travel time as the evaluation threshold without applying Equation 2.

Table 3 summarizes the evaluation results of route travel-time reliability for Route 2 and Route 3. Consider Route 2 as an illustration. Figure 4 displays the time-dependent reliability intervals for Route 2 under the no-train scenario and the one-train scenario. The x-axis shows the elapsed simulation time in seconds and the y-axis shows the travel time in minutes. The red straight

Table 2. Means and Standard Deviations of Route Travel Times in Minutes

Route	Scenario	Statistic	Five-minute interval for departure 600–900	900–1200	1200–1500	1500–1800	1800–2100	2100–2400	2400–2700	2700–3000	Average
Route 1	No-train	Mean	6.2	5.6	5.5	5.4	5.7	6.0	6.3	6.6	5.9
		SD	1.4	1.5	1.4	1.4	1.7	1.5	1.7	1.9	1.6
Route 2	One-train	Mean	7.1 *	7.4 *	4.4	2.7	3.4	3.5	3.6	2.9	4.4
		SD	2.6 *	1.3 *	1.5	1.5	1.3	1.7	1.3	1.3	1.6
	No-train	Mean	4.3	3.3	3.3	3.0	3.4	3.4	3.6	2.9	3.4
		SD	1.4	1.4	1.3	1.4	1.5	1.7	1.3	1.5	1.4
Route 3	One-train	Mean	13.3 *	11.0 *	10.3	10.0	9.9	8.9	7.4	7.6	9.8
		SD	1.8 *	1.6 *	1.6	1.5	1.1	1.4	1.4	1.7	1.5
	No-train	Mean	7.4	7.1	7.8	7.6	7.4	6.9	7.3	7.6	7.4
		SD	1.6	1.7	1.7	1.5	1.4	1.7	1.6	1.8	1.6
Route 4	No-train	Mean	9.9	10.7	10.5	10.6	11.4	10.9	11.0	14.2	11.2
		SD	1.4	2.0	1.5	2.0	1.7	1.9	1.5	1.8	1.7

Note: Route 1 and Route 4 are minimally impacted by train traffic. Therefore, only no-train scenarios are listed in this table. Route 2 and Route 3 are the two routes impacted by HRGCs. Cells with (*) indicate the intervals where the trains is present at the intersection impacted by HRGCs. Cells with underlined values indicate situations where 1) the impacted periods after a train has left the HRGC, and 2) the difference between the means for one-train and no-train scenarios is greater than one minute. SD = standard deviation.

line is the evaluation threshold, and the green dotted line in Figure 4 bounded with squares indicates the presence of a train at the HRGC near the intersection of 33rd Street and Cornhusker Highway on Route 2. The reliability of Route 2 is lowered to 49% from 72% due to the presence of the train.

Scenario 2: Impact on OD Level Reliability

The evaluation of the OD service is based on the major routes serving a given OD pair. The four routes in Figure 2 are further separated into section components as shown in Figure 5*a*, so that the OD network can be analyzed as a complex system with parallel and series sub-systems, as shown in Figure 5*b*. The shared sections of different routes are defined as "components." The independent sections of each route are defined as a "sub-route." The reliability intervals and evaluation thresholds for each component and sub-route are calculated based on Equations 1 and 2. The evaluation thresholds and the resulting reliability metrics are summarized together for no-train and one-train scenarios in Table 4.

In reliability engineering, the reliability of a parallel system and a series system can be calculated using Equation 3 *(25)*:

$$R(Parallel\,system) = 1 - \prod_{j=1}^{k}\left(1 - R_j\right) \tag{3a}$$

$$R(Series\,system) = \prod_{j=1}^{k} R_j \tag{3b}$$

where:

k = the number of components/sub-routes in the system.

R_j = the reliability of jth components/sub-routes connecting the OD, $j = 1, \ldots, k$.

The overall OD network reliability for the network in Figure 2 is 0.93 for the no-train scenario. In the one-train scenario, the overall OD reliability is decreased to 0.89. This type of analysis can quantify the impact of train traffic on reliability for an OD pair of interest, and enable reliability evaluation at the network level. It also provides a metric of reliability in cost-benefit studies by enabling increasing reliability to be considered in benefit-cost studies for infrastructure improvements such as viaduct construction.

Concluding Remarks

This paper presents the metrics and methodology for the reliability evaluation of a bi-modal transportation network, and exemplifies the methodology using a test

Table 3. Route Reliability Evaluation Results

Route No.	Evaluation threshold (min)	Reliability for one-train scenario (1)	Reliability for no-train senario (2)	Degradation ([2]-[1])/(2)
Route 2	4	49%	72%	32%
Route 3	9	23%	95%	76%

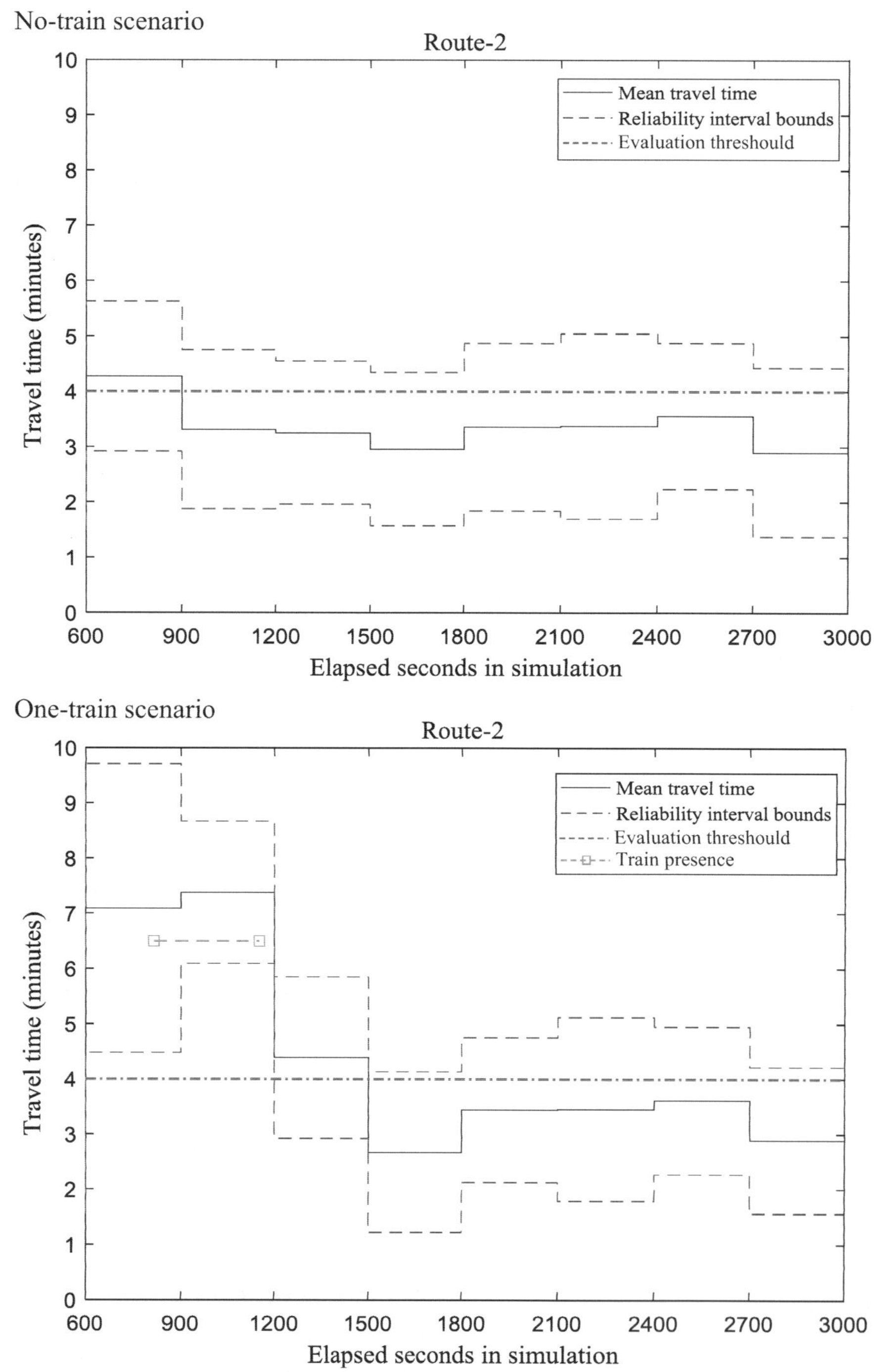

Figure 4. Time-dependent route travel time reliability intervals for Route 2.

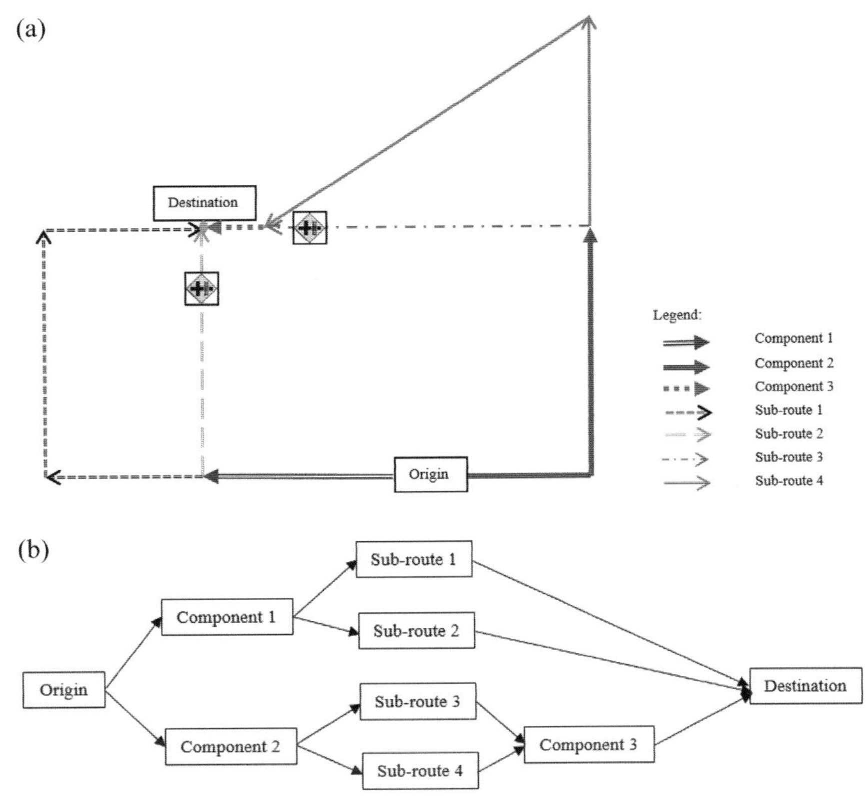

Figure 5. Defining the studied OD network as a complex system.

Table 4. Reliability of Each Component in the OD Network

Components	Distance (km)	t_f (min)	Number of signals	t_d (min)	t_i	E_r	Reliability (R) 0-train	1-train
Component 1 (c1)	0.45	0.48	1.00	0.13	0.61	0.91	0.92	0.93
Component 2 (c2)	2.77	3.24	3.00	0.38	3.61	5.42	0.96	0.98
Component 3 (c3)	0.16	0.14	1.00	0.13	0.26	0.40	0.71	0.71
Sub-route 1 (s1)	3.49	3.33	5.00	0.63	3.95	5.93	0.71	0.78
Sub-route 2 (s2)	1.81	1.95	3.00	0.38	2.32	3.48	0.76	0.40
Sub-route 3 (s3)	1.71	1.84	3.00	0.38	2.21	3.32	0.56	0.14
Sub-route 4 (s4)	3.20	3.56	6.00	0.75	4.31	6.46	0.53	0.63
Routes/OD			Calculation				Reliability 0-train	1-train
Route 1 & 2 (rt1&2)			$R_{rt1\&2} = R_{c1} \cdot [1 - (1 - R_{s1}) \cdot (1 - R_{s2})]$				0.86	0.81
Route 3 & 4 (rt3&4)			$R_{rt3\&4} = R_{c2} \cdot [1 - (1 - R_{s3}) \cdot (1 - R_{s4})] \cdot R_{c3}$				0.53	0.47
OD			$1 - (1 - R_{rt1\&2}) \cdot (1 - R_{rt3\&4})$				0.93	0.89

Note: The results are in two decimal digits because some components are shorter than 1 km in length, and the corresponding travel times are less than one minute.

network with HRGCs located in Lincoln, Nebraska. It should be noted that although the methodology and associated metrics were demonstrated with simulation data, the approach may be generalized to other networks, using simulated or empirical data, regardless of the presence of HRGC.

The arrival time interval information can be provided via a traveler traffic information system to assist drivers' route choices or to notify drivers of the impacted arrival time when a train is present in the corridor. In addition to regular drivers, this information will especially benefit the vehicles with higher priority on travel time reliability such as school buses, hazardous goods vehicles, delivery trucks, etc., and the routes with higher percentages of these vehicles.

The presence of train traffic increased both the mean and standard deviation of travel times on the routes passing through an HRGC so that they became less attractive choices compared with their alternatives, which were longer in terms of physical lengths. The duration of the impacted period depends on the train speed, traffic demand at the intersection, and the intersection operation condition. Given the evaluation thresholds defined in this study, the reliability of Route 3 is degraded by 76% while the reliability of Route 2 is degraded by 32% for the analysis period. This result can assist system operators in identifying the priority of proposed improvement projects.

In addition, this paper illustrates the concept of the OD-based reliability metrics. The roadway network was modeled as a combination of parallel systems and series systems. The proposed approach allows transportation agencies to quantify the impact of train traffic on a bimodal network in a more comprehensive way. The new metrics may augment or replace traditional metrics such as the exposure factor.

Future research topics include: 1) studying the impact of travel time reliability on the route choice behavior of drivers, which in turn affects the parameters of traffic information and management systems (*26–29*), 2) evaluating the impact of different train preemption logics on reliability based on high-resolution traffic event data (*30, 31*), and 3) applying the proposed methodology on real-time probe vehicle travel-time data provided by cost-effective technologies such as the GPS and Bluetooth (*32, 33*).

Author Contributions

The authors confirm contribution to the paper as follows: study conception and design, analysis and interpretation of results: Wu, Rilett; draft manuscript preparation: Wu, Rilett, Chen. All authors reviewed the results and approved the final version of the manuscript.

References

1. *Railroad-Highway Grade Crossing Handbook*. U.S. Department of Transportation Federal Highway Administration, Washington, D.C., 2007.
2. Zhao, L., and L. R. Rilett. Dynamic Model of Driver Approaching Behaviors at Highway-Rail Grade Crossings. Presented at 96th Annual Meeting of the Transportation Research Board, Washington, D.C., 2017.
3. Khattak, A., and L. Zheng. Pedestrian and Bicyclist Violations at Highway-Rail Grade Crossings. *Transportation Research Record: Journal of the Transportation Research Board*, 2011. 2250: 76–82.
4. Tung, L.W., and A. Khattak. Distracted Motor Vehicle Driving at Highway–Rail Grade Crossings. *Transportation Research Record: Journal of the Transportation Research Board*, 2015. 2476: 77–84.
5. Zhao, S., and A. Khattak. Injury Severity in Crashes Reported in Proximity of Rail Crossings–The Role of Driver Inattention. *Journal of Transportation Safety & Security*, 2017.
6. Khattak, A. Reducing Undesirable Actions of Motor Vehicle Drivers at Railroad–Highway Grade Crossings. *Transportation Research Record: Journal of the Transportation Research Board*, 2007. 2030: 54–58.
7. Khattak, A. Investigation of Train Warning Times and Gate Violations. *Transportation Research Record: Journal of the Transportation Research Board*, 2014. 2458: 104–109.
8. Khattak, A., and L.-W. Tung. Severity of Pedestrian Crashes at Highway-Rail Grade Crossings. *Journal of the Transportation Research Forum*, Vol. 54. No. 2, 2015. pp. 91–100.
9. Tung, L.-W. C. *The Investigation of Distracted Driving at Highway-Rail Grade Crossings*. PhD dissertation. Department of Civil Engineering, University of Nebraska-Lincoln, Lincoln, NE, 2014.
10. Yue, H., and E. Jones. Archiving Capability of Patio-temporal Data in Different Highway Railroad Grade Crossing (HRGC) Databases. *Proc., 20th Annual Intelligent Transportation System Conference*, Houston, TX, May 2012.
11. U.S. DOT, FHWA, FTA. *Status of the Nation's Highways, Bridges, and Transit: 2002 Conditions and Performance Report, Chapter 26*, Washington, D.C., 2002.
12. *Administration of State and Federal Highway-Rail Grade Crossing Safety Projects*. Nebraska Administrative Code Title 415:47, Nebraska Department of Roads, Lincoln, NE, July 2011.
13. Warner, J., E. S. Park, D. Lee, and A. Weissmann. Development of New Highway-Railroad Grade Crossing Priority Formula for Texas. Presented at 95th Annual Meeting of the Transportation Research Board, Washington, D.C., 2016.
14. Rodriguez, A., and A. Danaher. Railroad-Highway Grade Crossing Analysis for Corridor Planning Projects. Presented at 95th Annual Meeting of the Transportation Research Board,Washington, D.C., 2016.
15. Chaudhry, A., and T. Gates. Estimating Queue Lengths at Signalized Intersections Near Highway-Railroad Grade Crossings. Presented at 96th Annual Meeting of the Transportation Research Board, Washington, D.C., 2017.
16. *Manual on Uniform Traffic Control Devices: 2009 Edition*. U.S. Department of Transportation Federal Highway Administration, Washington, D.C., 2009.
17. Cho, H. *Preemption Strategy for Intersections near Highway-Railroad Crossings*. PhD dissertation, Texas A&M University, College Station, TX, 2003.
18. Chen, Y., and L. R. Rilett. Train Data Collection and Arrival Time Prediction System for Highway-Rail Grade

Crossings. *Transportation Research Record: Journal of the Transportation Research Board*, 2017. 2608: 36–45.

19. Chen, Y., and L. R. Rilett. Integrated Traffic Signal Control Along Corridor with Multiple Intersections Near Highway-Rail Grade Crossings. *Proc., 2013 Joint Rail Conference*, Knoxville, TN, USA, 2013.
20. Chen, Y. *An Adaptive Corridor-wide Signal Timing Optimization Methodology for Traffic Networks with Multiple Highway-Rail Grade Crossings*. PhD dissertation, Department of Civil Engineering, University of Nebraska-Lincoln, Lincoln, NE, 2015.
21. Appiah, J., and R. Rilett. Joint Estimation of Dynamic Origin-Destination Matrices and Calibration of Microsimulation Models Using Aggregate Intersection Turn Count Data. Presented at 89th Annual Meeting of the Transportation Research Board, Washington, D.C., 2010.
22. Wu, Z. *Measuring Reliability in Dynamic and Stochastic Transportation Networks*. PhD dissertation. Department of Civil Engineering, University of Nebraska-Lincoln, Lincoln, NE, 2015.
23. Elizabeth, G. J., A. Khattak, and L. R. Rilett. The University Nebraska- Lincoln's Highway Rail Grade Crossing Test Bed System. Presented at 88th Annual Meeting of the Transportation Research Board, Washington, D.C., 2009.
24. Fu, L., and L. R. Rilett. Expected Shortest Paths in Dynamic and Stochastic Traffic Networks. *Transportation Research Part B*, Vol. 32, No. 7, 1998, pp. 499–516.
25. Billinton, R., and R. Allan. *Reliability Evaluation of Engineering Systems*. Plenum Press, New York, NY, 1992.
26. Yue, H., L. R. Rilett, and P. Z. Revesz. Spatio-temporal Traffic Video Data Archiving and Retrieval System. *GeoInformatica*, Vol. 20, No. 1, 2015, pp. 59–94.
27. Zhao, H., D. Sun, H. Yue, M. Zhao, and S. Cheng. Using CSTPNS to Model Traffic Control CPS. *IET Software*, Vol. 11, No. 3, 2017.
28. Yue, H., and R. Yang. Development of Intelligent Transportation Systems and Plan of Integrated Information System. *Journal of Wuhan University of Technology*, Vol. 29, No. 4, 2005, pp. 560–563.
29. Yue, H., and P. Z. Revesz. TVICS: An Efficient Traffic Video Information Converting System. *Proc., 19th International Symposium on Temporal Representation and Reasoning (TIME'12)*, pp. 141–148, Leicester, UK, 2012.
30. Verghese, V., C. Liu, S. C. Subramanian, L. Vanajakshi, and A. Sharma. Development and Implementation of a Model-Based Road Traffic-Control Scheme. *Journal of Computing in Civil Engineering*, Vol. 31, No. 3, 2017, pp. 04016063.
31. Liu, C., A. Sharma, E. Smaglik, and S. Kothuri. TraSER: A Traffic Signal Event-Based Recorder. *SoftwareX*, Vol. 5, 2016, pp. 156–162. https://doi.org/10.1016/j.softx.2016.09.001
32. Quiroga, C. A., and D. Bullock. Travel Time Studies with Global Positioning and Geographic Information Systems: An Integrated Methodology. *Transportation Research Part C: Emerging Technologies*, Vol. 6, No. 1–2, 1998, pp. 101–127.
33. Zhou, H., R. Benz, and A. Voigt. Bluetooth Travel Time/Speed Data Analysis for Houston-Galveston Regional Transportation Study. Presented at 95th Annual Meeting of the Transportation Research Board, Washington, D.C., 2016.

The Standing Committee on Highway/Rail Grade Crossings (AHB60) peer-reviewed this paper (18-04762).

Article

TRR
JOURNAL OF THE TRANSPORTATION RESEARCH BOARD

Transportation Research Record
2018, Vol. 2672(10) 12–27

DOI: 10.1177/0361198118792763
journals.sagepub.com/home/trr
SAGE

Rails-Next-to-Trails: A Methodology for Selecting Appropriate Safety Treatments at Complex Multimodal Intersections

Anna Bovbjerg Alligood[1], Manali Sheth[1], Anne Goodchild[1], Edward McCormack[1], and Polina Butrina[1]

Abstract

There are more than 212,000 at-grade railroad crossings in the United States. Several feature paths running adjacent to the railroad tracks, and crossing a highway; they serve urban areas, recreational activities, light rail station access, and a variety of other purposes. Some of these crossings see a disproportionate number of violations and conflicts between rail, vehicles, and pedestrians and bikes. This research focuses on developing a methodology for appropriately addressing the question of treatments in these complex, multimodal intersections. The methodology is designed to be able to balance a predetermined, prescriptive approach with the professional judgment of the agency carrying out the investigation. Using knowledge and data from the literature, field studies, and video observations, a framework for selecting treatments based on primary issues at a given location is developed. Using such a framework allows the agency to streamline their crossing improvement efforts; to easily communicate and inform the public of the decisions made and their reasons for doing so; to secure stakeholder buy-in prior to starting a project or investigation; to make sure that approach and selected treatments are more standardized; and to ensure transparency in the organization to make at-grade crossings safer for pedestrians and bicyclists, without negatively impacting trains or vehicles.

There are approximately 212,000 at-grade railroad crossings in the United States (*1*). Most of these crossings consist of one or more sets of railroad tracks being crossed by a public highway. As the presence of transit rail, especially light rail, increases in our cities, and the nationwide efforts to install multi-use paths next to railroad tracks are successful—The Rails-to-Trails Conservancy—a new level of complexity to at-grade railroad crossings is added. These paths or trails will often run adjacent to the railroad tracks, to serve urban areas, recreational activities, light rail station access, and a variety of other purposes. In most cases they will also be crossing the public roadway close to the railroad tracks, establishing a second point of intersection with the highway and potential point of conflict, as seen in Figure 1. Red Xs mark the two primary points of conflict—the conflict between the pedestrian (yellow) and the vehicle (blue), which in turn leads to the second conflict between the vehicle (blue) and the train (black) in situations where the vehicle is stopping to avoid the pedestrian. Even in cases where a train is not present, having vehicles stopping on tracks is undesirable.

In the state of Oregon, as well as in many other U.S. states, it is illegal for a vehicle to stop on railroad tracks—"Obstructing the intersection" (*3*); but at the same time, it is also illegal to not stop for pedestrians, cyclists, and other users (non-vehicular users) who are crossing in front of the vehicle on a path, sidewalk, or railroad crossing (*4*). When a non-vehicular user is present in, for example, a crosswalk and approached by a car traversing the railroad tracks, the driver of the vehicle is forced to break one of the two laws or put themselves or other road users in danger. To avoid striking the pedestrian who is crossing in the intersection, this will most frequently result in the vehicle dwelling on the tracks to wait for the pedestrian to pass. Aside from being illegal, this poses a threat to the driver, who is in a train's path and may not be able to move out of the way in the event that a train, which would be unable to stop, approaches. As intersections become more complex in urban areas, with higher volumes and an increased level of information that needs to be processed by traffic participants,

[1]University of Washington, Seattle, WA

Corresponding Author:
Address correspondence to Anna Bovbjerg Alligood:
alligood.anna@gmail.com

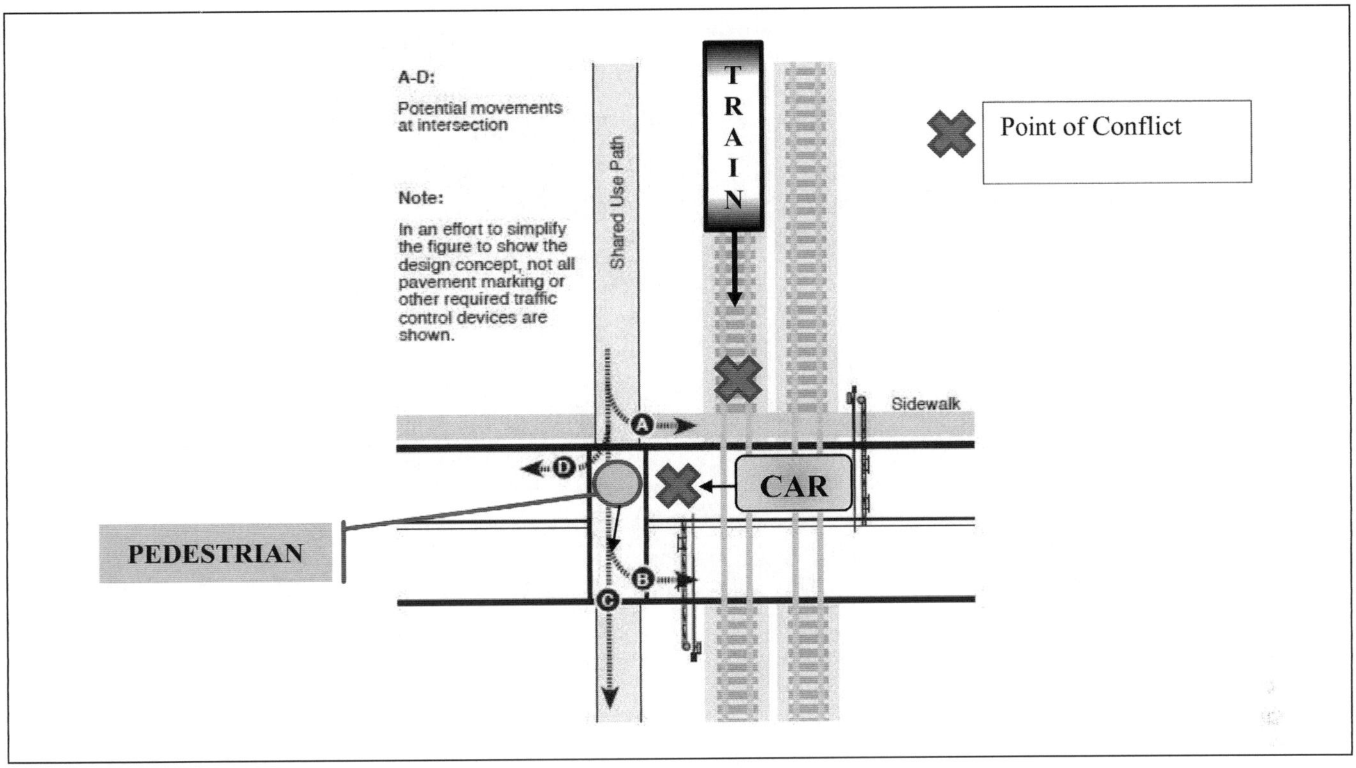

Figure 1. Conflicts at shared-use paths adjacent to the railroad tracks and perpendicular to the highway (2).

there is an increased probability of unintentional non-compliance, leading to conflicts between all types of users.

To adequately address safety concerns at this type of crossing, it is useful to regard these locations as single, complex multimodal intersections. Much literature is available to guide decision-making when it comes to applying treatments to highway–railroad crossings, and also for highway–sidewalk/path crossings, but limited guidance is available for these complex intersections where at least three different types of users are present. There is no standard or accepted methodology for applying treatments that improve safety and passability for all users, while addressing the legal issues previously described. Furthermore, these different types of users and infrastructure are represented by a variety of different stakeholders, both public and private, whose interests—which are potentially conflicting—should all be considered when attempting to mitigate issues or install new treatments at a railroad crossing. These can include railroad organizations, pedestrian/bike advocacy groups, cities and counties, and federal and state agencies, such as the FRA and state Departments of Transport (DOTs).

This paper addresses this gap by presenting a methodology for selecting the most appropriate treatment at a complex multimodal crossing. This is useful as it ensures more transparency in the process, which becomes more streamlined and predictable, to support collaboration with stakeholders. The following section describes existing practice and reviews the available literature. Following that the primary issues and ways of addressing them are presented, followed by conclusions and suggestions for areas of future research.

Existing Practice and Literature Review

There is no standard on how to address or treat at-grade highway–railroad crossings with a path running adjacent to the railroad tracks. The publications that transportation and traffic engineers and planners usually refer to for guidance do not explicitly cover this issue, though most of them mention the specific concerns for pedestrians and bicycles at these locations (*5*, *6*). This means that states, counties, and cities are often left to their own devices when it comes to selecting appropriate treatments. In the best cases, it means that the responsible agency will develop their own specific methodology for addressing this type of intersection and publish these as handbooks. In some cases, this means that the responsible agency will just treat the location as two separate entities, possibly leading to an increase in the confusion for users and in unintentional non-compliance, as their unique influence on each other is not considered. In

Figure 2. Two connecting trailheads on both sides of a highway are separated by a sign prohibiting pedestrian crossings.

other cases, as observed in various locations in Oregon, it may be decided to eliminate the pedestrian crossing, resulting in two unconnected trailheads on either side of a highway and, with that, an increase in the occurrence of illegal pedestrian and bicycle behavior. An example of such a crossing can be seen in Figure 2.

The first example of guidance handbooks on multimodal intersections was published by the Florida Department of Transportation (FDOT) in 1996 and contains experience and knowledge from an extensive field survey of more than 60 locations (*7*). More recently, a structured, flowchart-based methodology was developed by the Minnesota Department of Transportation (MnDOT), who published an extensive research project in 2013, which provides guidance on applicable treatments for complex railroad intersections (*8*). While both publications are informative, neither provide specific recommendations for the best possible solution addressing concerns such as efficacy, capital investment, and maintenance, as well as appearance and acceptance by its various users. The MnDOT lists a wide range of solutions or treatments and does not distinguish between different issues at a crossing, other than by the proxy of its characteristics, and therefore does not specifically address the condition at a specific crossing. This paper presents an approach that narrows down the number of recommendations that are made for an individual crossing, based on its identified primary issues, to be able to make the best possible infrastructure investment within the allocated budget.

This research project was conducted for the Oregon Department of Transportation Rail and Public Transport Division, and reflects this with respect to preferences and approved devices. To be applied in other districts, regional approvals and preferences should be considered. This paper does not examine all possible devices and treatments available nationally.

Primary Issues at Complex Intersections

In the pursuit of a methodology that can be used to select appropriate treatments, the first step was to identify and describe these railroad crossings and the primary categories of issues at them. As previously mentioned, both AASHTO and the *Manual on Uniform Traffic Control Devices* (MUTCD) suggest that this type of complex intersection is handled on a case-by-case basis (*5*, *6*). Treating crossings on a case-by-case basis, without any framework or guidance to inform decision-making and process, is time-consuming and can be perceived as unfair or inappropriate by stakeholders or the public. This paper suggests a methodology for ameliorating this situation.

The methodology presented in this paper seeks to identify some of the primary issues to streamline the process. These include issues with the built environment or infrastructure, lack of information for path users, and lack of information for drivers. These issues were identified using two different methodologies: first, a thorough literature review into available railroad treatments was conducted, followed by in-person observations and counts, which were further corroborated by video surveillance. Finally, issues were categorized and

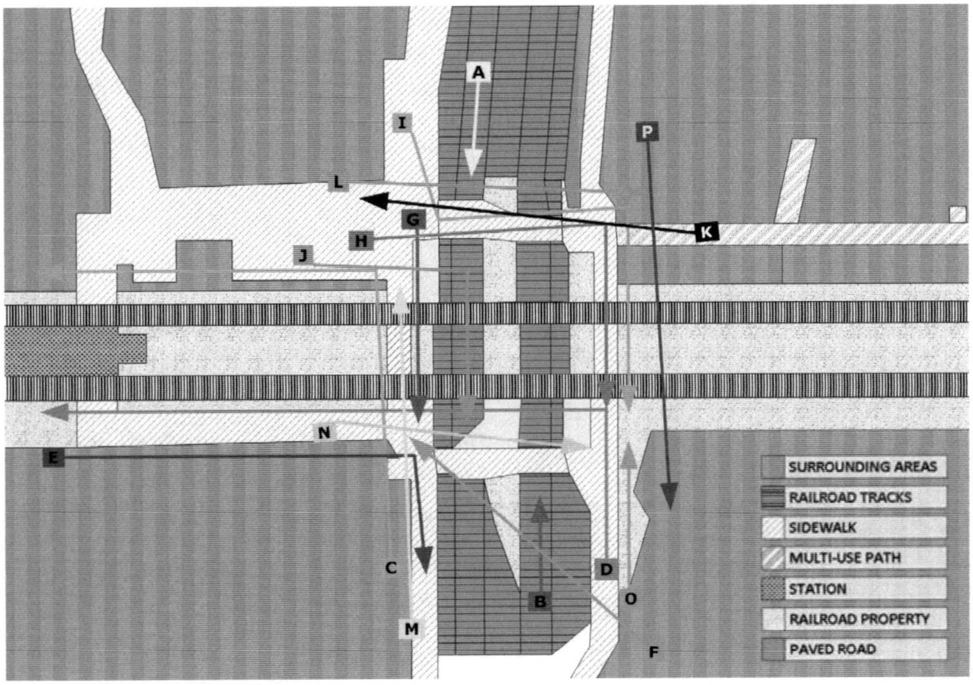

Figure 3. Example sketch of pedestrian and bike movements at a light rail crossing.

appropriate treatments were identified, which are described later in this paper.

Literature Review

In one of the earlier publications on having trails next to railroad tracks, the danger of the crossings was not specifically discussed, even though it was pointed out that most crossings in the U.S. have some form of warning device, other than the required crossbucks. Passive warning signs are not always enough for drivers to be aware of the upcoming railroad and path crossing (*9*). The following research pertains to pedestrians crossing the tracks perpendicularly. As there is limited research, one must look to existing knowledge about behavior and compliance at railroad crossings, though it is not directly applicable. In one study two active devices were compared, and it was found that the addition of automated gates to blinking flashers had the potential to reduce the percentage of drivers violating the warning signs from 67% to 38% (*10*). A paper published in 2013 investigated warning devices and signs for pedestrians and cyclists. They compared stated and actual behavior by pedestrians and cyclists and found that many participants were engaged in other activities while crossing, which interfered with their awareness of the tracks. Active signs were noticed more than passive signs, and the use of gates lowered rates of violation. People who crossed tracks more often generally displayed safer behavior than people who seldom crossed tracks. Pedestrian violate relatively more in urban areas than in rural areas. Larger groups were more likely to violate than were one or two people (*11*). This was confirmed by another study, which found that children under the age of eight expressed more risky behavior and induced risky behavior among others (*12*).

Field Studies

The second part of the research leading to identifying the primary issues was conducted as a field study of seven different railroad locations in the state of Oregon with the goal of collecting data and gaining more perspective about the issues at these crossings and their roots. With recommendations from the Rail Public Transit Division, seven study sites with different characteristics were chosen. The locations were visited and observed for several hours during summer months and counts were completed for all approaches and movements. Video was also collected for at least 24 h at each crossing.

Detailed counts were extracted from the video for both morning and afternoon peaks at all locations, for all approaches and movements. These were separated into pedestrian/cyclist and vehicle/train movements. Sketches were constructed showing the movements associated with the counts. An example of pedestrian and bike movements at a light rail crossing can be seen in Figure 3. This figure shows a location and the observed

Table 1. Example of Video Counts Related to Figure 3

	Video counts		
Movements	July 6, 2016 8 am–9 am	July 5, 2016 5 pm–6 pm	Average/hour
Pedestrians	68	51	59.5
Movement A	0	0	0
Movement B	0	0	0
Movement C	12	2	7
Movement D	11	0	5.5
Movement E	2	0	1
Movement F	1	0	0.5
Movement G	2	5	3.5
Movement H	0	0	0
Movement I	0	4	2
Movement J	0	0	0
Movement K	19	9	14
Movement L	12	18	15
Movement M	1	4	2.5
Movement N	6	5	5.5
Movement O	1	2	1.5
Movement P	1	2	1.5
Bikes	20	27	23.5
Movement A	2	10	6
Movement B	11	5	8
Movement C	1	0	0.5
Movement D	0	0	0
Movement E	0	1	0.5
Movement F	0	1	0.5
Movement G	0	2	1
Movement H	0	1	0.5
Movement I	0	0	0
Movement J	0	1	0.5
Movement K	2	4	3
Movement L	4	1	2.5
Movement M	0	1	0.5
Movement N	0	0	0
Movement O	0	0	0
Movement P	0	0	0

movements at this location. The letters correspond to the counts in Table 1. This enables the user to quickly identify the most common movements, which here is movement K from the multi-use path to the sidewalk and station area.

Description of Primary Issues

Using the knowledge gained from the field visits, reviewing the recorded video, and the available literature, the research group identified and described three overarching causes of problems: the built environment, lack of path user information, and lack of driver information. The two latter categories are concerned primarily with human behavior—our actions, skills, and knowledge—whereas the first category is purely concerned with the physical infrastructure and road design, and best practice within these two areas. The built environment category can be directly impacted by engineering and planning, design, and decisions. In the following, the three different categories are described, along with examples of problems that can be identified under each category. This list is not intended to cover all possible issues at at-grade railroad crossings, but encompasses the primary concerns that were observed through field studies and in the literature, and which were found to be the most useful for practitioners.

The Built Environment

The purpose of our transportation infrastructure is to facilitate movements that are safe and efficient. It does so through structure—such as road design, medians, and sidewalks—and through information, such as signage and pavement markings. When the built environment is lacking content, such as visibility or adequate travel paths, the infrastructure is not fulfilling its purpose of safely and effectively accommodating its users. This leads to undesirable situations for everyone, including those participating in the traffic, but also the responsible engineers and planners. The primary issues from the built environment are as follows:

- Speed: The posted speed limits are too high for the intended road utilization and type.
- Crossing design: The railroad tracks are elevated such that it makes drivers focus more on traversing the tracks and potentially decreases visibility of other road users and traffic control devices.
- Railroad crossing and path distance: The path and the railroad tracks are located either too close to each other or too far apart, making the crossing harder to negotiate.
- Stop line: The distance between the stop line and the tracks, and/or the stop line and the stop line of the opposite direction, and/or the stop line and the path is inappropriate.
- Insufficient crossing infrastructure: Pedestrians are not accommodated through shortest path routing, and therefore choose shortcuts to decrease their travel distance. This includes cutting across areas that are not intended for pedestrians, crossing diagonally, crossing on a track platform, walking on property, and generally bolting across to minimize their travel path, even if reasonable accommodation is available.
- Transit stop: Transit stops are located too close to the railroad crossing.
- Road/street infrastructure: Lack of grade separation or other form of structure between, for

example, the road and the sidewalk can lead to cars unintentionally driving on the sidewalk area, which can lead to conflicts with non-vehicular users.

- Visibility: Inadequate visibility caused by vegetation, buildings, or lack of street light (*13*).

Lack of Path User Information

This category is concerned with the actions and behavior exhibited by the users of the path running adjacent to the tracks, including pedestrians, cyclists, skateboarders, and a variety of other users. The type of path that runs adjacent to railroad tracks is typically a higher-speed path where the users are undisturbed by the surrounding traffic. When they do reach an intersection that means a change in their environment, they need to safely navigate it. When users are unprepared for an upcoming crossing, they can potentially end up in dangerous situations. It is important to ensure that path users are adequately informed of the upcoming crossing or break in their path and can be prepared to proceed safely. The identified primary issues from lack of path user information include the following:

- Speed: The layout and general use of a multi-use path leads to high bike speeds. Bikes may especially be likely to proceed through a crossing when already traveling at a high speed, even more so if on a primarily commuter-oriented path or high-speed trail.
- Signage: There is a lack of signage for bikes and pedestrians surrounding the crossing, leaving them unaware of upcoming crossings or rules of the road. This can also refer to pavement markings in substandard condition.
- Non-compliance: There is a high non-compliance rate of existing treatments and a lack of consequences for non-compliance (*13*).

Lack of Driver Information

This category concerns the users of the highway who are crossing both the railroad tracks and the path crossing. While pedestrians, cyclists, and others using facilities running perpendicular to the tracks and path also technically belong to this category, they are not explicitly handled in this research and do not seem to pose a significant problem. As with the previous category, it is vital that drivers approaching a crossing have adequate information, knowledge, and prior notification to be able to safely traverse the intersection. The primary issues from lack of driver information are as follows:

- Negotiation: If a railroad crossing is inappropriately spaced from a path, the driver will often treat both locations as two separate crossings, and this separation affects how they negotiate each crossing as two different obstacles rather than as one complex crossing. It is for this reason that crossings are generally not placed at curves, as this distracts the driver from paying adequate attention to both the railroad crossing and the curve.
- Vehicle speed: The actual speeds are too high compared to posted speed limits and for the intended road utilization and type.
- Signage: There is a lack of adequate signage to inform drivers of upcoming path or rail crossings, and of rules of the road about stopping for pedestrians or the speed limit, for example. This pertains to the railroad crossing itself, but also and especially to the path layout and the possibility of encountering pedestrians/cyclists. This also includes pavement markings being in substandard conditions (*13*).

The three categories with their 14 subcategories together describe the primary issues and an overall picture of potential conflicts at a complex intersection. These mechanisms should all be considered when attempting to prevent undesirable situations or behavior at an at-grade multimodal crossing. The following section will describe the developed methodology.

Addressing Primary Issues

The developed methodology is directed toward agencies attempting to select cost-effective treatments, which addresses primary issues at complex intersections. It assumes that it has already been established that the crossing in question is of concern to public safety. For this purpose, it also assumes that the responsible agency has information about the crossing and/or a methodology for collecting such. The primary issues are then decided based on existing knowledge and categorized according to the three primary categories and their subcategories presented in the previous section.

Outlining the Process

By combining the identified issues described in the previous section with the developed catalog of available treatments, it was possible to develop potential solutions for different situations at a variety of crossings. The process for selecting the treatment is as follows:

1. Type of rail: refers to whether the location carries heavy rail or light rail.
2. Identified issues: These are identified at the agency's discretion from existing or newly

collected data, such as number of trains per day, nearby activities, recent incidents or annual average daily traffic (AADT), field visits, video recordings, or other sources. It is generally recommended that for each crossing no more than five primary issues are selected. For most crossings it seems that between two and four focus areas are the most appropriate to sufficiently cover the issues at the crossing without "over-treating."

3. School locations: Typically, traffic engineers and planners are especially concerned when a location has young children. Khattak and Luo found that children under the age of eight were involved in "excessive gate-related violations in the absence of older crossing users" (*12*). For this reason, recommendations are different and the treatments generally more severe when the crossing is located within 0.5 miles of an elementary or middle school (K–8), or at the responsible agency's discretion.
4. Recommendations for locations: Generally, one primary recommendation is made for each location as identified in the previous three points, with one or more supplemental or secondary recommendations. These supplemental recommendations are marked with either "OR," meaning the two recommendations can be combined but should not necessarily be; or with "AND," meaning that the recommendations should be combined for best results.

Applying the Methodology

Using the process outlined enables the user to identify the primary issues and a selection of appropriate solutions, using the individual prescriptive tables. As the issues and/or parameters of a crossing increases and/or changes, so does the suggested solutions at the agency's discretion. The issues are selected based on a combination of field observations, video surveillance, public comments, previous history of incidents, and engineering judgment. Once the primary issues are identified and the characteristics of the crossings described, the appropriate solutions are selected and sketched. An example of the selection of treatments is given in the following sections. The prescriptive table for heavy rail is shown in Table 2.

Case Study: SE Spokane St, Portland

The crossing at SE Spokane St is a heavy rail crossing located in a mostly residential area in Portland, OR. The path, Springwater on the Willamette, running adjacent to the tracks has high volumes of pedestrians and bicycles throughout the day, as seen in Table 3. The AADT in 1993 was 3,644. The 2016 estimate is 6,000–8,000. On average, this crossing has two train events per day. The highway/railroad crossing is equipped with MUTCD R1-1, R15-1, W10-1, 8B-7, and has a stop line at all approaches. The path/highway crossing is equipped with MUTCD R1-1 and a stop line. A schematic of the crossing at SE Spokane St can be seen in Figure 4.

Based on field observations and video surveillance, the primary issues were identified and described using the proposed methodology as outlined above.

- Stop line: The stop line for westbound vehicles is placed 20 ft from the nearest track and >40 ft from the multi-use path. As the sight distance is categorized as semi-blind because of interior, vegetation, and topography, this results in drivers not being able to fully see the path and the tracks and their respective activity without crossing the stop line. Because of the downhill slope, the angle of the track, and the way the path is designed, this can be difficult to mitigate. It does lead to cars slowly entering onto the tracks and dwelling, and these issues should be ameliorated.
- Speed: This path is heavily commuter-oriented and from observations approximately half of its users are very familiar with the path and the crossing. It is a wide, well-maintained path, further increasing the speeds at which cyclists travel. The straightness of the path also leads to higher speeds and less awareness of the crossing.
- Non-compliance: Both trailheads are equipped with a smaller, lower-positioned version of a stop sign, asking especially cyclists to stop before entering the intersection.
- Negotiation: This intersection seems to be negotiated by the drivers as two separate intersections rather than a single complex intersection: The path, roads, and tracks are spaced far apart and there is limited visibility. Adding the slopes and angles to the mix leads to a situation that is highly complex, as there are many different approaches by different types of users. The unpredictability of the path users also seems to lead to further confusion for the drivers.

Based on the identified and described primary issues, the treatments were selected as shown in Table 4. This

Table 2. Table of Treatments for Heavy Rail

	Category	Treatment	Description
The built environment: design speed	Primary	Reassess speed limits	Reassessing the posted speed limits will encourage more drivers to keep to an appropriate speed.
	Secondary	Ensure that speed limits are posted and visible	Visibly showing and reminding drivers of the speed limits will potentially further enhance the effect of lowering speed limits.
	Primary, nearby school	Include crossing area in school zone	Compliance with speed limits are very good in school zones, which can be utilized in this situation.
	Secondary, nearby school	Reassess posted speed limits *or* Ensure that speed limits are posted and visible	Reassessing the posted speed limits will encourage more drivers to keep to an appropriate speed. *See above*
The built environment: vertical crossing design	Primary	Add signage	When drivers are not automatically aware of the road or of obstructions, signage can help draw their attention toward certain things.
	Secondary	Install rapid rectangular flashing beacons (RRFBs)	RRFBs are generally a good way of allowing pedestrians to make others aware of their presence, without continuously disturbing traffic when no pedestrians are present. This enforces the driver's attention toward path users and enables them to stop or slow down earlier.
	Primary, nearby school	Raised crosswalks	Raised crosswalks provide structure to the crossing, which ensures a focus on both the railroad–highway crossing and the path crossing. Furthermore, it supports users in crossing legally, as it nudges them toward walking on the raised crosswalk as opposed to next to it or diagonally. Raised crosswalks can present problems for certain classes of vehicles.
	Secondary, nearby school	Automatic RRFBs	*See above*
The built environment: horizontal crossing design	Primary	Move path closer to railroad tracks	Moving the path closer to the tracks makes drivers more likely to negotiate the crossing as one complex crossing as opposed to two separate crossings, which in turns make them more aware of activity on both segments of the complex crossing.
	Secondary	Signage *or* Variable message signs	*See above* Good compliance rates have been seen with variable signs that are appropriately designed.
	Primary, nearby school	Move path closer to railroad tracks *or* Install RRFBs	*See above* *See above*
	Secondary, nearby school	Install automatic RRFBs *or* Traffic signal	*See above* Conventional traffic signals are expensive and a last resort option for controlling traffic. When used in combination with automatic gates, traffic lights can in some instances lead to issues with negotiation, as users forget to check both the automatic gates and traffic signal at a location.

(continued)

Table 2. (continued)

	Category	Treatment	Description
The built environment: stop line	Primary	Move stop line closer to railroad tracks	Moving the stop line closer to the railroad tracks provides better visibility and ability to access the crossing for the driver. There can be some issues with right-of-way with moving the stop line closer.
	Secondary	Raised crosswalks	*See above*
	Primary, nearby school	Move stop line closer to railroad tracks	*See above*
		or	
		Raised crosswalks	*See above*
	Secondary, nearby school	Raised crosswalks	*See above*
		and	
		Dynamic enveloping *and*	Dynamic enveloping makes users more aware of the crossing as a point of conflict, slows down users, and enforces the location as one single complex crossing and not two separate crossings. It is especially a good choice in crossings with high train volumes or in wide crossings.
		Conflict zone traffic paint ("conflict paint")	Conflict paint is used for marking smaller areas of conflict than dynamic enveloping would. The two can also be used together to improve attention to the crossing overall and to some areas specifically. Note: interim approval
The built environment: insufficient crossing infrastructure	Primary	Supply crossing options *or*	If there is a path present on both sides of a highway, there should be a legal way for a user to cross this highway, within a reasonable distance from the path itself.
		Relocate crossing *and*	If there is a path present on both sides of a highway, there should be a legal way for a user to cross this highway, within a reasonable distance from the path itself.
		Add pavement markings *or*	Pavement markings provide structure and information to users about appropriate negotiation.
		Zebra stripes	Pavement markings provide structure and information to users about appropriate negotiation.
	Secondary	Dynamic enveloping	*See above*
	Primary, nearby school	Install pedestrian refuge *or*	A pedestrian refuge allows the user to cross the road in two or more segments. This gives structure to the crossing and is especially recommended in very wide crossings where RRFBs or similar are too costly for the purposes.
		Install RRFBs	*See above*
	Secondary, nearby school	Raised crosswalks	*See above*
		and	
		Dynamic enveloping	*See above*
		and	
		Conflict paint	*See above*
		or	
		Overcrossing *or* Undercrossing	Overcrossings and undercrossings are expensive options that are especially recommended close to schools, where children and teenagers often must navigate the crossing. They are generally a nuisance to cyclists and most other users, especially under ADA, but can be a last resort. *See above*

(continued)

Table 2. (continued)

	Category	Treatment	Description
The built environment: transit stop	Primary	Move transit stop *or*	Moving a transit stop further away from the crossing will have fewer people suddenly start running as they see their bus or train arrive or leave.
		Eliminate stop	Eliminating the transit stop does the same as moving the stop and may be a better option if the location is already adequately served by transit or a nearby stop for the same route.
	Secondary	*Refer to insufficient crossing infrastructure*	N/A
	Primary, nearby school	Move transit stop *or*	*See above*
		Eliminate stop	*See above*
	Secondary, nearby school	*Refer to insufficient crossing infrastructure*	N/A
The built environment: road/street infrastructure	Primary	Implement physical separation *or*	Physical separation keeps users in the desired areas of travel and therefore decreases the potential for conflicts and incidents.
		Pre-made concrete separation blocks	Using pre-made, ready-to-use sections is a faster and more economical way of providing physical separation as described above.
	Secondary	Pavement markings *or*	*See above*
		Zebra stripes	*See above*
	Primary, nearby school	Implement physical separation *or*	*See above*
		Quick curb	*See above*
	Secondary, nearby school	Raised crosswalks *or*	*See above*
		Bollards *or*	Obstructions work well to provide structure, and slow down and guide users toward the areas where they are preferred to walk or bike. Bollards are good because they allow for bike users and ADA users to continue to use the facility, compared to using fencing. Some find that bollards can present a fixed object hazard to cyclists.
		Fencing	Fencing should only be used in cases where it is paramount that users do not attempt to cross undesirably.
The built environment: visibility	Primary	Add signage	*See above*
	Secondary	Refer to horizontal crossing design *or*	N/A
		Maintenance of vegetation *or*	Unmaintained vegetation can potentially decrease visibility.
		Add street lights	Street lights are especially important in denser urban areas with transit running outside of daylight hours available nearby.
	Primary, nearby school	Add signage *and*	*See above*
		Refer to horizontal crossing design	N/A
	Secondary, nearby school	Maintenance of vegetation *or*	Unmaintained vegetation can potentially decrease visibility.
		Add street lights	*See above*

(continued)

Table 2. (continued)

	Category	Treatment	Description
Lack of path user information: trail speed	Primary	Speed treatment	Many different types of speed treatments are available. They are especially useful on high-volume routes used primarily for commuting. Even if the treatment chosen is ADA-compliant, it should not be implemented on routes with frequent ADA-use.
	Secondary	Add signage	*See above*
	Primary, nearby school	Speed treatment	*See above*
	Secondary, nearby school	Obstructions	Obstructions work well to provide structure, slow down users, and push them toward the areas where they are preferred to walk or bike. A variety of different options are available depending on the location.
Lack of path user information: signage	Primary	Add signage	*See above*
	Secondary	Tactile warning surfaces	This solution can provide critical information, especially to ADA users, and enable access
	Primary, nearby school	Add signage *or*	*See above*
		Variable message signs	*See above*
	Secondary, nearby school	In-pavement marker *or*	This solution is especially useful in urban areas, areas without street lights, and areas with high vehicle speeds.
		Automatic RRFBs	*See above*
Lack of path user information: non-compliance	Primary	Obstructions	*See above*
	Secondary	Variable message signs	*See above*
	Primary, nearby school	Obstructions	*See above*
	Secondary, nearby school	Educational initiatives *and*	Especially near schools it can be important to strengthen the understanding of the dangers of trains to children.
		Variable message signs	*See above*
Lack of driver information: negotiation	Primary	Dynamic enveloping *and*	*See above*
		Conflict paint	*See above*
	Secondary	Signage *or*	*See above*
		Traffic lights	*See above*
	Primary, nearby school	Raised crosswalks *and*	*See above*
		Dynamic enveloping *and*	*See above*
		Conflict paint	*See above*
	Secondary, nearby school	RRFBs *or*	*See above*
		Automatic RRFBs *or*	*See above*
		Traffic lights	*See above*
Lack of driver information: vehicle speed	Primary	Speed treatment *or*	Many different types of speed treatments are available. They are especially useful on high-volume routes with low truck rates.
		Pedestrian hybrid beacon (PHB) *or*	PHBs are generally a good way of allowing pedestrians to make others aware of their presence, without continuously disturbing traffic when no pedestrians are present. This enforces the driver's attention toward path users and enables them to stop or slow down earlier.
		RRFB	*See above*

(continued)

Table 2. (continued)

	Category	Treatment	Description
	Secondary	Conflict paint *and* Dynamic enveloping	*See above* *See above*
	Primary, nearby school	Speed treatments *and* In-pavement marker *or* HAWK *or* RRFB	*See above* This solution is especially useful in urban areas, areas without street lights, and areas with high vehicle speeds. *See above* *See above*
	Secondary, nearby school	Install active speed sign	This reminds motorists of their speed as it compares with the posted speed limit.
Lack of driver information: signage	Primary	Add signage *or* Add pavement markings	*See above* Pavement markings provide structure and information to users about appropriate negotiation.
	Secondary	Refer to insufficient crossing infrastructure	N/A
	Primary, nearby school	Add signage *or* Add pavement markings *and* Refer to insufficient crossing infrastructure	*See above* *See above* N/A
	Secondary, nearby school	Speed treatments *or* In-pavement markings *or* RRFBs	*See above* This solution is especially useful in urban areas, areas without street lights, and areas with high vehicle speeds. *See above*

Note: This table does not include all possible types of treatments and they may be excluded for several reasons, such as that the treatment is not commonly used, is not permitted in Oregon, or is experimental. Several treatments can be interchanged at the responsible agency's discretion.

Table 3. Counts from the Observations and Video Surveillance

Mode of transport	July 7, 2016, 8 am–9 am	July 6, 2016, 5:15 pm–6:15 pm	July 6, 2016, 3:23 pm–4:38 pm
Cars	136	281	439
Pedestrians	55	68	54
Bikes	154	305	250
Trains	0	0	0

table is based on Table 2 but is operationalized to make it easier for the agency using it to mark and review selected treatments. Table 4 is a condensed version of the original table and the full version can be found in the supporting publication (*13*).

The existing stop line is located 20 ft away from the crossing but should be moved slightly further away from the tracks to increase sight distance. The trail carries high volumes of pedestrians and bikes throughout the day, and on average more pedestrians and bikes than motor vehicles enter the crossing. It is in the city's interest to continue to prioritize and support the high volumes of pedestrians and bikes. A raised crosswalk addresses inadequate vertical design issues by increasing the driver's focus on the highway–path crossing, as opposed to solely on the highway–railroad crossing.

The combination of a raised crosswalk, conflict paint, and dynamic envelope further strengthens the

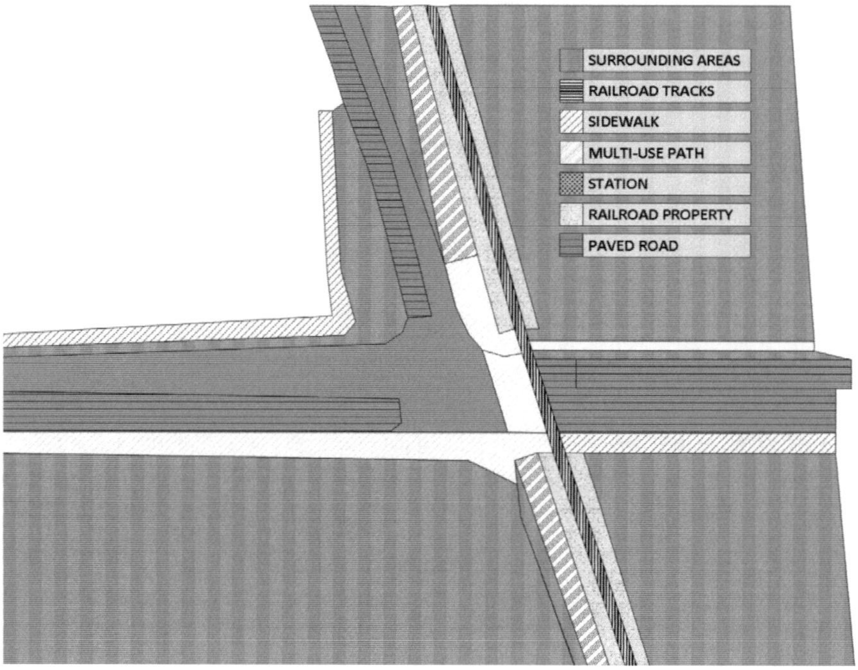

Figure 4. Schematic of the SE Spokane St crossing.

perception of the crossing as a single, complex crossing. This enables motor vehicles to better assess the crossing before entering and therefore negotiate it more safely. This can potentially lead to a decrease in conflicts between users and in the observance of illegal behavior. Further suggestions include adding a slight uphill slope on both sides of the raised crosswalk to nudge bicyclists to decrease their speed as they are approaching. This enables other users to access the raised crosswalk more easily. The changes should be made to ensure compliance with Americans with Disabilities Act (ADA) and Proposed Rights-of-Way Guidelines (PROWAG) requirements. Figures 5 and 6 show the proposed changes.

Conclusions

This research investigated and utilized several different methods and approaches to develop a methodology that can serve to support decision-making when selecting appropriate treatments in multimodal, complex intersections. The proposed methodology is designed to be able to balance a predetermined, prescriptive approach with the professional judgment of the agency carrying out the investigation. It allows the practitioner to utilize collected information about a complex multimodal intersection, apply it to a predetermined set of specifications using engineering judgment, and by that come up with a set of treatments that has previously been found to address the issues identified.

Currently there is not a standardized methodology for selecting appropriate treatments at complex railroad crossings, and they are either treated as two separate crossings or by following guidelines specific to the responsible DOT. Because complex intersections are generally different from each other, having a standardized method that does not allow for engineering judgment is not a preferred option. However, having a standardized set of guidelines that can support the work done by rail divisions and local governments across the country can improve outcomes when selecting treatments. A previously agreed-on methodology allows the relevant agency to streamline their crossing improvement efforts; to easily communicate and inform the public of the decisions made and their reasons for doing so; to secure stakeholder buy-in before starting a project or investigation, which will in turn lead to better outcomes; to make sure that approach and selected treatments are more standardized; and to ensure transparency in the organization to make at-grade crossings safer for pedestrians and cyclists, without negatively impacting trains or vehicles.

Table 4. Selected Treatments for This Location

	Crossings not near schools		Crossings near schools	
The built environment	Primary recommendation	Secondary recommendation(s)	Primary recommendation	Secondary recommendation(s)
☐ Speed	☐ Lower posted speed limits	☐ Ensure that speed limits are posted and visible	☐ Include crossing area in school zone	☐ Lower posted speed limits *or* ☐ Ensure that speed limits are posted and visible
☐ Vertical crossing design	☐ Add signage	☐ Install RRFBs	☐ Raised crosswalks	☐ Install automatic RRFBs
☐ Horizontal crossing design	☐ Move path closer to or further away from railroad tracks	☐ Signage *or* ☐ Variable message signs	☐ Move path closer to or further away from railroad tracks *or* ☐ Install RRFBs	☐ Install automatic RRFBs *or* ☐ Install traffic lights
☒ Stop line	☒ Move stop line closer to railroad tracks	☒ Raised crosswalks	☐ Move stop line closer to railroad tracks *or* ☐ Raised crosswalks	☐ Raised crosswalks *and* ☐ Dynamic enveloping *and* ☐ Conflict paint
☐ Insufficient crossing infrastructure	☐ Add signage *and* ☐ Supply crossing options *and* ☐ Add pavement markings *or* ☐ Relocate crossing *and* ☐ Add pavement markings	☐ Dynamic enveloping	☐ Pedestrian refuge *or* ☐ Install RRFBs	☐ Raised crosswalks *and* ☐ Dynamic enveloping *and* ☐ Conflict paint *or* ☐ Overcrossing *or* ☐ Undercrossing
☐ Transit stop	☐ Move stop *or* ☐ Eliminate stop	Refer to *insufficient crossing infrastructure*	☐ Move stop *or* ☐ Eliminate stop	Refer to *insufficient crossing infrastructure*
☐ Road/street infrastructure	☐ Implement physical separation *or* ☐ Quick curb	☐ Pavement markings *or* ☐ Marked crosswalk	☐ Implement physical separation *or* ☐ Quick curb	☐ Raised crosswalks *or* ☐ Bollards *or* ☐ Fencing
☐ Visibility	☐ Add signage	Refer to *horizontal crossing design* *or* ☐ Maintenance of vegetation *or* ☐ Add street lights	☐ Add signage *and* Refer to *horizontal crossing design*	☐ Maintenance of vegetation *or* ☐ Add street lights
Lack of path user information	Primary recommendation	Secondary recommendation(s)	Primary recommendation	Secondary recommendation(s)
☒ Speed	☒ Speed treatment	☐ Add signage	☐ Speed treatment	☐ Obstructions
☐ Signage	☐ Add signage	☐ Tactile warning surfaces	☐ Add signage *or* ☐ Variable message signs	☐ In-pavement marker *or* ☐ Automatic RRFBs
☒ Non-compliance	☒ Obstructions *or* ☐ Variable signs	Refer to *insufficient crossing infrastructure*	☐ Obstructions *or* ☐ Educational initiatives *and* ☐ Variable signs	Refer to *insufficient crossing infrastructure*

(continued)

Table 4. (continued)

	Crossings not near schools		Crossings near schools	
Lack of driver information	Primary recommendation	Secondary recommendation(s)	Primary recommendation	Secondary recommendation(s)
☒ Negotiation	☒ Dynamic enveloping *and* ☒ Conflict paint	☐ Signage *or* ☐ Traffic lights	☐ Raised crosswalks *and* ☐ Dynamic enveloping *and* ☐ Conflict paint	☐ RRFBs *or* ☐ Automatic RRFBs *or* ☐ Traffic lights
☐ Vehicle speed	☐ Speed treatment *or* ☐ PHBs *or* ☐ RRFBs	☐ Conflict paint *and* ☐ Dynamic enveloping	☐ Speed treatment *and* ☐ In-pavement marker *or* ☐ PHBs *or* ☐ RRFBs	☐ Active speed sign
☐ Signage	☐ Add signage *or* ☐ Add pavement markings	Refer to *insufficient crossing infrastructure*	☐ Add signage *or* ☐ Add pavement markings *and* Refer to *insufficient crossing infrastructure*	☐ Speed treatments *or* ☐ In-pavement markings *or* ☐ RRFBs

Figure 5. Proposed changes to SE Spokane St.

Acknowledgments

This research was funded by the Oregon Department of Transportation's Research Section. The authors would like to thank the ODOT Rail Public Transit Division for their support and expertise throughout the project. The project was furthermore supported by the FHWA and by the Valle Scholarship and Scandinavian Exchange Program.

Author Contributions

The authors confirm contribution to the paper as follows: study conception and design: AG, EM, ABA; data collection: ABA, PB, MS; analysis and interpretation of results: ABA, AG, EM; draft manuscript preparation: ABA. All authors reviewed the results and approved the final version of the manuscript.

Figure 6. Proposed changes to SE Spokane St.

References

1. US FRA. *Highway-Rail Crossing & Trespassing Fact Sheet*. U.S. Department of Transportation, Washington, D.C., 2016.
2. Birk, M. L., A. Ferster, M. G. Jones, P. K. Miller, G. M. Hudson, J. Abrams, and D. Lerch. *Rails-with-Trails: Lessons Learned*. U.S. Department of Transportation, Washington, D.C., 2002.
3. Oregon Department of Transportation. Rules of the Road for Drivers. In *Vehicle Code Book*. Oregon Department of Transportation, Salem, OR, 2015.
4. Oregon Department of Transportation. *Oregon Pedestrian, Bicycle and Driver Rules: Selected Statutes*. Oregon Department of Transportation, Salem, OR, 2016.
5. FHWA. *Manual on Uniform Traffic Control Devices*. Federal Highway Administration, Washington, D.C., 2009.
6. AASHTO. *A Policy on Geometric Design of Highways and Streets*, 6th ed. American Association of State Highway and Transportation Officials, Washington, D.C., 2011.
7. FDOT. *Trail Intersection Design Handbook*. Florida Department of Transportation, Tallahassee, FL, 1996.
8. Noyce, D. A., Z. Li, J. Ash, and G. Khan. *Best Practices Synthesis and Guidance in At-Grade Trail Crossing Treatments*. Minnesota Department of Transportation, St. Paul, MN, 2013.
9. Birk, M. L., A. Ferster, M. G. Jones, P. K. Miller, G. M. Hudson, J. Abrams, and D. Lerch. *Rails-with-Trails: Lessons Learned—Literature Review, Current Practices, Conclusions*. U.S. Department of Transportation, Washington, D.C., 2002.
10. Meeker, F., D. Fox, and C. Weber. A Comparison of Driver Behavior at Railroad Grade Crossings with Two Different Protection Systems. *Accident Analysis & Prevention*, Vol. 29, No. 1, 1997, pp. 11–16.
11. Metaxatos, P., and P. S. Sriraj. *Pedestrian/Bicyclist Warning Devices and Signs at Highway–Rail and Pathway–Rail Grade Crossings*. Illinois Center for Transportation, Rantoul, IL, 2013.
12. Khattak, A., and Z. Luo. Pedestrian and Bicyclist Violations at Highway–Rail Grade Crossings. *Transportation Research Record: Journal of the Transportation Research Board*, 2011. 2250: 76–82.
13. Bovbjerg, A., M. Sheth, A. Goodchild, and E. McCormack. *Multi-Modal Intersections: Resolving Conflicts Between Trains, Motor Vehicles, Bicyclists and Pedestrians*. Report No. FHWA-OR-18-04. Oregon Department of Transportation, Salem, OR, 2017.

The Standing Committee on Highway/Rail Grade Crossings (AHB60) peer-reviewed this paper (18-04794).

Article

TRR
JOURNAL OF THE TRANSPORTATION RESEARCH BOARD

Transportation Research Record
2018, Vol. 2672(10) 28–37

DOI: 10.1177/0361198118780885
journals.sagepub.com/home/trr
SAGE

Modeling the Probability of Hazardous Materials Release in Crashes at Highway–Rail Grade Crossings

Amirfarrokh Iranitalab[1], Yashu Kang[1], and Aemal Khattak[1]

Abstract

Crashes at Highway–Rail Grade Crossings (HRGCs) that involve a truck or a train carrying hazardous materials (hazmat) expose people and the environment to potentially severe consequences of hazmat release. This research involved statistical modeling of the probability of hazmat release from trucks and/or trains in crashes at HRGCs to identify factors associated with hazmat release. The Federal Railroad Administration (FRA) HRGC crash dataset (2007–2016) yielded two subsets of crashes: 1) those involving hazmat-carrying trucks, and 2) those involving hazmat-carrying trains. Results from a logistic regression model using data subset 1 (crashes involving hazmat-carrying trucks) with hazmat release/no release as the response variable showed that standard flashing signal lights, railroad crossbucks, and railroad classes II and III (relative to railroad class I) were associated with lower hazmat release probability from hazmat-carrying trucks. Hazmat release probability from trucks was higher with freight train involvement. Results from a logistic regression model using data subset 2 (crashes involving hazmat-carrying trains) revealed that hazmat release probability from trains was lower with warmer temperature. However, the probability of release from trains was greater with railroad class II (relative to railroad class I), type of highway user (different types of trucks and motorcycle relative to automobiles), and weather conditions (fog, sleet or snow, relative to clear). A comparison of the results from this study with HRGC crash severity studies highlighted the importance and usefulness of this study.

Surface transportation carries a substantial volume of hazardous materials (hazmat) transported across the United States. Trucks constitute more than 50% of the total annual hazmat transportation in the United States (*1*), and more than 2.6 million carloads of hazmat was transported by rail in 2016 in North America (*2*). Hazmat-carrying trains and trucks potentially expose people and the environment to infrequent but potentially severe consequences of crashes that result in release of hazmat. For example, 2581 hazmat-carrying train incidents were reported to the Federal Railroad Administration (FRA) in the 2012–2016 time period, while only 84 of those incidents led to hazmat release (*3*). However, the consequences of a train crash and release of chlorine gas in Graniteville, South Carolina in 2005 resulted in 9 fatalities, hundreds of injuries, evacuation of approximately 5400 people, and monetary costs exceeding $6.9 million (*4*). While these crashes may occur anywhere on the transportation system, crashes at highway–rail grade crossings (HRGCs), where the two modes of transportation meet, may lead to hazmat release from either trucks or trains, or both. Identifying the contributing factors to hazmat release in HRGC crashes involving a hazmat-carrying truck or train is important for setting policies and for making more informed public safety related decisions.

With a focus on crashes at HRGCs involving hazmat-carrying trucks and/or trains, the research objective was to identify the effects of highway users' characteristics, truck/train attributes, environment, land-use and HRGC traits on the probability of hazmat release from trucks or trains in these crashes. The FRA's HRGC crash dataset (2007–2016) yielded two crash data subsets: 1) crashes involving hazmat-carrying trucks, and 2) crashes involving hazmat-carrying trains. Logistic regression models were estimated using each data subset with hazmat release/no release as the response variable. Both models provided useful information about the presence and magnitude of effects of explanatory variables on hazmat release from these two transportation modes. Based on

[1]Department of Civil Engineering and Nebraska Transportation Center, University of Nebraska-Lincoln, Lincoln, NE

Corresponding Author:
Address correspondence to Aemal Khattak: khattak@unl.edu

the results, this paper presents recommendations for countermeasures and for policies toward decreasing hazmat release in HRGC crashes.

Organization of the remainder of this paper is as follows. The next section presents a review of literature on safety of hazmat transportation and HRGCs. Methodology, data and variables, and modeling results are the ensuing sections of this paper. Discussion, conclusions and a list of references complete this paper.

Literature Review

A significant number of studies have focused on safety of hazmat transportation and safety of HRGCs. This section presents a review of the relevant results from those studies. Although findings regarding safety of hazmat transportation and safety of HRGCs are helpful in studying crashes leading to hazmat release at HRGCs, the review of published literature did not uncover studies specifically focused on the safety of hazmat transportation at HRGCs by trucks and trains.

Safety of Hazmat Transportation

Studying safety of hazmat transportation is commonly based on risk-based approaches due to the rarity of hazmat release incidents. The usual definition of risk involves a set of components including occurrence probability of a hazmat-involved incident, exposure measures (e.g., traffic in terms of truck-miles and train-miles, population, environment, etc.), probability of hazmat release in an incident, and consequences of the release. A number of studies focused on the methods for quantification of these components or using risk in different approaches towards reduction of occurrence probability and severity of hazmat incidents. Countermeasures regarding hazmat transportation risk reduction recommended in these studies included routing strategies (*1*, *5–8*), hazmat transportation network design (*9–12*), tank car safety design enhancement in rail transportation (*4*, *13–15*), hazmat-carrying train speed reduction (*16*) and emergency response improvements (*1*, *17*).

Some studies applied optimization techniques for reducing the risk of hazmat release in crashes. Zografos and Androutsopoulos (*1*) proposed a decision support system for evaluating routes and implemented the method using hazmat distribution within the area of Thriasion Pedion, Greece. By combining transportation network modeling and risk assessment, Glickman et al. (*5*) concluded that significant risk reductions were achievable with relatively minor alterations in routing patterns. Verter and Kara (*9*, *10*) investigated the possibility of designing a transportation network for transportation of hazmat to minimize population exposure. Similarly, Erkut and Gzara (*11*) modeled the transportation network design problem as a bi-level network flow formulation, and Erkut and Alp (*12*) considered designating hazmat routes in and through a major population center. Saat and Barkan (*13*), and Barkan et al. (*14*, *15*) presented different approaches to tank car safety design via optimization to address the tradeoff between safety design and tank car weight.

Nayak et al. (*18*) presented a number of methods for quantifying the probability of accidents and severity of impacts of hazmat accidents in rail transportation. Saccomanno and El-hage (*19*, *20*) focused on tank car derailment, a common cause of hazmat release in train incidents. Liu et al. (*21*) analyzed number of derailed tank cars as a measure of severity of tank car derailments in freight-train accidents. Verma et al. (*22*) proposed a bi-objective model for planning and managing rail-truck intermodal hazmat transportation.

Safety at Highway-Rail Grade Crossings

The majority of studies on safety of HRGCs focused on analysis of crash frequency and severity at these transportation junctions. Raub (*23*) examined the performance of four specific warning device classes (crossbucks only, STOP signs, flashing lights, and gates) at HRGCs and compared their effects on crash frequency. Hu et al. (*24*) studied and identified factors associated with crash injury severity at HRGCs, which included number of daily trains, number of daily trucks, highway separation, an obstacle detection device, and approaching crossing marks. Other factors that were associated with crash frequency and severity at HRGCs include highway motor vehicle driver's age and behavior, traffic volume, and weather conditions (*25*, *26*). Zhao and Khattak (*27*) found that greater number of highway lanes at HRGCs, the presence of standard flashing-light signals and clear weather decreased the likelihood of severe injuries. Zhao et al. (*28*) showed that higher train speed, female pedestrians and commercial land use were associated with more severe injuries in pedestrian-train crashes at HRGCs. Fan et al. (*29*) identified pick-up trucks, concrete, and rubber surfaces associated with more severe crashes at HRGCs, while truck-trailers, snow and fog, and higher daily traffic volumes were more likely to be observed in less severe crashes.

The contributing factors to crash frequency and severity at HRGCs were relatively consistent in the reviewed studies. However, these results do not necessarily hold in describing probability of hazmat release, which warrants the investigation of hazmat-related crashes at HRGCs. The next section presents the research methodology and a description of the modeling technique.

Methodology

This research involved estimation of two logistic regression models for capturing possible impacts and associations of explanatory variables on the probability of hazmat release from hazmat-carrying trucks (Truck Model) and trains (Train Model) in crashes reported at HRGCs. Occurrences of hazmat release or no release from trucks and trains respectively were binary response variables in the two logistic regression models. Explanatory variables included HRGC-related traits, train and highway user characteristics, type of crash, and environmental and land-use characteristics. The following paragraphs describe the logistic regression model, estimation methods and the tools for model interpretation used in this study.

Logistic regression models are the common binary response regression models (*30, 31*). These models are a type of generalized linear model with the Bernoulli distribution assumption for the response variable and cumulative density function of a logistic probability distribution as the link function. In binary response models, the quantity that is estimated is the probability of success, π. In the context of this paper, π_i is probability of hazmat release in the ith HRGC crash from trucks and trains in the Truck and Train models, respectively. If $x_{i1}, \ldots, x_{ip}$ are p explanatory variables measured on the ith observation (i^{th} HRGC crash), in logistic regression models π_i is defined as in Equation 1.

$$\pi_i = \frac{exp(\beta_0 + \beta_0 x_{i1} + \ldots + \beta_p x_{ip})}{1 + exp(\beta_0 + \beta_0 x_{i1} + \ldots + \beta_p x_{ip})} \quad (1)$$

In this equation $\beta_0, \ldots, \beta_p$ are the logistic regression parameters or coefficients of the explanatory variables. Equation 1 can be rewritten as shown in Equation 2. The left side of Equation 2 is the natural logarithm for the odds of success (hazmat release) and the right side of this equation is a linear combination of the regression parameters with the explanatory variables, often referred to as linear predictors. This transformation of π_i is referred to as the logit transformation (*31*).

$$log\left(\frac{\pi_i}{1-\pi_i}\right) = \beta_0 + \beta_0 x_{i1} + \ldots + \beta_p x_{ip} \quad (2)$$

Model interpretation in this study used the odds ratios (OR). For a c-unit increase in a continuous explanatory variable, x, the OR is defined as shown in Equation 3. This measure is interpreted as "the odds of hazmat release from trucks or trains change by OR times for every c-unit increase in x, holding the other variables constant". For a categorical explanatory variable, the value of c is 1 and the interpretation is "the odds of hazmat release from trucks or trains change by OR times as large for $x = 1$ than for $x = 0$, holding the other variables constant" (*30, 31*).

$$\text{OR} = \frac{\text{Odds}_{x+c}}{\text{Odds}_x} = e^{c\beta_i} \quad (3)$$

Maximum Likelihood (ML) is the most common estimation method for logistic regression models. In some cases this estimator might lead to biased results, e.g., when an outcome of the response variable is rare, or presence of complete or quasi separation (an explanatory variable separates the data between 0 and 1 for the response variable (*31*)). The bias-reduction method developed by Firth (*32*) is a solution for this issue. In this study quasi separation was present in the Truck Model (detected by observing abnormally large estimated standard errors for some of the coefficients (*31*) in case of using ML), and values of 1 (hazmat release) were rare in the response variable of the Train Model (0.13%). Therefore, in both models ML was replaced with the Firth estimation method.

Data and Variables

Ten-year U.S. HRGC accident/incident data (2007–2016) and HRGC history inventory data were obtained from FRA safety database, which included 21,761 total number of reported accidents (*3*). Crashes were matched with the inventory dataset based on unique HRGC identification number and approximate date of crash. Two subsets of crashes were extracted from this dataset: 1) crashes with hazmat-carrying highway users (trucks) containing 75 crashes, and 2) crashes with hazmat-carrying trains, including 3726 crashes. Truck Model and Train Model were estimated using these two subsets, respectively. Table 1 presents the variables and their statistics for the truck subset, while Table 2 shows similar information for the train subset.

Modeling Results

Two logistic regression models were estimated for hazmat release from trucks (Truck Model) and trains (Train Model) in crashes at HRGCs. Variable selection was based on Corrected Akaike's Information Criterion (AICc). Some variables in both models were not statistically significant (at $\alpha = 0.10$ level), but were retained in model specifications, since they contributed to the models according to AICc (via describing small proportions of variations in the response variable and affecting other parameters of the models) (*31*). Table 3 shows the modeling results including estimated coefficients, standard errors, OR, and 90% profile likelihood ratio confidence intervals for OR. The significance of estimated coefficients with 90% confidence can be judged by looking at the OR confidence intervals (if each interval does not contain 1, hypothesis of equality of the coefficient with zero is rejected). Figure 1 shows the point estimates and profile likelihood ratio confidence intervals of OR, for better observation and easier interpretation.

Table 1. Variables and Statistics of the Hazmat-Carrying Truck Data Subset

Variable	Variable name	Values and statistics
Response variable		
Hazmat release	HAZREL	1 = No (45.33%), 2 = Yes (54.67%)
Explanatory variables		
Highway user characteristics		
Type of vehicle	TYPVEH	1 = Truck (20.00%), 2 = Truck-trailer (72.00%), 3 = Pick-up Truck (08.00%)
Vehicle speed	VEHSPD	Mean = 8.3467, Variance = 164.5268
Driver age	DRIVAGE	Mean = 47.3467, Variance = 169.0944
Driver gender	DRIVGEN	1 = Male (98.67%), 2 = Female (1.33%)
Train characteristics		
Railroad class	TYPRR	1 = Class I (81.33%), 2 = Class II (01.33%), 3 = Class III (17.33%)
Freight train	FREIGHT	1 = No (21.33%), 2 = Yes (78.67%)
Train speed	TRNSPD	Mean = 32.4110, Variance = 326.5510
Number of cars	NBRCARS	Mean = 51.5135, Variance = 1615.2670
Crash characteristics		
Type of crash	TYPACC	1 = Train struck highway user (92.00%), 2 = Train struck by highway user (08.00%)
Environment and land-use characteristics		
Temperature	TEMP	Mean = 63.1600, Variance = 473.4335
Weather	WEATHER	1 = Clear (64.00%), 2 = Cloudy (30.67%), 3 = Rain (05.33%), 4 = Fog, sleet, snow (00.00%)
Visibility	VISIBLTY	1 = Dawn (06.67%), 2 = Day (70.67%), 3 = Dusk (08.00%), 4 = Dark (14.67%)
Type of land use	DEVELTYPE	1 = Open space (41.33%), 2 = Residential (09.33%), 3 = Commercial (10.67%), 4 = Industrial (30.67%), 5 = Other (08.00%)
HRGC characteristics		
Cantilever flashing-light signals	CANTALIVERFLS	1 = No (94.67%), 2 = Yes (05.33%)
Standard flashing-light signals	STANDARDFLS	1 = No (69.33%), 2 = Yes (30.67%)
Bells	BELLS	1 = No (58.67%), 2 = Yes (41.33%)
Crossbucks	CROSSBUCKS	1 = No (32.00%), 2 = Yes (68.00%)
Gates	GATES	1 = No (70.67%), 2 = Yes (29.33%)
Highway traffic signal	HWYTRFICSIG	1 = No (98.67%), 2 = Yes (01.33%)
Audible	AUDIBLE	1 = No (80.00%), 2 = Yes (20.00%)
Stop sign	STOPSIGNS	1 = No (64.00%), 2 = Yes (36.00%)
Other control devices	OTHER	1 = No (82.67%), 2 = Yes (17.33%)
Public/Private HRGC	PUBLIC	1 = No (29.33%), 2 = Yes (70.67%)

In hazmat-carrying truck crashes at HRGCs, with 90% confidence and holding all the other variables constant except the variable being interpreted, it can be said that presence of standard flashing-light signals decreased the odds of hazmat release from trucks by an amount between 0.0475 to 0.4716 times, relative to its absence. Railroad crossbucks decreased the odds of release from trucks by 0.0461 to 0.4065 times and public crossings increased these odds by 1.6148 to 15.8892 times, compared with private crossings. Railroad classes II and III decreased the odds of release from trucks by amounts between 0.0013 and 0.9781 and between 0.0496 and 0.4631, respectively, relative to railroad class I. Freight trains increased truck release odds by 1.9958 to 17.4551 times, compared with non-freight trains. Crossing control devices introduced as "Other control devices" (in Table 1) decreased the odds of release from trucks by 0.0907 to 0.8836 times, compared with absence of these control devices. Sufficient statistical evidence was not available to support the existence of effects of any other variables that were considered in this study on the release of hazmat from trucks in HRGC crashes.

In hazmat-carrying train crashes at HRGCs, again, with 90% confidence and holding all the other variables constant except the variable being interpreted, railroad class II increased the odds of hazmat release from trains by 1.3266 to 62.4336 times, relative to railroad class I. Railroad class III did not have any significant difference from railroad class I regarding the probability of hazmat release from trains. Type of highway user changed the probability of hazmat release from trains: trucks, truck-trailers and pick-up trucks increased the odds of release by an amount between 1.6463 to 57.1876 times, compared with automobiles; crashes with motorcycles, other motor vehicles and other objects increased these odds by 2.1248 to 112.5912 times, relative to automobiles; hazmat release probability did not change in crashes with vans, buses and school buses, and pedestrians relative to automobiles. An increase in temperature by 5° F decreased the odds of hazmat release from trains by an amount

Table 2. Variables and Statistics of the Hazmat-Carrying Train Data Subset

Variable	Variable names	Values and statistics
Response variable		
Hazmat release	HAZREL	1 = No (99.86%), 2 = Yes (0.13%)
Explanatory variables		
Highway user characteristics		
Type of vehicle	TYPVEH	1 = Auto (42.53%), 2 = Truck/truck-trailer/pick-up truck (40.55%), 3 = Van/bus/school bus (03.43%), 4 = Pedestrian (04.85%), 5 = Motorcycle/other (08.64%)
Vehicle speed	VEHSPD	Mean = 7.2321, Variance = 128.9675
Driver age	DRIVAGE	Mean = 42.3601, Variance = 299.6472
Driver gender	DRIVGEN	1 = Male (75.46%), 2 = Female (24.54%)
Train characteristics		
Railroad class	TYPRR	1 = Class I (86.23%), 2 = Class II (02.60%), 3 = Class III (11.16%)
Train speed	TRNSPD	Mean = 32.7384, Variance = 243.8642
Number of cars	NBRCARS	Mean = 69.1539, Variance = 1322.8460
Crash characteristics		
Type of crash	TYPACC	1 = Train struck highway user (82.54%), 2 = Train struck by highway user (17.46%)
Environment and land-use characteristics		
Temperature	TEMP	Mean = 60.8345, Variance = 491.5737
Weather	WEATHER	1 = Clear (69.78%), 2 = Cloudy (20.33%), 3 = Rain (05.60%), 4 = Fog, sleet, snow (04.29%)
Visibility	VISIBLTY	1 = Dawn (05.17%), 2 = Day (57.11%), 3 = Dusk (05.52%), 4 = Dark (23.18%)
Type of land use	DEVELTYPE	1 = Open space (28.35%), 2 = Residential (21.08%), 3 = Commercial (26.79%), 4 = Industrial (16.30%), 5 = Other (07.48%)
HRGC characteristics		
Cantilever flashing-light signals	CANTALIVERFLS	1 = No (80.69%), 2 = Yes (19.31%)
Standard flashing-light signals	STANDARDFLS	1 = No (54.30%), 2 = Yes (45.70%)
Bells	BELLS	1 = No (41.88%), 2 = Yes (58.12%)
Crossbucks	CROSSBUCKS	1 = No (32.00%), 2 = Yes (68.00%)
Gates	GATES	1 = No (70.67%), 2 = Yes (29.33%)
Highway traffic signal	HWYTRFICSIG	1 = No (97.61%), 2 = Yes (02.39%)
Audible	AUDIBLE	1 = No (63.88%), 2 = Yes (36.12%)
Stop sign	STOPSIGNS	1 = No (78.98%), 2 = Yes (21.02%)
Other control devices	OTHER	1 = No (83.75%), 2 = Yes (16.25%)
Public/Private HRGC	PUBLIC	1 = No (12.82%), 2 = Yes (87.18%)

between 0.7990 to 0.9863 times. Fog, sleet, and snow increased the odds of release by 1.3229 to 24.0584 times, relative to clear weather. There was not enough statistical evidence toward the existence of any impacts or association of any other variables on the release of hazmat from trains in HRGC crashes.

Discussion and Conclusion

Standard flashing-light signals, railroad crossbucks and "other crossing control devices" (as is in Table 1) were effective in reducing the probability of hazmat release from trucks in truck-train HRGC crashes. The use of such control devices is recommended at HRGCs with high hazmat-carrying truck traffic. Prioritization of safety countermeasures implementation may be given to public HRGCs since hazmat release is more probable at these locations relative to private crossings. Freight trains were associated with higher probability of hazmat release from trucks. This finding is reasonable as freight trains are usually longer and heavier relative to other (e.g., passenger) trains. HRGCs with more frequent passage of trains that belong to railroad classes II and III, and less frequent passage of freight trains were safer for hazmat-carrying trucks. Routes that minimize the interaction between these trucks with class I railroads and freight trains may be preferred and considered in the route selection of hazmat-carrying trucks.

Hazmat-carrying class II railroads were more vulnerable in HRGC crashes relative to class I railroads in terms of hazmat release. Extra train hazmat safety consideration is recommended for hazmat-carrying trains on routes with HRGCs that carry high volumes of trucks, truck-trailers and pick-up trucks, and also motorcycle and other vehicles (relative to automobiles). With the exception of motorcycles, different types of trucks and

Table 3. Results of Truck and Train Logistic Regression Models

Variable	c	Truck model					Train model				
		Estimated coefficient	Standard error	OR	OR 90% confidence interval		Estimated coefficient	Standard error	OR	OR 90% confidence interval	
					Lower level	Upper level				Lower level	Upper level
(Intercept)	NA	0.1276	0.6314	NA	NA	NA	−7.52731	1.44417	NA	NA	NA
TYPACC2	1	−1.5033	1.0194	0.2224	0.0409	1.0216	0.94787	0.66771	2.5802	0.8604	7.738
STANDARDFLS	1	−1.8406	0.7061	0.1587	0.0475	0.4716	—	—	—	—	—
CROSSBUCKS	1	−1.9297	0.6706	0.1452	0.0461	0.4065	—	—	—	—	—
PUBLIC	1	1.5747	0.7058	4.8294	1.6148	15.8892	—	—	—	—	—
TYPRR2	1	−2.6878	2.4092	0.0680	0.0013	0.9781	2.20836	1.17077	9.1008	1.3266	62.4336
TYPRR3	1	−1.8332	0.6919	0.1599	0.0496	0.4634	−0.84804	1.12983	0.4283	0.0668	2.7466
FREIGHT	1	1.7275	0.6682	5.6267	1.9958	17.4551	NA	NA	NA	NA	NA
OTHER	1	−1.2336	0.7053	0.2912	0.0907	0.8836	—	—	—	—	—
TYPVEH2	1	—	—	—	—	—	2.27244	1.07845	9.7031	1.6463	57.1876
TYPVEH3	1	—	—	—	—	—	2.32472	1.51188	10.2238	0.8503	122.9214
TYPVEH4	1	NA	NA	NA	NA	NA	2.31776	1.50243	10.1529	0.8577	120.1858
TYPVEH5	1	NA	NA	NA	NA	NA	2.73871	1.20683	15.467	2.1248	112.5912
TEMP	5	—	—	—	—	—	−0.02382	0.01281	0.8877	0.799	0.9863
WEATHER2	1	—	—	—	—	—	1.00773	0.68463	2.7394	0.8883	8.4473
WEATHER3	1	—	—	—	—	—	0.80258	1.16608	2.2313	0.3278	15.1896
WEATHER4	1	—	—	—	—	—	1.73017	0.88173	5.6416	1.3229	24.0584
DEVELTYPE2	1	—	—	—	—	—	−0.42363	1.22511	0.6547	0.0873	4.9111
DEVELTYPE3	1	—	—	—	—	—	0.42381	0.89897	1.5278	0.3482	6.7025
DEVELTYPE4	1	—	—	—	—	—	1.18478	0.84209	3.27	0.8185	13.0645
DEVELTYPE5	1	—	—	—	—	—	1.53165	0.97012	4.6258	0.9379	22.8138

Note: — = not used in the final model; NA = not applicable.

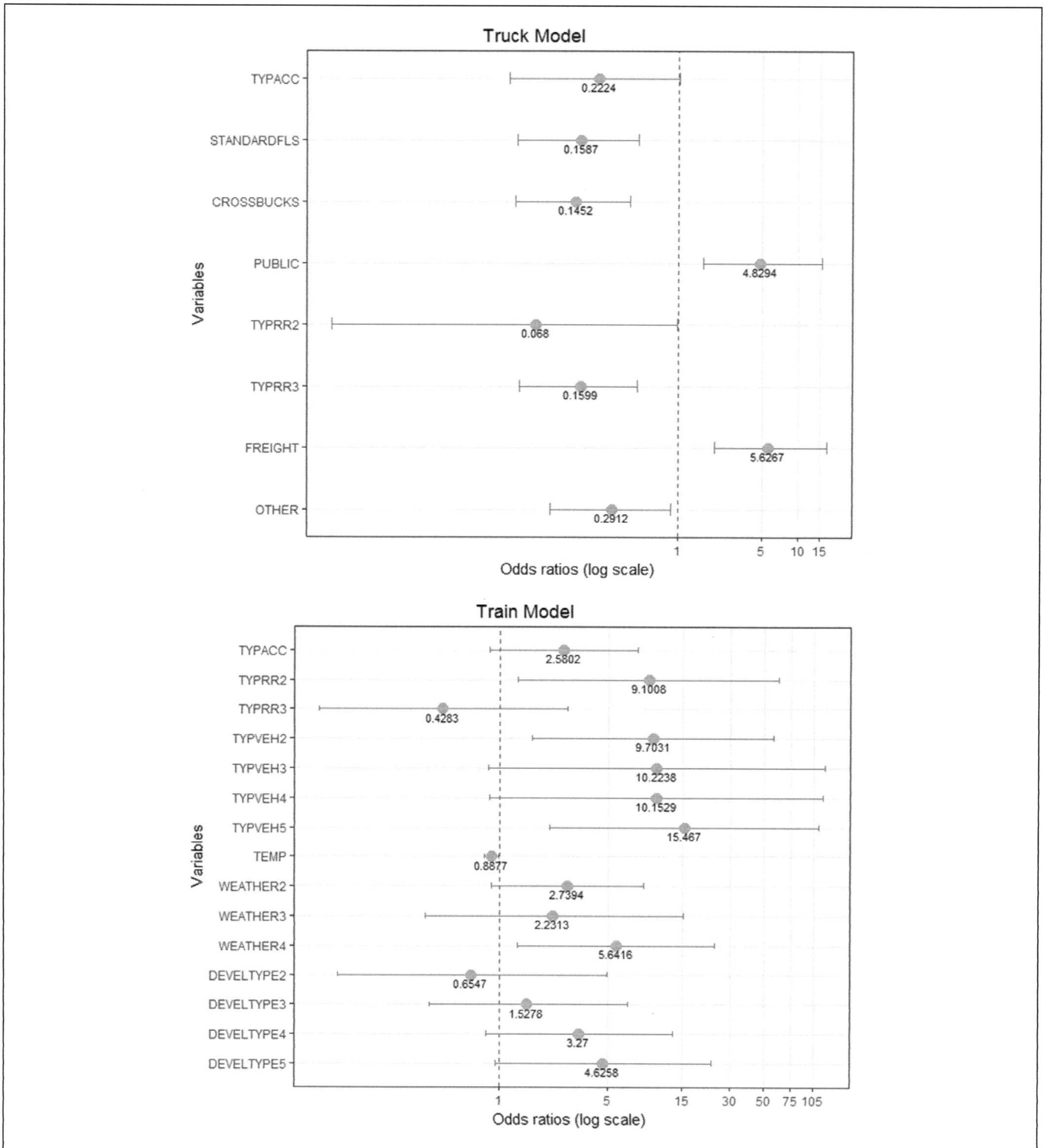

Figure 1. OR and profile likelihood ratio confidence intervals for truck and train models.

other vehicles (e.g., recreational vehicles) are heavier than automobiles on average, leading to potentially more severe collisions and higher probability of hazmat release from trains. Since higher temperature and presence of fog, sleet or snow were associated with smaller and larger probability of hazmat release, respectively, weather considerations are recommended in shipping of hazmat by rail and may be used in relevant policy-making.

Table 4. Comparison of the Results of Six HRGC Crash Severity Studies with Hazmat Release Results

Variables	Crash severity						Hazmat release	
	(Hao & Daniel (*25*))	(Haleem & Gan (*35*))	(Eluru et al. (*36*))	(Zhao et al. (*28*))	(Kang & Khattak (*37*))	(Zhao & Khattak (*27*))	Highway user	Train
TYPACC2	—	Increase	Increase	Increase	Increase	Increase	NS	NS
STANDARDFLS	—	—	NS	NS	Increase	—	Decrease	NS
CROSSBUCKS	—	—	Decrease	NS	Increase	—	Decrease	NS
PUBLIC	—	—	—	—	—	—	Increase	NS
TYPRR2	—	—	—	—	—	—	Decrease	Increase
TYPRR3	—	—	—	—	—	—	Decrease	NS
FREIGHT	—	—	—	Increase	—	Increase	Increase	NA
OTHER	—	—	NS	—	—	—	Decrease	NS
TYPVEH2	Increase	Decrease	NS	NA	Decrease	Decrease	NS	Increase
TYPVEH3	Increase	Decrease	Decrease	NA	NS	—	NS	NS
TYPVEH4	NA	NS	—	NA	—	—	NA	NS
TYPVEH5	Increase	NS	—	NA	Increase	—	NA	Increase
TEMP	—	—	Decrease	NS	—	—	NS	Decrease
WEATHER2	Increase	NS	NS	NS	Increase	—	NS	NS
WEATHER3	Increase	NS	Decrease	Increase	NS	—	NS	NS
WEATHER4	Increase	Decrease	Decrease	Increase	Decrease	—	NS	Increase
DEVELTYPE2	Decrease	NS	—	NS	NS	—	NS	NS
DEVELTYPE3	Decrease	NS	—	Increase	NS	—	NS	NS
DEVELTYPE4	Decrease	Decrease	—	NS	Decrease	—	NS	NS
DEVELTYPE5	Decrease	NS	—	NS	NS	—	NS	NS
OFFPEAK	Decrease	—	—	—	—	—	—	—
VEHSPD	Increase	—	—	NA	Increase	Increase	NS	NS
VISIBLTY1	Increase	—	—	—	NS	NS	NS	NS
VISIBLTY3	Increase	—	—	—	Increase	Decrease	NS	NS
VISIBLTY4	Increase	—	—	—	Increase	NS	NS	NS
TRNSPD	Increase	Increase	Increase	Increase	Increase	Increase	NS	NS
DRIVAGE	Increase	Increase	Increase	Increase	Increase	Increase	NS	NS
NONPAVED	Increase	—	—	—	—	—	—	—
AADT	Increase	Decrease	—	NS	Decrease	—	—	—
DRIVGEN	—	Increase	Increase	Increase	Increase	Increase	NS	NS
BELLS	—	Decrease	—	Increase	—	—	NS	NS
GATES	—	—	Increase	NS	NS	—	NS	NS
HWYTRFICSIG	—	—	NS	—	NS	—	NS	NS
AUDIBLE	—	—	NS	—	NS	—	NS	NS
STOPSIGNS	—	—	Increase	—	Decrease	—	NS	NS
NBRCARS	—	—	—	—	Increase	—	NS	NS

Note: — = not considered in study; NS = not significant; NA = not applicable.

Both models may be used as a part of a risk assessment framework. The framework may include at least two steps: models that predict the occurrence of crashes at HRGCs, based on variables such as highway and rail traffic, land use, control devices, etc. (e.g., (*33*, *34*)); and models, such as those estimated in this study, that predict the probability of hazmat release from trucks, trains, or both, given the occurrence of a crash. The product of these two probabilities can provide a hazmat risk measure for each HRGC, useful to serve as a prioritization tool for countermeasure implementation or resource allocation.

As was mentioned in the literature review, a significant number of papers studied injury severity of HRGC crashes and used this criteria to evaluate control devices and other related factors at HRGCs. It was also mentioned that there was a lack of research on hazmat release crashes reported at HRGCs. The question that may arise is whether the factors that increase/decrease crash severity at HRGCs are consistent with the factors that increase/decrease the probability of hazmat release (positive correlation between crash severity and hazmat release). This consistency may question the importance of this study. To investigate this possibility, Table 4 summarized the results of six studies regarding crash severity at HRGC, and the results of this study for comparison. It should be noted that, although some variables were defined differently in some studies, the final results were consistently reported in this table.

This comparison shows that there were variables that affected crash severity in different studies, almost consistently, but were not associated with hazmat release probability, such as type of crash, train speed, driver age and gender, vehicle speed and some types of land-use. The effects of types of highway user, temperature and weather on crash severity were inconsistent throughout the severity papers. Although these variables were not significant in hazmat release from trucks, some of them affected hazmat release from trains, but not necessarily in the same way (direction) as in the crash severity. The positive effects of standard flashing-light signals, railroad crossbucks and other control devices on hazmat release from trucks were not observed in the majority of the crash severity literature. Public/private HRGCs and type of railroad were not considered in the reviewed crash severity studies, although they were associated with hazmat release. Freight trains increased crash severity and the probability of hazmat release from trucks in crashes at HRGCs. In general, with an exception of some cases, crash severity modeling results were not consistent with hazmat release modeling outcomes, indicating that policies and countermeasures based on crash severity studies may not be relevant to decreasing hazmat release in crashes at HRGCs. Thus, this underscores the necessity of investigating hazmat release in crashes at HRGCs.

The large proportion of missing values in potentially important variables in the dataset and, consequently, not using those variables in the model specifications was a limitation in this research. These variables included details about HRGC control devices, actions of highway users during crashes, sight obstructions, type of hazmat, roadway conditions, etc. For future studies, researchers may address this issue by using more complete datasets (i.e., datasets with few missing values). Other modeling methods can be utilized for analyzing hazmat-related crashes at HRGCs that might lead to further insights. Short-term and long-term costs and damages of hazmat release at HRGCs may be studied to prioritize countermeasures and policies regarding public safety improvements at HRGCs.

References

1. Zografos, K. G., and K. N. Androutsopoulos. A Decision Support System for Integrated Hazardous Materials Routing and Emergency Response Decisions. *Transportation Research Part C: Emerging Technologies*, Vol. 16, No. 6, 2008, pp. 684–703.
2. Association of American Railroads, Bureau of Explosives. *Annual Report of Hazardous Materials Transported by Rail: 2016*. Bureau of Explosives, Washington, D.C., 2016.
3. Federal Railroad Adminsitration Office of Safety Analysis. *Railroad Safety Data*. Federal Railroad Administration. http://safetydata.fra.dot.gov/OfficeofSafety/publicsite/downloads/downloads.aspx. Accessed May 1, 2017.
4. Liu, X., M. R. Saat, and C. P. L. Barkan. Integrated Risk Reduction Framework to Improve Railway Hazardous Materials Transportation Safety. *Journal of Hazardous Materials*, Vol. 260, 2013, pp. 131–140.
5. Glickman, T. S., E. Erkut, and M. S. Zschocke. The Cost and Risk Impacts of Rerouting Railroad Shipments of Hazardous Materials. *Accident Analysis and Prevention*, Vol. 39, No. 5, 2007, pp. 1015–1025.
6. Abkowitz, M., and P. D.-M. Cheng. Developing a Risk/Cost Framework for Routing Truck Movements of Hazardous Materials. *Accident Analysis and Prevention*, Vol. 20, No. 1, 1988, pp. 39–51.
7. Saccomanno, F. F., and A.-W. Chan. Economic Evaluation of Routing Strategies for Hazardous Road Shipments. *Transportation Research Record: Journal of the Transportation Research Board*, 1985. 1020: 12–18.
8. List, G. F., P. B. Mirchandani, M. A. Turnquist, and K. G. Zografos. Modeling and Analysis for Hazardous Materials Transportation: Risk Analysis, Routing/Scheduling and Facility Location. *Transportation Science*, Vol. 25, No. 2, 1991, pp. 100–114.
9. Kara, B. Y., and V. Verter. Designing a Road Network for Hazardous Materials Transportation. *Transportation Science*, Vol. 38, No. 2, 2004, pp. 188–196.
10. Verter, V., and B. Y. Kara. A Path-based Approach for Hazmat Transport Network Design. *Management Science*, Vol. 54, No. 1, 2008, pp. 29–40.

11. Erkut, E., and F. Gzara. Solving the Hazmat Transport Network Design Problem. *Computers & Operations Research*, Vol. 35, No. 7, 2008, pp. 2234–2247.
12. Erkut, E., and O. Alp. Designing a Road Network for Hazardous Materials Shipments. *Computers & Operations Research*, Vol. 34, No. 5, 2007, pp. 1389–1405.
13. Saat, M. R., and C. P. L. Barkan. Generalized Railway Tank Car Safety Design Optimization for Hazardous Materials Transport: Addressing the Trade-off between Transportation Efficiency and Safety. *Journal of Hazardous Materials*, Vol. 189, No. 1, 2011, pp. 62–68.
14. Barkan, C. P. L., S. V. Ukkusuri, and S. Travis Waller. Optimizing the Design of Railway Tank Cars to Minimize Accident-caused Releases. *Computers & Operations Research*, Vol. 34, No. 5, 2007, pp. 1266–1286.
15. Barkan, C. P. L. Improving the Design of Higher-capacity Railway Tank Cars for Hazardous Materials Transport: Optimizing the Trade-off Between Weight and Safety. *Journal of Hazardous Materials*, Vol. 160, No. 1, 2008, pp. 122–134.
16. Kawprasert, A., and C. Barkan. Effect of Train Speed on Risk Analysis of Transporting Hazardous Materials by Rail. *Transportation Research Record: Journal of the Transportation Research Board*, 2010. 2159: 59–68.
17. Berman, O., V. Verter, and B. Y. Kara. Designing Emergency Response Networks for Hazardous Materials Transportation. *Computers & Operations Research*, Vol. 34, No. 5, 2007, pp. 1374–1388.
18. Nayak, P. R., D. B. Rosenfield, J. H. Hagopian, A. D. Lillie, and A. Park. *Event Probabilities and Impact Zones for Hazardous Materials Accidents on Railroads*. Report DOT/FRA/ORD-83/20. FRA, U.S. Department of Transportation, 1983.
19. Saccomanno, F. F., and S. M. El-Hage. Minimizing Derailments of Railcars Carrying Dangerous Commodities through Effective Marshaling Strategies. *Transportation Research Record: Journal of the Transportation Research Board*, 1989. 1245: 34–51.
20. Saccomanno, F. F., and S. M. El-Hage. Establishing Derailment Profiles by Position for Corridor Shipments of Dangerous Goods. *Canadian Journal of Civil Engineering*, Vol. 18, No. 1, 1991, pp. 67–75.
21. Liu, X., M. R. Saat, X. Qin, and C. P. L. Barkan. Analysis of U.S. Freight-train Derailment Severity Using Zero-truncated Negative Binomial Regression and Quantile Regression. *Accident Analysis and Prevention*, Vol. 59, 2013, pp. 87–93.
22. Verma, M., V. Verter, and N. Zufferey. A Bi-objective Model for Planning and Managing Rail-truck Intermodal Transportation of Hazardous Materials. *Transportation Research Part E: Logistics and Transportation Review*, Vol. 48, No. 1, 2012, pp. 132–149.
23. Raub, R. A. Examination of Highway-rail Grade Crossing Collisions Over 10 Years in Seven Midwestern States. *Institute of Transportation Engineers Journal*, Vol. 76, No. 4, 2006, p. 16.
24. Hu, S. R., C. S. Li, and C. K. Lee. Investigation of Key Factors for Accident Severity at Railroad Grade Crossings by Using a Logit Model. *Safety Science*, Vol. 48, No. 2, 2010, pp. 186–194.
25. Hao, W., and J. Daniel. Severity of Injuries to Motor Vehicle Drivers at Highway-Rail Grade Crossings in the United States. *Transportation Research Record: Journal of the Transportation Research Board*, 2013. 2384: 102–108.
26. Russo, B., P. Savolainen, and P. Tarmo. Examination of Factors Affecting Frequency and Severity of Crashes at Rail-Grade Crossings. Presented at 92nd Annual Meeting of the Transportation Research Board, Washington, D.C., 2013.
27. Zhao, S., and A. Khattak. Motor Vehicle Drivers' Injuries in Train–Motor Vehicle Crashes. *Accident Analysis and Prevention*, Vol. 74, 2015, pp. 162–168.
28. Zhao, S., A. Iranitalab, and A. Khattak. A Clustering Approach to Injury Severity in Pedestrian-Train Crashes at Highway-Rail Grade Crossings. *Journal of Transportation Safety and Security*, 2018. https://doi.org/10.1080/19439962.2018.1428257
29. Fan, W. D., M. R. Kane, and E. Haile. Analyzing Severity of Vehicle Crashes at Highway-Rail Grade Crossings: Multinomial Logit Modeling. *A Message from the JTRF Co-General Editors 3*, Vol. 54, No. 2, 2015, p. 39.
30. Agresti, A., and M. Kateri. Categorical Data Analysis. In *International Encyclopedia of Statistical Science* (M. Lovric, ed.), Springer Berlin Heidelberg, Berlin, Heidelberg, pp. 206–208.
31. Bilder, C. R., and T. M. Loughin. *Analysis of Categorical Data with R*. CRC Press, Boca Raton, FL Florida, 2014.
32. Firth, D. Bias Reduction of Maximum Likelihood Estimates. *Biometrika*, Vol. 80, No. 1, 1993, pp. 27–38.
33. Oh, J., S. P. Washington, and D. Nam. Accident Prediction Model for Railway-highway Interfaces. *Accident Analysis and Prevention*, Vol. 38, No. 2, 2006, pp. 346–356.
34. Yan, X., S. Richards, and X. Su. Using Hierarchical Tree-based Regression Model to Predict Train-vehicle Crashes at Passive Highway-rail Grade Crossings. *Accident Analysis and Prevention*, Vol. 42, No. 1, 2010, pp. 64–74.
35. Haleem, K., and A. Gan. Contributing Factors of Crash Injury Severity at Public Highway-railroad Grade Crossings in the U.S. *Journal of Safety Research*, Vol. 53, 2015, pp. 23–29.
36. Eluru, N., M. Bagheri, L. F. Miranda-Moreno, and L. Fu. A Latent Class Modeling Approach for Identifying Vehicle Driver Injury Severity Factors at Highway-railway Crossings. *Accident Analysis and Prevention*, Vol. 47, 2012, pp. 119–127.
37. Kang, Y., and A. Khattak. Cluster-Based Approach to Analyzing Crash Injury Severity at Highway–Rail Grade Crossings. *Transportation Research Record: Journal of the Transportation Research Board*, 2017. 2608: 58–69.

The Standing Committee on Highway/Rail Grade Crossings (AHB60) peer-reviewed this paper (18-03375).

Article

TRR
JOURNAL OF THE TRANSPORTATION RESEARCH BOARD

Injury Severity of Truck Drivers in Crashes at Highway-Rail Grade Crossings in the United States

Transportation Research Record
2018, Vol. 2672(10) 38–47

DOI: 10.1177/0361198118781183
journals.sagepub.com/home/trr
SAGE

Waleed A. Khan[1] and Aemal J. Khattak[1]

Abstract

The physical and operational characteristics of large trucks distinguish them from other types of vehicles in terms of facility design needs and safety requirements. A critical node in the surface transportation network is the highway-rail grade crossing (HRGC) because it represents a conflict point between different modes of transportation. The focus of this research was to identify factors related to different injury severity levels of truck/truck-trailer drivers in crashes reported at HRGCs. This study utilized a mixed logit model to investigate injury severity of those drivers and relied on 2007–2014 Federal Railroad Administration (FRA) crash and inventory data involving trucks/truck-trailers. Results showed that truck/truck-trailer drivers' injuries in crashes reported at HRGCs were positively associated with train speed, when train struck the road user (truck/truck-trailer), when the driver "went around crossing gates", older drivers, crashes reported in rural areas, and crashes at crossings with a minimum crossing angle of 60–90 degrees. Presence of crossbucks, gates, track obstructions, and HRGCs located within 500 feet of a highway were associated with relatively less severe driver injuries. The paper provides recommendations for safety improvements at HRGCs and recommendations for future research.

The 2012 Commodity Flow Survey (CFS-2012), jointly conducted by the U.S. Census Bureau and the Bureau of Transportation Statistics (BTS), estimated shipping of 11.3 billion tons of freight valued at more than $13 trillion over the U.S. freight transportation system and generating 3.3 trillion ton-miles of travel in 2013 (*1*). Freight transportation by roads dominated the nation's movement of freight for value and tonnage, accounting for 73.1% of the value ($10.1 trillion) and about 71% of weight (8.1 billion tons). Truck and rail each accounted for 1.2 trillion ton-miles. Single mode truck was the dominant mode of freight transportation, accounting for at least 60% of the total value of shipments for 43 states in the U.S. According to Freight Facts and Figures 2015 (*2*), total yearly shipments will likely increase to 28.5 billion tons, with domestic shipments of about 23 billion tons by 2040. Truck and train traffic will increase in the future due to the expected growth in the demand for freight. This will likely increase the risk of conflicts between these two modes of transport thereby exacerbating a multimodal safety issue.

According to the Federal Railroad Administration (FRA) crash data (*3*), heavy-vehicle crashes at highway-rail grade crossings (HRGCs) range from 14% to 17% of the total crashes reported during 2007–2014. In 2014, of the 2,293 crashes at grade crossings, 390 (17%) involved heavy vehicles (truck/truck-trailer) on public crossings with 19 truck driver fatalities constituting 7.2% of the total fatalities reported at HRGCs. Figure 1 presents details of heavy-vehicle involved crashes over the period 2007–2014.

Truck-involved crashes at HRGCs are important to investigate because of the potential severity of crash outcomes at these locations and due to the potential disruption of both the highway and the rail networks. Current safety research is mostly focused on crashes reported at non-HRGC locations while published research on crashes reported at HRGCs is not specifically focused on trucks. Trucks have unique characteristics compared with other motor vehicles in terms of size, weight, and acceleration characteristics. However, a review found relatively limited literature on truck-involved crashes at HRGCs and therefore the topic requires attention. This research advances the understanding of factors related to driver injury severity of heavy vehicles (trucks/truck-trailers) at HRGCs by investigating the effects of HRGC design characteristics, vehicle attributes, driver characteristics, and environmental conditions on reported injury severities.

This paper consists of five sections. Section 1 is this introduction. Section 2 presents a review of relevant published literature. Section 3 describes the data including descriptive statistics for the dataset. Section 4 presents a brief modeling

[1]Department of Civil Engineering and Nebraska Transportation Center, University of Nebraska-Lincoln, Lincoln, NE

Corresponding Author:
Address correspondence to Aemal J. Khattak: khattak@unl.edu

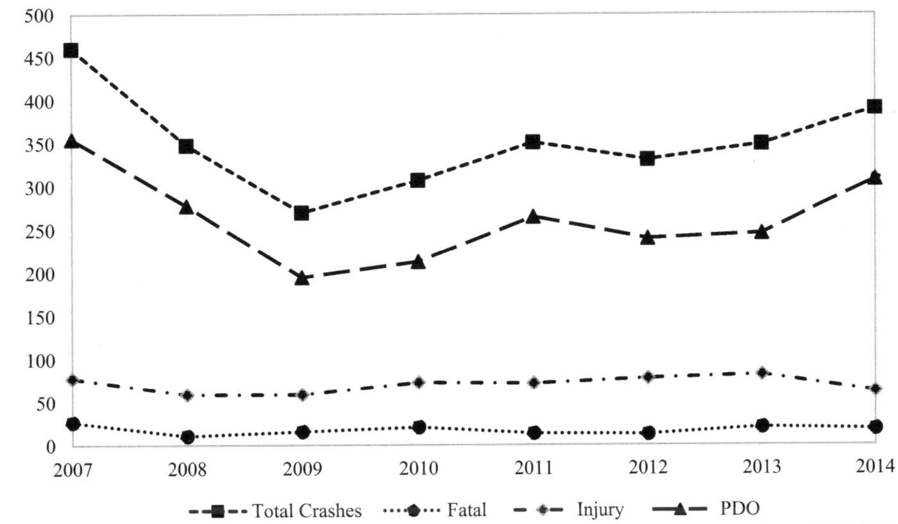

Figure 1. Heavy-vehicle crashes at HRGCs in the U.S. (2007–2014).

background and its results, and discussion of the different variables associated with driver injury severity of trucks/truck-trailers at HRGCs. Section 5 summarizes this study, presents conclusions from the analysis, provides recommendations for safety improvements at HRGCs, and proposes future research.

Literature Review

This literature review covers injury severity of different road users at HRGCs and various factors associated with it. It also covers assorted statistical models used to identify factors related to crash injury severity.

Injury Severity at Highway-Rail Grade Crossings

Research focused on crash injury severity at HRGCs mostly utilized multinomial logit (MNL), ordered probit (OP), ordered logit (OL) and mixed logit (mixed generalized logit) models to identify different factors associated with crash injury severity. Hu et al. (*4*) conducted a study in Taiwan using 592 HRGCs by estimating a generalized logit model and reported that the likelihood of more severe crash injuries increased with an increase in the number of trucks and daily trains; they reported highway obstacle and separation detection devices associated with more severe crashes. Eluru et al. (*5*) estimated a latent segmentation based OL model using the FRA crash data (1997–2006). Their results indicated that time of crash, presence of snow and/or rain, motor vehicle driver age, driver behavior before the crash and motor vehicle role in the crash were key factors influencing injury severity.

Hao and Daniel (*6*) determined different factors influencing drivers' injury severity at HRGCs using an OP model by utilizing FRA crash data (2002–2011). Factors related to higher injury severity of motor vehicle drivers included adverse weather conditions, low visibility, train and vehicle speeds greater than 50 mph, highways with annual average daily traffic (AADT) over 10,000, crashes reported in open areas, and crashes involving trucks and semi-trailers.

Russo and Savolainen (*7*) used an OL model using the FRA crash data to identify different factors of rail, highway, traffic and driver characteristics associated with the frequency and injury severity of HRGCs crashes. Factors found positively associated with more severe injuries included females, motor vehicle drivers aged over 60 years, drivers that did not stop at crossings and trains with speeds greater than 60 mph.

Fan and Haile (*8*) used a MNL model to identify factors that increased crash injury severity at HRGCs by using the FRA crash data (2005–2012). Drivers aged 25 years and older, pick-up trucks and crossing surfaces with concrete or rubber were related to more severe crashes. Foggy and snowy weather conditions, truck-trailers, certain land development types and higher AADT were associated with less severe crashes. Hao and Daniel (*9*) identified different factors affecting driver injury severity of vehicle driver at HRGC under different weather conditions using a mixed logit model. Their results showed that injury severity was more prevalent in crashes involving vehicles or trains with higher speeds. Light conditions and unpaved surfaces also increased injury severity.

Some literature on heavy vehicles at HRGCs was found and reviewed. In Australia, the trend of heavy-vehicle involvement in rail-involved crashes has shown significant potential for catastrophic outcomes (*10*); the risk may be increasing with growing numbers of longer and heavier freight vehicles using the road network coupled with increased train services and higher speeds. Another study conducted in Australia (*11*) focused on understanding design and behavior issues that affect HRGC safety by conducting a series of group discussions with train and truck drivers. This study concluded that vehicle driver visibility (line of sight and angles of approach) and effective vehicle clearance (impeded acceleration, length of carriage maneuverability) were affected by the configuration of HRGCs. However, driver compliance towards violation of crossing protocols was often due to saving time or due to high familiarity with the crossing.

Ishak et al. (*12*) introduced Petri nets, a graphical and mathematical modeling tool in assessing risk at HRGCs when heavy vehicles were passing through intersecting areas. Factors associated with heavy-vehicle collisions at HRGCs included traffic level of service (LOS), percentage of heavy vehicles and the distance of grade crossing to or from the nearest intersection. Truck drivers' violations of traffic laws increased crash risk.

A Nebraska study (*13*) on the frequency and type of crossing gate related violations by truck drivers showed that violations increased at HRGCs with longer time between onset of flashing lights and train arrivals and with greater truck traffic at HRGCs. As well, the study reported most gate violations occurring during nighttime. Liu et al. (*14*) conducted a safety analysis of truck-involved crashes to identify factors associated with drivers' behavior prior to collisions. Their results showed that truck-involved crashes at HRGCs equipped with gates were generally less severe compared to those occurring at crossings without gates. Truck drivers at crossings without gates more likely failed to make an appropriate stop, or proceeded after a short stop, or stopped on crossing before crash occurrence.

Using an OP model, Hao et al. (*15*) reported speed control as a measure to reduce truck drivers' injury severities at HRGCs. They also reported area type, traffic volume, roadway pavement, weather condition, and visibility as factors influencing truck driver injury severity at HRGCs. Using an OL model, Fan et al. (*16*) reported that truck-trailers were associated with less severe injuries (among other factors) in train-vehicle crashes.

Potential Modeling Approaches

Injury severity data may be considered as nominal or ordinal and relevant modeling techniques may be used. Frequently used nominal models include MNL, nested logit (NL) and mixed logit models (RPL model), while generalized ordered logit (GOL) model, OL, and OP models are commonly used ordinal models. The modeling approach for injury severity depends on the quality and quantity of data available for the analysis. A number of data characteristics and their limitations have been identified in the past that may be critical in development and application of a statistical model.

Some of the commonly used models for modeling injury severities in the past decade are OL/OP model (*17–19*), MNL (*20–22*) and NL (*21*). Abdel-Aty (*23*) compared OP, MNL and NL models, in addition to identifying different factors associated with injury severity at intersections and roadway sections. The OP model was comparatively simple and produced better results in terms of the model's goodness of fit and number of significant variables entered in the model specification. Abdel-Aty and Abdel Wahab (*24*) compared results from an Artificial Neural Network (ANN) and an OP model and showed more accurate predictions from the ANN model.

A study by Haleem and Abdel-Aty (*25*) on traffic crash injury severity at un-signalized intersection concluded that the binary probit model showed better goodness of fit compared to the disaggregated OP and NL models. Yasmin and Eluru (*26*) compared different ordered and unordered response models for driver injury severity of crashes. Nominal response models were MNL, NL and order generalized extreme value logit (OGEV) while ordinal response models included OL and GOL models. They reported that OGEV and NL models reduced to simple MNL models but the GOL model performed better in terms of data fit than OL and MNL models. The authors examined the performance of GOL and MNL models and their results indicated the GOL model as an equivalent ordered response model to the MNL model for ordinal discrete variables.

Yasmin et al. (*27*) investigated a better model for injury severity of pedestrians by comparing three order response models: OL model, latent segmentation based ordered logit model (LSOL) and GOL. Their results indicated that LSOL performed better than the GOL and OL model for identifying factors associated with different injury severity levels of pedestrians. The effect of sample size on model estimation was investigated by Ye and Lord (*28*). The three models estimated in the study were OP, MNL and RPL. The results indicated that RPL model required the largest sample size while the OP model required the smallest. In terms of model interpretations, the RPL model performed better than the MNL model.

Zhao and Khattak (*29*) used the FRA crash data to identify different variables associated with driver injury severity in train-motor vehicle crashes at HRGCs. They compared OP, MNL and RPL models to identify a suitable model to explore factors related to different injury severity levels. They concluded that the RPL model and the MNL model performed better for injury severity analysis of motor vehicle drivers involved in crashes at HRGCs.

In summary, the majority of reviewed research was focused on general highway traffic that included passenger cars, trucks, buses, motorcycles, etc. HRGC safety studies of specific types of vehicles were relatively few and heavy-vehicle safety at grade crossing appears to be somewhat under-explored and there is a gap in research of injury severity of truck/truck-trailer drivers in crashes reported at HRGCs. A mixed logit model, also known as RPL, was selected in this study due to its wide acceptance and better performance in past studies. The RPL model handles the

Table 1. FRA Reported Crashes at HRGCs by Vehicle Types (2007–2014)

Type of vehicle	At-grade crossing			Driver injury severity levels of total reported crashes		
	Total reported crashes	Type of crossing				
		Public	Private	PDO	Injury	Fatal
Auto	7465	6833	632	4972	1955	538
Truck	1159	877	282	750	313	96
Truck-trailer	2699	1929	770	2189	431	79
Pick-up truck	2454	2042	412	1542	683	229
Van	601	502	99	383	172	46
Bus	23	22	1	18	5	0
School bus	4	4	0	4	0	0
Motor cycle	50	49	1	22	15	13
Other/other motor vehicle	1293	1036	257	715	395	183
Pedestrians	1039	1006	33	NA	NA	NA
Total	16787	14300	2487	–	–	–

correlations in crash data much more efficiently and bias in results is less likely.

Data Processing

The dataset utilized in this study was extracted from the FRA HRGC inventory and crash databases (*3*). This section describes the reduction of the extracted dataset into the final dataset used for model estimation.

Data Source

Originally started in 1975, the FRA national HRGC inventory database includes both current and historical records with 80,000–100,000 crossings updated per year. The database contains three major data files; HRGC accident file, HRGC history file and HRGC inventory file. The unique crossing ID serves as the linkage amongst these three files. The HRGC accident database provides a history of all crashes reported at HRGCs and the surrounding conditions at the time of crash. It contains details such as speed of train and vehicle involved in the crash, type of train, type of materials carried (by freight vehicles), type of vehicle, crash circumstances, time of day, environmental conditions, and driver attributes etc.

The HRGC inventory file provides current crossing inventory information, which reflects the current state of each crossing with reference attributes. The HRGC history file reflects changes made to crossings including a reason for the update and an effective date for the update. This history file contains previous records of every crossing before any changes were made to it; this is helpful in getting inventory information prior to any changes. The inventory file contains information such as AADT, active and passive warnings, warning type, area type, geometric characteristics and coordinates of the crossings.

In order to get inventory information for the year a crash occurred, both highway-rail crossing inventory and highway-rail crossing history files were utilized. The data was substantially checked and cleaned for consistency. Some accident file IDs were missing in the HRGC inventory file; in such cases the accidents were removed from the final dataset.

Data Reduction

Initially, crashes at HRGCs were extracted from HRGC accidents database for the period 2007–2014. The unique ID numbers between the three datasets were then used to extract inventory information for each accident/incident. The total number of accidents/incidents were 17,635 and this number included all types of crashes reported at crossings. Heavy-vehicle (truck/truck-trailer) involved crashes at public grade crossings were then extracted from the dataset, which contributed to about 15.9% (2,806) of the total (17,635) accidents/incidents reported at HRGCs for 2007-2014. Table 1 presents the details of reported crashes and the different driver injury severity levels with respect to each vehicle type. The different severity levels noted in Table 1 are based on total reported crashes at HRGCs with a total of 16,787 crashes (i.e., crashes on public and private crossings). During data reduction some IDs did not match across the files or in some instances information on injury severity was missing. In such cases, the observations were omitted from the final dataset resulting in 2,664 observations in the final dataset.

Data Description

The dependent variable, i.e. injury severity, consisted of three severity levels: property damage, injury and fatal injury. These three levels of injury were coded as 0 (property damage, PDO), 1 (injury, INJ) and 2 (fatal). The final dataset consisted of 2,005 PDOs (75.26%), 525 INJ (19.7%) and 134(5%) fatal crashes.

The parameters used in modeling were mostly related to HRGC geometric characteristics, traffic-related variables such as different types of passive and active controls, truck/truck-trailer driver attributes, environmental aspects, and some crash specific details such driver behavior and crash circumstances. A total of 94 independent variables were tested for statistical association with injury severity of truck drivers. Table 2 presents details of some of the variables used in model estimation.

Statistical Method

This section provides a background to the RPL model, its estimation and results.

Random Parameter Logit Model

Because driver injury severity is discrete, a discrete outcome model, such as the RPL, was selected for this study. The RPL offers relaxation of the independence of irrelevant alternatives (IIA) property, which states that the ratio of the choice probabilities of any pair of alternatives is independent of the presence or absence of any other alternative in a choice set. An implication of IIA is that all pairs of alternatives are equally similar or dissimilar. This assumes that all information in the random components is identical in quantity for the set of attributes that are not observed and the relationship between pairs of alternatives and hence across all alternatives (*30*).

The RPL model is a relatively recent development for the analysis of discrete data (*31*). The random parameter model addresses a weakness of standard MNL by allowing parameter values to vary across observations. For the derivation and application of the standard MNL model, it is assumed that parameters are fixed across all observations. When this assumption is incorrect, the parameter estimates and outcome probabilities are inconsistent (*32*).

Random parameter logit model appropriately accounts for the possibility of variation of different parameters across individual observations. Following the work presented by McFadden and Train (*31*) to develop the RPL approach, consider a function determining discrete outcome probabilities as

$$T_{in} = \beta_i X_{in} + \varepsilon_{in} \tag{1}$$

where

β_i = vector of estimable parameter for discrete outcome i,

X_{in} = vector of the observable characteristics (covariates) that calculate discrete outcomes for observation n, and

ε_{in} = disturbance term.

The standard MNL form can be written as

$$P_i(j) = \frac{\exp(\partial_j + \beta_l X_{ij})}{\sum_{j=1}^{J} \exp(\partial_j + \beta_j X_{ij})} \tag{2}$$

Where, P_i (j) = the probability of observation i having discrete outcome j (j denoting all possible outcomes for observation n). By defining a mixed model (with a mixing distribution) whose outcome probabilities are defined as Pi (j) with

$$P_i\ (j) = \int P_i(j) f(\beta|\varphi) d\beta \tag{3}$$

Where $f(\beta|\varphi)$ represents the density function of β and φ, and refers to the mean and variance of the density function. By putting the values of Equation 2 in Equation 3 we get

$$P_i\ (j) = \int \frac{\exp(\partial_j + \beta_l X_{ij})}{\sum_{j=1}^{J} \exp(\partial_j + \beta_j X_{ij})} f(\beta|\varphi) d\beta \tag{4}$$

Equation 4 indicates that the mixed logit probabilities P_i (j) are the weighted average of the standard MNL probabilities with the weights determined by the density function. In case of $f(\beta|\varphi)$ =1, the model reduces to simple MNL. The term β of Equation 2, can now account for observation-specific variations of the effect of x on outcome probabilities, with the density function used to $f(\beta|\varphi)$ determine β. Different types distribution (normal, uniform, triangular distribution) can be used as a density function for β. RPL probabilities are thus a weighted average of different values of β across different observations where some elements of parameter vector β are random parameters and some are fixed.

RPL Model Estimation and Results

In the RPL model estimation, all independent variables were first assumed random and both the uniform and normal distributions were tested. The random parameters that were not statistically significant at 90% significance level (p-value = 0.10) for both uniform and normal distributions were then kept as fixed parameters in the model specification and examined. Standard Halton Sequence (SHS) intelligent draws were selected for the RPL model estimation. Initially the number of Halton draws and iterations were set to 40 to identify significant random and fixed parameters at a significance level of 90% (p-value = 0.10). The final model was then revised by increasing the number of draws (SHS) and maximum iterations to 300. Table 3 presents results of the final RPL model and Table 4 presents the estimated marginal effects of the variables significant at 90% significance level.

For injury crash level in the RPL model, indicator for availability of gates at HRGCs followed a normal random

Table 2. Descriptive Statistics for the Variables Incorporated in the RPL Model

Variable Type	Description and coding	Frequency Yes = 1 No = 0	Mean	Standard deviation
Dependent variable				
Truck driver injury severity	PDO	2005	–	–
	Injured	525	–	–
	Fatal	134	–	–
Independent variables				
Vehicle characteristics				
Vehicle type	Truck	832	0.31	0.46
	Truck-trailer	1832	0.69	0.46
Vehicle speed (mph)		NA	7.79	11.54
Hazardous materials carried	Yes	711	0.27	0.44
	No	1950		
Railway characteristics				
Train speed (mph)		NA	30.67	18.36
Primary obstruction of track view	Yes=1	97	0.04	0.19
	No=0	2566		
Driver attributes				
Driver age equal or below 22 years		NA	45.70	13.61
Driver age between 22 and 55 years				
Driver age 55 years or above				
Driver gender	Male	2472	0.96	0.04
	Female	96	0.19	0.19
Truck driver behavior/action of highway user	Driver went around the gates	199	0.07	0.26
	Standing RR equipment/ did not stop	1364	0.51	0.50
	Stopped on crossing	784	0.29	0.46
	Went around/ through temporary barricade	317	0.12	0.32
Traffic control characteristics				
Active controls				
Gates available (indicator)	Yes=1	1091	0.92	1.19
	No=0	1573		
Is track signaled	Yes=1	1430	0.55	0.50
	No=0	1183		
Highway traffic signal controlling crossing	Yes=1	74	0.03	0.16
	No=0	2590		
Nearby highway intersection have traffic signals	Yes=1	271	0.19	0.39
	No=0	1163		
Train detection system indicator	Yes=1	1369	0.52	0.50
	No=0	1247		
Indicator for availability of bells	Yes=1	1375	0.52	0.50
	No=0	1289		
Passive controls				
Stop sign available	Yes=1	383	0.26	0.66
	No=0	2281		
Crossbuck assembly available indicator	Yes=1	1894	0.71	0.45
	No=0	770		

Note: NA = not applicable.

distribution implying that the variable may vary from one crash to another crash. All other independent variables were included as fixed parameters in the model. Fourteen parameters (including random parameter) were statistically significant at the 90% significance level. Parameters that increased the likelihood of crash severity at 90% significance level were: vehicle and train speed, crashes occurring in rural area, crossing angle of 60–90°, driver age above 55 years,

Table 3. RPL Model Results

Random parameter logit model				
Log likelihood function	−1343.55			
Chi squared	2047.9			
McFadden pseudo R-squared	0.4325			
Akaike information criterion (AIC)	2729.1			
Significance level	0			
No. of observations	2155			
Truck driver injury severity at HRGCs	**Coefficient**	**Standard error**	**z**	**Prob. Z>Z**
Injury severity level: injury				
Constant	−2.63981	0.27001	−9.78	0.000
Speed of train	0.037	0.00454	8.61	0
Speed of vehicle (truck/truck-trailer)	0.0178	0.00508	3.5	0.0005
Motorist behavior: motorist went around gates	1.2	0.30791	3.9	0.0001
Truck indicator in crash	0.89554	0.12869	6.96	0
Smallest crossing: 60–90°	0.3908	0.16848	2.32	0.0203
Crossbuck	−0.1286	0.06205	−2.07	0.0382
Rural area	0.2951	0.1388	2.13	0.0335
Indicator for primary obstruction of track view	−0.7269	0.3942	−1.84	0.0652
Highway near intersection (500ft)	−0.2401	0.121	−1.98	0.0472
Indicator if track is signaled	−0.2818	0.1325	−2.13	0.0334
Random parameter in utility functions				
Indicator for gates availability	−1.3834	0.5311	−2.6	0.0092
Distns. of Standard deviation				
Indicator for gates availability	1.3207	0.7514	1.76	0.0788
Injury severity level: fatal				
Constant	−8.04	0.5666	−14.2	0
Speed of train	0.06122	0.0063	9.68	0
Speed of vehicle (truck/truck-trailer)	0.02893	0.0085	3.4	0.0007
Rural area	0.8334	0.2571	3.24	0.0012
Position of vehicle: vehicle moving over crossing	1.4523	0.3211	4.53	0
Age of driver: above 55 years	0.675	0.2099	3.22	0.0013
Truck indicator in crash	1.4	0.2081	6.73	0
Circumstances of crash: rail equipment struck highway user	0.6236	0.3441	1.81	0.07

crash involving trucks at HRGCs, motorist behavior; went around the gates (violation), and crash circumstances in which train struck the roadway user (truck/truck-trailer). However, primary obstruction of track view, crossings within 500-ft of a highway, presence of gates and indicator for "if track is signalized" were negatively associated with injury severity of truck/truck-trailer drivers in crashes at HRGCs.

Results and Discussion

Thirteen parameters were statistically significant with different truck/truck-trailer drivers' injury severity levels in train–heavy-vehicle crashes at HRGCs. Results showed that train speed and the vehicle speed were positively associated with injury severity at the 99% significance level. Both findings were similar to those reported in reviewed literature (*9*, *12*, *14*, *29*) and were reasonable as higher speeds are commonly associated with more severe injuries. Truck-involved crashes at HRGCs significantly increased the likelihood of a more severe crash. The total number of truck-train crashes consisted of 9.5% fatal and 27.8% injury crashes, whereas truck-trailer had a higher number of total crashes (68.7%) but consisted of 3% fatal and 16% injury crashes.

Statistically significant HRGC geometric characteristics associated with truck/truck-trailer driver injury severity included; crossings at rural roads, crossing angle 60–90°, intersecting roadway within 500-ft of the crossing, and primary obstruction of the track view. Crashes reported in the rural areas were more severe. About 58% of the total crashes (2,594) were reported in rural areas in which the total number of injury and fatal crashes were 24% and 7.2% respectively. Primary obstruction of track view and roadway located within 500-ft of the crossing were negatively associated with severity level injury. In the analyzed data, no fatal or injury crashes were reported when truck view was obstructed. Similarly, 17.3% and 4.6% of total crashes

Table 4. Marginal Effects of the Statistically Significant Variables

Variables significant at 90% significance level	RPL	
	Injury	Fatal
Vehicle characteristics		
Speed of train	0.1845	0.1273
Speed of vehicle (truck/ truck-trailer)	0.0255	0.0175
Truck indicator in crash	0.051	0.036
Driver attributes		
Motorist behavior: the motorist went around the gates	0.0134	N/S
Age of driver (above 55 years)	N/S	0.011
Crash specific characteristics		
Circumstances of crash: rail equipment struck highway user	N/S	0.0269
Position of vehicle: vehicle moving over crossing	N/S	0.0611
Traffic characteristics		
Indicator if track is signaled	–0.0214	N/S
Indicator for gates availability	–0.0224	N/S
Crossbuck	–0.0293	N/S
Geometric characteristics		
Rural area	0.0289	0.0324
Smallest crossing: $60^0 - 90^0$	0.0485	N/S
Indicator for primary obstruction of track view	–0.0023	N/S
Highway near intersection (500ft)	–0.0199	N/S

Note: N/S = not significant at the 10% level. All other values are statistically significant.

reported at crossings within 500-ft of intersecting roadway resulted in injuries and fatalities, respectively; whereas 23.7% and 6% of total crashes occurring at crossing not within 500-ft of roadway were injury and fatal respectively. This explains the negative sign associated with the two variables for injury category of severity level in the estimated model.

Different active and passive traffic controls were examined in model estimation, and crossbuck, gates installed, and if the track was signaled reduced the likelihood of more injury severity. The model results showed that the availability of gates decreased the likelihood of a severe crash of heavy-vehicle drivers at HRGCs; this finding is consistent with previous studies (*14*). The parameter representing the presence of gates followed a normal random distribution implying that it varied from crash to crash. This parameter was negatively associated with injury category of severity level.

Driver attributes that increased injury severity of truck drivers included drivers aged above 55 years and when they went around the gates while crossing. Crash specific characteristics that increased the likelihood of crash injury severity were: when trains struck road user and vehicle moving over the crossing i.e., the vehicle failed to stop for an oncoming train. Both findings appear reasonable and consistent with each other. Drivers are vulnerable to more severe injuries when the train strikes the road user. About 62% of total crashes (2007–2014 crashes) were reported when road user was moving over the crossing, in which 26% were injury and 7.3% were fatal crashes.

Summary and Conclusion

Truck and truck-trailer crashes at HRGCs reported between 2007 and 2014 were utilized in this study with 2,664 observations. Thirteen independent variables were statistically significantly associated with the dependent variable at the 90% significance level (alpha = 0.10). Statistically significant parameters positively associated with injury severity included speed of train and road user (truck/truck-trailer), truck-train crash, driver behavior "went around the gates", drivers older than 55 years, crashes reported in rural areas, and crashes at minimum crossing angle of 60–90°. Crash specific characteristics increasing the likelihood of fatalities included when train struck heavy vehicle and when the vehicle was moving over the crossing.

Higher speeds were associated with more severe injuries. This finding is reasonable and consistent with previous injury severity research. The total number of truck-train crashes reported were comparatively lower (Table 2) than trailer-train crashes. However, truck-train crashes constituted 9.5% fatal and 27.8% injury crashes. Whereas truck-trailer crashes at HRGC consisted of 3% fatal and 16% injury crashes (2007–2014).

The dataset included 200 crashes resulting from the truck/truck-trailer driver going around crossing gates resulting in about 28% injury and 9.6% fatal crashes. About 58% of the total crashes were reported in rural areas, which consisted about 7.2% fatal and 23.8% injury crashes. Trucks/truck-trailers crossing rails and failing to make a stop for the oncoming train resulted in more severe injuries. Examples of such instances include truck drivers unaware of oncoming trains due to poor visibility, the absence of appropriate traffic warnings, and driver inattention. Trucks/truck-trailers struck by a train while moving over the crossing consisted of 26% injury and 7.3% fatal crashes. Older drivers and when train struck the road user (truck/truck-trailer) increased the likelihood of a more severe crash. This finding is reasonable and consistent with injury severity of motor vehicles at HRGC (*29*).

Variables that significantly decreased the likelihood of a severe crash were: crossings with gates, if the track was signaled, presence of crossbuck, and if the track was obstructed. The variables representing the presence of gates followed a normal random distribution implying that it varied from crash to crash. This parameter was negatively associated with injury category of severity levels.

Conclusion

This research was undertaken with the objective to identify factors associated with injury severity of truck/truck-trailer drivers in crashes reported at HRGCs. Based on the results, the following conclusions are offered:

- Higher vehicular speeds at HRGCs increase truck/truck-trailer driver injury severity; slower speeds will improve the safety of HRGCs.
- Injuries were more severe when trains struck trucks/truck-trailers; perhaps making trucks more conspicuous to trains may afford the train engineer some extra time to slow down and reduce crash severity.
- Drivers of trucks/truck-trailers passing around gates were more severely injured; the deployment of countermeasures (e.g., four-quadrant gates and other physical barriers) will likely improve HRGC safety.
- The presence of crossbucks reduced injury severity of truck/truck-trailer drivers and ensuring the installation and maintenance of crossbucks will contribute to HRGC safety improvement.

Limitations and Future Research

This research investigated different factors associated with injury severity of drivers of trucks/truck-trailers and did not take into account the most severe injury in crashes. Results may be different if the most severe injury was considered. While drivers' behavior is an important attribute in HRGC safety, truck drivers' characteristics such as health/fitness, illnesses, financial and educational levels, driving experience, past traffic citations, were not available for inclusion in this study. Future research may explore how to collect such data to analyze a richer dataset. Future studies can consider other types of models and analyses. This research utilized the unordered response of the dependent variable; future studies may consider ordered response models such as OP and GOL etc. Other methods such as ANN and different data mining techniques were used in the past (*33–35*). Such methods may be used to investigate truck drivers' injury severities in crashes reported at HRGCs. This study focused on public HRGCs and future studies may include private HRGCs.

References

1. U.S. Department of Transportation, Federal Highway Administration. *Freight Analysis Framework - FHWA Freight Management and Operations* [Online]. http://www.ops.fhwa.dot.gov/freight/freight_analysis/faf/. Accessed November 2, 2016.
2. U.S. Department of Transportation. *Freight Facts and Figures 2015*. http://www.rita.dot.gov/bts/sites/rita.dot.gov.bts/files/FFF_complete.pdf. Accessed August 11, 2016.
3. Federal Railroad Administration Office of Safety Analysis. *FRA Safety Data*. 2016. http://safetydata.fra.dot.gov/OfficeofSafety/Default.aspx. Accessed August 12, 2016.
4. Hu, S.-R., C.-S. Li, and C.-K. Lee. Investigation of Key Factors for Accident Severity at Railroad Grade Crossings by Using a Logit Model. *Safety Science*, Vol. 48, No. 2, 2010, pp. 186–194.
5. Eluru, N., M. Bagheri, L. F. Miranda-Moreno, and L. Fu. A Latent Class Modeling Approach for Identifying Vehicle Driver Injury Severity Factors at Highway-Railway Crossings. *Accident Analysis & Prevention*, Vol. 47, 2012, pp. 119–127.
6. Hao, W., and J. Daniel. Motor Vehicle Driver Injury Severity Study Under Various Traffic Control at Highway-Rail Grade Crossings in the United States. *Journal of Safety Research*, Vol. 51, 2014, pp. 41–48.
7. Russo, B., and P. T. Savolainen. Examination of Factors Affecting Frequency and Severity of Crashes at Rail-Grade Crossings. In 92nd Annual Meeting of the Transportation Research Board Compendium of Papers, Washington, D.C., 2013.
8. Fan, W. D., M. R. Kane, and E. Haile. Analyzing Severity of Vehicle Crashes at Highway-Rail Grade Crossings: Multinomial Logit Modeling. *Journal of the Transportation Research Forum*, Vol. 54, No. 2, 2015, pp. 39–56.
9. Hao, W., and J. Daniel. Driver Injury Severity Related to Inclement Weather at Highway-Rail Grade Crossings in the United States. *Traffic Injury Prevention*, Vol. 17, No. 1, 2015, pp. 31–38.
10. Australian Transport Safety Bureau. *Railway Level Crossing Safety Bulletin*. ATSB Rail Safety Bulletin, Canberra, Australia 2008.
11. Davey, J., A. Wallace, N. Stenson, and J. Freeman. The Experiences and Perceptions of Heavy Vehicle Drivers and Train Drivers of Dangers at Railway Level Crossings. *Accident Analysis & Prevention*, Vol. 40, No. 3, 2008, pp. 1217–1222.
12. Ishak, S. Z., S. Somenahalli, and W. L. Yue. An Assessment of Heavy Vehicle Safety at Level Crossing Using Petri Nets: South Australia Case Studies. *Proceedings of the Eastern Asia Society for Transportation Studies*, Vol. 8, 2011, p. 359.
13. Khattak, A. J., and M. Gao. *Truck Safety at Highway-Rail Grade Crossings*. Report MATC-UNL-225. Final report to the Mid-America Transportation Center, 2012.

14. Liu, J., A. J. Khattak, S. H. Richards, and S. Nambisan. What are the Differences in Driver Injury Outcomes at Highway-Rail Grade Crossings? Untangling the Role of Pre-Crash Behaviors. *Accident Analysis & Prevention*, Vol. 85, 2015, pp. 157–169.
15. Hao, W., C. Kamga, X. Yang, J. Ma, E. Thorson, M. Zhong, and C. Wu. Driver Injury Severity Study for Truck Involved Accidents at Highway-Rail Grade Crossings in the United States. *Transportation Research Part F: Traffic Psychology and Behaviour*, Vol. 43, 2016, pp. 379–386.
16. Fan, W., L. Gong, and E. Haile. Severity Analysis of Vehicle Crashes on Highway-Rail Grade Crossings: Ordered Response Logit Modeling. *Advances in Transportation Studies*, 2016, Vol. 38, pp. 5–20.
17. O'Donnell, C., and D. Connor. Predicting the Severity of Motor Vehicle Accident Injuries Using Models of Ordered Multiple Choice. *Accident Analysis & Prevention*, Vol. 28, No. 6, 1996, pp. 739–753.
18. Kockelman, K. M., and Y.-J. Kweon. Driver Injury Severity: An Application of Ordered Probit Models. *Accident Analysis & Prevention*, Vol. 34, No. 3, 2002, pp. 313–321.
19. Kweon, Y.-J., and K. M. Kockelman. Overall Injury Risk to Different Drivers: Combining Exposure, Frequency, and Severity Models. *Accident Analysis & Prevention*, Vol. 35, No. 4, 2003, pp. 441–450.
20. Carson, J., and F. Mannering. The Effect of Ice Warning Signs on Ice-Accident Frequencies and Severities. *Accident Analysis & Prevention*, Vol. 33, No. 1, 2001, pp. 99–109.
21. Lee, J., and F. Mannering. Impact of Roadside Features on the Frequency and Severity of Run-Off-Roadway Accidents: An Empirical Analysis. *Accident Analysis & Prevention*, Vol. 34, No. 2, 2002, pp. 149–161.
22. Khorashadi, A., D. Niemeier, V. Shankar, and F. Mannering. Differences in Rural and Urban Driver-Injury Severities in Accidents Involving Large-Trucks: An Exploratory Analysis. *Accident Analysis & Prevention*, Vol. 37, No. 5, 2005, pp. 910–921.
23. Abdel-Aty, M. A. Analysis of Driver Injury Severity Levels at Multiple Locations using Ordered Probit Models. *Journal of Safety Research*, Vol. 34, No. 5, 2003, pp. 597–603.
24. Abdel-Aty, M. A., and H. T. Abdelwahab. Predicting Injury Severity Levels in Traffic Crashes: A Modeling Comparison. *Journal of Transportation Engineering*, Vol. 130, No. 2, 2004, pp. 204–210.
25. Haleem, K, and M. A. Abdel-Aty. Examining Traffic Crash Injury Severity at Unsignalized Intersections. *Journal of Safety Research*, Vol. 41, No. 4, 2010, pp. 347–357.
26. Yasmin, S., and N. Eluru. Evaluating alternate discrete outcome frameworks for modeling crash injury severity. *Accident Analysis & Prevention,* Vol. 59, 2013, pp. 506-521.
27. Yasmin, S., N. Eluru, and S. V. Ukkusuri. Alternative Ordered Response Frameworks for Examining Pedestrian Injury Severity in New York City. *Journal of Transportation Safety & Security*, Vol. 6, No. 4, 2014, pp. 275–300.
28. Ye, F., and D. Lord. Comparing Three Commonly Used Crash Severity Models on Sample Size Requirements: Multinomial Logit, Ordered Probit and Mixed Logit Models. *Analytic Methods in Accident Research*, Vol. 1, 2014, pp. 72–85.
29. Zhao, S., and A. Khattak. Motor Vehicle Drivers' Injuries in Train–Motor Vehicle Crashes. *Accident Analysis & Prevention*, Vol. 74, 2015, pp. 162–168.
30. Hensher, D. A., J. M. Rose, and W. H. Greene. *Applied Choice Analysis: A Primer*. Cambridge University Press, 2005.
31. McFadden, D., and K. Train. Mixed MNL Models for Discrete Response. *Journal of Applied Econometrics*, Vol. 15, No. 5, 2000, pp. 447–470.
32. Washington, S. P., M. G. Karlaftis, and F. Mannering. *Statistical and Econometric Methods for Transportation Data Analysis*. CRC Press, Boca Raton, Fla., 2010.
33. Abdelwahab, H. T., and M. A. Abdel-Aty. Development of Artificial Neural Network Models to Predict Driver Injury Severity in Traffic Accidents at Signalized Intersections. *Transportation Research Record: Journal of the Transportation Research Board*, 2001. 1746: 6–13.
34. Chang, L.-Y., and H.-W. Wang. Analysis of Traffic Injury Severity: An Application of Non-Parametric Classification Tree Techniques. *Accident Analysis & Prevention*, Vol. 38, No. 5, 2006, pp. 1019–1027.
35. Chimba, D., and T. Sando. Neuromorphic Prediction of Highway Injury Severity. *Advances in Transportation Studies*, Vol. 19, No. 1, 2009, pp. 17–26.

The Standing Committee on Highway/Rail Grade Crossings (AHB60) peer-reviewed this paper (18-05356).

Article

TRR
JOURNAL OF THE TRANSPORTATION RESEARCH BOARD

Transportation Research Record
2018, Vol. 2672(10) 48–61

DOI: 10.1177/0361198118794067
journals.sagepub.com/home/trr
SAGE

Direct and Indirect Effects of Nebraska Motor Vehicle Drivers' Characteristics on Inattentive Driving at Highway-Rail Grade Crossings

Shanshan Zhao[1] and Aemal J. Khattak[2]

Abstract

This research investigated the direct and indirect effects of the characteristics of motor vehicle drivers in Nebraska, including their socioeconomic characteristics, safe driving knowledge, and attitudes to inattentive driving at highway–rail grade crossings (HRGCs). A random statewide mail survey of drivers aged 19 years or older provided the data analyzed in this research. The use of a structural equation model (SEM) enabled identification of direct and indirect effects among the variables. Conclusions were that drivers' attitudes toward safety issues at HRGCs had both direct and indirect effects on drivers' inattentive behavior. Drivers' intent to violate or obey driving regulations indirectly influenced driver inattention. Their overall knowledge of safely negotiating HRGCs did not directly affect inattentive driving, but it indirectly affected inattentive driving through interference with drivers' intent to violate or obey regulations. Drivers' perceptions of delay, safety, clarity, and reliability indirectly affected their inattentive behavior. Variables associated with fewer instances of inattentive driving included: positive attitudes toward safety issues at HRGCs; lower intent to violate (or greater intent to obey) driving rules at HRGCs; less frequent use of HRGCs; higher perceptions of the safety, reliability, and so forth of local HRGCs; older drivers ($\geq$60); better knowledge of safely negotiating HRGCs; lower educational levels; male drivers; lower income households (<US$30 k per year); and longer residency in the current city. Safety information dissemination, high household income (>US$100 k per year), and younger drivers had direct effects on drivers' safe driving knowledge and could lead to higher overall knowledge levels of safely negotiating HRGCs.

Introduction

Motor vehicle drivers' inattention to the driving task includes drivers engaging in and being distracted by secondary tasks, drowsiness, fatigue, and daydreaming among other things. As defined by Regan et al. (*1*), driver inattention is "insufficient or no attention to activities critical for safe driving." Inattention can easily lead to driving errors such as decision-making errors (*2, 3*). A driver's cognitive distraction can be as risky as visual and manual distractions. When engaged in cognitive tasks, drivers make fewer inspections of the instruments and mirrors inside the vehicle and pay less attention to traffic lights outside the vehicle (*4*). The consequences of inattentive driving can be especially dangerous at rail crossings where careful scanning for train presence and watching for warning signals are important. Although not directly pertaining to highway–rail grade crossings (HRGC), Grace et al. (*5*) reported increased chances of traffic rule violations at intersections when drivers were involved with secondary tasks.

Motor vehicle driver inattention contributed to a high proportion of roadway crashes. According to the U.S. Railroad Safety Statistics 2005, about 41% of all reported rail crossing crashes were due to driver inattention (*6, 7*). Young et al. (*8*) reported that motor vehicle drivers' unintentional violations, which are potentially caused by inattention and low awareness of risks (*8–11*), account for about half of all crashes at rail crossings in Australia. Research has shown an association between driver inattention and more severe crash injuries (*12*).

Despite the possibility of significant harm, driver inattention at rail crossings is frequent. The naturalistic driving studies (*13, 14*) of the Federal Railroad

[1]Connecticut Transportation Safety Research Center, Connecticut Transportation Institute, University of Connecticut, Storrs, CT
[2]Department of Civil Engineering, University of Nebraska-Lincoln, Lincoln, NE

Corresponding Author:
Address correspondence to Shanshan Zhao: shanshan.h.zhao@uconn.edu

Administration (FRA) reveal that on average light vehicle and heavy truck drivers at rail crossings engaged in secondary tasks 46.7% and 21% of the driving time, respectively. At passive grade crossings, driver distraction was a primary cause or a contributing factor in 20% of all cases (*15*, *16*). A survey of 4,402 drivers in Australia revealed that 25% had engaged in risky behavior at rail level crossings and motor vehicle driver inattentiveness was the most significant risk factor (*7*). Tung and Khattak (*17*) investigated motor vehicle driving distraction in vicinity of HRGCs using data collected from video recordings and reported that about one-third of the observed motor vehicle drivers were distracted.

Research shows that the most common sources of inattention at HRGCs are talking to or looking at passengers, light vehicle drivers using cellphones (*14*), and heavy truck drivers using cellphones, smoking or lighting cigarettes (*13*). Stereo systems and passengers were sources of in-vehicle distractions while highway traffic was a common external distraction (*15*). In addition to cellular phone use, interacting/talking with passengers, distraction from outside the vehicles, and adjusting in-vehicle equipment, cognitive distraction was also a major distraction (*9*, *11*).

Regarding other reasons that can lead to driver inattention at rail crossings, Åberg (*18*) concluded that drivers' previous experience of the absence of trains at crossings affected their motivation to acquire information at the crossing and the impulse to look for trains increased as the number of trains at the crossing increased. Driver inattention can be a result of drivers' tendency not to expect a train; drivers seemed to underestimate the number of trains passing a crossing. Absence of trains reinforced this low expectancy of trains each time the driver passed the crossing (*15*).

Overall, significant research has addressed drivers' inattention and distraction in general highway settings but research regarding the contribution of these factors to rail crossing safety is limited (*16*). Naturalistic driving studies (*8*, *13*, *14*), fixed-site observational studies (*17*, *18*), crash report analyses (*9*, *15*), and questionnaire surveys or focus group interviews (*18*, *19*) have been employed in investigating drivers' inattentive driving or its impact at HRGCs. However, surveys focused particularly on driver inattention and distraction at HRGCs are sparse. In particular, the effects of driver characteristics – their intent to obey or violate driving regulations, their knowledge of negotiating HRGCs, and their experience of negotiating local rail crossings – on inattentive driving at HRGCs have not been explored during the past two decades.

Knowledge of direct and indirect relationships between different driver characteristics and inattentive driving is useful as it can provide important insights into improving safety at HRGCs. Therefore, the objective of this research was to investigate the direct and indirect effects between different characteristics of drivers and inattentive driving at HRGCs; these characteristics included drivers' safety knowledge levels, their attitudes toward safe driving at HRGCs, and socioeconomic attributes. A statewide random-sample self-report survey of drivers aged 19 years and older residing in Nebraska provided the data analyzed in this research. The analysis utilized a structural equation model (SEM).

Method

Data Collection

The survey questionnaire designed for distribution to households in Nebraska sought data on a variety of HRGC safety related issues, respondent driving behavior, attitudes, and socioeconomic attributes. This survey questionnaire consisted of 31 questions. The Bureau of Sociological Research (BOSR) of the University of Nebraska-Lincoln helped administer the survey (i.e., mail the survey questionnaire, send reminders to non-responders, and receive and code completed questionnaires) in July/August 2015. A random sample of 2,500 households in Nebraska were requested to participate in the survey. The BOSR received 980 survey questionnaires from the respondents for an overall response rate of 39.2%. Nebraska has a population of around 1.9 million and 736,613 households. According to FRA, there are about 9,965 HRGCs in Nebraska. Table 1 lists the survey questions pertinent to this research.

Analytical Methods

Analysis of the direct and indirect effects of various driver characteristics on inattentive driving at HRGCs utilized a SEM, which is a hybrid between analysis of variance (ANOVA)/regression and factor analysis. In SEM, the response variable in one equation may appear as a predictor in another equation; one variable could influence another variable reciprocally directly or indirectly through intermediaries. A SEM requires two inputs – the qualitative causal assumptions from the researcher and the empirical data used for the research. SEM then results in two consequences of the two inputs – the causal relationships and statistical measures of fit for the assumptions (*20*).

A SEM has two parts: the measurement model and the structural model (*21*). The measurement model builds the relationships between latent variables (factors) and their manifest indicators. The structural model relates both latent and manifest variables. Specifically, the structural model describes three types of relationships; the

Table 1. Survey Questions Pertinent to This Research

Local rail crossings

1. As a motor vehicle driver, please indicate your agreement or disagreement with the following statements (Strongly agree, Agree, Neutral, Disagree, Strongly disagree)
  1a. I believe motorist delays at rail crossings in my city (the city of your residence at the time of this survey) are excessive.
  1b. I feel unsafe when driving at rail crossings in my city.
  1c. I feel traffic signs and pavement markings at rail crossings in my city are confusing.
  1d. I doubt the reliability of the train warning devices (e.g., flashing lights, bells, gates, etc.) at the rail crossings in my city.
  1e. I've never received information on rail crossing safety.

Use and knowledge of rail crossings
  3. What type of personal motor vehicle do you drive on a daily basis most often? (Passenger car, Pick-up truck, Minivan, Motorcycle, Other (specify), Do not drive a personal motor vehicle daily)
  4. What type of work motor vehicle do you drive on a daily basis most often? (Passenger car, Pick-up truck, Minivan, Motorcycle, Other (specify), Do not drive a personal motor vehicle daily)
  5. During the past 14 days, how often did you drive across rail crossings?
    _______ times during the past 14 days.
  7. Based on your experience, how many trains do you think pass through this crossing *(the most frequently used crossing you mentioned in Question 6)* daily?
    _______ trains pass through on a daily basis.

Questions 8–16 ask your knowledge of safely negotiating a rail crossing (notice choices were omitted in this manuscript to save space).
  8. What does a crossbuck sign require a driver to do when approaching a rail crossing?
  9. Railroad companies post an emergency 1-800 number at crossings. The purpose of this number is to *(check all that apply).*
  10. What should a motor vehicle driver do when approaching a rail crossing and the crossing lights start flashing?
  11. What should a motor vehicle driver do if the crossing lights start flashing after he/she has started to cross the tracks?
  12. At a rail crossing that is designated as a Quiet Zone indicated by "No Train Horn", the train will...
  13. What should a motor vehicle driver do if his/her vehicle stalls on a rail crossing?
  14. Which of the following may be considered a motor vehicle violation at a gated rail crossing? *(Check all that apply).*
  15. What should a motor vehicle driver do at a gated rail crossing if the gates do not open after a train has passed?
  16. Which of the following vehicles must stop at all rail crossings unless the crossing is abandoned, exempted, or a flagman is present? *(Check all that apply).*

Activities and experiences while driving across rail crossings

17. Indicate how often you participated in each of the following activities while driving across rail crossings (Always, Often, Sometimes, Rarely, Never).
  17a. Look left and right to check for trains when approaching a rail crossing.
  17b. Drive across a rail crossing when the train warning devices (e.g., lights, bells, etc.) were activated.
  17c. Drive across a rail crossing when the gates were descending, ascending or in a level position.
  17d. Stop and check for trains when there is a STOP sign at the crossing.
  17e. Talk to other passengers in the vehicle while driving across a rail crossing.
  17f. Eat or drink while driving across a rail crossing.
  17g. Talk on a cell phone while driving across a rail crossing (including using hands-free arrangements).
  17h. Text or use Apps on a cellphone or other electronic device while driving across a rail crossing.
  17i. Reach for objects inside the vehicle (e.g., food, phone, map, etc.) while driving across a rail crossing.
  17j. Adjust any in-vehicle equipment (e.g., radio, heater/air conditioning, windows, etc.) while driving across a rail crossing.
  17k. Distracted by a person, object or event (e.g., accident) outside of the vehicle while driving across a rail crossing.
  17l. Mentally not focused on the driving task while driving across a rail crossing.
  17m. Smoking cigarettes while driving across a rail crossing.
  17n. Other distraction (e.g., personal grooming) while driving across a rail crossing.

18. As a motor vehicle driver, please indicate your agreement or disagreement with the following statements (Strongly agree, Agree, Neutral, Disagree, Strongly disagree).
  18a. I believe safety is a significant issue at rail crossings.
  18b. I do not like to wait for passing trains at rail crossings.
  18c. I like to accelerate my vehicle and quickly get across whenever train warning devices get activated.
  18d. I routinely stop when train warning devices are active even if I have a chance to cross the tracks before train arrival.
  18e. I regret stopping when train warning devices were active and I had a chance to get across before arrival of the train at the crossing.
  18f. I like to drive across the tracks after a train has passed even though warning devices may still be active.
  18g. I ensure that all warning devices have stopped after the passage of a train before I drive across the tracks.
  18h. I like to drive around/between fully lowered gates when I can.
  18i. I support technology that will block cellphone signals at rail crossings (except for emergency calls) to reduce distracted driving.
  18j. I support stronger law enforcement at rail crossings.
  18k. I am familiar with Operation Lifesaver.
  18l. I would like to receive information on rail crossing safety.
  18m. Playing "chicken," intentionally stopping a vehicle on a rail crossing in front of an oncoming train, is fun.

(continued)

Table 1. (continued)

Local rail crossings
General information (answering choices were omitted here to save space)
23. How long have you lived in your city (the city of your residence at the time of this survey)?
25. How long have you been a licensed driver?
26. What is your gender?
27. What is your age group?
28. What is your highest level of education?
30. What is your approximate annual household income (i.e., combined for all household members)?

relationships among latent variables, the relationships among observed variables, and the relationships between latent variables and observed variables that are not factor indicators. A set of regression equations describe these relationships. Linear regression equations used for continuous latent variables that are dependent variables, linear regression equations for continuous observed dependent variables, censored normal or censored-inflated normal regression equations used for censored observed dependent variables, Poisson or negative binomial regression equations for count observed dependent variables, logistic or probit regression equations for categorical observed dependent variables, etc. (*21*).

Besides the manifest variables that are directly observed and measured and the latent variables that are not directly measured, in a SEM, variables that are not influenced by other variables in the model are called exogenous variables (represented by x); variables that are influenced by other variables in the model are called endogenous variables (represented by y). In a SEM, γ represent the structural parameters relating an endogenous to an exogenous variable, and β are for structural parameters relating one endogenous variable to another, whereas ζ represent the disturbance terms.

Results

Patterns in Response

Table 2 provides summary statistics for questions included in the survey questionnaire. Table 2*a* shows that respondents generally believed that the traffic signs and pavement markings at their local rail crossings were clear and not confusing, that the rail crossings were safe, and that the train warning devices such as flashing lights, bells, gates, etc., were reliable. Most of the complaints came from excessive delays at rail crossings and lack of safety information on local rail crossings. Table 2*b* presents the percentages of different types of motor vehicles used by the respondents; most of the respondents drove passenger cars (including SUVs) for both personal and work purposes, followed by pickup trucks.

Table 2*c* shows that the majority (76%) of the respondents used a HRGC at least once during the two-week period before responding to the questionnaire. This shows that most of the respondents were frequent users of HRGCs and were likely to recall their experience of using HRGCs. Table 2*d* summarizes the expected number of trains at the local HRGC which the respondent used most frequently. Table 2*e* indicates that respondents generally took correct actions at rail crossings with active traffic control devices, but many respondents did not fully understand signage at rail crossings, the risks of certain violations, and the actions needed in cases of emergencies.

Table 2*f* shows that over 82% of the respondents did not cross upon activation of the warning devices or gates. Most respondents seldom texted or used apps at rail crossings. A majority of the drivers responding kept from reaching for objects in the vehicle (66.2%), talking on a phone (53.7%), mental distraction (53.0%), and adjusting in-vehicle objects (51.3%). Relatively few drivers were distracted by outside objects, eating or drinking when crossing, or talking to passengers.

Table 2*g* indicates the majority of the respondents agreed that safety is a significant issue at rail crossings (83.2% = 54.7% + 28.5%). Over 54% supported technologies that can block cellphone signals at rail crossings (except for emergency calls) to reduce distracted driving. About 58% of the drivers supported stronger law enforcement toward rule violations at HRGCs. Respondents seemed to know relatively little about public information programs dedicated to reducing collisions, injuries, and fatalities at HRGCs (only 21.9% acknowledged they knew), such as Operation Lifesaver. Only 23.6% respondents indicated a desire to receive information on rail crossing safety. The survey revealed that although respondents did not like to wait for trains to pass, most of them did not accelerate vehicles to go across the tracks when warning devices were activated.

Missing values in Table 2 arise from survey respondents not responding to some of the questions (called item nonresponse). One way to deal with item

Table 2. Summary Statics in Responses

(*a*) Driver perceptions of local rail crossings (in percentage %)

Aspects of perceptions	Strongly agree	Agree	Neutral	Disagree	Strongly disagree	Not answered
Q1a. Excessive delays	4.9	11.1	24.2	35.3	18.2	6.3
Q1b. Unsafe	1.3	5.2	12.9	47.0	27.0	6.5
Q1c. Confusing signs and markings	0.6	2.1	10.1	55.9	24.6	6.6
Q1d. Unreliable warning devices	1.8	5.8	12.0	49.6	23.9	6.8
Q1e. No safety info received	13.8	28.3	17.7	22.6	10.6	7.1

(*b*) Types of vehicles (in percentage %)

Vehicle type	Passenger car	Pick-up truck	Minivan	Motorcycle	Other	Not drive	Not answered
Q3. Personal motor vehicle	67.2	16.3	6.5	0.3	0.6	1.7	7.2
Q4. Work motor vehicle	14.2	11.3	2.2	-	4.4	64.7	4.5

(*c*) Frequency of HRGC use (in percentage %)

Q5. Use frequency of rail crossings (times/day)	None	0<freq.≤7	7<freq.≤14	14<freq.≤28	freq.>28	Not answered
Percentage %	17.1	34.3	15.7	13.9	11.9	7.0

(*d*) Estimated daily train passages at local HRGCs

Q7. Expected train passages/day	None	0<freq.≤10	freq.>10	Unknown or not answered
Percentage %	9.6	38.0	20.6	31.8

(e) Questions testing drivers' knowledge of driving at rail crossings (in percentage %)*

	Choices				
Questions	A	B	C	D	Not answered
Q8. Meaning of crossbuck signs	23.8	**45.2**	23.2	1.3	6.5
Q9. Use of railroad 1-800 number	**73.5**	**30.7**	**58.3**	18.5	3.9
Q10. Actions when lights flashing	0.2	5.1	**90.7**	0.1	3.9
Q11. Actions when lights start flashing while crossing	0.5	**92.1**	2.3	1.0	4.0
Q12. Meaning of Quiet Zone	9.3	3.6	**66.7**	15.5	4.9
Q13. Actions when stalled on tracks	0.2	7.9	**84.3**	1.2	6.4
Q14. Considered of violations	**77.6**	**91.8**	**65.0**	1.7	3.8
Q15. Actions when gates did not ascend immediately after train passed	1.0	**91.2**	0.2	3.3	4.3
Q16. Vehicles must stop at rail crossings	**95.3**	**79.1**	**81.4**	1.3	3.4

(*f*) Participation in attentive and inattentive driving activities (in percentage %)

Activities	Always	Often	Sometimes	Rarely	Never	Not answered
Q17a. Look left and right to check for trains	70.9	13.4	5.8	3.1	3.2	3.7
Q17b. Cross when warning devices activated	0.6	0.1	1.7	11.1	82.4	4.0
Q17c. Cross when gates descending, ascending or leveled	0.5	0.3	1.2	7.0	86.0	4.9
Q17d. Stop at STOP signs	77.9	8.3	2.1	1.0	5.4	5.3
Q17e. Talk to passengers	2.3	11.2	36.6	19.4	26.0	4.4
Q17f. Eat or drink	1.0	5.4	24.4	23.4	41.5	4.2
Q17g. Talk on a phone	0.6	4.1	19.6	18.1	53.7	4.0
Q17h. Text or use apps	0.2	0.7	3.8	8.9	82.6	3.9
Q17i. Reach for objects	0.3	1.3	8.7	19.7	66.2	3.8
Q17j. Adjust in-vehicle equipment	0.5	2.9	13.9	27.6	51.3	3.9
Q17k. Distracted by outside object	0.1	1.5	13.1	36.1	44.9	4.3
Q17l. Mental distraction	1.0	1.4	9.4	30.9	53.0	4.4
Q17m. Smoke cigarettes	0.7	2.9	5.2	2.7	84.5	4.1
Q17n. Other form of inattention	0.1	0.1	3.4	14.8	77.2	4.4

(continued)

Table 2. (continued)

(g) Attitudes and intentions regarding safe driving at rail crossings (in percentage %)

Questions	Strongly agree	Agree	Neutral	Dis-agree	Strongly disagree	Not answered
Q18a. Safety at rail crossings is important	54.7	28.5	9.5	4.7	1.0	1.6
Q18b. Do not like to wait for trains to pass	10.8	32.4	28.9	13.0	12.7	2.2
Q18c. Like to accelerate to cross through when warning devices are activated	1.8	2.4	6.1	33.9	53.9	1.8
Q18d. Routinely stop when warning devices are activated even if there is a chance to cross	48.9	34.1	5.6	3.5	6.0	1.9
Q18e. Regret for stopping for trains when there is a chance to cross	2.6	5.9	13.5	34.7	41.5	1.8
Q18f. Like to cross after train passage but warning devices still active	1.0	1.2	3.4	35.0	57.6	1.8
Q18g. Ensure warning devices off before crossing	57.2	34.3	3.0	1.7	2.0	1.7
Q18h. Like to drive around fully lowered gates	0.9	0.1	0.3	16.1	80.7	1.8
Q18i. Support technology that blocks cell phone signals at rail crossings	33.5	21.1	20.5	11.2	11.6	2.0
Q18j. Support stronger law enforcement	29.2	29.1	27.9	7.2	4.4	2.2
Q18k. Familiar with Operation Lifesaver	10.6	11.3	21.3	26.5	26.0	4.2
Q18l. Would like to receive info on rail crossing safety	8.9	14.7	34.2	21.4	17.7	3.1
Q18m. Feel it is fun to play "chicken" at rail crossings	1.2	0.0	0.3	3.7	93.2	1.6

(h) Demographic characteristics

Q23. Residency in current city	Mean=27.61	Standard Deviation=20.95
Answers to the following questions	Yes	No
Q25. Licensed driver for more than 10 years	84.80%	6.43%
Q26. Female driver	51.63%	39.29%
Q27. Driver age <30	9.39%	88.88%
Q27. Driver age ≥30 and <60	42.96%	55.31%
Q27. Driver age ≥60	38.88%	59.39%
Q28. Up to high school education	19.80%	76.63%
Q28. Up to associate degree education	29.59%	66.84%
Q28. Up to bachelor's degree education	24.69%	71.73%
Q28. Higher than bachelor's degree education	15.31%	81.12%
Q30. Household annual income (in US$) <30,000	16.73%	72.45%
Q30. Household annual income (in US$) ≥30,000 and <60,000	24.29%	64.90%
Q30. Household annual income (in US$) ≥60,000 and <100,000	21.43%	67.76%
Q30. Household annual income (in US$) ≥100,000	19.69%	69.49%

Note: Bold font indicates correct answers. Sample size: 980.

nonresponse is to discard the entire record but discarding the entire response from a respondent with a few item nonresponses may significantly reduce the sample size. Another way to compensate for item nonresponse is imputation, which means assigning a value (e.g., the mean) for the missing data (*22*). However, use of the same value for each missing datum artificially reduces variance of the variable with missing data. This research adopted pairwise deletion to deal with item nonresponse, which is a compromise between the two aforementioned methods. In pairwise deletion, each correlation between each pair of variables is calculated from all cases that have valid data on those two variables, even though there might be missing data for other variables of the same cases.

Direct and Indirect Effects Analysis using Structural Equation Modeling

Investigation of direct and indirect effects of motor vehicle drivers' characteristics on their inattentive driving behavior at HRGCs utilized a SEM. An overall risk

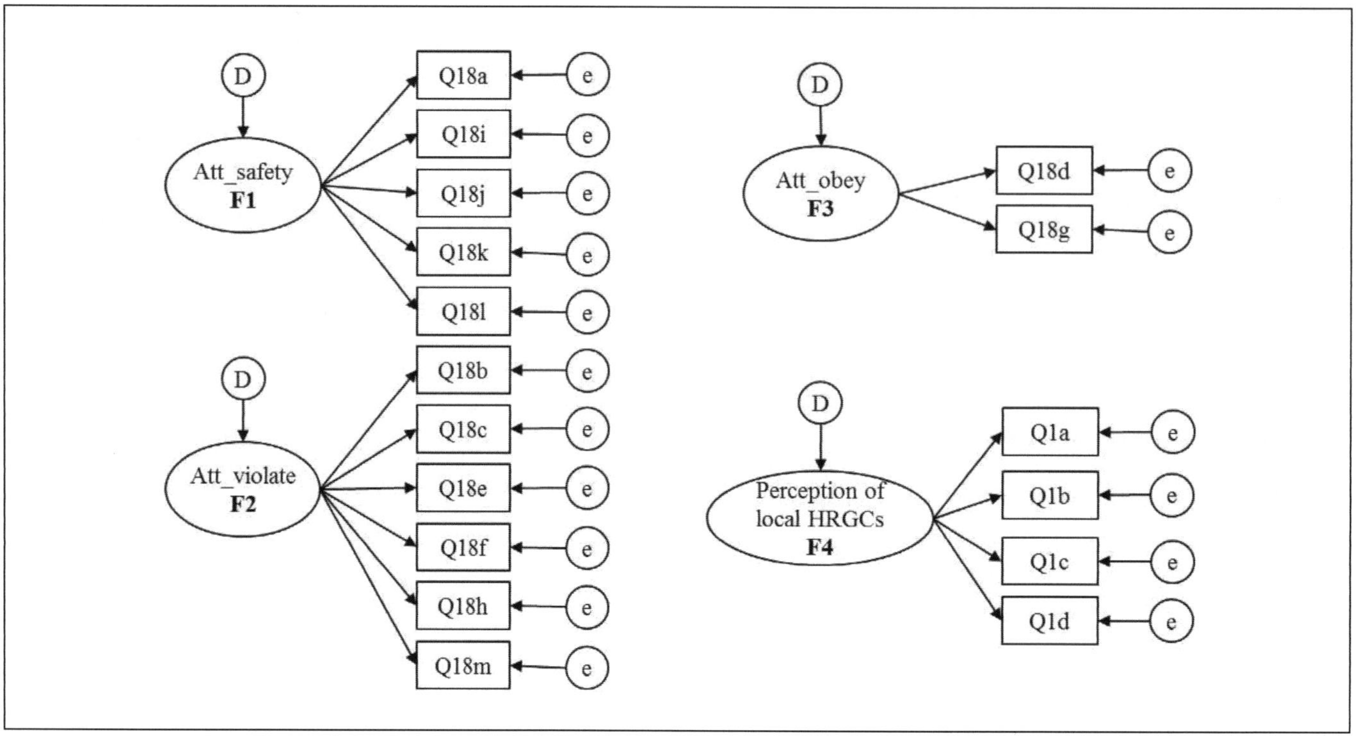

Figure 1. Measurement model.
*Q18a–m are questions 18a–m in the survey; Q1a–d are questions 1a–d in the survey.

score with regard to driving inattention was calculated for each respondent based on his/her responses to questions 17e through 17n (questions 17a through 17d were not directly asking about drivers' inattentive behavior and thus were not integrated into the overall risk score). Participants' knowledge of driving safely at HRGCs was assessed in questions 8 through 16, the responses to which were aggregated into one overall safe driving knowledge score theoretically ranging from 0 to 27 points with each correct response worth 3 points. The overall knowledge score was then categorized into four levels: level 1 (overall knowledge score of 0–17), level 2 (overall knowledge score of 18–20); level 3 (overall knowledge score of 21–23); and level 4 (overall knowledge score of 24–27). The thresholds between levels were determined by the fact that the sampled population had scores mostly clustered between 18 and 27.

A theoretical SEM model including the assumed direct and indirect effects was estimated. The proposed SEM was separated into the "measurement model" and the "structural model" and is presented in Figures 1 and 2. As shown in Figure 1, the measurement model illustrates the mapping of observed measures onto the theoretical latent factors: (1) factor 1 (i.e., F1) – drivers' attitude toward safety at HRGCs; (2) factor 2 (i.e., F2) – drivers' intent to violate rules at HRGCs; (3) factor 3 (i.e., F3) – drivers' habit of obeying rules at HRGCs; and (4) factor 4 (i.e., F4) – drivers' perceptions of local HRGCs. Questions 18a and 18i through 18l measured factor 1; Questions 18b, 18c, 18e, 18f, 18h, and 18m measured factor 2; Questions 18d and 18g measured factor 3; and Questions 1a through 1d measured factor 4. No correlations were assumed between the latent factors.

In Figure 2, the structural models show the direct and indirect causal and correlational links between the latent factors as well as other observed variables that are not part of the measurement model. The uni-directional arrows indicate direct effects assumptions and the bi-directional arrows reflect correlation assumption between two variables. The lack of an arrow from one variable to another indicates an assumption that no direct or indirect causal relationship or correlation exists between the two. Drivers' overall knowledge level of safely negotiating HRGCs (represented by Q8–16), attitude toward safety issues at HRGCs (represented by F1), intent to violate rules at HRGCs (represented by F2), habit of obeying rules at HRGCs (represented by F3), and perceptions of delay, safety, clarity, and reliability of local HRGCs (represented F4) were assumed to have direct effects on inattentive driving behavior at HRGCs (represented by 17e–n). Drivers' attitudes toward safety (F1) also affects their rule violating (F2) and obeying intent (F3). The latter two were assumed to share some underlying common reasons that are not revealed by factors considered here, and the relationship was represented by a bi-directional curved arrow in Figure 2. Drivers' perceptions of local

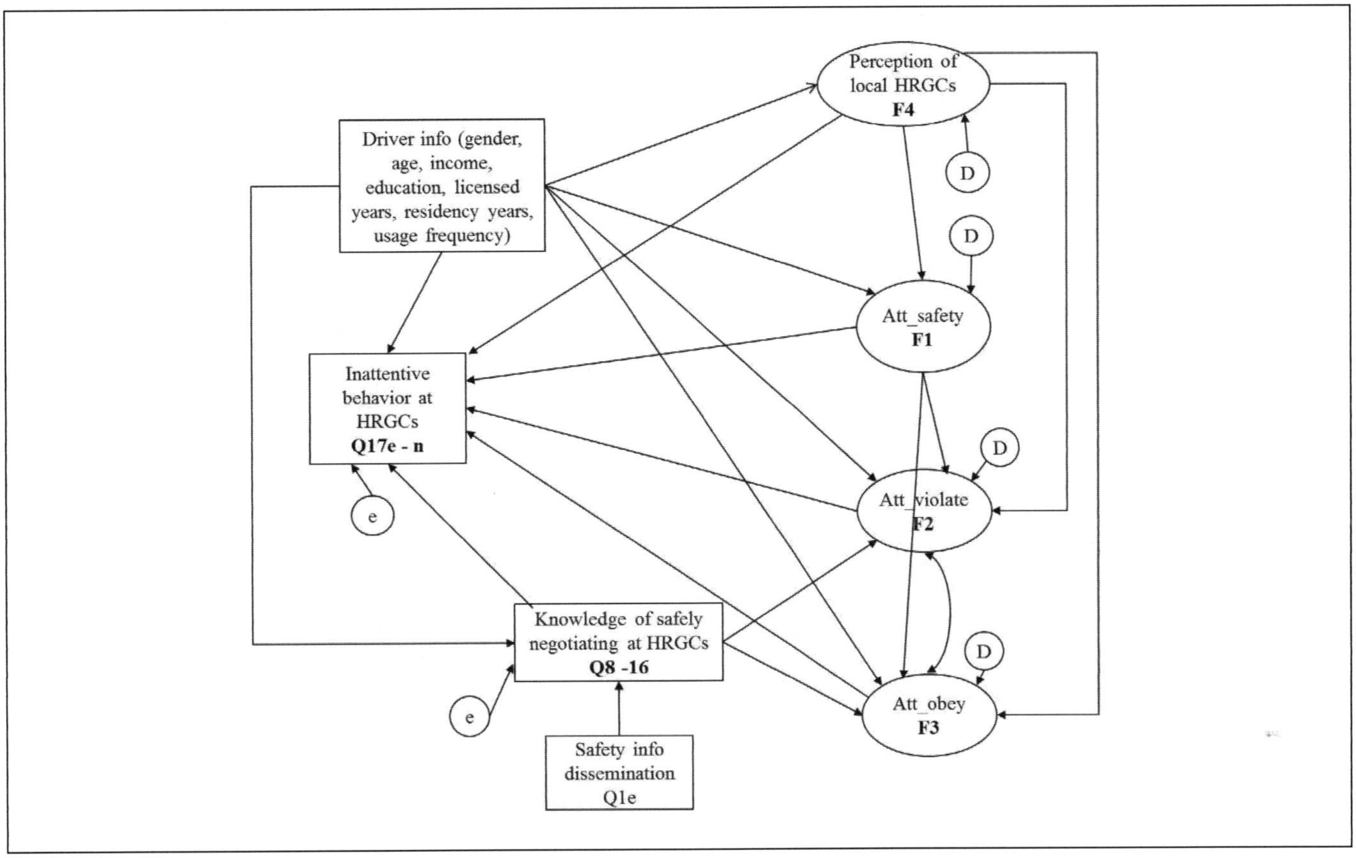

Figure 2. Structural model.

crossing conditions (F4) were assumed to have a direct effect on their higher or lower attitudes toward safety (F1) and intent to violate (F2) and obey rules (F3). Question 1e asked for drivers' exposure to information on rail crossing safety (Q1e) and it was assumed to affect their knowledge at HRGCs (Q8–16). The overall knowledge level (Q8–16), on the other hand, affects drivers' intent to violate rules (F2) or obey rules (F2) at HRGCs. Finally, driver attributes including gender, age, household income, education level, years of being a licensed driver, residency years in the current city, and frequency of using HRGCs were all assumed to have some direct effects on higher or lower levels of all the other factors.

The above SEM model was estimated in R (an open source programming language and software environment for statistical computing and graphics) using the "lavaan" package (*23*, *24*) and the model fit criteria including CFI (comparative fit index), RMSEA (root mean square error of approximation), and SRMR (standardized root mean squared residual) were used to evaluate the model result. As suggested by previous research on SEM (*25*), CFI > 0.95, RMSEA < 0.05, and SRMR < 0.06 ˜ 0.08 indicate good model fit. The result of the proposed SEM structure (above) showed poor model fit (CFI < 0.90) and suggested re-specification of the model. The model re-specification process combined adding parameters (e.g., adding correlations, regressors) that improved model fit and deleting parameters that were not statistically significant at the 90% level. The modification index (MI) was computed for each fixed (at zero) parameter. The value of a given MI reflects the minimum amount that the chi-square statistic is expected to decrease if the parameter is set free (*26*). Thus, a large MI may indicate significant improvement in model fit if that particular parameter is freed for estimation instead of fixed at zero. Notably, any modification in the structured model requires theoretical justification. A model with good fit was reached after several modifications to the original proposed model. Figure 3 presents the final model; this model achieves a CFI = 0.969, RMSEA = 0.029, and SRMR = 0.074, indicating a good fit.

By comparing the final model in Figure 3 with the original proposed model in Figures 1 and 2, it can be seen that key relations between the latent variables (F1 through F4) and their relations to the inattentive behavior variable remain unchanged. An arrow missing from the overall knowledge level variable to the inattentive

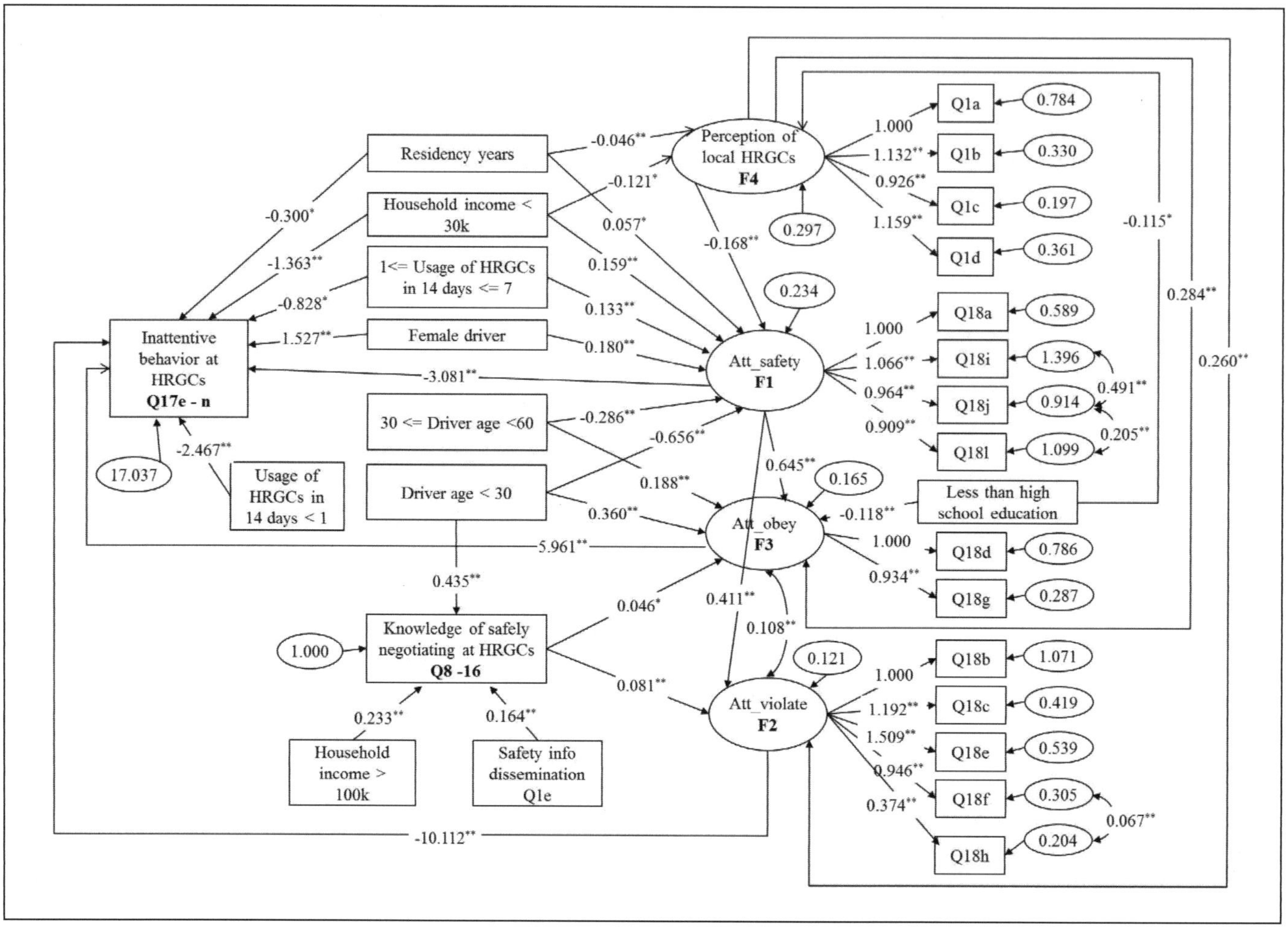

Figure 3. Final SEM model.
**Statistically significant at 95%; *statistically significant at 90%. Sample size = 660.

behavior variable indicated that the former does not have a direct effect on the latter, which is also true for the missing arrow from F4 to inattentive driving behavior. Driver related characteristics, including gender, age, income, education, licensed years, residency years, and HRGC use frequency were found significant in some of the relations, but not all. Removing the non-significant arrows from these variables to the key variables (such as the latent variables and the response variables including inattentive behavior and knowledge level) were not considered a violation of the theoretical assumptions because they were tentatively included in the first place. For the measurement models of the latent factors, a couple of the measures were removed (Q18k and Q18m) while three extra correlations between the measures were added – correlations between Q18i and Q18j, Q18j and Q18l, Q18f and Q18h.

Table 3 presents the parameter estimates for the structural models on driver inattention and overall knowledge levels, including the unstandardized estimates (the "estimate" column), standard error of the estimates, z-value of the estimates, p-value, and standardized estimates (the "std.lv" column for standardized solutions when only latent variables are standardized and the "std.all" column for standardized solutions when both latent and observed variables are standardized). The unstandardized estimates kept the scaling information of the variables and can only be interpreted with reference to the scales of the variables. The standardized estimates are non-scaling and comparable, which may help pick up more important factors and relationships. Standardized estimates with absolute values greater than 0.50 indicate a "large" direct effect, values around 0.30 indicate a "medium" direct effect and values less than 0.10 indicate a "small" effect (*27*).

The "lavaan" package estimates the ordinal response (recall that drivers' knowledge was categorized into four ordinal levels) in SEM using ordered probit regression (though this can be changed to logit). Table 3 presents their thresholds.

Table 3. Standardized and Unstandardized Coefficients for the Structural Equation Model

Structural models:		Estimate	Std.Err	z-value	$P(>\|z\|)$	Std.lv	Std.all
Inattentive driving behavior at HRGCs (Q17e–n) ~							
(Low to high scores indicating negative to positive attitudes)	Attitude toward safety issues at HRGCs (F1)	−3.081	0.739	−4.171	0.000	−1.775	−0.322
	Attitude toward violating traffic rules at HRGCs (F2)	−10.112	2.086	−4.848	0.000	−4.380	−0.793
	Attitude toward obeying traffic rules at HRGCs (F3)	5.961	1.594	3.740	0.000	3.180	0.576
(1=Yes 0=No)	Use of HRGCs <1 in the past 2 weeks	−2.467	0.585	−4.220	0.000	−2.467	−0.157
(1=Yes 0=No)	Use of HRGCs ≥1 and ≤7 in the past 2 weeks	−0.828	0.449	−1.842	0.065	−0.828	−0.073
(1=Yes 0=No)	Female	1.527	0.408	3.739	0.000	1.527	0.137
(1=Yes 0=No)	Household annual income <US$30,000	−1.363	0.585	−2.330	0.020	−1.363	−0.095
(Categories: L1= <5 yrs, L2=5–15 yrs, L3=15–25 yrs, L4=25–35 yrs, L5=>35 yrs	Residency in current city	−0.300	0.163	−1.833	0.067	−0.300	−0.082
Knowledge of safely negotiating HRGCs (Q8to16cat) ~							
(1= strongly agree, 5=strongly disagree)	Never receive safety info on rail crossing safety (Q1e)	0.164	0.038	4.305	0.000	0.164	0.201
(1=Yes 0=No)	Household annual income ≥US$100,000	0.233	0.116	1.998	0.046	0.233	0.097
(1=Yes 0=No)	Driver age <30	0.435	0.167	2.599	0.009	0.435	0.139
Thresholds:							
	Q8to16cat|t1	−0.586	0.193	−3.042	0.002	−0.586	−0.566
	Q8to16cat|t2	0.117	0.194	0.600	0.548	0.117	0.113
	Q8to16cat|t3	0.919	0.197	4.671	0.000	0.919	0.888

Discussion

The interpretation of the SEM results relies on the concepts of direct, indirect, and total effects. A direct effect is the impact that one variable (e.g., exogenous) has directly on another variable (e.g., endogenous). For example, through the SEM model results, it can be seen that gender has a direct effect on a driver's involvement with inattentive driving at HRGCs (represented by the uni-directional arrow in Figure 3, symbolized as female → Q17e–n). An indirect effect represents the effect of one variable on another variable through mediating variables. For example, gender was found to have an indirect effect on inattentive driving through F1 (as shown by the uni-directional arrows flowing from female to F1 and F1 to Q17e–n in Figure 3, symbolized as female → F1 → Q17e–n). The total effect is the summation of the direct and indirect effect. The total effect for gender (female) on involvement with inattentive driving would be the sum of the direct and the indirect effects. Table 4 summarizes the direct, indirect, and total effects of variables on overall driver inattention risk and overall knowledge level revealed by the SEM in the section "Direct and Indirect Effects Analysis Using Structural Equation Modeling"; the tables also show the 95% confidence intervals for those effects.

Table 4*a* shows that a driver's attitude toward safety issues at HRGCs has a direct effect on his or her involvement with inattentive driving behavior at HRGCs – a driver's more positive attitude reduces inattentive driving. This attitude also has some indirect effects on inattentive driving by affecting the driver's intent to violate or obey rules at HRGCs. The indirect effect has the same sign as the direct effect. Notice the indirect effect in this relation is not statistically significant. In summary, variables that only have direct effects on inattentive driving and can reduce inattentive driving include: lower intent to violate rules at HRGCs, greater intent to obey the rules at HRGCs, and less frequent use of HRGCs. Variables that only have indirect effects and can reduce inattentive driving are: higher perceptions of safety, reliability etc., of local HRGCs; older drivers (≥60); higher knowledge of safely negotiating HRGCs; and a lower educational level. Variables that have both direct and indirect effects and reduce inattentive driving include: positive attitudes toward safety issues at HRGCs, occasional HRGC use (1–7 times per 2 weeks), male drivers, lower income households (<US$30 k per year), and

Table 4a. Direct, Indirect, and Total Effects on Inattentive Driving Involvement at HRGCs

Variables	Effect	Estimate	Std.Err	z-value	P(>|z|)	95% Conf.int	
						Lower	Upper
Attitude toward safety issues at HRGCs (F1)	Direct	−3.081	0.739	−4.171	0.000	−4.529	−1.633
	Indirect	−0.318	0.607	−0.523	0.601	−1.508	0.872
	Total	−3.399	0.544	−6.245	0.000	−4.465	−2.333
Attitude toward violating traffic rules at HRGCs (F2)	Direct	−10.112	2.086	−4.848	0.000	−14.201	−6.023
	Indirect	—	—	—	—	—	—
	Total	−10.112	2.086	−4.848	0.000	−14.201	−6.023
Attitude toward obeying traffic rules at HRGCs (F3)	Direct	5.961	1.594	3.740	0.000	2.837	9.085
	Indirect	—	—	—	—	—	—
	Total	5.961	1.594	3.740	0.000	2.837	9.085
Use of HRGCs <1 in the past 2 weeks	Direct	−2.467	0.585	−4.220	0.000	−3.614	−1.320
	Indirect	—	—	—	—	—	—
	Total	−2.467	0.585	−4.220	0.000	−3.614	−1.320
Use of HRGCs ≥1 and ≤7 in the past two weeks	Direct	−0.828	0.449	−1.842	0.065	−1.708	0.052
	Indirect	−0.452	0.226	−2.000	0.046	−0.895	−0.009
	Total	−1.280	0.471	−2.716	0.007	−2.203	−0.357
Female	Direct	1.527	0.408	3.739	0.000	0.727	2.327
	Indirect	−0.612	0.211	−2.905	0.004	−1.026	−0.198
	Total	1.075	0.450	2.389	0.017	0.193	1.957
Household annual income <US$30,000	Direct	−1.363	0.585	−2.330	0.020	−2.510	−0.216
	Indirect	−0.497	0.262	−1.895	0.058	−1.011	0.017
	Total	−1.859	0.604	−3.077	0.002	−3.043	−0.675
Residency in current city	Direct	−0.300	0.163	−1.833	0.067	−0.619	0.019
	Indirect	−0.177	0.081	−2.192	0.028	−0.336	−0.018
	Total	−0.477	0.162	−2.952	0.003	−0.795	−0.159
Perception of local HRGCs (F4)	Direct	—	—	—	—	—	—
	Indirect	−0.367	0.290	−1.267	0.205	−0.935	0.201
	Total	−0.367	0.290	−1.267	0.205	−0.935	0.201
Driver age <30	Direct	—	—	—	—	—	—
	Indirect	4.141	0.657	6.304	0.000	2.853	5.429
	Total	4.141	0.657	6.304	0.000	2.853	5.429
Driver age ≥30 and <60	Direct	—	—	—	—	—	—
	Indirect	2.090	0.430	4.858	0.000	1.247	2.933
	Total	2.090	0.430	4.858	0.000	1.247	2.933
Knowledge of safely negotiating HRGCs (Q8to16cat)	Direct	—	—	—	—	—	—
	Indirect	−0.538	0.143	−3.755	0.000	−0.818	−0.258
	Total	−0.538	0.143	−3.755	0.000	−0.818	−0.258
Less than high school education	Direct	—	—	—	—	—	—
	Indirect	−0.661	0.362	−1.829	0.067	−1.371	0.049
	Total	−0.661	0.362	−1.829	0.067	−1.371	0.049

Table 4b. Direct, Indirect, and Total Effects on Knowledge Level of Safely Negotiating HRGCs

Variables	Effect	Estimate	Std.Err	z-value	P(>|z|)	95% Conf.int	
						Lower	Upper
Never receive safety info on rail crossing safety (Q1e)	Direct	0.164	0.038	4.305	0.000	0.090	0.238
	Indirect	—	—	—	—	—	—
	Total	0.164	0.038	4.305	0.000	0.090	0.238
Household annual income ≥US$100,000	Direct	0.233	0.116	1.998	0.046	0.006	0.460
	Indirect	—	—	—	—	—	—
	Total	0.233	0.116	1.998	0.046	0.006	0.460
Driver age <30	Direct	0.435	0.167	2.599	0.009	0.108	0.762
	Indirect	—	—	—	—	—	—
	Total	0.435	0.167	2.599	0.009	0.108	0.762

longer residency tenure in the current city of the respondent's residence.

Notice that this research revealed that gender (female) had opposite direct and indirect effects on inattentive driving – directly, the SEM results revealed that females generally were more involved in inattentive driving; while indirectly, the model results indicated that females had a more positive attitude toward safety issues at HRGCs, which reduced their involvement with inattentive driving. Although the final total effect of "female" is still negative (i.e., the combined effects shown by the SEM results indicated that females were more involved in inattentive driving at HRGCs), the direct effect is mitigated by the opposite indirect effect. This finding conforms with previous research that reported female drivers to be slightly more likely to report being distracted or observed in more inattentive driving (*28*, *29*). However, generalization of this finding to the larger population requires caution since the results are based on the data available for this research. On the contrary, when the direct and indirect effects have the same sign, the effect of that factor on the inattentive behavior is reinforced because the direct and indirect effects have the same direction.

Table 4*b* shows that safety information dissemination, high household income (>US$100 k per year), and younger drivers only have direct effects on safety knowledge level and can lead to higher safety knowledge levels. Notably, these estimates are based on ordinal probit regression results and thus represent the increasing probability of knowledge falling into the j level versus the probability of falling into the (j– 1) level, where j is any of the second, third or fourth level of a driver's overall knowledge that were introduced in the section "Direct and Indirect Effects Analysis Using Structural Equation Modeling".

Conclusion

The objective of this research was to investigate direct and indirect effects of drivers' characteristics including their driving knowledge levels, their attitudes toward safety at HRGCs and socioeconomic attributes on their inattentive driving at HRGCs. SEM was used as the analysis tool assuming a series of direct and indirect effects. This theoretical structure was tested using the collected survey data, but resulted in an unacceptable model fit. By removing non-significant relations and adding relations with large modification indices (MI) and ensuring sensible removal and additions, the model was modified to achieve a superior fit, where CFI>0.95 and SRMR<0.08. The SEM revealed a relatively complete direct and indirect effects flow chart between driver factors and their knowledge, behavior, and intent. The direct, indirect, and total effects that the numerous exogenous variables had on the endogenous variables (i.e., drivers' involvement with inattentive driving, overall knowledge level, latent variables of attitude toward safety issues at HRGCs, intent to violate rules and obey rules, and perceptions of local HRGCs) and the causal relations between the endogenous variables were estimated.

Drivers' attitudes toward safety issues at HRGCs had both direct and indirect effects on drivers' inattentive driving behavior at HRGCs. Drivers' intent to violate or obey regulations indirectly affected their inattentive behavior. No evidence of a direct relationship between drivers' overall knowledge level of safely negotiating HRGCs and inattentive driving was uncovered, but the former indirectly affected the latter through interfering with drivers' intent to violate/obey regulations. No evidence was found that drivers' perceptions of delay, safety, clarity, and reliability directly affected their inattentive driving behavior. However, those factors had indirect effects on inattentive driving by influencing drivers' attitudes toward safety and intent to violate/obey rules. Demographic information of the drivers such as residency tenure in the current city of residence, income, gender, age, education, and frequency of rail crossing use had some direct or indirect effects on drivers' inattentive behavior.

Above all, inattentive driving behavior at HRGCs was directly and indirectly affected by drivers' attitudes toward safety issues at HRGCs. Drivers' knowledge of correct HRGC negotiation appeared to only indirectly affect inattentive driving behavior through influencing drivers' intent to violate or obey regulations. Therefore, compared with increasing drivers' knowledge, focusing on drivers' attitudes seems to be a more promising strategy to reduce inattentive driving at HRGCs. Although, changing drivers' attitudes and driving styles is not an easy task and reducing inattentive driving at HRGCs by attempting to change drivers' attitudes would be a protracted process.

Generalization of the conclusions to a larger driving population requires consideration of several research limitations. Firstly, the research used data pertaining to Nebraska only and generalization to states substantially different than Nebraska may be questionable. Secondly, data collected through the survey pertained to drivers that were aged 19 years or older. Therefore, the research findings and conclusions do not apply to drivers younger than 19 years, who may behave differently than drivers aged 19 years or older. Thirdly, there are inevitable biases associated with stated preference (self-reported) data. Although it is often difficult to gauge and control, biases exist to various degrees in self-reported surveys. Participants in this survey may not have answered truthfully for a variety of reasons that are difficult to control. The sample may also be biased because people who chose to participate are likely to be the ones that have stronger opinions on the topic. While a careful review of the collected data did not reveal any obvious biases, caution is warranted when referencing these findings and conclusions. Despite the limitations, the research contributes to the body of knowledge by investigating potential factors associated with inattentive driving behavior at HRGCs, which may not be possible without relying on stated preference data. The quantification of the effects that drivers' characteristics, attitudes, and knowledge have on their driving behavior can help program sponsors develop more targeted and effective educational, technological or regulatory programs that aim at reducing inattentive driving and improving HRGC safety.

Author Contributions

The authors confirm contribution to the paper as follows: study conception and design: Shanshan Zhao, Aemal J. Khattak; data collection: Shanshan Zhao, Aemal J. Khattak; analysis and interpretation of results: Shanshan Zhao, Aemal J. Khattak; draft manuscript preparation: Shanshan Zhao, Aemal J. Khattak. All authors reviewed the results and approved the final version of the manuscript.

References

1. Regan, M. A., C. Hallett, and C.P. Gordon. 2011. Driver Distraction and Driver Inattention: Definition, Relationship and Taxonomy. *Accident Analysis and Prevention*, Vol. 43, No. 5, 2011, pp. 1771–1781.
2. Young, K. L., and P. M. Salmon. Examining the Relationship between Driver Distraction and Driving Errors: A Discussion of Theory, Studies and Methods. *Safety Science*, Vol. 50, No. 2, 2012, pp. 165–174.
3. Young, K. L., P. M. Salmon, and M. G. Lenné. At the Cross-Roads: An On-Road Examination of Driving Errors at Intersections. *Accident Analysis and Prevention*, Vol. 58, 2013, pp. 226–234.
4. Harbluk, J. L., Y. I. Noy, P. L. Trbovich, and M. Eizenman. An On-Road Assessment of Cognitive Distraction: Impacts on Drivers' Visual Behavior and Braking Performance. *Accident Analysis and Prevention*, Vol. 39, No. 2, 2007, pp. 372–379.
5. Ashley, G., O. A. Osman, S. Ishak, and J. Codjoe. Effect of Secondary Tasks and Driver Behavior on Crash and Near-Crash Risk: Naturalistic Driving Study. Presented at 96th Annual Meeting of the Transportation Research Board, Washington, D.C., 2017.
6. Federal Railroad Administration. *Railroad Safety Statistics 2005 Annual Report*. Federal Railroad Administration, Washington, D.C., 2006.
7. Searle, A., L. Di Milia, and D. Dawson. *An Investigation of Risk-Takers at Railway Level Crossings*. Report No. 2. CRC for Rail Innovations, Brisbane, Australia, 2012.
8. Young, K. L., M. G. Lenné, V. Beanland, P. M. Salmon, and N. A. Stanton. Where Do Novice and Experienced Drivers Direct Their Attention on Approach to Urban Rail Level Crossings? *Accident Analysis and Prevention*, Vol. 77, 2015, pp. 1–11.
9. Caird, J. K., J. I. Creaser, C. J. Edwards, and R. E. Dewar. *A Human Factors Analysis of Highway-Railway Grade Crossing Accidents in Canada*. Report No. TP-13938E. Transport Canada, Calgary, 2002.
10. Freeman, J., and A. Rakotonirainy. Mistakes or Deliberate Violations? A Study into the Origins of Rule Breaking at Pedestrian Train Crossings. *Accident Analysis and Prevention*, Vol. 77, 2015, pp. 45–50.
11. Salmon, P. M., G. J. M. Read, N. A. Stanton, and M. G. Lenné. The Crash at Kerang: Investigating Systemic and Psychological Factors Leading to Unintentional Non-Compliance at Rail Level Crossings. *Accident Analysis and Prevention*, Vol. 50, 2013, pp. 1278–1288.
12. Zhao, S., and J. K. Aemal. Injury Severity in Crashes Reported in Proximity of Rail Crossings–The Role of Driver Inattention. *Journal of Transportation Safety & Security* (just-accepted), Vol. 10, No. 6, 2017, pp. 507–524.
13. Ngamdung, T., and M. DaSilva. *Driver Behavior Analysis at Highway-Rail Grade Crossings using Field Operational Test Data – Heavy Trucks*. Report No. DOT-VNTSC-FRA-12-01. Federal Railroad Administration. Office of Research and Development, Washington, D.C., 2012.
14. Ngamdung, T., and D. DaSilva. *Driver Behavior Analysis at Highway-Rail Grade Crossings using Field Operational Test Data – Light Vehicles*. Report No. DOT-VNTSC-FRA-12-02. Federal Railroad Administration. Office of Research and Development, Washington, D.C., 2013.
15. NTSB. *Safety Study: Safety at Passive Grade Crossings. Volume 1: Analysis*. NTSB/SS-98/02. National Transportation Safety Board, Washington, D.C., 1998.
16. Yeh, M., and J. Multer. *Driver Behavior at Highway- Railroad Grade Crossings: A Literature Review from 1990-2006*. Report No. DOT/FRA/ORD-08/03. U.S. DOT Volpe Center, Cambridge, Mass., 2008.
17. Tung, L. C., and A. J. Khattak. Distracted Motor Vehicle Driving at Highway–Rail Grade Crossings. *Transportation Research Record: Journal of the Transportation Research Board*, 2015. 2476: 77–84.
18. Åberg, L. Driver Behavior at Flashing-Light, Rail-Highway Crossings. *Accident Analysis and Prevention*, Vol. 20, No. 1, 1998, pp. 59–65.
19. Wallace, A. *Motorist Behavior at Railway Level Crossings: The Present Context in Australia*. PhD dissertation. Centre for Accident Research and Road Safety, Queensland University of Technology, 2008.
20. Bollen, K. A., and J. Pearl. Eight Myths about Causality and Structural Equation Models. In *Handbook of Causal Analysis for Social Research*. Springer Netherlands, 2013, pp. 301–328.
21. Muthén, L., and B. Muthén. Confirmatory Factor Analysis and Structural Equation. In *MPlus Statical Analysis with Latent Variables (User's Guide)*. Los Angeles, CA, 1998, pp. 43–78.
22. Brick, J. M., and G. Kalton. Handling Missing Data in Survey Research. *Statistical Methods in Medical Research*, Vol. 5, No. 3, 1996, pp. 215–238.

23. Rosseel, Y. *The lavaan Tutorial*. Department of Data Analysis. Ghent University, 2015.
24. Rosseel, Y., D. Oberski, J. Byrnes, L. Vanbrabant, V. Savalei, E. Merkle, M. Hallquist, M. Rhemtulla, M. Katsikatsou, M. Barendse, and M. Chow. *Package 'lavaan': Latent Variable Analysis*. R Package Version 0.5-20, 2015.
25. Hu, L., and P. M. Bentler. Cutoff Criteria for Fit Indexes in Covariance Structure Analysis: Conventional Criteria versus New Alternatives. *Structural Equation Modeling: A Multidisciplinary Journal*, Vol. 6, No. 1, 1999, pp. 1–55.
26. Hox, J. J., and T. M. Bechger. An Introduction to Structural Equation Modeling. *Family Science Review*, Vol. 11, 1998, pp. 354–373.
27. Suhr, D. The Basics of Structural Equation Modeling. Presented at SAS User Group of the Western Region of the United States (WUSS), Irvine, Calif., 2006.
28. Heck, K. E., and R. M. Carlos. Passenger Distractions among Adolescent Drivers. *Journal of Safety Research*, Vol. 39, No. 4, 2008, pp. 437–443.
29. Goodwin, A. H., R. D. Foss, S. S. Harrell, and N. P. O'Brien. *Distracted Driving Among Newly Licensed Teen Drivers*. AAA Foundation for Traffic Safety, Washington, D.C., 2012.

The Standing Committee on Highway/Rail Grade Crossings (AHB60) peer-reviewed this paper (18-05424).

Article

TRR
JOURNAL OF THE TRANSPORTATION RESEARCH BOARD

Transportation Research Record
2018, Vol. 2672(10) 62–73

DOI: 10.1177/0361198118790849
journals.sagepub.com/home/trr
SAGE

Evaluation of Railway Ballast Permeability Using Machine Vision–Based Degradation Analysis

Haohang Huang[1], Maziar Moaveni[1], Scott Schmidt[2], Erol Tutumluer[1], and John M. Hart[3]

Abstract

Railway ballast degrades progressively as a result of accumulated traffic primarily through abrasion and particle breakage. Degraded ballast may cause reduced lateral and longitudinal stability, ineffective drainage, and excessive settlement of track structures, all of which would adversely affect the performance of ballasted track. Traditional methods of ballast degradation assessment involve time-consuming field sampling and laboratory sieve analysis; moreover, determining the level of track performance deterioration at which ballast maintenance is best considered still remains challenging. This paper investigates the permeability of railway ballast through laboratory testing and provides insight into its field drainage capacity under degraded condition using an innovative approach of field imaging. Constant head permeability tests were conducted on clean and degraded ballast samples which indicated nonlinear power-curve trends, especially for clean ballast, of unit flow amount with its hydraulic gradient. Imaging-based degradation analysis using machine vision technology was also performed on clean and degraded in-service ballast to correlate Fouling Index (FI) from laboratory sieving with Percent Degraded Segments (PDS) obtained from the recently developed image segmentation algorithm. Accordingly, a new Permeability Index (PI) is introduced in this paper to define ballast permeability in the form of a bilinear model developed from the machine vision–based ballast degradation analysis. Based on the findings of this study, a two-stage ballast cleaning process for determining the timeframe of ballasted track maintenance considering its drainage capacity is proposed.

Railway ballast, comprising uniformly graded coarse aggregate particles, serves as an essential structural and functional component of railway track. Different functions of ballast include maintaining track geometry and structural stability, providing load distribution and track resilient capacity, and facilitating drainage (*1*). The progressive degradation of in-service ballast primarily occurs through particle abrasion and breakage, or through external fouling agents, such as subgrade soil or coal dust intrusion, which fill the voids in ballast with finer particles. This phenomenon when combined with the presence of high moisture content may result in rapid deterioration of track structure with reduced shear strength properties, impeded drainage, and excessive yet differential track settlement (*2–4*). As a result, the cost of periodic maintenance and long-term track renewal activities would increase.

Despite considerable research and practitioner experience related to railway maintenance, the stage at which the ballast layer requires the type of maintenance activity (such as shoulder cleaning, undercutting, ballast renewal, etc.) corresponding to various degradation levels and drainage capacity has not been adequately investigated. Generally, the decision on "when" and "how" to do ballast maintenance is based on field visual inspection which is subjective. Furthermore, the concerns of ballast drainage are customarily considered in railway design process rather than in long-term maintenance practice. Consequently, problematic drainage issues of degraded ballast and their locations may remain undetected until they are visible. This may raise concerns particularly when ballasted track is part of demanding shared

[1]Department of Civil and Environmental Engineering, University of Illinois at Urbana-Champaign, Urbana, IL
[2]Nicholson Construction Company, Kalamazoo, MI
[3]University of Illinois at Urbana-Champaign, Beckman Institute for Advanced Science and Technology, Computer Vision and Robotics Laboratory, Urbana, IL

Corresponding Author:
Address correspondence to Erol Tutumluer: tutumlue@illinois.edu

corridors where heavy freight trains and higher speed passenger trains are jointly operated.

Previous studies on clean and degraded railway ballast permeability commonly adopted a linear discharge velocity (unit amount of flow) versus hydraulic gradient relationship that had been originally proposed by Darcy (*5*), and evaluated the hydraulic conductivity at different hydraulic gradients in vertical constant head permeameter tests (*6–11*). Cole calculated hydraulic gradients in railway trackbed from flow nets and found the typical hydraulic gradients present in the field to be in the range of 0.1–1.0 (*12*). Studies on rock fill dam permeability also investigated the flow characteristics of large aggregate particles. According to Taylor (*13*), turbulent flow through porous media has a discharge velocity–hydraulic gradient relationship that does not follow Darcy's Law. Later, Wilkins found the relationship to be a nonlinear power curve (*14*):

$$v = C \cdot i^n \tag{1}$$

where

v is the discharge velocity,

C is an empirical constant depending on the fluid viscosity and the mean particle hydraulic radius defined by Taylor (*13*),

i is the hydraulic gradient, and

n is an empirical exponent typically found to be 0.54.

Schmidt et al. (*15*) recently conducted constant head permeability tests and validated the power-curve nature of the discharge velocity–hydraulic gradient relationship to correctly represent the permeability of uniformly graded ballast coarse aggregates.

For evaluating levels of ballast degradation in the field, Selig and Waters proposed the Fouling Index (FI) (*1*):

$$\mathrm{FI} = P_4 + P_{200} \tag{2}$$

where P_4 and P_{200} represent the weight percentages of particles passing the No. 4 (4.75 mm) and the No. 200 (0.075 mm) sieves, respectively. However, challenges related to collecting adequate amounts of field sampled ballast and the spatial variability of ballast conditions in a typical track structure need to be addressed in determining representative ballast particle size distributions from sieve analyses. For example, ballast samples obtained from different track locations such as shoulder, crib, and so on, can introduce extreme variability in FI values.

Emerging ballast inspection technologies based on electromagnetic waves and imaging methods have been recently implemented to replace traditional ballast sampling and sieving procedures for determining the level of ballast degradation. Al-Qadi et al. developed efficient analytical tools for reducing noise interference and better implementation techniques for the Ground Penetration Radar (GPR) applied to detect ballast degraded zones (*16*). Zarembski et al. coupled the field data using Laser Imaging Detection and Ranging (LIDAR) with GPR data to visualize the longitudinal and vertical profile of a ballast layer (*17*). Anbazhagan et al. utilized seismic surface waves to characterize shear strength properties of ballast fouled by various sources of materials (*18*). Moaveni et al. applied computer vision techniques to identify ballast particles based on shape and size characteristics, and proposed implementation of imaging-based inspection methods for evaluating field degraded ballast conditions (*19, 20*). Despite these encouraging developments in nondestructive techniques for ballast inspection and assessment of in-service ballast condition, in-situ drainage or hydraulic properties in terms of ballast permeability at different degradation levels have not been studied properly.

Objective

This paper presents an innovative approach to quantitatively relate parameters obtained from a recently developed imaging-based ballast field inspection technique to ballast saturated hydraulic conductivity or permeability behavior. The objective is to establish imaging-based degradation analysis data in the field to be linked to the ballast permeability, which will be essential for determining the timeframe of ballasted track maintenance. Therefore, the findings presented here will provide useful information to railway track maintenance decision makers seeking to improve track drainage and at the same time monitor the type and severity of ballast degradation.

Ballast Inspection System Based on Machine Vision

A machine vision–based ballast inspection system (MV-BIS) is designed and programmed to acquire field images of ballast cross sections, process those images using image processing techniques, and quantify the corresponding ballast degraded conditions. All stages of the ballast imaging and sampling are illustrated in Figure 1. Two types of ballast cross sections including vertical trench cross sections and horizontal shoulder cut sections are inspected and imaged as part of field image acquisition activities. After the working surface is flattened, an adjustable camera mounting system is positioned for consistent image acquisition of various locations of the trench wall. Note that a 1-in. (25.4 mm) white-colored calibration ball is placed on the area of interest to ensure a consistent spatial resolution (i.e., number of pixels per

Figure 1. Flowchart of field ballast imaging and sampling processes.

unit length) by adjusting the uniformity of the pixel input into the next imaging process step.

Laboratory sieve analysis and imaging-based ballast degradation analysis are then conducted on the collected ballast samples and associated images. The goal of the sieve analysis is to determine the particle size distribution for collected samples, thereby generating ground truth datasets for validating the results from image processing outputs.

The imaging-based degradation analysis utilizes a watershed segmentation algorithm as the core of the image processing module, which has been proven to be more efficient than other leading algorithms in the presence of mutually touching particles in a trench cross section or aggregate stockpile views (*21*). After extracting the grayscale image of field ballast cross section by preprocessing, the algorithm separates out individual segments. As railroad ballast particles are generally expected to be in convex or nearly convex shape, the segments are individually subjected to a convex-hull test, filtering out nonparticle segments and identifying true ballast particles. This test first constructs for each segment its convex hull and computes the ratio between the number of pixels of the segment and that of the convex hull (see Figure 2).

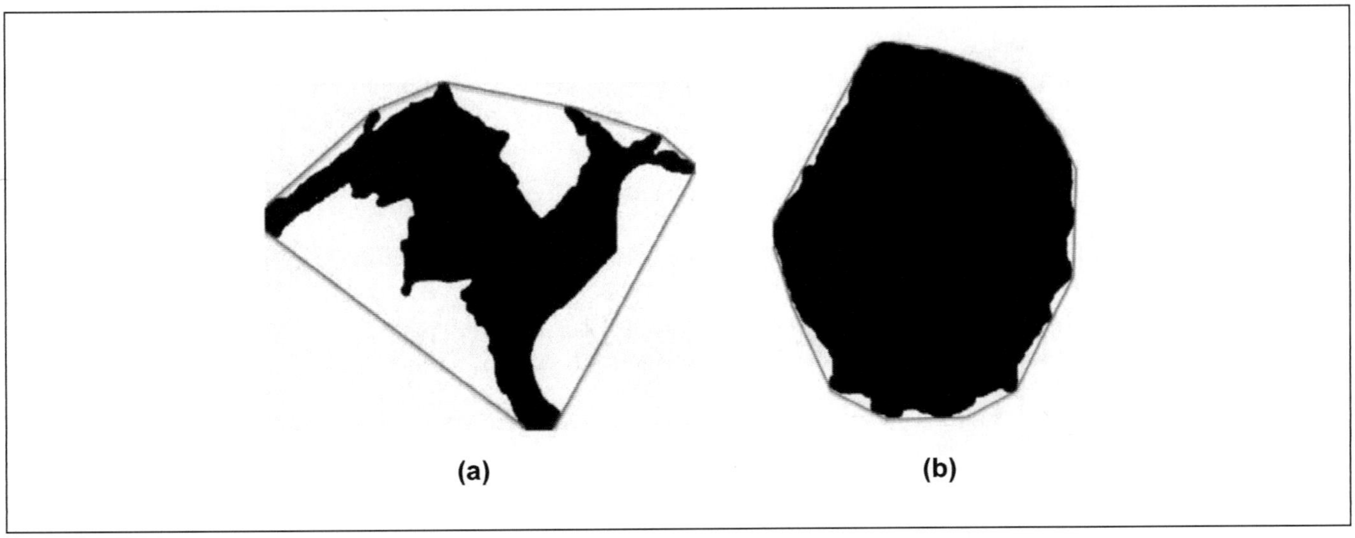

Figure 2. (*a*) Convex hull constructed for a false-segment (non-ballast particle) and (*b*) Convex hull constructed for a true-segment (ballast particle).

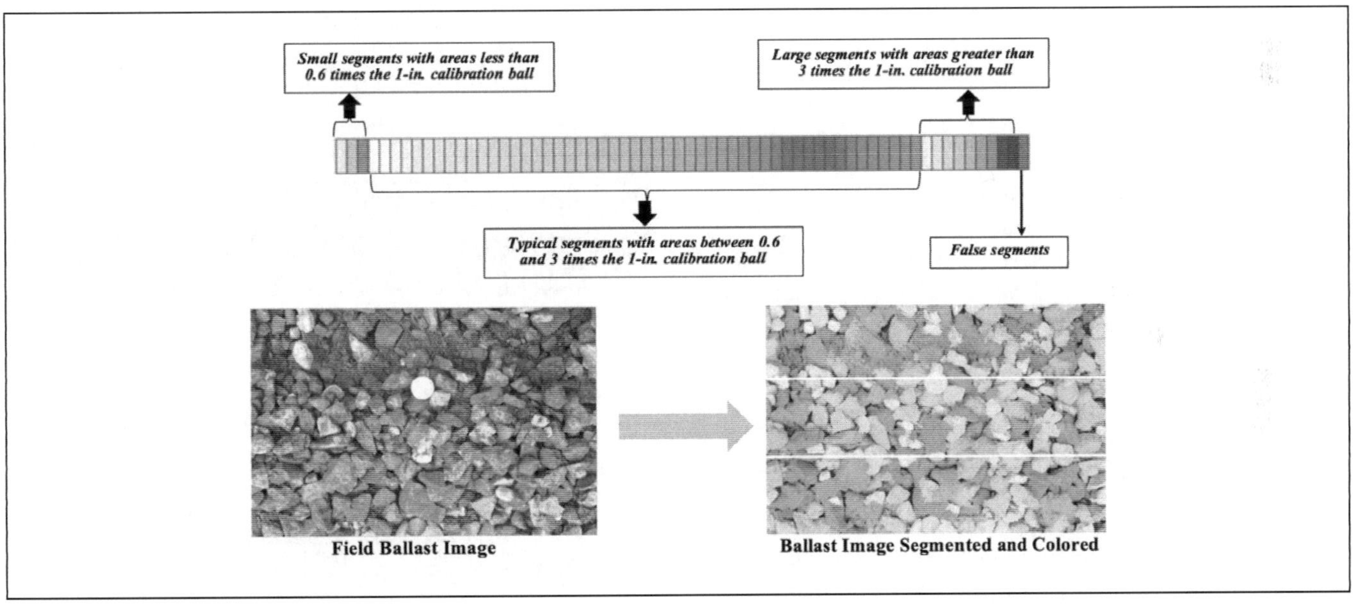

Figure 3. Segment size categorized and colored using a color-coded key.

More details about the image processing techniques used for ballast segmentation can be found elsewhere (*22*).

Subsequently, Percent Degraded Segments (PDS) is calculated using an area-based approach by categorizing all true segments in the image. With reference to the area of the 1-in. (25.4 mm) calibration ball, the segments are categorized and colored into three size bands including typical, small, and large. Figure 3 shows an example of an image after segmentation and coloring using a color-coded key. The "typical" category represents average size ballast particles, the "small" category represents severely degraded particles, and the "large" represents oversized areas with particles too fine to be identified individually, such as fine-grained soils. The two latter categories are therefore labeled as Degraded Segments, with the segment-size thresholds set as 60% and 300% respectively.

The equation for calculating PDS is

$$\text{Define}: J = \left\{ 1 \leq i \leq n | T_s < \frac{S_i}{b} < T_l \right\}$$
$$\text{PDS}(\%) = 1 - \frac{\sum_{j \in J} S_j}{A} \times 100 \tag{3}$$

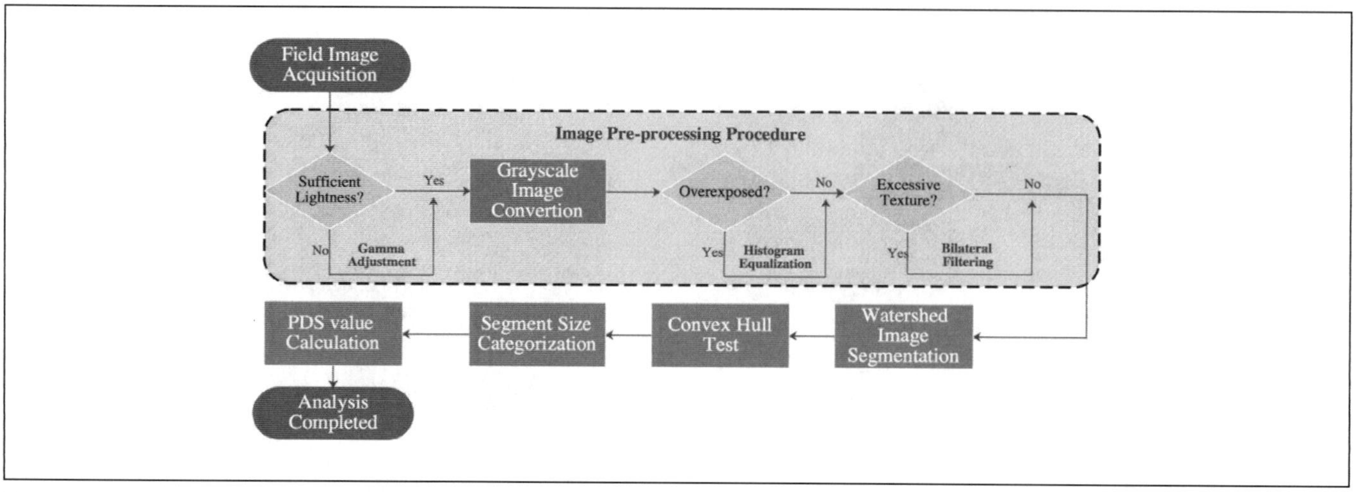

Figure 4. Flowchart of the imaging-based degradation analysis including preprocessing procedure.

where

$T_s = 60\%, T_l = 300\%$ represents the chosen segment-size threshold for small and large particles respectively,

$\{S_i | 1 \leq i \leq n\}$ denotes the set of all segments obtained,

$\{S_j | j \in J\}$ denotes the designated segments that are within the small and large segment-size thresholds,

b is the area of the 1-in. (25.4 mm) diameter calibration ball, and

A is the total area of the image.

The flowchart of the entire imaging-based ballast degradation analysis, including several preprocessing image enhancement techniques such as gamma adjustment, histogram equalization, and bilateral filtering (*23*) is shown in Figure 4.

Description of University of Illinois Constant Head Aggregate Permeameter (UI-CHAP)

The UI-CHAP as a constant head permeameter has been designed for testing large aggregates such as railway ballast in an effort to overcome the specimen size challenges of small constant head permeameters currently available. For the UI-CHAP equipment, an actual photo, a labeled flow diagram, and a diagram with dimensions are all provided in Figure 5. Note that the permeability of the ballast samples reported in this paper is referred to as "the horizontal permeability" which is caused by the specimen flow arrangement of the UI-CHAP (see Figure 5, *b* and *c*).

The flow of water through the UI-CHAP begins in the upstream tower and flows downward until it turns 90° and proceeds horizontally through the ballast sample box. The piezometers located on either side of the ballast sample box are used to measure the total head that exists at the entrance and exit of the ballast sample to the

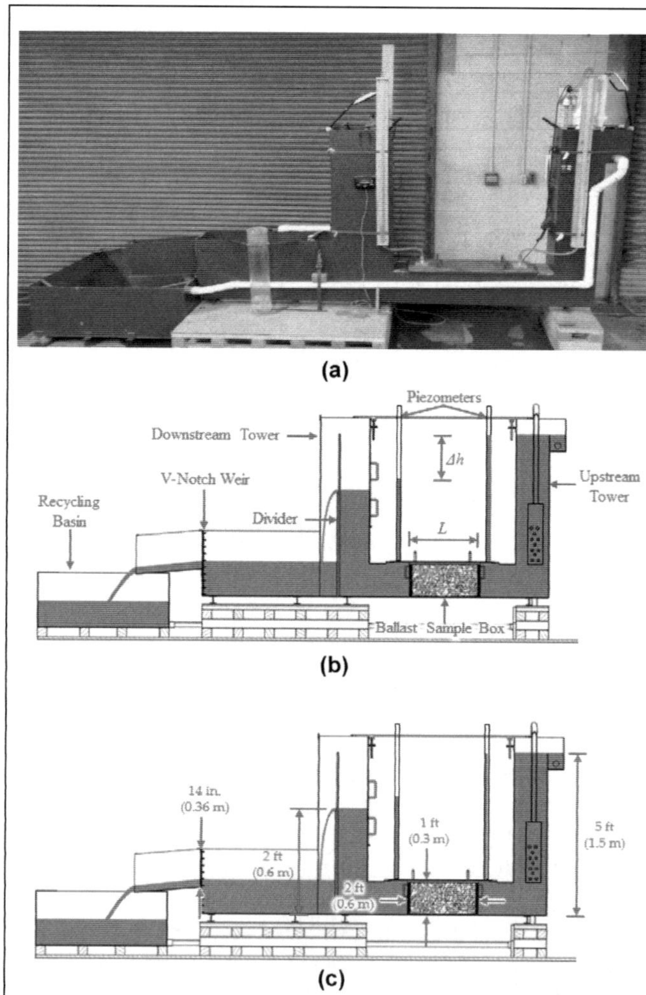

Figure 5. (*a*) Photo, (*b*) labeled diagram, and (*c*) dimensions of University of Illinois Constant Head Aggregate Permeameter (UI-CHAP).

nearest millimeter. The UI-CHAP consists of two systems: the right-hand system (from the upstream tower to the downstream tower divider) measures the total head difference, and the left-hand system (from the downstream tower divider to the recycling basin) is used to measure the constant flow rate of water through the ballast sample. More details about UI-CHAP can also be found elsewhere (*15*, *24*).

Each sample tested is prepared by sieving and mixing the aggregates to achieve desired gradations. Then, the sample is compacted in the ballast sample box in several lifts using a vibratory roller. The sample box is 30.5 cm × 30.5 cm × 61 cm (1 ft. × 1 ft × 2 ft) and is placed within a sealed channel after the compaction (see Figure 5, *b* and *c*). The sample box contains a metal mesh and geosynthetic filter to prevent fine material loss on both ends, and the channel sealing consists of plumber putty.

Materials and Research Methodology

The database used in the study included a total of 25 ballast samples and their associated field images captured at different degradation levels. The information from samples is compiled and reordered here with the original image ID designations, as shown in the first two columns in Table 1. Fourteen ballast samples were collected from vertical trench cross sections at the Transportation Technology Center, Inc. (TTCI) Facility for Accelerated Service Testing (FAST) located in Pueblo, Colorado, designated as Source 1; the other 11 samples were collected from horizontal shoulder cut sections at a revenue service line of BNSF Railway located near Kansas City, Missouri, designated as Source 2.

The full gradation curves of these 25 ballast samples are presented in Figure 6*a*, in which they are sorted from the lowest to the highest FI values. In the UI-CHAP permeability tests, a total of 13 ballast samples and one sand sample were prepared to match desired gradations as given in Figure 6*b*; for gradations at FI = 1.6, 3.1, 7.4, 14.0, and 39.0 two identical samples were prepared to verify the repeatability of the test results. Note that the material used during the sample preparation for two of the gradations (FI = 2.0 and FI = 10.3) was from Source 2, and the rest were from Source 1. Samples at FI = 1.6, 3.1, 7.4, 14.0, and 39.0 had been previously tested by Schmidt et al. (*15*), who had also conducted replicate tests on these materials to investigate the repeatability of permeability measurements.

According to Table 1 and Figure 6*a*, FI values for all 25 samples vary from 4.75 to 46.04, which covers a wide range of ballast gradations from clean to heavily degraded conditions. Accordingly, the computed PDS values range from 34.87% to 87.09%. The imaging-based degradation analysis described in the previous section was applied to all the 25 field images of the ballast cross sections. For each image designation listed in Table 1, the output PDS value from Equation 3 was compared with the associated Selig's FI value obtained through sieve analysis and Equation 2 for the corresponding sample. The PDS values and FI values were reported to an accuracy of two decimal points. FI values were plotted against PDS results and statistical regression analysis was conducted between these two parameters.

Table 1. Results of FI and PDS in Imaging-Based Ballast Degradation Analysis

Ballast sample name (-)	Image ID designation (-)	FI (-)	PDS (%)
Source1-Sample1	1-1050-C1	11.34	50.94
Source1-Sample2	1-1050-C2	10.6	50.14
Source1-Sample3	2-1308-C1	16.9	57.82
Source1-Sample4	2-1308-C2	14.19	53.82
Source1-Sample5	3-1354-C1	22.81	61.8
Source1-Sample6	3-1354-C1-Wall2	17.78	59.01
Source1-Sample7	3-1354-I1	20.11	61.28
Source1-Sample8	3-1354-C2-Wall2	22.98	70.04
Source1-Sample9	3-1396-C1	20.79	55.25
Source1-Sample10	3-1396-C2	20.07	62.53
Source1-Sample11	4-1460-C1	17.25	58.05
Source1-Sample12	4-1460-C2	13.71	52.98
Source1-Sample13	5-1557-C1	16.26	61.21
Source1-Sample14	5-1557-C2	8.87	50.31
Source2-Sample1	7-8-9	4.79	49.8
Source2-Sample2	20-21-22	46.04	81.19
Source2-Sample3	34-35-36	26.81	68.39
Source2-Sample4	38-39-40	28.92	68.39
Source2-Sample5	39-40-41	5.03	34.87
Source2-Sample6	50-51-52	25.46	74.62
Source2-Sample7	63-64-65	35.3	87.09
Source2-Sample8	75-76-77	4.75	35.24
Source2-Sample9	78-79-80	5.25	42.64
Source2-Sample10	81-82-83	33.32	74.11
Source2-Sample11	87-88-89	13.94	50.68

Note: - indicates parameter is unitless.

The results of imaging-based ballast degradation analysis are shown in Table 1 and Figure 7. Figure 7 shows the results of statistical analysis and the regression FI-PDS model achieved with a coefficient of determination R^2 equal to 0.86, which indicates a strong linkage between FI and PDS values and can be used as an efficient tool for imaging-based assessment of ballast degradation in the field.

The 13 laboratory ballast samples at different degradation levels and one sand sample were tested in the laboratory to determine their constant head permeability properties using the UI-CHAP. For each test, the discharge velocity was computed and plotted against hydraulic gradient. As shown in the first column in Table 2, no filter was used to test clean ballast samples

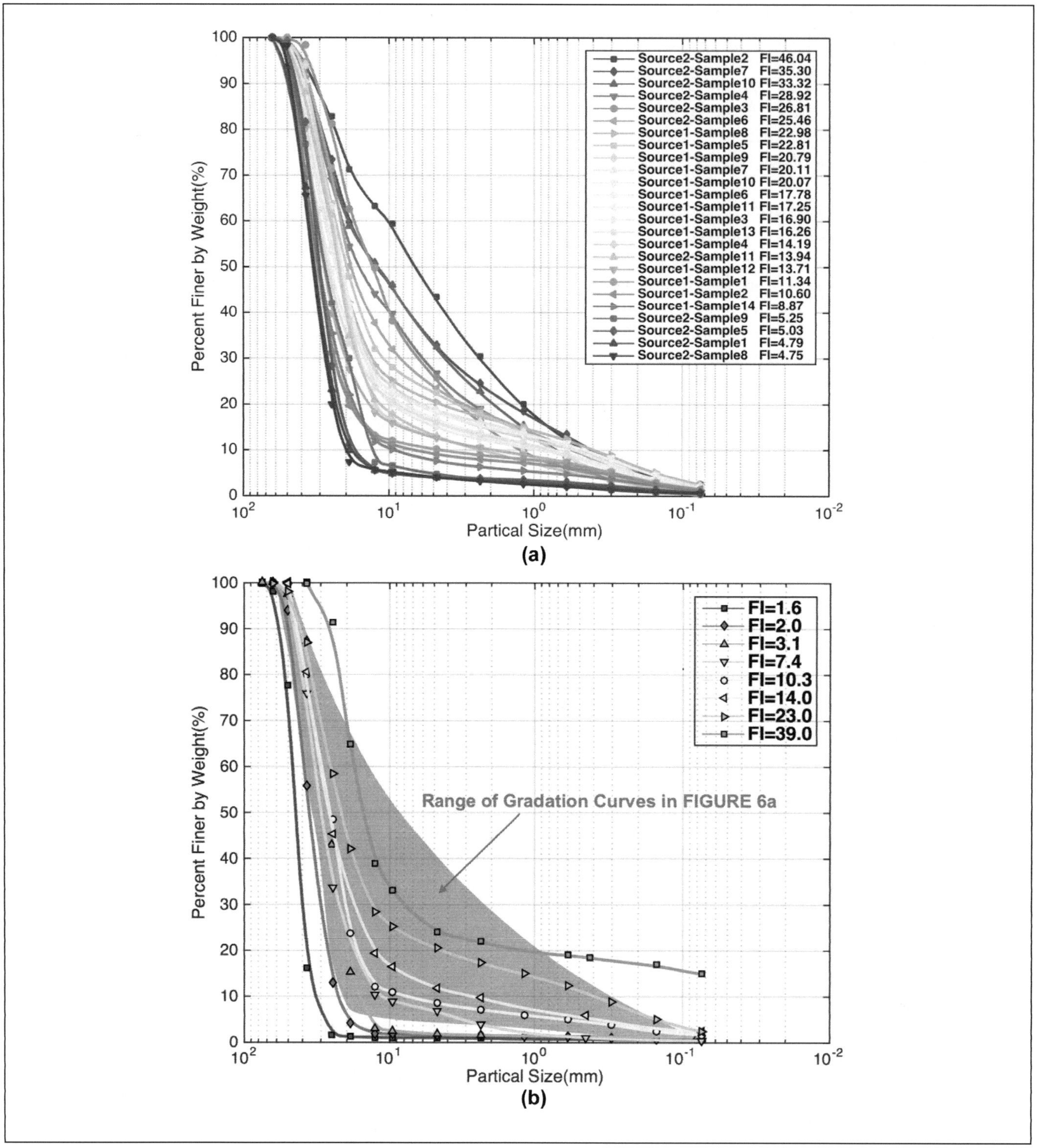

Figure 6. (*a*) Gradation curves for all samples in ballast degradation analyses and (*b*) Gradation curves for ballast samples tested in the permeameter test.

Table 2. Results of UI-CHAP Constant Head Permeameter Tests

Samples ID (-)	Backcalculated PDS* (%)	Eq. (1): $v = C \cdot i^n$ Coefficient *C*, PI (-)	Variation of tests (%)	Eq. (1): $v = C \cdot i^n$ Exponent *n* (-)	Coefficient of determination R^2 (-)
FI = 1.6-1, no filter	36.20	15.86	5.5	0.51	0.999
FI = 1.6-2, no filter	36.20	16.74		0.54	0.999
FI = 2.0	36.74	12.20	na	0.50	0.999
FI = 3.1-1, no filter	38.21	13.13	0.3	0.55	0.999
FI = 3.1-2, no filter	38.21	13.17		0.54	0.999
FI = 7.4-1	43.97	1.20	28.3	0.52	0.999
FI = 7.4-2	43.97	0.86		0.66	0.998
FI = 10.3	47.46	0.24	na	0.72	0.982
FI = 14.0-1	52.82	0.28	10.7	0.52	0.998
FI = 14.0-2	52.82	0.31		0.59	0.999
FI = 23.0	64.88	0.21	na	0.68	0.996
FI = 39.0-1	86.32	0.18	50.0	0.84	0.985
FI = 39.0-2	86.32	0.09		0.54	0.979
Sand	100.00	0.04	na	1.00	0.983

Notes: * Refer to Figure 7; Eq. = Equation; - indicates parameter is unitless; na = not applicable.

at FI = 1.6 and 3.0, whereas all of the other samples were tested with a geotextile fabric.

Based on the previous findings of Wilkins (*14*) and Schmidt et al. (*15*), the velocity–gradient relationship is expected to be nonlinear and a constant hydraulic conductivity value does not exist. Therefore, it was decided that the empirical constant in the power-curve model shown in Equation 1, *C*, should be used as a critical parameter that captures different magnitudes of ballast permeability. This constant can be obtained by fitting the nonlinear power curve (i.e., flow trend line). Note that parameter *C* is defined as the Permeability Index (PI) in this paper.

Because of the indirect relationship between permeameter test results and imaging-based PDS, the FI values of the permeameter samples were first backcalculated to PDS values based on the positive correlation between FI and PDS. The PI was then plotted against the backcalculated PDS results to investigate the ballast permeability change at different degradation levels.

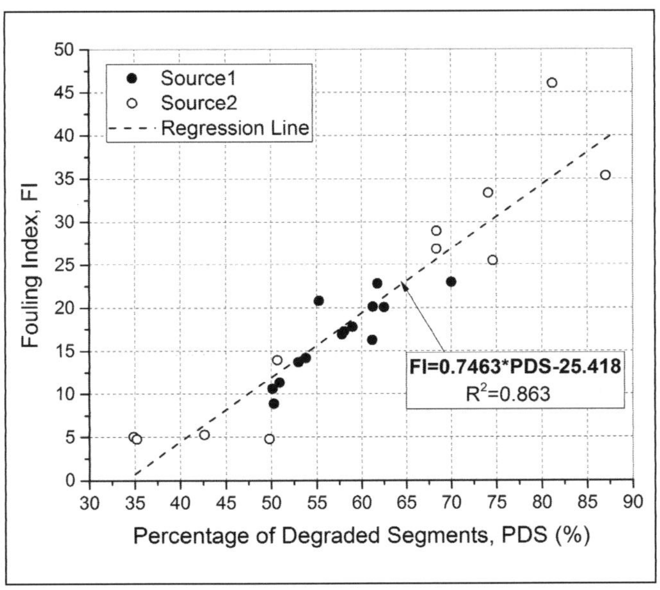

Figure 7. FI versus PDS for all samples.

Results and Discussion

The results of ballast permeability tests are illustrated in Figure 8. Furthermore, the PI and the exponent *n* obtained by fitting the nonlinear flow trend line in the form of Equation 1 are tabulated in Table 2. The tendency of significant permeability decrease at moderately degraded conditions is observed. Lastly, based on the validated correlation between FI-PDS values and the permeability test results, a bilinear relationship between PI and PDS values is proposed, as illustrated in Figure 9.

Figure 8 shows the nonlinear velocity gradient relationship observed from the 13 ballast samples tested.

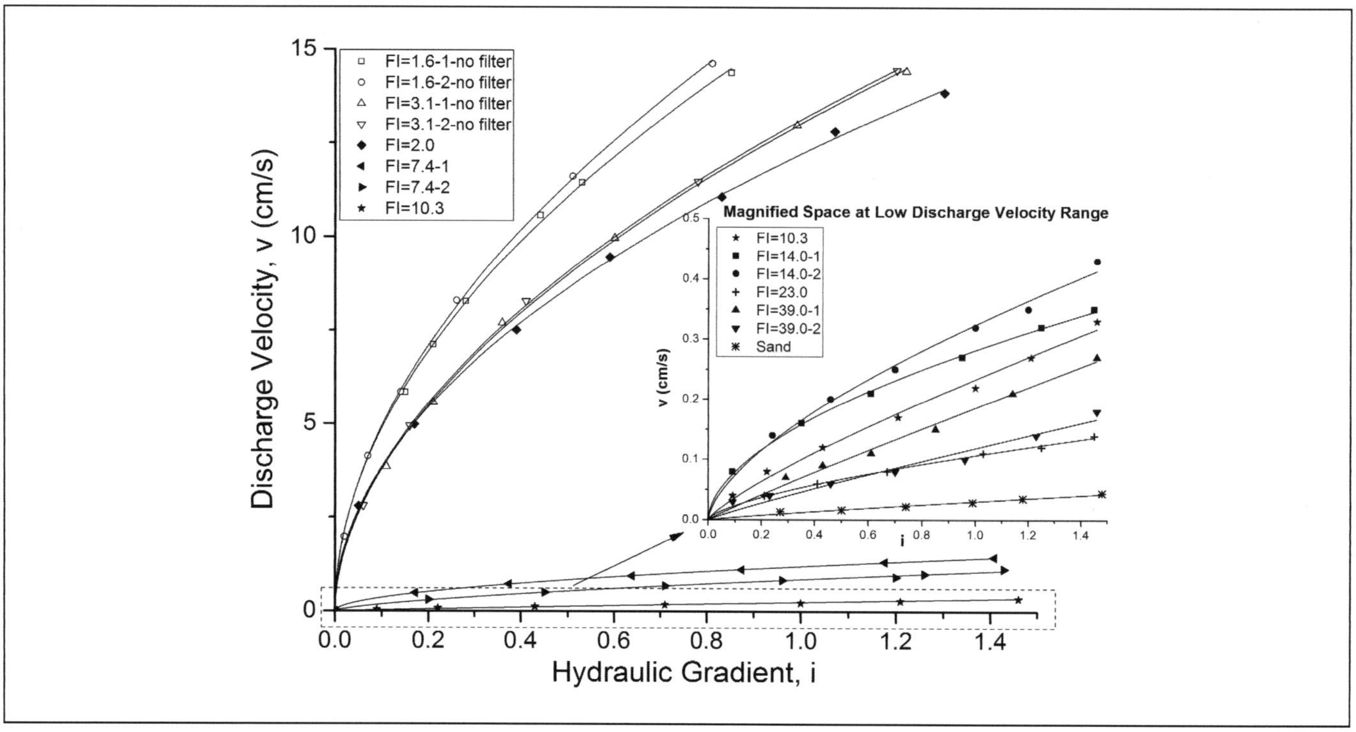

Figure 8. Discharge velocity versus hydraulic gradient for samples tested in the permeameter.

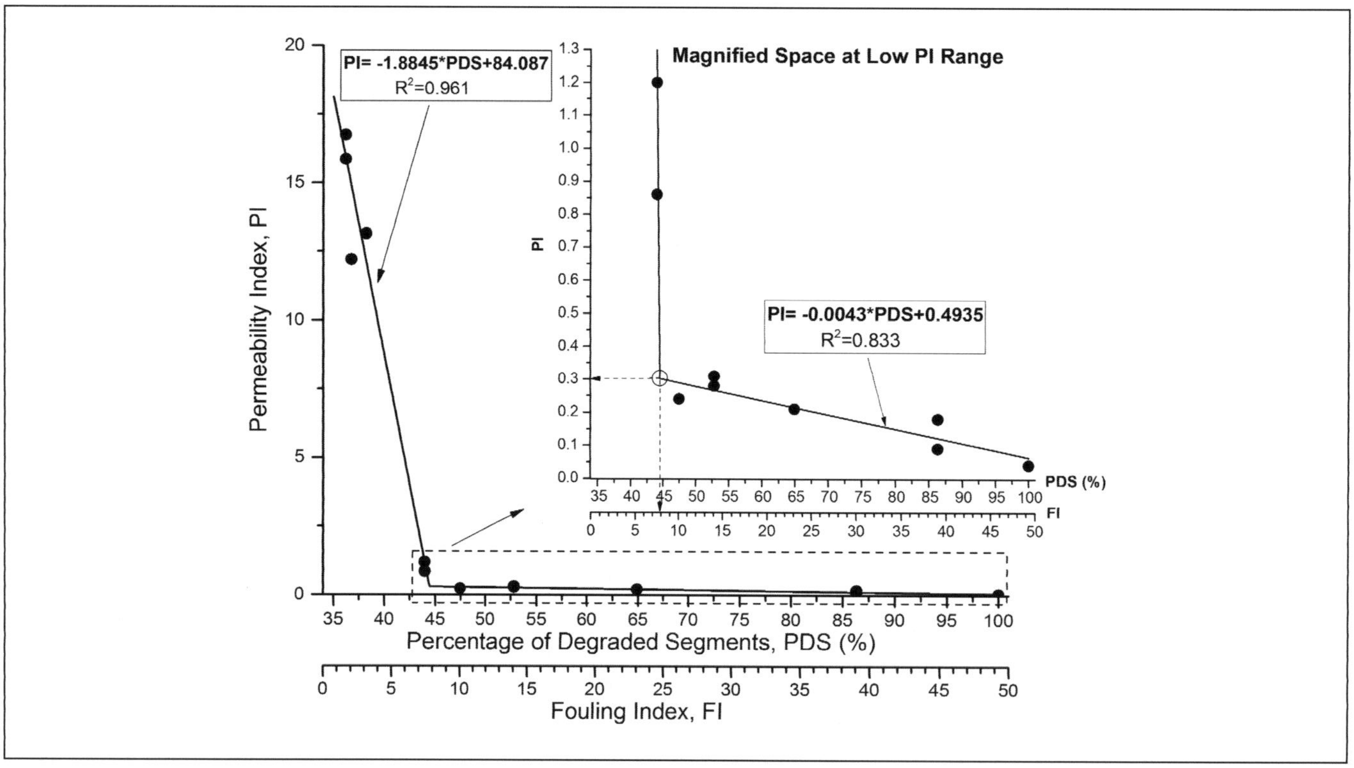

Figure 9. PI versus PDS and FI plot for samples in permeameter test.

The flow trend lines show that the permeability decreases and approaches the line of sand as the ballast samples become more degraded, and the shapes of the curves also become similar to the linear trend observed for sand, as was previously discussed by Schmidt et al. (*15*). The rapid decrease in the PI (see Table 2) indicates that the greatest loss in permeability occurs during the degradation process from clean to moderately degraded condition. The PI values at clean condition are more than tenfold that at moderately or heavily degraded conditions. This could be explained by the internal structure of the compacted ballast sample, which is a stable skeleton of large particles with voids partially filled with fine particles. As the level of degradation increases from clean to moderately degraded, the fine particles start to fill up the voids, thus forming less permeable paths that restrict the flow and decrease the permeability significantly. Furthermore, the saturated flow could mobilize the degraded fine material to yield variable permeability by reshaping and rearranging those paths. The rearrangement of fine material could thereby introduce more randomness in the pattern and distribution of flow paths. Therefore, when higher contents of fine material are present, the measured PI values of tests replicated show more variability. As shown in the fourth column of Table 2, the PI values of original test and replicate test differ within 6% for both FI = 1.6 and FI = 3.0 samples, but at FI = 7.4, 14.0 and 39.0 they differ to the extent of 28.3%, 10.7% and 50.0%, respectively.

Figure 8 also illustrates the effect of geosynthetic filter restriction on the flow characteristics. The tests with and without filter restriction could represent different field drainage conditions encountered in practice. Those tests conducted without filter (e.g., FI = 1.6 and 3.0) simulate clean or newly placed ballast resting on the top, shoulders, or slopes of the trackbed. The fine materials of ballast in these locations are not subjected to strong movement restriction in flow path, such that any degraded material could be washed out of the ballast as the degradation progresses. By contrast, the remaining samples with geosynthetic filter (e.g., FI = 2.0 and FI = 7.4 to FI = 39.0) represent ballast located within the trackbed where the degraded material is closely retained by the subballast, subgrade soil, and surrounding fine particles. Additionally, the result of the FI = 2.0 sample shows that the filter restriction decreases the permeability slightly; nevertheless, the PI value corresponding to FI = 2.0 (clean sample, with filter) still has the same order of magnitude as FI = 1.6 and 3.0 (clean sample, without filter). This indicates that for clean ballast, different field conditions described above could slightly influence its permeability.

Figure 9 proposes a bilinear relationship between the newly defined ballast permeability indicator PI in this study and the imaging-based ballast degradation indicator PDS. As shown in Figure 6*b*, the gradations of ballast samples tested in UI-CHAP are mostly in the range—and follow the general trend—of the gradations collected in MV-BIS. Therefore, the PDS values of these permeameter samples can be backcalculated from their corresponding FI values based on the validated FI-PDS model shown in Figure 7. Then, the PI values of those samples are plotted against the corresponding PDS and FI values, as shown in Figure 9. According to the distinct order of magnitude of the PI values, the data points are separated into two groups (i.e., Group 1: FI = 1.6–7.4, and Group 2: FI = 10.3–39.0 and sand), and linear regression analysis is conducted for both groups. As expected, the first group of clean ballast yields a high coefficient of determination $R^2 = 0.96$; the second group, with more randomness in permeability, has a coefficient of determination $R^2 = 0.83$. By connecting the two regression lines, a critical point at which PI = 0.3 and PDS = 45% (FI = 8) can be found for the bilinear PI-PDS model. An intuitive interpretation of the PI value at this critical point is that the ballast has a hydraulic conductivity of 0.3 cm/s at a hydraulic gradient equal to 1.0. This critical point indicates that the ballast permeability decreases significantly from clean condition to a moderately degraded condition at a PDS value of 45% or FI value of 8, then generally maintains its low values with a slow rate of decrease. In other words, railway engineering problems caused by insufficient ballast drainage could potentially be detected in an early stage of degradation when the FI value exceeds this threshold, and the threshold in terms of PDS value indicates a more efficient imaging-based tool for such evaluation process in field. Note that this finding is only valid in the range of the limited number of samples tested and analyzed in this study. Therefore, this bilinear relationship should not be extrapolated unless a significant number of tests are conducted on a comprehensive sets of ballast samples.

Conclusions and Recommendations

The following conclusions can be drawn related to the railway ballast degradation and permeability correlation established from the imaging-based degradation analysis data and the constant head permeability test results conducted within the scope of this study:

- A comprehensive review of the literature showed that determining the in-situ drainage properties of railway ballast in a quantitative manner has not been well studied. No previous study was found which relates field images of ballast with their corresponding permeability behavior.
- An imaging-based index called PDS was validated and successfully implemented in this study to

distinguish different levels of ballast degradation from field captured cross section images. The image processing algorithm showed a strong correlation with commonly used Selig's FI values obtained from laboratory sieve analysis results.

- Constant head permeability tests conducted on ballast samples at different degradation levels verified that ballast has a nonlinear power-curve trend of flow characteristics for the discharge velocity related to hydraulic gradient. The tests performed with and without filter could therefore represent the field drainage conditions of (*a*) used ballast within the trackbed where restricted flow occurs when the fine materials are structurally retained, and (*b*) clean or newly placed ballast resting on locations where free surface flow occurs when the fine materials can be washed out.
- A new parameter called Permeability Index (PI) was introduced to represent ballast drainage capacity. A bilinear relationship observed between PI and imaging-based PDS and laboratory FI results successfully captured the decrease in permeability as ballast degradation increases.

The findings of this study show promising future potential of the developed image processing technique for replacing subjective visual inspection methods or time-consuming ballast sampling processes. Accordingly, this innovative technology is envisioned for full development to provide automated means of track condition monitoring and installation on portable platforms (e.g., ballast shoulder cleaners) to obtain degradation levels periodically along the entire track network. As for field ballast permeability assessment, railway engineers will have a better tool at their discretion to decide the process of ballast cleaning. The breakpoint of the bilinear model in Figure 9 could be used to envision a two-stage ballast cleaning procedure considering its permeability: a preliminary in-situ cleaning effort (e.g., flushing, air-blowing, vibratory settling, etc.) to remove degraded materials before the ballast permeability decreases significantly, and a final thorough cleaning process (shoulder cleaning, undercutting and renewal) when the ballast reaches overall severely degraded condition. The two-stage ballast cleaning procedure could potentially maintain ballast drainage capacity, extend the life cycle of in-service railway ballast, and reduce the long-term cost of ballast maintenance. Finally, the proposed methods in this research are envisioned to provide ballast service life prediction as well as technology implementation in the management of ballast maintenance.

Note that practical drainage behavior of railway ballast, however, is more complicated because of the dynamic nature of train loadings on ballasted track. The transient hydraulic responses of ballast (e.g., fine material pumping, excess pore water pressure, etc.) under dynamic cyclic loading have not been properly investigated yet. Therefore, further laboratory and in-situ tests as well as theoretical analysis need to be performed on this subject.

Acknowledgments

The Association of American Railroads (AAR), TRB Innovations Deserving Exploratory Analysis (IDEA) Rail Safety program funded by Federal Railroad Administration, BNSF Railway, and Union Pacific Railroad have provided financial support for the development of a machine vision–based ballast inspection system and the UI-CHAP permeameter equipment described in this paper. The authors greatly appreciate the contributions and technical support received from Rail Transportation and Engineering Center (RailTEC) and Illinois Center for Transportation (ICT) at the University of Illinois in Urbana-Champaign. Moreover, the authors would like to recognize the help and support that was provided by Dennis Morgart, William (Zach) Dombrow, and Mike Wnek from BNSF Railway, and David Davis, Mike McHenry, Dingqing Li, and Colin Basye from TTCI.

Author Contributions

The authors confirm contribution to the paper as follows. Study conception and design: Erol Tutumluer, Haohang Huang, Maziar Moaveni, Scott Schmidt, John M. Hart; data collection: Scott Schmidt, Haohang Huang, Maziar Moaveni, John M. Hart; analysis and interpretation of results: Haohang Huang, Scott Schmidt; draft manuscript preparation: Haohang Huang, Maziar Moaveni. All authors reviewed the results and approved the final version of the manuscript.

References

1. Selig, E. T., and J. M. Waters. *Track Geotechnology and Substructure Management*. Thomas Telford Publications, London, 1994.
2. Chiang, C. C. *Effects of Water and Fines on Ballast Performance in Box Tests*. Master of Science Degree Project Report No. AAR89-366P. Department of Civil Engineering, University of Massachusetts, Amherst, 1989.
3. Han, X., and E. T. Selig. Effects of Fouling on Ballast Settlement. *International Heavy Haul Railway Conference*, 1997, pp. 257–268.
4. Ebrahimi, A., J. M. Tinjum, and T. B. Edil. Protocol for Testing Fouled Railway Ballast in Large-Scale Cyclic Triaxial Equipment. *Geotechnical Testing Journal*, Vol. 35, No. 5, 2012, pp. 796–804. https://doi.org/10.1520/GTJ103846.
5. Darcy, H. *The Public Fountains of the City of Dijon. Appendix—Note D. Henry Darcy and His Law*. https://bae.okstate.edu/faculty-sites/Darcy/index.htm. Accessed July 7, 2016.
6. Parsons, B. *Hydraulic Conductivity of Railroad Ballast and Track Substructure Drainage*. Master's thesis. Department of Civil Engineering, University of Massachusetts, Amherst, 1990.

7. Tennakoon, N., B. Indraratna, C. Rujikiatkamjorn, S. Nimbalkar, and T. Neville. The Role of Ballast Fouling Characteristics on the Drainage Capacity of Rail Substructure. *Geotechnical Testing Journal*, Vol. 35, No. 4, 2012, pp. 1–12. https://doi.org/10.1520/GTJ104107.
8. Rahman, A. J. *Permeability, Resistivity, and Strength of Fouled Railroad Ballast*. Master's thesis. Civil, Environmental, and Architectural Engineering Department, University of Kansas, 2013.
9. Danquah, W. O., G. S. Ghataora, and M. N. Burrow. The Effect of Ballast Fouling on the Hydraulic Conductivity of the Rail Track Substructure. Proceedings of XV Danube-European Conference on Geotechnical Engineering (DECGE), Austrian Society for Soil Mechanics and Geotechnical Engineering, Vienna, Austria, 2014, pp.997–1002.
10. Paiva, C., M. Ferreira, and A. Ferreira. Ballast Drainage in Brazilian Railway Infrastructures. *Construction and Building Materials*, Vol. 92, 2015, pp. 58–63. https://doi.org/10.1016/j.conbuildmat.2014.06.006.
11. Su, Z., H. Huang, and G. Jing. Experimental Analysis Permeability Characteristics of Fouling Railway Ballast. *Metallurgical and Mining Industry*, Vol. 7, No. 9, 2015, pp. 992–997.
12. Cole, B. *Analysis of Railroad Track Substructure Drainage*. PhD dissertation. Department of Civil Engineering, University of Massachusetts, Amherst, 1989.
13. Taylor, D. W. *Fundamentals of Soil Mechanics*. John Wiley & Sons, New York, 1948, pp. 97–123.
14. Wilkins, J. K. Flow of Water Through Rock Fill and Its Application to the Design of Dams, *New Zealand Engineering*, 1955, pp. 382–387.
15. Schmidt, S., S. Shah, M. Moaveni, B. J. Landry, E. Tutumluer, C. Basye, and D. Li. Railway Ballast Permeability and Cleaning Considerations. *Transportation Research Record: Journal of the Transportation Research Board*, 2017. 2607: 24–32.
16. Al-Qadi, I. L., W. Xie, R. Roberts, and Z. Leng. Data Analysis Techniques for GPR Used for Assessing Railroad Ballast in High Radio-Frequency Environment. *Journal of Transportation Engineering*, Vol. 136, No. 4, 2010, pp. 392–399. https://doi.org/10.1061/(ASCE)TE.1943-5436.0000088.
17. Zarembski, A. M., G. T. Grissom, and T. L. Euston. Use of Ballast Inspection Technology for the Prioritization, Planning and Management of Ballast Delivery and Placement. Proceedings of the Annual Conference, American Railway Engineering and Maintenance-of-way Association, Indianapolis, IN, 2013, pp. 1062–1097.
18. Anbazhagan, P., B. Indraratna, C. Rujikiatkamjorn, and L. Su. Using a Seismic Survey to Measure the Shear Modulus of Clean and Fouled Ballast. *Geomechanics and Geoengineering*, Vol. 5, No. 2, 2010, pp. 117–126.
19. Moaveni, M., S. Wang, J. Hart, E. Tutumluer, and N. Ahuja. Evaluation of Aggregate Size and Shape by Means of Segmentation Techniques and Aggregate Image Processing Algorithms. *Transportation Research Record: Journal of the Transportation Research Board*, 2013. 2335: 50–59.
20. Moaveni, M., Y. Qian, I. I. Qamhia, E. Tutumluer, C. Basye, and D. Li. Morphological Characterization of Railroad Ballast Degradation Trends in the Field and Laboratory. *Transportation Research Record: Journal of the Transportation Research Board*, 2016. 2545: 89–99.
21. Vincent, L., and P. Soille. Watersheds in Digital Spaces: An Efficient Algorithm Based on Immersion Simulations. *IEEE Transactions on Pattern Analysis & Machine Intelligence*, Vol. 13, No. 6, 1991, pp. 583–598.
22. Moaveni, M., E. Tutumluer, S. Wang, J. M. Hart, N. Ahuja, and M. McHenry. Evaluating In-Service Railroad Ballast Condition Using Machine Vision Based Inspection Systems. Presented at 96th Annual Meeting of the Transportation Research Board, Washington, D.C., 2017.
23. Gonzalez, R. C., and R. E. Woods. *Digital Image Processing*. Prentice Hall, N.J., 2002.
24. Schmidt, S. *Railway Ballast Permeability*. Master's thesis. Department of Civil and Environmental Engineering, University of Illinois at Urbana-Champaign, 2017.

The Standing Committee on Railway Maintenance (AR060) peer-reviewed this paper (18-04780).

Article

TRR
JOURNAL OF THE TRANSPORTATION RESEARCH BOARD

Transportation Research Record
2018, Vol. 2672(10) 74–84

DOI: 10.1177/0361198118793501
journals.sagepub.com/home/trr
SAGE

Support Condition and Traffic Loading Patterns Influencing Laboratory Determination of Under Ballast Mat Bedding Modulus and Insertion Loss

Arthur de O. Lima[1], Marcus S. Dersch[1], Erol Tutumluer[1], J. Riley Edwards[1], and Yu Qian[2]

Abstract

In recent years, noise and vibration concerns have grown as environmental regulations and requirements impose greater responsibilities on infrastructure owners. Under ballast mats, rubber elastomers inserted below the ballast or concrete slab, have been widely deployed and studied in Europe, but the amount of research to date in North America is limited. Current testing practices for obtaining component level properties of under ballast mats are based on European practices and loading environments. Moreover, these procedures use experimental setups, which are often not representative of field loading conditions. With this in mind, the research presented in this paper investigates static bedding modulus properties of three under ballast mats by varying support and loading conditions to simulate revenue-service field scenarios involving both ballasted and concrete slab track. Performance prediction indicators such as insertion loss (related to vibration reduction) were also evaluated using prediction models that required the use of the experimentally obtained bedding modulus results as inputs. Results showed a difference of up to 33% in bedding modulus results among the support conditions tested. Additionally, the insertion loss was changed by up to 1.8 dB. Traffic pattern simulations also demonstrated a sharp rate of stiffening due to static preload conditioning as well as a gradual rate of asymptotic stiffening with accumulated loading cycles. This finding further identifies the need to quantify a revenue service "working range" stiffness for the component.

Environmental requirements related to noise and vibration disturbances near new and existing rail lines, especially in populated areas, have become consistently stricter (*1*). This has driven the industry to seek alternatives to mitigate such disturbances. Under ballast mats (UBMs), also referred to as ballast mats, are one of multiple methods (i.e., floating slabs, under tie pads, mass spring systems, etc.) used to reduce the propagation of vibrations in the track structure (*2*).

Under ballast mats are pads made from an elastic material (recycled tire rubber, Ethylene Propylene Diene Monomer (EPDM) rubber, Polyurethane foam, etc.) and installed below the ballast layer of a ballasted track structure or under the concrete slab in a slab track design. They have been shown to be most effective in mitigating frequencies between 30 and 200 Hz (*3*, *4*). This is the frequency range that causes the most human discomfort (*5*). Furthermore, frequencies above this threshold attenuate quickly into the adjacent ground and are not generally considered to be problematic.

Bedding modulus is a well-established property used to characterize UBMs. It is defined as the amount of force required to displace a unit area sample by a unit deflection, and is calculated as the tangent, or secant modulus, at a specific stress value in the stress/displacement curve.

Under ballast mats are typically designed and manufactured to achieve a specific insertion loss – ratio of signal levels (i.e., vibration amplitudes) before and after the installation of a filter (i.e., UBM) in units of decibels – depending on the specified operating environment (e.g., freight, passenger, open track, slab track, etc.). This

[1]Rail Transportation and Engineering Center (RailTEC), Department of Civil and Environmental Engineering, University of Illinois at Urbana-Champaign, Champaign, IL
[2]Department of Civil and Environmental Engineering, The University of South Carolina, Columbia, SC

Corresponding Author:
Address correspondence to Arthur de O. Lima: aolima@illinois.edu

performance parameter is determined from prediction models relying on inputs from the characteristics of track structure, loading environment, and materials (*6–8*). Bedding modulus is one such parameter and has great importance in predicting performance levels. Thus, a proper understanding and use of this input property is essential for an accurate interpretation of revenue-service track performance.

European countries and rail agencies have used and/or studied UBMs for many decades for both passenger and freight services (*7, 9–13*). Meanwhile, in North America, Class I railroads have primarily deployed UBMs on ballasted bridge decks (concrete or steel) and tunnels with limited research being conducted to date. While the uses for UBMs relating to reduction of noise and vibration are known, applications in freight railroads are mostly limited to the improvement of track transition performance by providing a reduction in track stiffness on the structure, thus reducing impact loading and differential settlements at the bridge abutments (*14–16*). Conversely, loading magnitudes in transit applications are less, thus research findings from European studies can be applied to understand the behavior of UBMs in North American transit applications (*17*).

To date, the German Deutsches Institut für Normung (DIN) 45673-5, titled "Mechanical vibration – Resilient elements used in railway tracks – Part 5: Laboratory test procedures for under-ballast mats" (*18*), (hereafter referred to as DIN), is the only standardized testing procedure available for the determination of UBM mechanical properties describing test procedures, setups, and loading characteristics to be employed.

Nevertheless, growing interest in North America for UBMs has established a demand for the development of uniform testing procedures to determine a bedding modulus value that can be representative of freight railroad loading environments. Currently, established laboratory testing procedures to quantify bedding modulus use two steel plates to apply loads and simple pre-load conditioning of the sample (*18*). It is hypothesized, however, that the bedding modulus value quantified in this manner may not fully represent revenue service conditions and therefore may lead to unrealistic estimations of insertion loss performance. Additionally, based on a literature review on the mechanical behavior of rubbers under load and similar findings in this research project (*19*), it is believed that loading patterns due to train traffic over a line may result in gradual changes in an UBM's bedding modulus as a result of strain-crystallization effects in the crystalline networks of the rubber (*20–22*). This necessitates studying UBM's revenue service working range.

This paper presents results from laboratory experiments performed in the Research and Innovation Laboratory (RAIL) at the University of Illinois at Urbana-Champaign (UIUC). These experiments focused on investigating the effects of varying support conditions, loading procedures, and sample conditioning on the UBM bedding modulus values as well as their resulting insertion loss. It is important to note that the values of static bedding modulus (C_{stat}) are used as a proxy to compare results within this study – using insertion loss estimations – between the different tests conducted and are not intended to be used as a real estimation of the component's performance due to its dynamic nature.

Methodology

Experimental Setup

Testing was conducted using the Pulsating Load Testing Machine (PLTM) frame at RAIL in the Harry Schnabel Jr. Geotechnical Laboratory at UIUC. The PLTM setup includes a 250-kN (55,000-lb) vertical actuator for load application. In the test setup, four potentiometers quantified loading plate vertical displacements. These potentiometers were located on each corner of the loading plate, which was attached to the actuator (Figure 1). This arrangement exceeds the recommended number of displacement gauges (three) specified by the DIN. Considerations were also made to ensure that the UBM sample was the only component to deform as the concrete block and frame were assumed to be rigid.

Experimental Test Matrix

An experimental test matrix was developed to investigate some of the questions pertaining to variability in bedding modulus as a function of support conditions. Three UBM designs with varying thicknesses and geometry were subjected to the testing procedures described below. Samples were labeled in sequential order as A, B, and C according to their maximum thicknesses, and were subjected to testing with quasi-static conditions based on loading regimes recommended by the DIN 45673-5 (*18*). Table 1 presents the general characteristics of the UBM samples tested.

Support Condition Effects. To better quantify how different support conditions may affect the values for bedding modulus, three support conditions were tested in the laboratory. A baseline value was obtained as specified within the DIN, and two other supports which better represent the UBM field service conditions were also employed. As discussed previously, this experiment was important since bedding modulus is a critical input into insertion loss prediction models, and a slight change in

Figure 1. (a) PLTM setup for bedding modulus testing of UBM samples; (b) detail of potentiometer arrangement.

Table 1. Under Ballast Mat Sample Characteristics

Label	Color	Thickness (mm [in.]) Minimum	Thickness (mm [in.]) Maximum	Sample size (mm [in.])	Construction
Type A		5 (0.197)	10 (0.394)	254 (10)	
Type B	Black	8 (0.315)	17 (0.670)	x	Profiled mat bonded to flat protective layer
Type C		7 (0.275)	25 (0.984)	254 (10)	

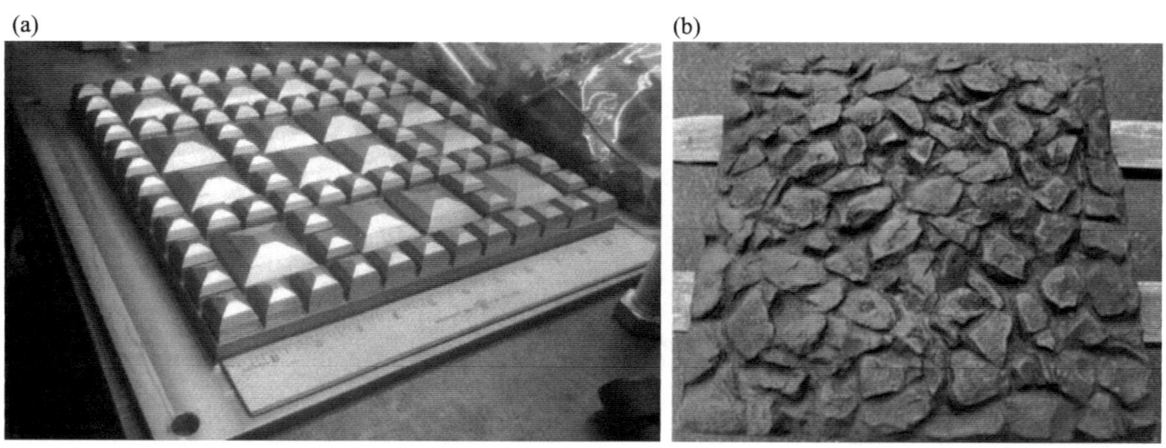

Figure 2. (*a*) UIUC manufactured GBP; (*b*) German DIN Ballast Plate (*24*).

bedding modulus values can influence the predicted performance.

In an effort to reduce the variability and increase the repeatability of laboratory testing, researchers adopted the European Geometric Ballast Plate (GBP) standardized testing apparatus (*23*) (Figure 2*a*). The GBP is designed to represent the same contact surface area as the original German DIN Ballast Plate (*24*) (Figure 2*b*), which is no longer available to researchers and testing laboratories.

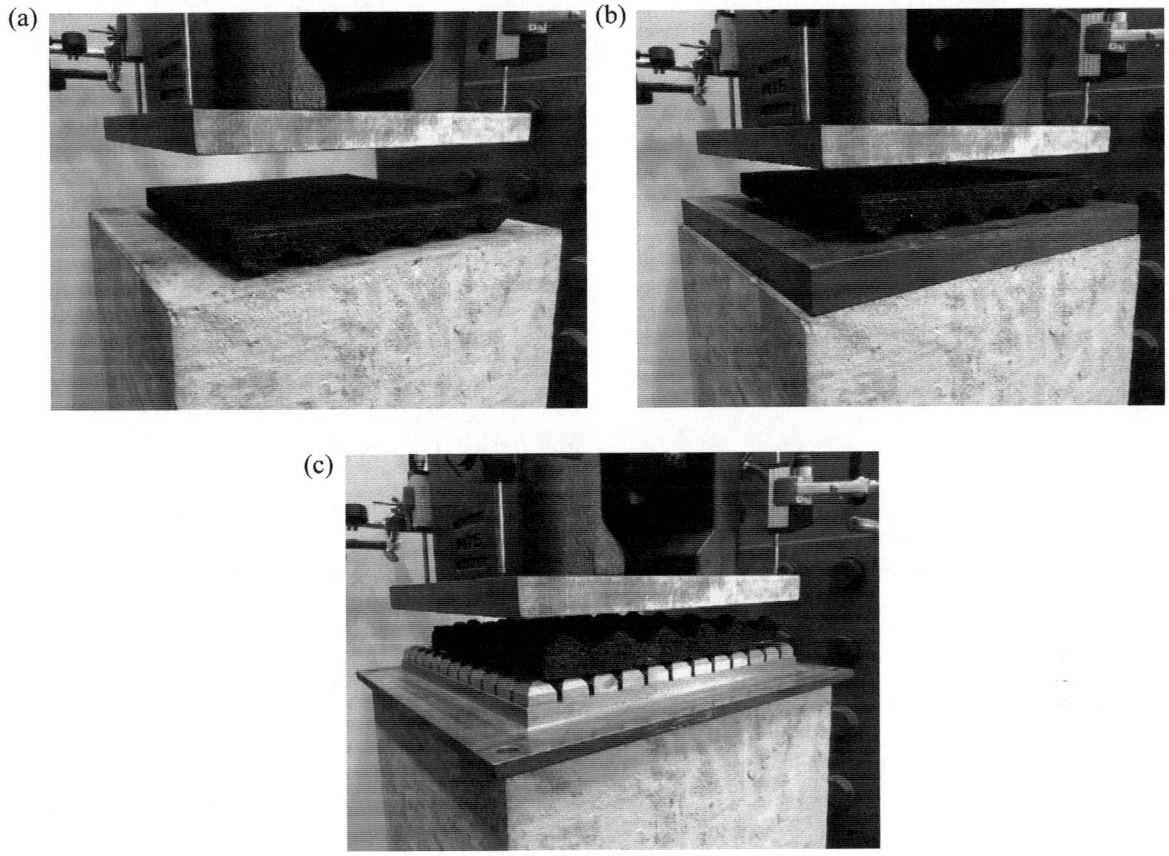

Figure 3. Support conditions used in the experiment: (*a*) concrete; (*b*) steel; and (*c*) GBP.

The GBP has a complex surface representing various ballast particles yet has symmetric geometry providing a uniform contact surface independent of specimen orientation. Nevertheless, the authors understand that results obtained using the GBP may differ from actual ballast particles; yet, the GBP provides increased repeatability by removing the intrinsic variability of the non-homogeneous nature of ballast particles.

Support conditions were selected with the objective of approximating the contact surface characteristics of field conditions to which the material would be subjected. Therefore, a 35.6 × 35.6 × 71.1 cm (14 × 14 × 28 in.) concrete block support (Figure 3*a*) was employed to represent applications on concrete bridge decks, tunnels or floating slab track. The European GBP (Figure 3*c*) (*23*) was used as a means to simulate the interface loading condition of railroad ballast; please note that the sample orientation changed to ensure the UBM was oriented properly with regards to the GBP (Figure 3*c*). In this case, a steel plate (Figure 3*b*) was adopted as the control setup of the recommended support condition in the German DIN 45673-5. For all cases, load was applied using a flat steel plate attached to the actuator head.

Traffic Pattern Effects. To quantify an UBM's stiffening due to loading, three separate test scenarios were developed. The first two scenarios were performed only on Type A samples and were considered to represent the behavior of the other samples in this study. The third testing scenario was performed on all three sample types.

First, a loading scenario was developed to quantify the change in component stiffness due to continuous loading. UBMs were first subjected to a preload and then to cycles of loading at 5 Hz. Measurements of static bedding modulus were obtained once every 50 cycles of loading for up to 1,500 cycles. Immediately after the 1,500 cycles were applied, the UBMs were held statically at maximum load for 10 minutes. After, the applied load was reduced to the initial preload level for sample rest during which time bedding modulus values were obtained at incrementally longer intervals of 10, 30, and 120 minutes.

The second loading scenario was developed to quantify the change in component stiffness due to initial conditioning of the UBM samples during the preload phase of testing. This was achieved by observing the changes in bedding modulus, for the same sample, over preload time intervals of 1, 2, 3, 5, 10, 30, 45, 60, and 120 minutes. All samples were allowed a period of unloaded rest for either two or six hours between preload applications.

The final loading scenario was developed to quantify the stiffening and recovery of the UBM samples during representative in-service loading conditions and how this stiffening would affect the bedding modulus of the UBM over time. This was achieved by developing revenue service traffic patterns for a rail transit line. Therefore, New York City Transit heavy rail vehicles were simulated assuming 32 axles per train consist and headways varying between 3, 5, and 10 minutes. Headways were varied to quantify the effect of traffic density/rest period on UBM bedding modulus.

Testing Procedures

All testing was based on procedures presented in the German DIN 45673-5 (*18*). This standard provided the authors with a baseline reference of established best practices in determining the bedding modulus of UBM samples. However, it was imperative that modifications to the procedures be made both to compare the bedding modulus values under European freight and passenger service loads and to better represent the North American loading environment.

Therefore, the Talbot equation (*25*) and Kerr's Boussinesq formulation (*26*) were used to relate axle load to ballast stress and determine the maximum magnitude of force to be applied on the specimen. Talbot's formulation determined the stress value for a typical axle load as it represented a more conservative scenario. For the 95th percentile of typical North American freight heavy axle load of 356 kN (80 kips) (*27*), a ballast stress of 262 kPa (38 psi) was determined for a ballast depth of 30.5 cm (12 in.) below the crosstie. Then, based on the UBM sample size, it was possible to determine the maximum load to be applied as 16.9 kN (3.8 kips).

For the investigation of support condition effects in accordance with DIN procedures, three cycles of quasi-static loads within the range of 0.9–16.9 kN (0.2–3.8 kips) were applied to each sample following a continuous loading and unloading rate of 0.01 $N/mm^2/s$ (1.45 psi/s). Measurements of force and displacements were obtained for the last complete loading cycle. Four test replicates were then conducted to assess the level of variability with the proposed test procedure.

During the investigation of traffic pattern effects, test procedures were modified to allow for more precise capture of the stiffening and recovery effects of traffic. The traffic loading range employed consisted in a preload/minimum load of 1.8 kN (0.4 kips) specified by the DIN and maximum load as described above. For the determination of the static bedding modulus at each stage, load ranges were maintained constant but only a single load cycle was performed. The tests were executed this way given the fact that the two initial conditioning cycles recommended by the DIN procedures could impart changes to the bedding modulus results and thus mask the true effects of traffic loading.

Insertion Loss Prediction Models

Various models to determine the expected performance of UBMs have been developed, one of which was developed by Wettschureck and Kurze (W&K) (*28*). W&K's model was found to provide satisfactory results when compared with field data on insertion loss available to the researchers and chosen for use throughout this research.

This theoretical impedance model consists of a unidimensional, single degree-of-freedom representation of the track structure in which three separate impedances are considered. First, the source impedance represents all track components present above the level of installation of the UBM, including ballast, crosstie, rail, and unsprung wheelset mass. Second, an individual impedance value depicts the UBM (i.e., filter) to be added to the structure. Lastly, the terminal impedance includes characteristics of the support in which the material is to be installed (i.e., either the subgrade or a concrete slab). The insertion loss equation used in the W&K model is given below (Equation 1). A more detailed presentation of the equations for the source and terminal impedance employed in this model was documented by Wettschureck (*7*):

$$\Delta L_e = 20\log\left|1 + \frac{\frac{j\omega}{s_M}}{\frac{1}{Z_i} + \frac{1}{Z_a}}\right| dB \quad (1)$$

where ΔL_e = insertion loss (dB);
j = imaginary unit;
ω = radian frequency (Rad);
s_M = UBM stiffness (N/m);
Z_i = source impedance; and
Z_a = terminal impedance.

Note that the UBM stiffness s_M in the above is determined using the following equation:

$$s_M = s''_M \times S_w \times (1 + jd_M) \quad (2)$$

where s''_M = dynamic bedding modulus (N/m^3);
Sw = effective load transfer area of the ballast-UBM interface as defined in (*28*) (m^2); and
d_M = loss factor of the UBM.

As noted previously, this study has employed the static bedding modulus as a proxy for UBM performance and so this was the value input as s''_M in Equation 2 above.

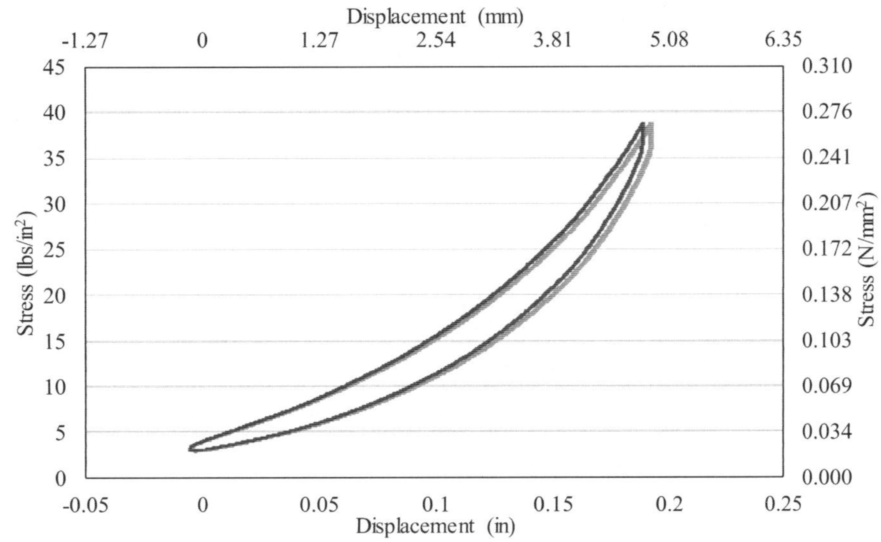

Figure 4. Hysteresis loops for sample Type C with steel support condition.

Results and Discussion

Measurements of displacement from all potentiometers were zeroed based on their initial recorded values and subsequently averaged to obtain the absolute deformation (i.e., UBM displacement). Force measurements were used to compute the stress by dividing the values by the total area of the specimen being tested. These stresses are presented in Figure 4, which shows hysteresis loops as overlapping curves for all four replicates of Type C sample tested with the steel support condition.

The hysteretic behavior of the UBM's elastic material is clearly observed in Figure 4 indicating the occurrence of energy dissipation due to internal friction in the material (*29*). This results in loss of strength in the unloading phase of the test providing lower stress values for a same strain measurement. Energy loss in the system may be represented by the work corresponding to the area engulfed by the loop (*30*).

Support Condition Effects

Based on the values presented in the hysteresis loop, the bedding modulus for each test was calculated as the secant modulus of each loop. A summary of the mean results from all repetitions is presented in Figure 5.

Consistency of the measurements was observed throughout all tests. Measurements of the variability were taken as the maximum absolute percentage deviation from the mean within a single test procedure, and these were found to be 1.7% for steel, 3.8% for concrete, and 2.3% for the GBP in all tests conducted as part of this experiment. Given the low variability within a given sample and support condition, averages of the replicates were chosen as a reasonable method to present the results.

As can be seen in Figure 5, there is a noticeable difference between all support conditions, with a trend of concrete consistently yielding the highest values of bedding modulus followed by steel and lastly by the GBP. Comparatively, concrete and steel provided very similar results with the GBP giving slightly lower values. The differences among the samples range from as low as 2.3% for Type C to 11.7% for Type A (Figure 5) between concrete and steel supports. One possible reason for such differences could be the effect of frictional forces between the UBM and the concrete surface microstructure. These frictional forces would induce a lateral confinement of the sample which, in turn, by Poisson's effect would impose additional restrictions to the vertical deformation of the material and result in reduced deformations for the same applied load.

Also, the GBP support displayed the smallest overall bedding modulus for all support conditions with values up to 33.3% lower than the results from the same tests performed on concrete or steel. This may be explained by the presence of the profiled surface of the plate, which provides space for the material to deform into, space that is not present in the flat surfaces of either concrete and steel supports. Nevertheless, statistical analysis conducted with a significance level of 0.05 demonstrated all results to be significantly different across support conditions.

Traffic Pattern Effects

Similarly, bedding modulus values were calculated for each step of the traffic loading pattern simulation procedures undergone by each sample as previously described.

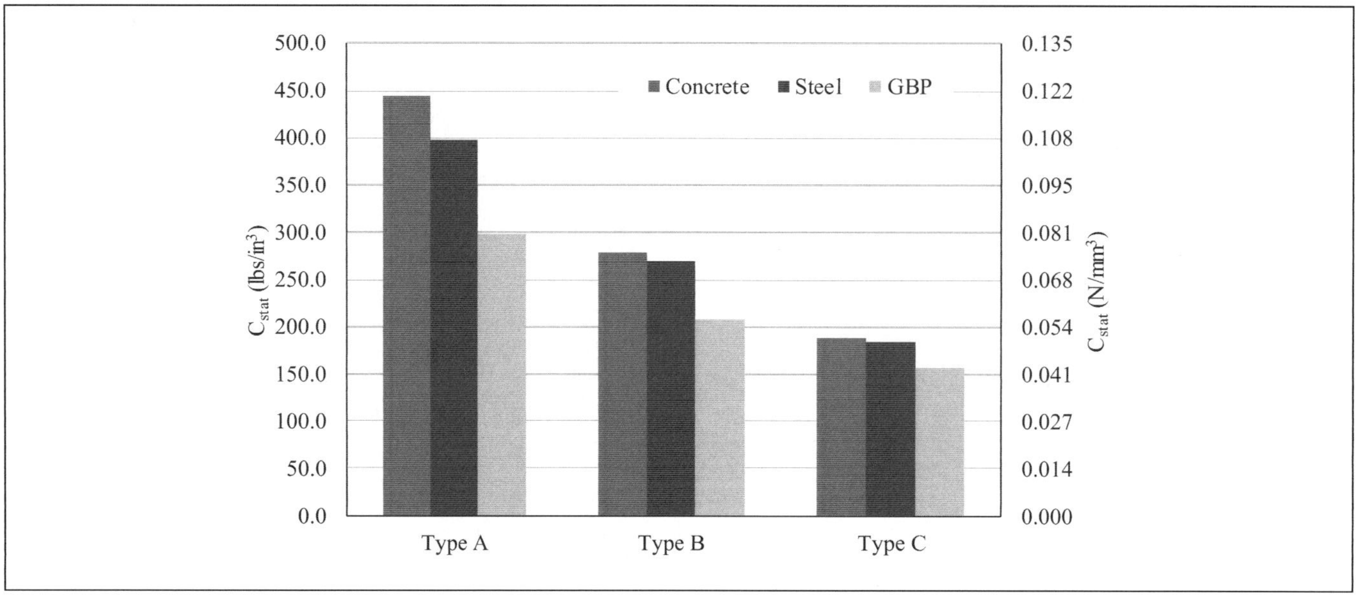

Figure 5. Static bedding modulus results for samples tested.

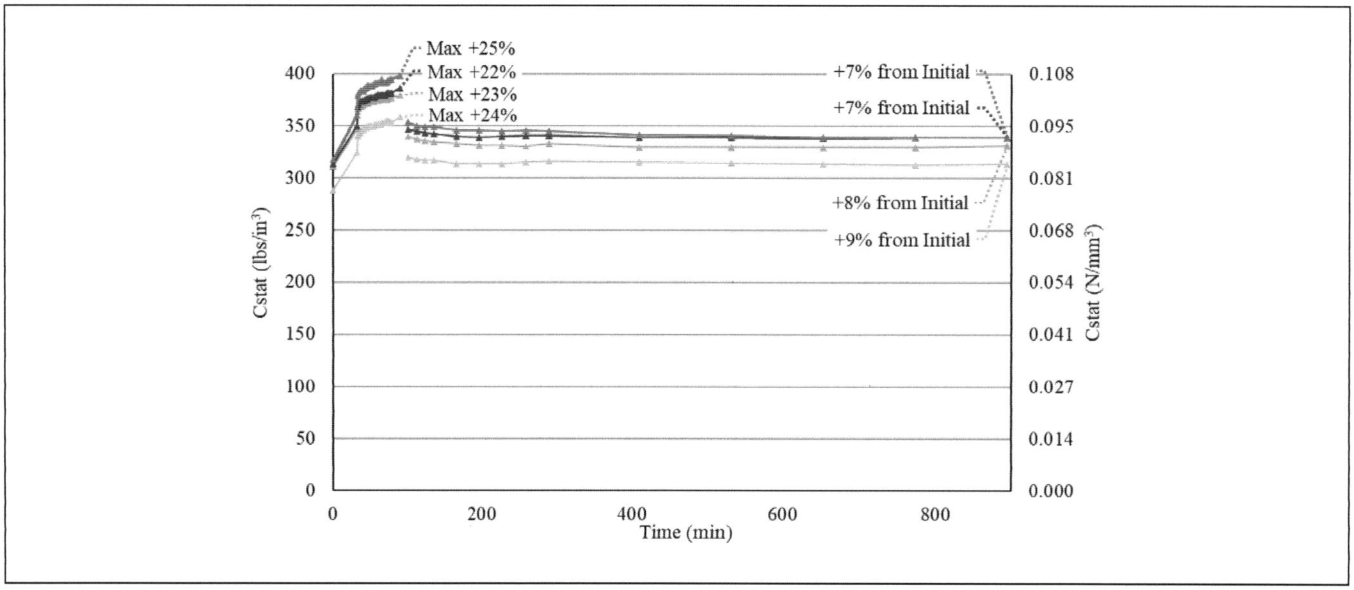

Figure 6. Continuous loading stiffening and preload recovery results.

Results from the first simulation are presented in Figure 6. A consistent increase in bedding modulus values with incremental loading cycles is clearly observed for the four Type A samples tested. A maximum increase of 25% was found after 1,500 cycles of loading plus 10 minutes of constant load. Further, it is possible to observe the rapid development of elastic recovery for all samples after only 10 minutes of rest under preload reaching stiffness values lower than the initial preload conditioning. It is worth noting the inflection point, which is present in all curves after around 200 minutes of testing. One possible explanation for the inflection point is that there is a change in the net balance between the recovery of the preload rest and the stiffening due to the loading imparted while measuring the sample's static bedding modulus. Moreover, samples continued to recover reaching an asymptote value of about 7%–9% higher than the initial condition after approximately 180 minutes of rest.

Figure 7 presents the effects of preload conditioning time on the results, showing that the initial stiffening of the sample occurs at a rate larger than the procedure is capable of detecting, and the continuance of such load

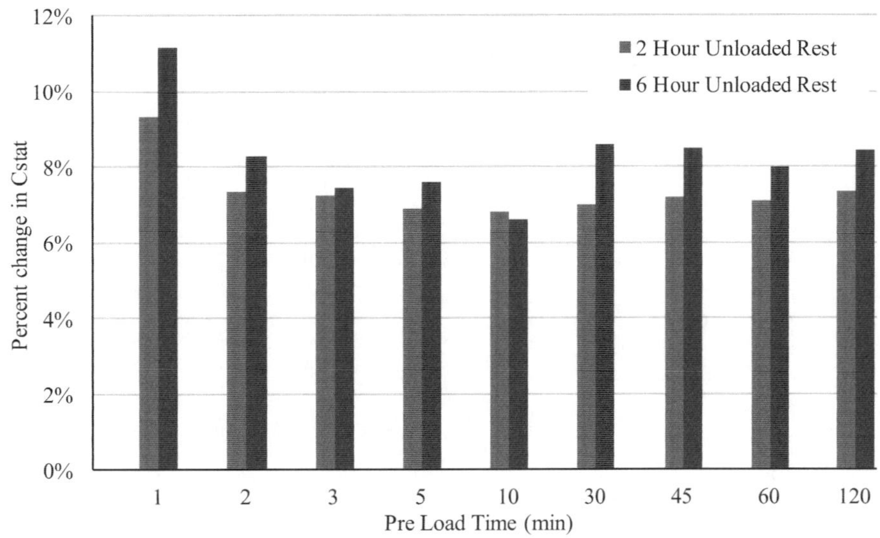

Figure 7. Effects of preload conditioning time results.

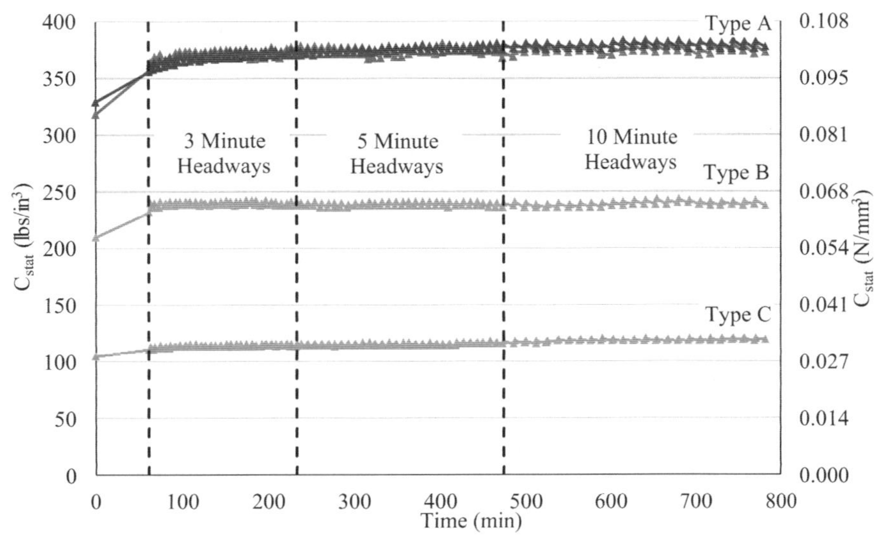

Figure 8. Revenue service traffic loading simulation results.

over the sample for extended periods of time does not generate significant changes to the sample stiffness. Moreover, similar trends were observed between the two tests conducted with different rest periods between preload applications but statistical analysis with a significance value of 0.05 concluded statistical difference of the results. This finding is consistent with the observed results previously described in which the recovery of the sample occurs within the first few minutes of sample unloading.

Finally, for all samples tested, results shown in Figure 8 demonstrate similar trends in behavior; an initial stiffening occurs due to the preload conditioning phase, followed by a gradual stiffening to an asymptotic value of the samples with an increased number of simulated train passes. Amplitudes of stiffness variations for each train pass are larger for the thinner sample (Type A) and reduce with the increase in sample thickness. Further, for a significance level of 0.05, stiffness

Table 2. One-Third Octave Band Frequency Insertion Loss Results for All Samples and Support Conditions Tested

One-third octave band frequency	Insertion loss prediction [dB]								
	Type A			Type B			Type C		
	Concrete	Steel	GBP	Concrete	Steel	GBP	Concrete	Steel	GBP
31.5	−1.9	−2.0	−1.9	−1.8	−1.8	−0.9	−0.5	−0.3	0.7
40	−1.1	−0.7	0.9	1.4	1.6	3.8	4.7	4.9	6.4
50	3.7	4.6	7.2	7.9	8.1	10.6	11.5	11.7	13.2
63	13.5	14.4	17.0	17.6	17.8	20.2	21.0	21.2	22.6
80	20.1	21.0	23.3	23.9	24.1	26.3	27.1	27.3	28.6
100	14.6	15.4	17.6	18.1	18.3	20.4	21.2	21.4	22.6
125	12.2	13.0	15.0	15.5	15.8	17.8	18.5	18.7	19.9
160	10.9	11.6	13.6	14.1	14.4	16.3	17.1	17.2	18.4
200	10.3	11.0	13.0	13.4	13.7	15.6	16.3	16.5	17.7
Average insertion loss difference	*−0.6* (Concrete vs. Steel)		*1.7* (GBP vs. Steel)	*−0.2* (Concrete vs. Steel)		*1.8* (GBP vs. Steel)	*−0.2* (Concrete vs. Steel)		*1.2* (GBP vs. Steel)

variation amplitudes were found to be different across train headways for all samples other than Type C.

The two results reported for Type A represent the same sample that was put through the procedure a second time after approximately one week of being completely unloaded. Results undoubtedly demonstrate the elastic recovery capabilities of the sample; once unloaded over time the sample could recuperate and present behavior like the original fresh specimen.

Insertion Loss Predictions

Bedding modulus results were also input into the W&K prediction model to determine the resulting insertion loss of the different support conditions employed. Values were calculated assuming the UBM was deployed on a ballasted concrete bridge deck for which track and substructure characteristics were chosen based on values obtained from the literature. Insertion loss results were determined for one-third octave band frequencies in the range of interest (i.e., 30–200 Hz). Table 2 presents the results from all samples and support conditions. Moreover, the bottom section of this table provides a comparison between the resulting insertion losses of concrete and GBP against the control steel as the average insertion loss difference.

For example, considering Type A, the prediction model calculations based on bedding modulus results from a test using concrete support yielded values of insertion loss 0.6 dB lower on average than the same calculation made based on bedding modulus obtained from the control. In contrast, insertion loss based on results from testing using the GBP was 1.7 dB larger on average than the control. This same trend was observed throughout all sample types.

Conclusion

Laboratory experiments were conducted to investigate the effects of test setup and loading procedural variations on the measurements of UBM performance parameters, i.e., static bedding modulus and insertion loss. Three different test support conditions were evaluated together with three different loading pattern simulations conducted to investigate the loading effects. Findings from the laboratory experiments are as follows:

- Testing procedures employed proved to exhibit high repeatability for a given sample and support condition, which was within 3.8% of the mean, with the steel providing the least variability between the three supports evaluated.
- Results presented showed a consistent reduction in bedding modulus for the GBP support, which produced results up to 33% lower than the other two support conditions. In contrast, results obtained from concrete and steel supports showed little difference but concrete tests consistently presented higher values.
- Even though consistently lower, GBP results provide evidence to support the proposed adoption of the GBP as a standard equipment for the testing of UBMs. Additionally, applications of this apparatus could extend to the fatigue testing of the material providing a much simpler setup when compared with current practices of implementing a ballast box.
- Results from simulated loading patterns demonstrated the gradual increase in bedding modulus values with the accumulation of loading cycles over the sample. However, recovery of sample

properties could be observed to develop at high rates (less than 10 minutes) after load was removed and the UBM could rest.

- Preload results also demonstrated the rapid generation of initial preconditioning stiffening due to a constant static load over the samples at a rate larger than the sensitivity of the test procedure.
- Results of the revenue-service traffic pattern simulations provided evidence of a proposed "working range" stiffness of UBMs. This was beyond the fact that no effect of traffic density could be observed.
- Predictions from a previously proposed model depicted the influences of the variation of bedding modulus to the studied performance parameters showing differences of up to 1.8 dB, on average, in the insertion loss calculations for the different results from the support condition investigation. This may have substantial effect depending on the level of mitigation needed for a specific application.

Acknowledgments

The authors would like to thank Progress Rail Services for providing funding for this project. Additional support was provided by the National University Rail (NURail) Center, a USDOT-OST Tier 1 University Transportation Center. The authors extend their gratitude to Tim Prunkard and the UIUC Machine Shop for their assistance in manufacturing the GBP and with equipment setup. We would also like to thank the students and staff from RailTEC for their assistance and advice, especially Arkaprabha Ghosh for his support with the development of the early stages of this research as well as Brendan Schmit and Kaila Simpson for their assistance with laboratory testing. J. Riley Edwards has been supported in part by the grants to the UIUC RailTEC from Canadian National Railway and Hanson Professional Services.

Author Contributions

The authors confirm contribution to the paper as follows: study conception and design: Arthur de Oliveira Lima, Marcus S. Dersch, J. Riley Edwards and Yu Qian; data collection: Arthur de Oliveira Lima and Marcus S. Dersch; analysis and interpretation of results: Arthur de Oliveira Lima, Marcus S. Dersch, and J. Riley Edwards; draft manuscript preparation: Arthur de Oliveira Lima, Marcus S. Dersch, J. Riley Edwards, Erol Tutumluer and Yu Qian. All authors reviewed the results and approved the final version of the manuscript.

References

1. Carl, E. H., D. A. Towers, and L. D. Meister. *Transit Noise and Vibration Impact Assessment*. Publication FTA-VA-90-1003-06. U.S. Department of Transportation, Federal Transit Administration, Office of Planning and Environment, Washington, D.C., 2006, p. 274.
2. *State of the Art Review of Mitigation Measures on Track*. Publication RIVAS_UIC_ WP3_D3_1_V01-3. International Union of Railways, Paris, France, 2011, p. 85.
3. Jones, C. J. C., and J. R. Block. Prediction of Ground Vibration from Freight Trains. *Journal of Sound and Vibration*, Vol. 193, No. 1, 1996, pp. 205–213.
4. Wilson, G. P., H. J. Saurenman, and J. T. Nelson. Control of Ground-Borne Noise and Vibration. *Journal of Sound and Vibration*, Vol. 87, No. 2, 1983, pp. 339–350.
5. Hussein, M. F. M. *Vibration from Underground Railways*. University of Cambridge, Cambridge, MA, 2004.
6. Auersch, L. Dynamic Axle Loads on Tracks With and Without Ballast Mats: Numerical Results of Three-Dimensional Vehicle-Track-Soil Models. *Proceedings of the Institution of Mechanical Engineers, Part F: Journal of Rail and Rapid Transit*, Vol. 220, No. 2, 2006, pp. 169–183. https://doi.org/10.1243/09544097F00105.
7. Wettschureck, R. Measures to Reduce Structure-Borne Noise Emissions Induced by above-Ground, Open Railway Lines. *Rail Engineering International*, Vol. 1, 1997, pp. 12–16.
8. Hanson, C. E., and H. L. Singleton. Performance of Ballast Mats on Passenger Railroads: Measurement vs. Projections. *Journal of Sound and Vibration*, Vol. 293, No. 3–5, 2006, pp. 873–877. https://doi.org/10.1016/j.jsv.2005.08.052.
9. Wettschureck, R. G. Vibration and Structure-Borne Sound Isolation by Means of Cellular Polyurethane (PUR) Elastomers. Presented at the SVIB Vibrationsdagen, Stockholm, Sweden, 1994.
10. Wettschureck, R. G., M. Heim, and M. Tecklenburg. Long-Term Efficiency of Ballast Mats Installed in the Rapid Transit Railway Tunner near the Philharmonic Hall of Munich Germany. *Proc., 10th International Congress on Sound and Vibration*, Stockholm, Sweden, 2003, pp. 403–410.
11. Wettschureck, R. G., M. Heim, and M. Tecklenburg. Long-Term Properties of Sylomer® Ballast Mats Installed in the Rapid Transit Railway Tunnel near the Philharmonic Hall of Munich, Germany. *Rail Engineering International*, Vol. 31, No. 4, 2002, pp. 6–11.
12. Sol-Sánchez, M., F. Moreno-Navarro, and M. Rubio-Gámez. The Use of Elastic Elements in Railway Tracks: A State of the Art Review. *Construction and Building Materials*, Vol. 75, 2015, pp. 293–305.
13. Wettschureck, R. G., F. Breuer, M. Tecklenburg, and H. Widmann. Installation of Highly Effective Vibration Mitigation Measures in a Railway Tunnel in Cologne, Germany. *Rail Engineering International*, No. 4, 1999, pp. 12–16.
14. Li, D., and L. Maal. Heavy Axle Load Revenue Service Bridge Approach Problems and Remedies. Presented at the Joint Rail Conference 2015, San Jose, CA, 2015.
15. Curt, M., and D. Otter. *Summary of Stress-State Reduction in Concrete Bridges*. Publication TD-13-025. Transportation Technology Center Inc., Pueblo, CO, 2013.
16. de O. Lima, A., M. S. Dersch, Y. Qian, E. Tutumluer, and J. R. Edwards. Laboratory Evaluation of under-Ballast Mat Effectiveness to Mitigate Differential Movement Problem in Railway Transition Zones. Presented at the

10th International Conference on the Bearing Capacity of Roads, Railways and Airfields, Athens, Greece, 2017.
17. Vuchic, V. R. *Urban Transit Systems and Technology*. John Wiley & Sons, Hoboken, NJ, 2007.
18. DIN 45673-5. *Mechanical Vibration - Resilient Elements Used in Railway Tracks - Part 5: Laboratory Test Procedures for under-Ballast Mats*. Deutsches Institut fur Normung, Berlin, Germany, 2013.
19. de O. Lima, A., M. S. Dersch, Y. Qian, E. Tutumluer, and J. R. Edwards. Laboratory Mechanical Fatigue Performance of under-Ballast Mats Subjected to North American Loading Conditions. Presented at the 11th International Heavy Haul Association Conference, Cape Town, South Africa, 2017.
20. Bauman, J. T. *Fatigue, Stress, and Strain of Rubber Components: A Guide for Design Engineers*. Hanser Publications, Munich, Cincinnati, OH, 2008.
21. Mars, W. V., and A. Fatemi. Factors That Affect the Fatigue Life of Rubber: A Literature Survey. *Rubber Chemistry and Technology*, Vol. 77, No. 3, 2004, pp. 391–412.
22. Woo, C. S., and H. S. Park. Useful Lifetime Prediction of Rubber Component. *Engineering Failure Analysis*, Vol. 18, No. 7, 2011, pp. 1645–1651. https://doi.org/10.1016/j.engfailanal.2011.01.003.
23. EN 16730:2016. *Railway Applications - Track - Concrete Sleepers and Bearers with under Sleeper Pads*. British Standards Institution, London, UK, 2016.
24. Rudolf Schilder. USP (Under Sleeper Pads): A Contribution to Save Money in Track Maintenance. Presented at the AusRAIL PLUS 2013, Driving the Costs Out of Rail, Canberra, Australia, 2013.
25. Talbot, A. N. *Second Progress Report: Special Committee on Stresses in Railroad Track*. American Railway Engineering Association, Chicago, IL, 1920.
26. Kerr, A. D. *Fundamentals of Railway Track Engineering*. Simmons-Boardman Books Inc., Omaha, NE, 2003.
27. Van Dyk, B. *Characterization of the Loading Environment for Shared-Use Railroad Superstructure in North America*. Master's thesis. University of Illinois at Urbana-Champaign, Champaign, IL, 2013.
28. Wettschureck, R., and U. J. Kurze. Insertion Loss of Ballast Mats. *Acustica*, Vol. 58, No. 3, 1985, pp. 177–182.
29. LaClair, T. J. Rolling Resistance. In *The Pneumatic Tire*. U.S. Department of Transportation, Washington, D.C., 2006, p. 699.
30. Hopkinson, B., and G. T. Williams. The Elastic Hysteresis of Steel. *Proceedings of the Royal Society of London A: Mathematical, Physical and Engineering Sciences*, Vol. 87, No. 598, 1912, pp. 502–511. https://doi.org/10.1098/rspa.1912.0104.

The Standing Committee on Railway Maintenance (AR060) peer-reviewed this paper (18-03123).

Article

TRR
JOURNAL OF THE TRANSPORTATION RESEARCH BOARD

Transportation Research Record
2018, Vol. 2672(10) 85–95

DOI: 10.1177/0361198118784382
journals.sagepub.com/home/trr
SAGE

Investigation of Relationship between Train Speed and Bolted Rail Joint Fatigue Life using Finite Element Analysis

Hao Yin[1], Yu Qian[2], J. Riley Edwards[1], and Kaijun Zhu[3]

Abstract

Reducing the allowable operating speed or imposing temporary speed restrictions are common practices to prevent further damage to rail track when defects are detected related to certain track components. However, the speeds chosen for restricted operation are typically based on past experience without considering the magnitude of the impact load around the rail joints. Due to the discontinuity of geometry and track stiffness at the bolted rail joints, an impact load always exists. Thus, slower speeds may not necessarily reduce the stresses at the critical locations around the rail joint area to a safe level. Previously, the relationship between speed and the impact load around the rail joints has not been thoroughly investigated. Recent research performed at the University of Illinois at Urbana-Champaign (UIUC) has focused on investigating the rail response to load at the joint area. A finite element model (FEM) with the capability of simulating a moving wheel load has been developed to better understand the stress propagation at the joint area under different loading scenarios and track structures. This study investigated the relationship between train speed and impact load and corresponding stress propagation around the rail joints to better understand the effectiveness of speed restrictions for bolted joint track. Preliminary results from this study indicate that the contact force at the wheel–rail interface would not change monotonically with the changing train speed. In other words, when train speed is reduced, the maximum contact force at the wheel–rail interface may not necessarily reduce commensurately.

Two neighboring rails need to be connected to provide a uniform running surface for trains. Using rail joints or welding rails (i.e., continuously welded rail) are the two main methods of joining the rails together. With the increasing popularity of continuously welded rail due to its many maintenance and service life benefits, the number of in-service bolted joints has reduced significantly, and rail joint research has also decreased as a result. However, many bolted joints remain in the track, especially in rail transit systems. Because of the unique loading environment in rail transit systems, such as high-frequency and high-repetition (i.e., number of load replications), defects associated with bolted rail joints still pose safety and operational challenges.

Rail-end bolt-hole cracks and upper fillet cracks are two of the major defects which can cause a rail break or even loss of rail running surface. Previous research has concluded that the stress concentration around the rail-end bolt-hole and the rail upper fillet areas are the primary reason for crack initiation and propagation (*1*–*3*). Without proper methods to identify the defects in the rail joints in a timely manner, the risk for damage to the track structure and/or derailments is higher (*4*, *5*).

To reduce the risk of accidents caused by potential failure of the track, temporary speed restrictions are typically applied to sections where defects are detected. In October 2000, over 1,800 emergency speed restrictions were imposed and a nationwide track investigation and replacement program was conducted in the United Kingdom after the Hatfield derailment (*6*). In February 2015, the Washington Metropolitan Area Transit Authority decided to slow down trains on some sections as a safety precaution to prevent incidents with rails that were potentially cracked or broken (*7*). Intuitively, to slow down trains would reduce the dynamic load on the rails and other track components. Due to the differences between track structures and operation practices, the

[1]Rail Transportation and Engineering Center – RailTEC, Department of Civil and Environmental Engineering, University of Illinois at Urbana-Champaign, Urbana, IL
[2]Department of Civil and Environmental Engineering, University of South Carolina, Columbia, SC
[3]Walter P. Moore and Associates, Inc., Washington, DC

Corresponding Author:
Address correspondence to Yu Qian: yuqian@sc.edu

speed restrictions among different freight railroads and transit agencies vary and are often based on past experience. Due to the discontinuity of geometry and track stiffness at the bolt rail joints, an impact load will always exist. Thus, slower operation speed may not necessarily reduce the stresses at the critical locations around the rail joint area to a safe level. Furthermore, the relationship between rail stresses at the joint area and operating speed has not been thoroughly investigated.

Recent research performed at the University of Illinois at Urbana-Champaign (UIUC) has focused on investigating the rail responses at the joint area. A finite element (FE) model has been developed to better understand the stress propagation at the joint area with different loading scenarios and track structures. This study investigated the relationship between train speed and the stresses around the rail-end bolt-hole and upper fillet areas, which were identified as the most critical locations (*8*), with the objective of better understanding the effectiveness of speed restrictions. The predicted fatigue life of rail joints under different train speeds was also studied. Results indicate that the stresses in critical rail locations were not proportional to train speed, which does not align with conventional wisdom. In other words, lower train speeds do not necessarily ease the stress concertation around the joint area and consequently extend the fatigue life of rail joints.

Objective and Scope

The objective of this study is to investigate the relationship between stress distributions and consequent fatigue life at the critical locations around the rail joint area and train speed. Specifically, stresses at the rail–wheel contact interface, the rail-end bolt-hole, and the rail-end upper fillet will be investigated with the objective of evaluating the effectiveness of speed restrictions. A FE model that was previously developed to study optimal joint bar configurations (*8*, *9*) was adapted to simulate moving wheel loadings with various train speeds. The fatigue life of the upper fillet area was also estimated with a fatigue life predictive model based on results from the FE analysis. Findings from this study can help to better understand the relationship between train speed and the fatigue life of rail joints and will aid in the refinement of future guidelines for speed restrictions to be more reflective of the stress state of the track and its components.

Numerical Simulation Approach

Commercially available software known as *Abaqus/CAE* was selected to perform the FE simulations. A linear FEM of a rail joint system that had been previously developed, calibrated, and validated was further refined to simulate the dynamic response of the rail joint system to the impact load caused by moving wheels. For analysis of the fatigue life, the commercially available fatigue life analysis software *fe-safe* was selected to perform the prediction. The loading history of the moving wheel passing the gap of the rail joint obtained from the dynamic FE analysis was then used as the input for the fatigue life prediction, the estimated fatigue life (total cycle number of wheels passing before damage) was obtained as the results of the fatigue life analysis. The procedure of bolted rail joint FE analysis and fatigue life analysis is illustrated in Figure 1.

Dynamic FE Analysis Model

In order to gain insight into the response of the rail joint due to the impact loading caused by each wheel pass, a

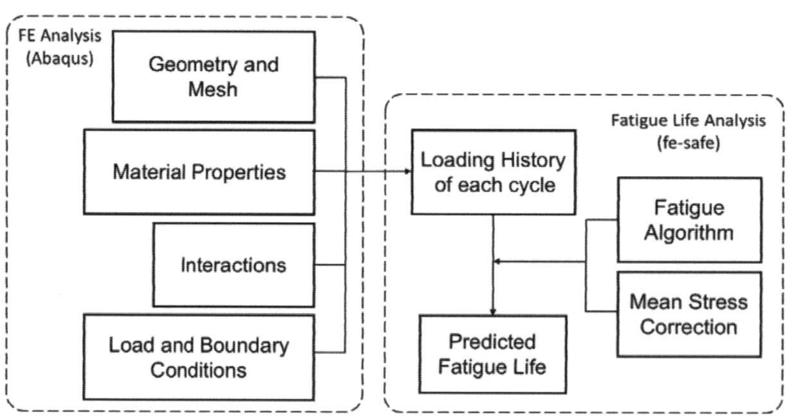

Figure 1. Procedure used for FE analysis of bolted rail joint and for fatigue life prediction.

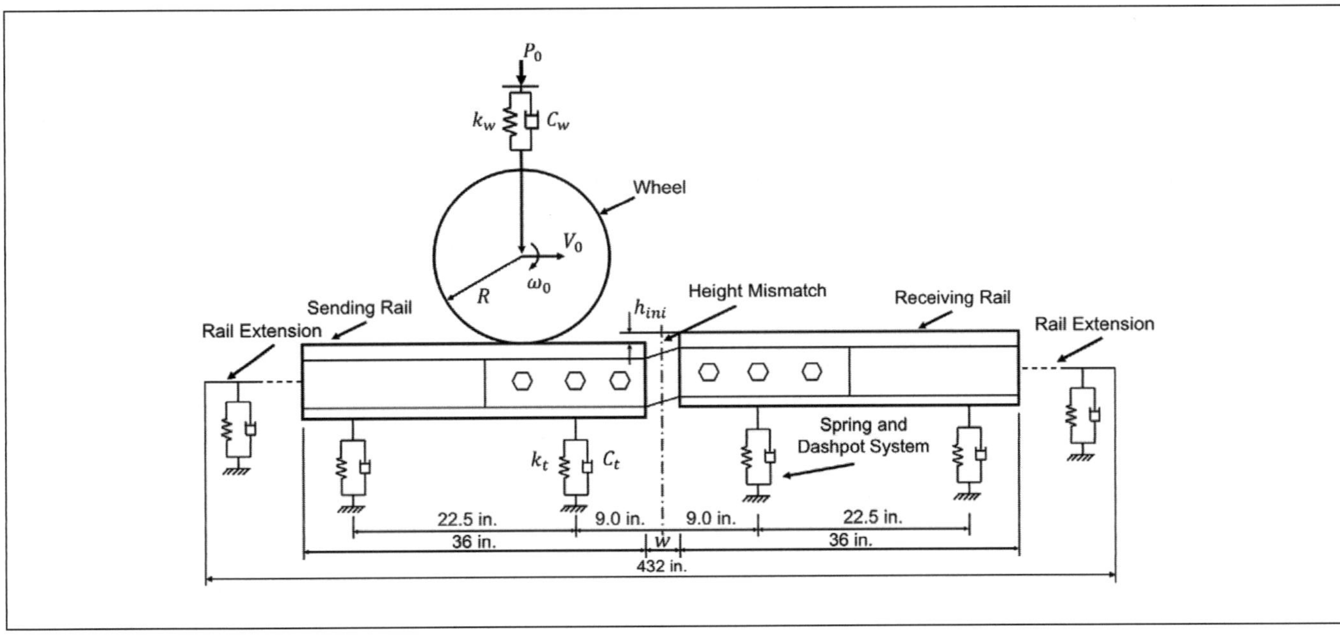

Figure 2. Schematic diagram of UIUC's FE model of bolted rail joint.

dynamic FE model was developed using *Abaqus/CAE* Explicit (Figure 2).

The 115RE rail and standard joint bars were selected to represent a typical joint used in rail transit systems in the United States. The centermost crosstie spacing was 18 in. (45.7 cm), and other crossties were spaced at 22.5 in. (57.2 cm) on center. The total length of each rail was 216 in. (548.6 cm), based on the sensitivity analysis of rail length published in an earlier publication (*8*), the length of each rail modeled with 3-D deformable solid elements set to 36 in. (91.4 cm), and the remaining 180 in. (457.2 cm) of each rail was simplified by assigning rail section properties to linear beam elements. The gap (w) between sending rail and receiving rail was set to $w = 0.125$ in. (0.318 cm), and the initial height mismatch (h_{ini}) between the sending rail and receiving rail was also introduced in this dynamic FE model to better simulate the geometric imperfections at the rail joints caused by poor assembly, ground settlement, etc. Based on a similar study of the mechanical responses to the height mismatch at the rail joint (*10*), a height mismatch of $h_{ini} = 0.005$ in. (0.013 cm) was selected to obtain the rail's response to the impact load when the wheel passed the gap. For the geometry of the wheel, the diameter of the wheel was set to $R = 17$ in. (43.2 cm), which is a typical size of railcar wheel used in heavy rail transit systems, such as the MTA New York City Transit Authority. Due to the fact that the behavior of the rail joint system was primarily studied in the vertical plane and the models were loaded vertically and symmetrically in the longitudinal direction of the rail, the railcar wheel was modeled as a cylinder without a flange. Figure 3 shows the components of the FE model generated in the simulation.

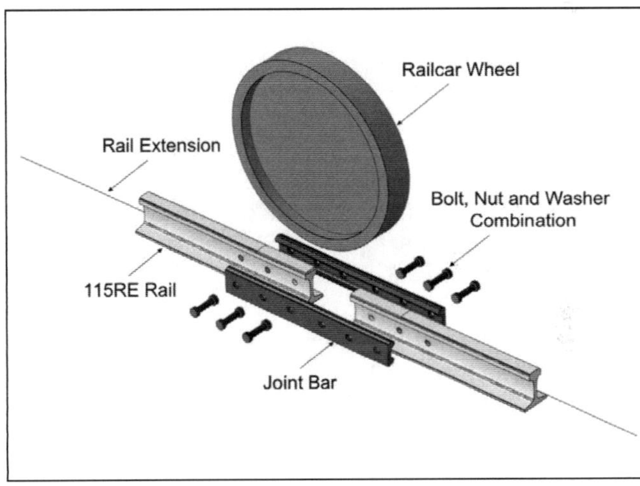

Figure 3. Components of the bolted rail joint assembly used in the dynamic FE model.

Material Properties. All the parts (i.e., wheel, rail, rail joint) were assumed to behave elastically in the dynamic FE analysis and a correction of long-term behavior of materials was performed in conjunction with the fatigue life analysis. The Young's modulus, Poisson's ratio, and the density of the wheel, rails, rail joints, and bolts were assigned as 29,000 ksi (199.9 GPa), 0.33, and 0.283 lb/in.^{3} (7,833.4 kg/m^3), respectively. The supporting system (e.g., crosstie, ballast, etc.) was represented in the model

Table 1. Coefficient of Friction (COF) Values Used in the FE Model

Frictional interaction	COF
Bolt–rail interface	0.20
Bolt–joint bar interface	0.20
Rail–joint bar interface	0.20
Rail–rail pad interface	0.30
Wheel–rail interface	0.15

by linear spring and dashpot elements, with details of the simplifications included in an earlier publication (*8*). k_t and C_t were the spring stiffness and damper coefficients, and the equivalent springs and dampers were those contributed from the crosstie, rail pad, ballast, subgrade, etc. Using a track modulus of 4,000 psi (27.58 MPa) provided by New York City Transit Authority (NYCTA), and results from previous research pertaining to equivalent springs and dampers, $k_t = 90,000$ lbf/in. (15,761 kN/m) and $C_t = 90$ lbf·s/in. (15.76 kN·s/m) were selected. Similarly, k_w and C_w were the spring stiffness and damper coefficient of springs representing the suspension system of a train car and $k_t = 1,000$ lbf/in. (175.13 kN/m) and $C_t = 0.8$ lbf·s/in. (0.14 kN·s/m) were selected, which are consistent with other studies (*10, 11*).

Contact Interactions. Contact interactions between components were formulated using surface-to-surface contact discretization, and a master–slave surface pair was defined for each contact pair. This contact formulation method prevents large and undetected penetrations from nodes on the master surface into the slave surface, providing more accurate stress and strain results compared with other methods (*12*). The basic Coulomb friction model with the penalty friction formulation was used to simulate the frictional force response at the contact interface. The maximum allowable frictional stress is related to contact pressure by the coefficient of friction (COF) between contacting bodies. The COFs of the contact pairs in the model were determined from the literature and are summarized in Table 1 (*13, 14*).

Load and Boundary Conditions. For loading conditions, since the stress distribution between the threaded bolt and nut is not the primary zone of interest in this study, the combination of the bolt, nut, and washer was simplified into a single component. The bolt torque moment was represented by bolt preload calculated with Equation 1 by the bolt torque moment and bolt diameter (*15*):

$$P_b = \frac{T}{KD} \quad (1)$$

where

P_b = bolt preload (lbf)
T = bolt torque moment (lbf·in.)
K = coefficient of the bolt torque moment (43.8–56.2)
D = bolt diameter (in.)

The bolts used for the 115RE rail joints had a diameter D of 1 in. (2.54 cm), the torque moment T was chosen as 4,425 lbf·in. (500 N·m), and $K = 45$ was selected based on previous research (*8*). Thus, the bolt preload P_b was calculated as 22,000 lbf (97.86 kN) per bolt. The axle load of 16,500 lbf (73.40 kN) from the train car was first applied to a spring element which represented the suspension, and then vertically passed to the wheel. For boundary conditions, the displacements of each component at lateral and longitudinal direction of the rail were limited since the behavior of the rail joint system was primarily studied in the vertical direction.

In addition, because the explicit solver was used for the dynamic FE analysis, the time increment size must be limited to a very small number to avoid numerical stability and convergence issues and, after a sensitivity study of the time increment size was conducted, 0.0001s/step was selected. All of the constants and variables that were considered in the dynamic FE model are summarized in Table 2.

Fatigue Life Analysis

The fatigue life analysis was performed primarily based on the load history and distribution of stresses calculated from the dynamic FE models. In addition to the FE analysis results, information on material properties, as well as the selection of the methods of fatigue algorithm and mean stress correction, were of great importance during the fatigue analysis. *fe-safe* was selected to perform the fatigue analysis for bolted rail joints taking into consideration the effects of various impact loads caused by various wheel speeds. The methodology used in this study is illustrated in Figure 4.

Loading History. The wheel–rail contact force history obtained from the dynamic FE analysis was used as the load history for each cycle of wheel passing and was input directly into *fe-safe*. This load history was utilized as the base load, and was factored using a load factor function. The estimated fatigue life could be considered as the total cycles of loading that the system has experienced before damage occurs, namely, the total number of wheels passing over the rail joint before damage initiates.

Material Properties of Fatigue Life Analysis. Based on a test report provided by NYCTA, the ultimate tensile strength of the steel used for 115RE rail was approximately 177.0 ksi (1,220 MPa) and strength at 10^7 cycles (fatigue limit)

Table 2. Constants and Variables for FE Model

Constants	
Crosstie spacing (center)	18 in. (45.7 cm)
Crosstie spacing	22.5 in. (57.2 cm)
Rail section	115RE Rail
Rail length	216 in. (548.6 cm) in total, 36 in. (91.4 cm) with 3D elements, 180 in. (457.2 cm) with 1D elements
Gap height mismatch, h_{ini}	0.005 in. (0.013 cm)
Gap width, w	0.125 in. (0.318 cm)
Joint bar design	Standard Joint Bar
Bolt preloading	22,000 lbf (97.86 kN) per bolt
Wheel radius	17 in. (43.2 cm)
Wheel load	16,500 lbf (73.40 kN)
Suspension spring stiffness	1,000 lbf/in. (175.13 kN/m)
Track modulus	4,000 psi (27.58 MPa)
Equivalent spring stiffness	90,000 lbf/in. (15,761 kN/m)
Time increment size	0.0001 s/step
Variables	
Train speed (Wheel rolling speed, no slippage)	5 mph (8.0 km/h)
	10 mph (16.1 km/h)
	20 mph (32.1 km/h)
	30 mph (48.3 km/h)
	40 mph (64.4 km/h)
	50 mph (80.5 km/h)
	60 mph (96.6 km/h)

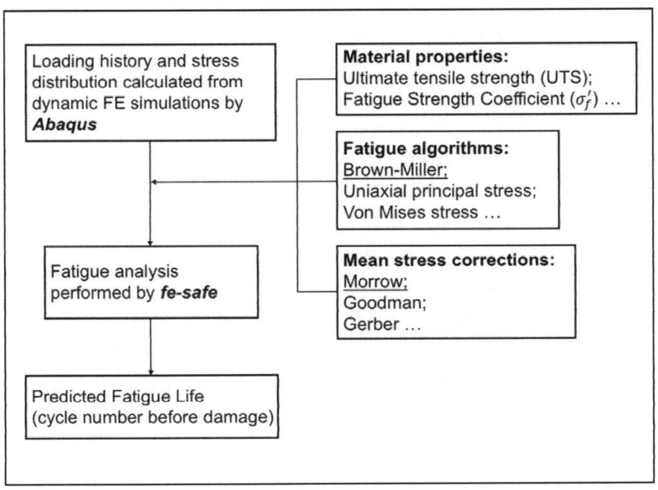

Figure 4. Methodology for fatigue life analysis.

was 61.5 ksi (424 MPa), which were two key parameters used for the fatigue life analysis. The fatigue limit represents a cyclic stress amplitude below which the material does not fail and could be cycled indefinitely (i.e., an infinite fatigue life). For ductile steel specifically, the fatigue limit is the strength of the material at 10^7 cycles of loading. In other words, if the steel structural system could experience at least 10^7 cycles of loading without cracking or other damage, it is assumed that no fatigue damage would occur under the same loading conditions (*16*).

Fatigue Analysis Algorithms. The Brown–Miller criterion was selected for this specific fatigue analysis, which gave the most realistic fatigue life estimates for ductile metals. The Brown–Miller equation suggests that the maximum fatigue damage occurs on the plane which experiences the maximum shear strain amplitude, and that damage is a function of both this shear strain amplitude ($\Delta\gamma_{max}/2$) and the normal strain amplitude ($\Delta\varepsilon_n/2$). Accordingly, different from the conventional strain-life equation (Equation 2), the Brown–Miller equation (Equation 3) alters the left-hand side of the equation with the addition of shear strain amplitude and normal strain amplitude (*17*).

$$\frac{\Delta\varepsilon}{2} = \frac{\sigma_f'}{E}(2N_f)^b + \varepsilon_f'(2N_f)^c \tag{2}$$

where
$\Delta\varepsilon/2$ = applied strain amplitude
$2N_f$ = endurance in reversals
σ_f' = fatigue strength coefficient
ε_f' = fatigue ductility coefficient
b = fatigue strength exponent
c = fatigue ductility exponent

$$\frac{\Delta\gamma_{max}}{2} + \frac{\Delta\varepsilon_n}{2} = C_1\frac{\sigma_f'}{E}(2N_f)^b + C_2\varepsilon_f'(2N_f)^c \tag{3}$$

where
$\Delta\gamma_{max}/2$ = shear strain amplitude

$\Delta\varepsilon_n/2$ = normal strain amplitude
C_1 = 1.65 (constant)
C_2 = 1.75 (constant)

The constants $C_1 = 1.65$ and $C_2 = 1.75$ were derived based on the assumption that cracks initiate on the plane of maximum shear strain. However, for complex variable amplitude loading, it was found that better agreement with test results was obtained by assuming that the most damaged plane was the one that produced the highest value of $(\Delta\gamma_{max}/2 + \Delta\varepsilon_n/2)$. or that case, constants C_1 and C_2 will have slightly different values on this plane. Nevertheless, the values shown in Equation 3 could be applied generally (*18*).

Mean Stress Corrections. Typically, it is common for a load history to have a non-zero mean stress, σ_m, which is defined in Equation 4. The fatigue performance would vary as the mean stress changes. The influence of mean stress can be characterized as the influence of stress amplitude, σ_a, the distance of minimum stress to maximum stress in a fatigue loading cycle (Equation 5).

$$\sigma_m = \frac{\sigma_{max} + \sigma_{min}}{2} \tag{4}$$

$$\sigma_a = \frac{\sigma_{max} - \sigma_{min}}{2} \tag{5}$$

where

σ_m = mean stress (psi)
σ_a = stress amplitude (psi)
σ_{max} = maximum stress (psi)
σ_{min} = minimum stress (psi)

Generally, it can be observed that, for mean stress, a tensile mean stress has a detrimental effect on endurance cycles N_f, whereas a compressive mean stress has a beneficial effect. For stress amplitude, the endurance cycles N_f increase as the applied stress amplitude σ_a decreases (*19*). To correct the influence of mean stress, the Morrow mean stress correction was adopted for the Brown–Miller criterion. After the application of the Morrow mean stress correction, the Brown–Miller equation (Equation 3) becomes Equation 6, with a corrected elastic term by subtracting the mean normal stress on the plane, $\sigma_{n,m}$ (*20*).

$$\frac{\Delta\gamma_{max}}{2} + \frac{\Delta\varepsilon_n}{2} = C_1 \frac{\left(\sigma_f' - \sigma_{n,m}\right)}{E}\left(2N_f\right)^b + C_2\varepsilon_f'\left(2N_f\right)^c \tag{6}$$

where

$\sigma_{n,m}$ = mean normal stress (psi)

Discussion of the Results

Critical outputs from the dynamic FE model, such as the wheel–rail contact force, Von Mises stress around rail-end bolt-hole, Von Mises stress at rail-end upper fillet, and the vertical displacement at rail end, were analyzed. Figure 5 shows examples of these parameters when the wheel was passing different locations around the joint calculated in the simulation at train speed of 20 mph (32.1 km/h).

The loading history of the vertical contact force at the wheel–rail interface when the wheel was moving at a speed of 20 mph (32.1 km/h) is shown in Figure 6. It should be noticed that the original data from the simulation was the time history of the wheel–rail contact force, and it was modified by changing the independent variable (x-axis) from the time to the relative wheel position on the rail surface. As such, the starting point was set to the left end of the joint bar and the ending point was set to the right end of the joint bar, as shown in the schematic drawings at the bottom of Figure 6. When the wheel was running on the sending rail approaching the gap, the wheel–rail contact force was relatively stable, around 16,500 lbf (73.4 kN), approximately the same value as the applied wheel load, with certain variation due to the wheel and track vibration. When the wheel rolled over the gap between the two rails, an unloading stage was observed. Once the wheel contacted with the second rail after passing the gap, a peak contact force (P1) of 40,832 lbf (181.6 kN), was recorded which was the response of the rail to the impact of the moving wheel. Another peak contact force (P2) showed up after P1, which was the response of the track system.

Figure 7 shows the mechanical response of the rail to the impact load due to the wheel rolling over the gap at various train speeds. Figure 7*a* plots all the peak wheel–rail contact force (P1) values for the different simulations with different train speeds. Note the first peak contact force, P1, is always higher than the second peak contact force, P2 (*21*). By comparing the P1 values at different operation speeds, it is clear that the magnitude of P1 was not related to train speed in a linear manner. In other words, reducing train speed from 60 mph (96.6 km/h) to 5 mph (8.0 km/h), the peak wheel–rail contact force did not reduce monotonically. When the operation speed was 60 mph (96.6 km/h), the value of P1 was 40,253 lbf (179.1 kN), when the operation speed reduced to 50 mph (80.5 km/h) and 40 mph (64.4 km/h), P1 reduced to 37,800 lbf (168.1 kN) and 36,500 lbf (162.4 kN), respectively. However, when the operation speed further reduced to 30 mph (48.3 km/h) and 20 mph (32.1 km/h), P1 increased to 41,916 lbf (186.5 kN) and 40,832 lbf (181.6 kN), respectively. This finding was counterintuitive, and the same trend was also observed for the maximum Von Mises stress around the bolt-hole and upper fillet area in Figure 7*b* and *c*.

According to conventional wisdom in the rail industry, contact force generally decreases monotonically with

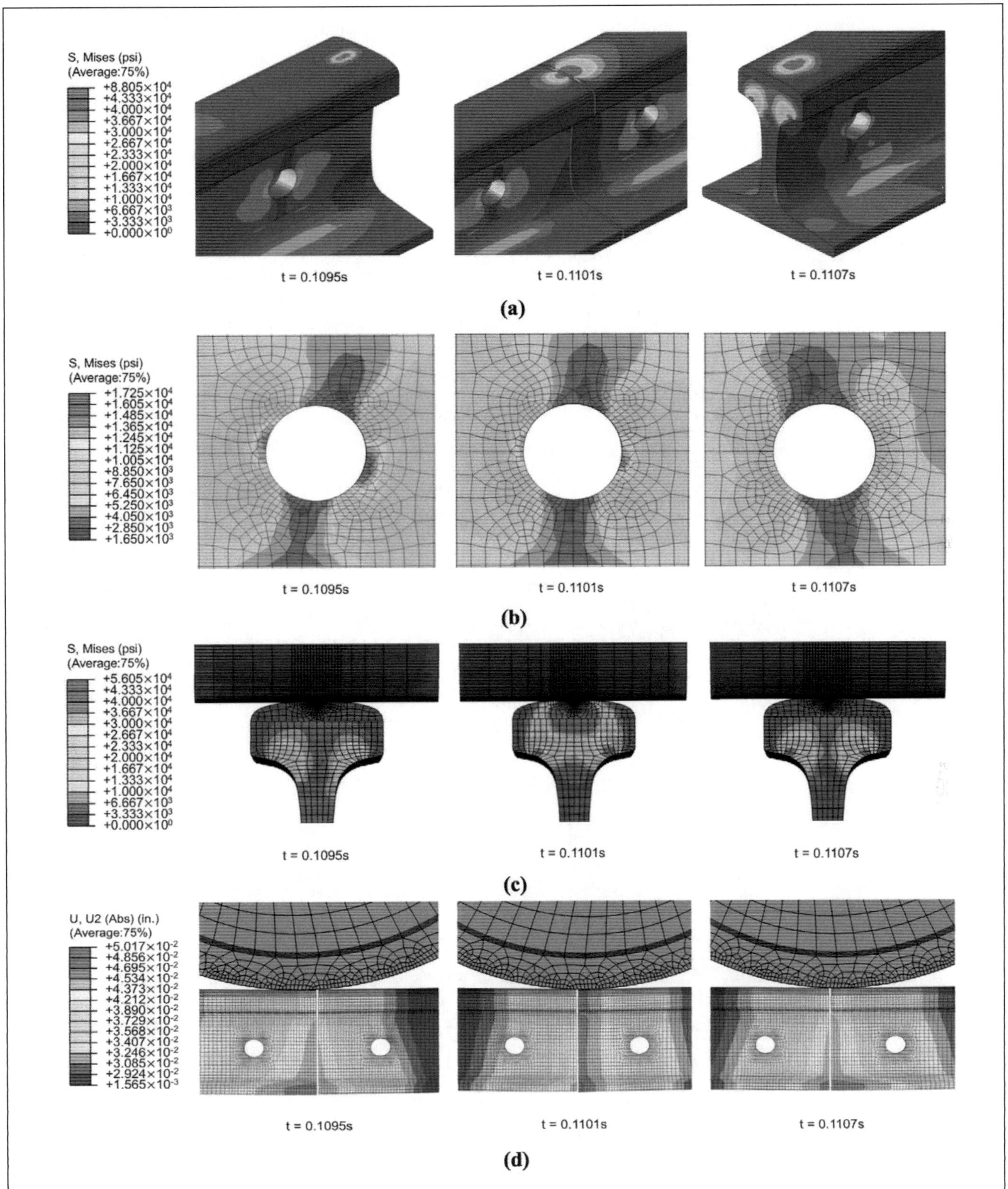

Figure 5. Examples of the simulation results at train speed of 20 mph (32.1 km/h): (*a*) wheel–rail contact patch; (*b*) Von Mises stress around rail-end bolt-hole; (*c*) Von Mises stress at rail-end upper fillet; (*d*) vertical displacement at rail end.

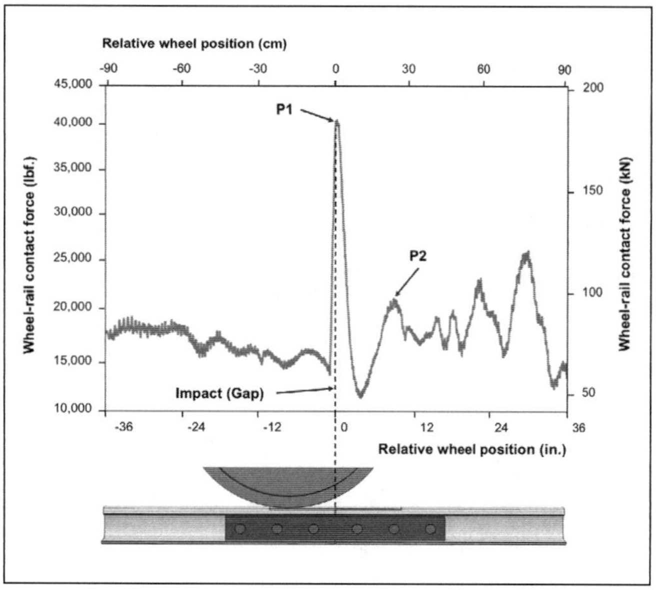

Figure 6. Contact force history of wheel–rail interface of bolted rail joint at train speed of 20 mph (32.1 km/h).

decreasing train speed (*21*), but the findings from this study shown in Figure 7 do not support this. In the literature, the concept that dynamic load increases as traveling speed increases is based on the well-established vehicle–track interaction theory, without considering the joints. However, there are two important differences between this study and existing literature: (a) the gap between the two rails and (b) the differential displacement of the two rails at the joint. Due to the gap between the two rails, the sending rail and the receiving rail will not have the same displacement at the same time. When the wheel is approaching the end of the sending rail, the displacement of the end of the sending rail increases. The displacement of the sending rail will cause the joint bar to move together. The displacement of the joint bar will then cause the displacement of the receiving rail. The sending rail will reach its maximum displacement when the wheel is on top of the end of the rail (*8*), right before the wheel rolls over the gap. However, the receiving rail will not reach the same displacement simultaneously. The differential displacement of the two rails will cause additional height mismatch (h_w) before the wheel hits the

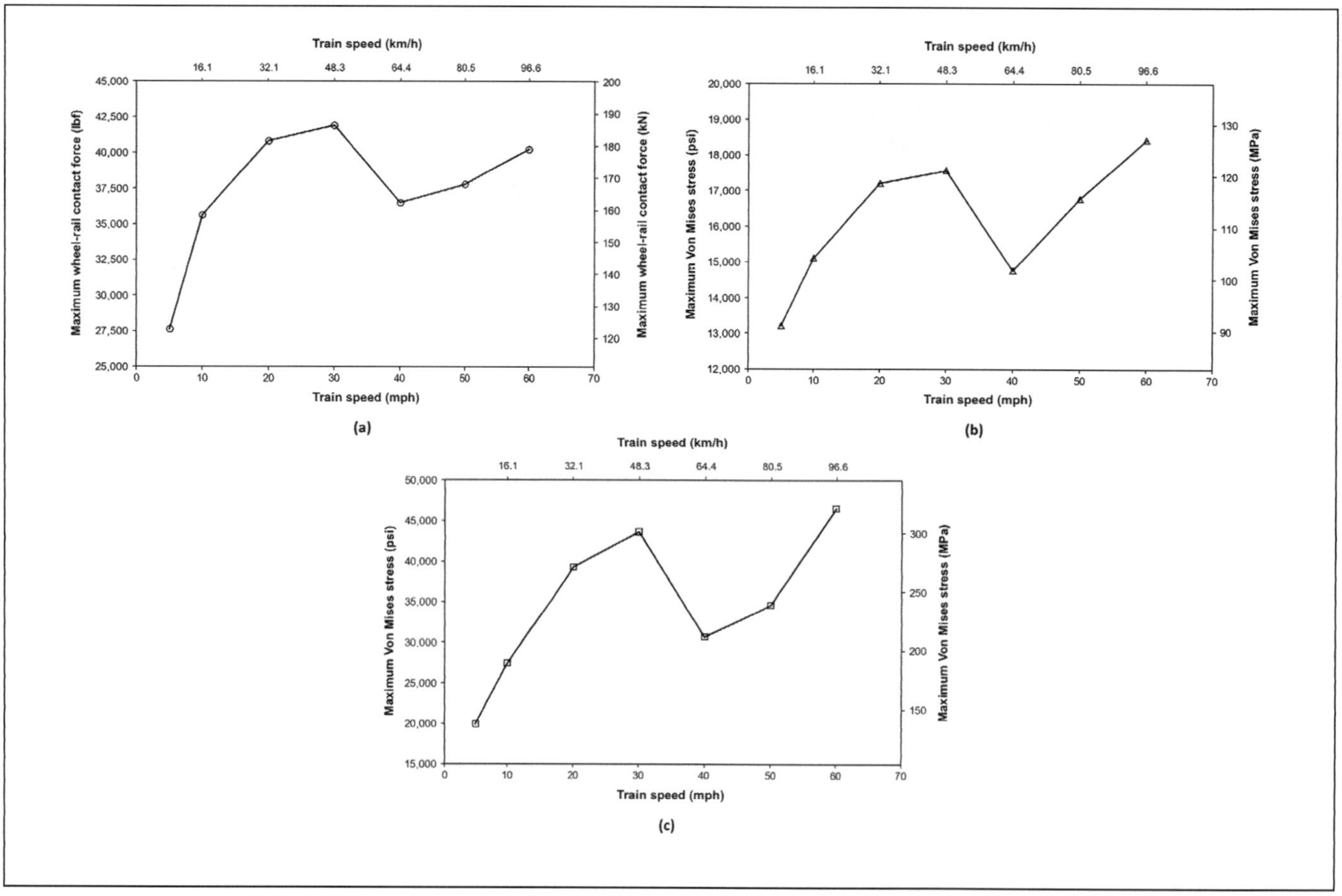

Figure 7. Mechanical responses of rail joint at various operation speeds: (*a*) maximum contact force at the wheel–rail interface; (*b*) maximum Von Mises stress around the rail-end bolt-hole; (*c*) maximum Von Mises stress at rail-end upper fillet.

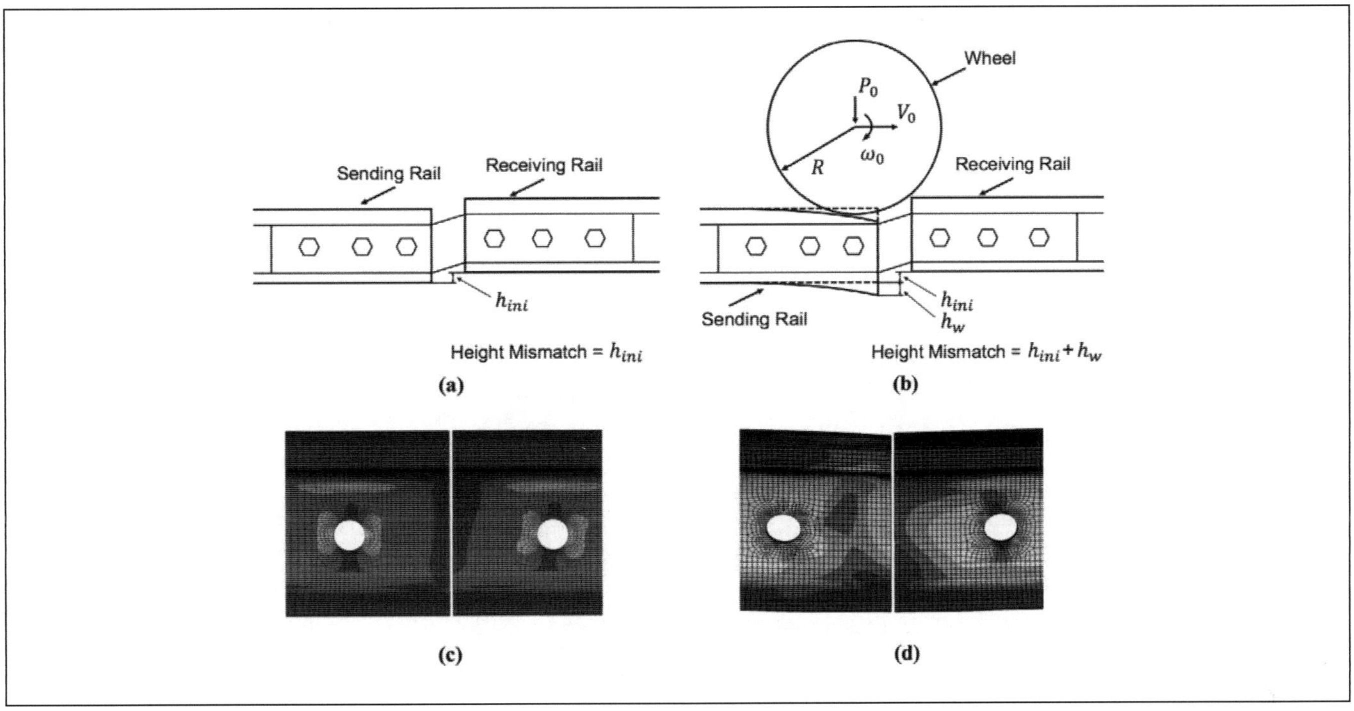

Figure 8. Schematic drawings and FEM examples of height mismatch caused by wheel.

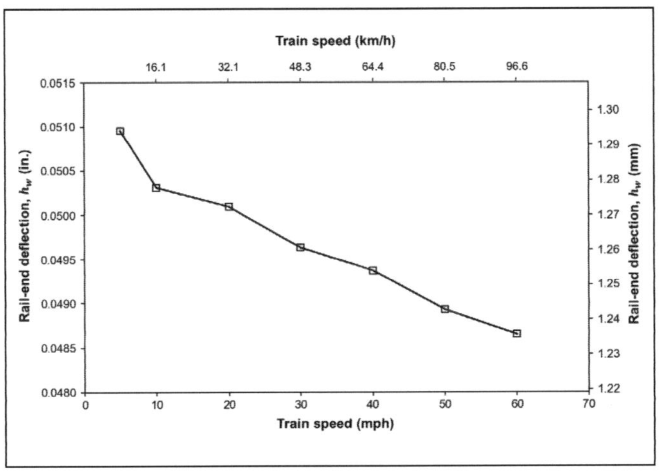

Figure 9. Rail height mismatch caused by wheel loading at various operation speeds, h_w.

receiving rail (Figure 8). Previous research has shown that the maximum contact force when the wheel hits the receiving rail increases as a function of height mismatch (*10*). Figure 9 shows that height mismatch increased when the speed decreased. Figures 8 and 9, when combined, show that when the operation speed reduced, the rail height mismatch would increase, and as a result, the maximum contact force could increase. Due to the rail height mismatch at the joint, and the relationship of the operation speed and the rail mismatch discussed above, the maximum contact force may not decrease monotonically as operation speed decreases, as illustrated again in Figure 10.

Based on the results shown in Figure 7*b* and *c*, the stresses calculated around the bolt-hole area were significantly smaller than the stresses around the upper fillet area, which was also shown in a previous study (*8*, 9). Based on this result, the rail-end upper fillet area was selected to perform the fatigue life analysis. Figure 11 presents the fatigue life of the upper fillet as predicted based on the loading history (see Figure 6 for example) with the same configurations but different train speeds simulated in this study. Assume trains continue to operate at a speed of 60 mph (96.6 km/h), the estimated fatigue life would be 6.6×10^5 wheel passes. If a speed restriction were issued, and the speed reduced to 40 mph (64.4 km/h) or 10 mph (16.1 km/h), the estimated fatigue life would increase to 4.2×10^6 or 2.9×10^6 wheel passes, an increase of 536% and 339%, respectively. However, if the speed was reduced to 30 mph (48.3 km/h) or 20 mph (32.1 km/h), the estimated fatigue life would decrease to 4.3×10^5 or 1.7×10^5 wheel passes, a reduction of 74% and 35%, respectively. Also, the trend line of estimated fatigue life shows that the fatigue life at rail-end upper fillet was highly correlated with mechanical responses of rail (Figure 7), and the estimated fatigue life was negatively correlated with the impact load applied to the rail joint (i.e. maximum wheel–rail contact force).

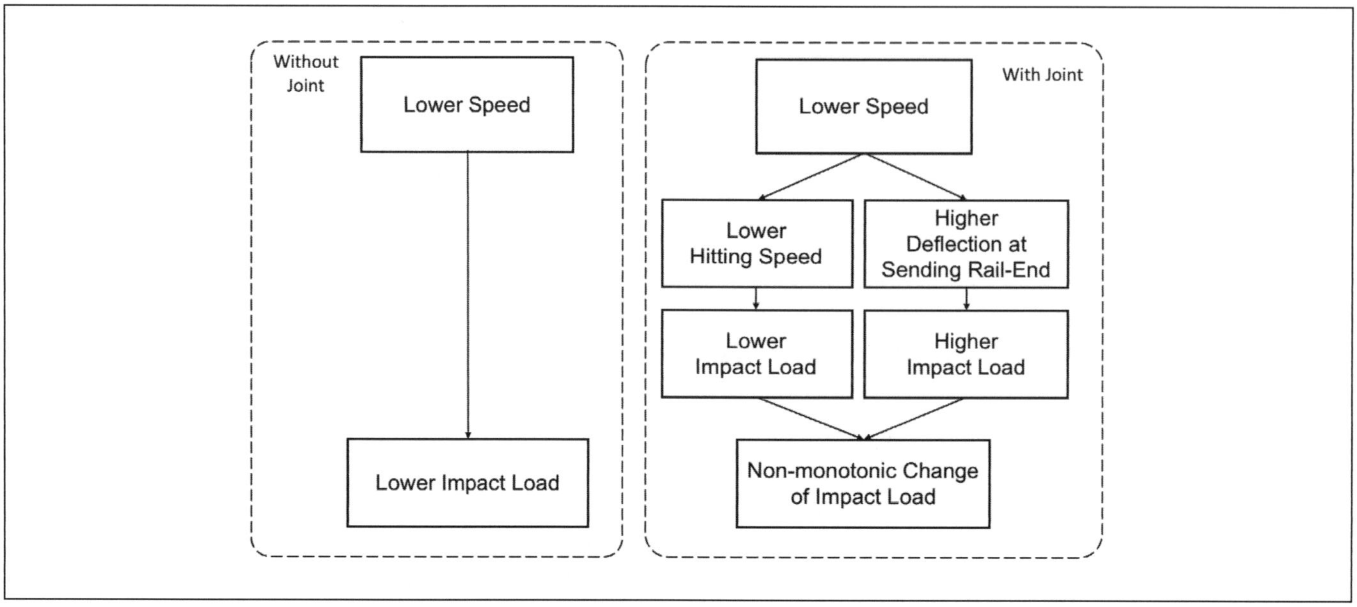

Figure 10. Relationship between operation speed and contact force.

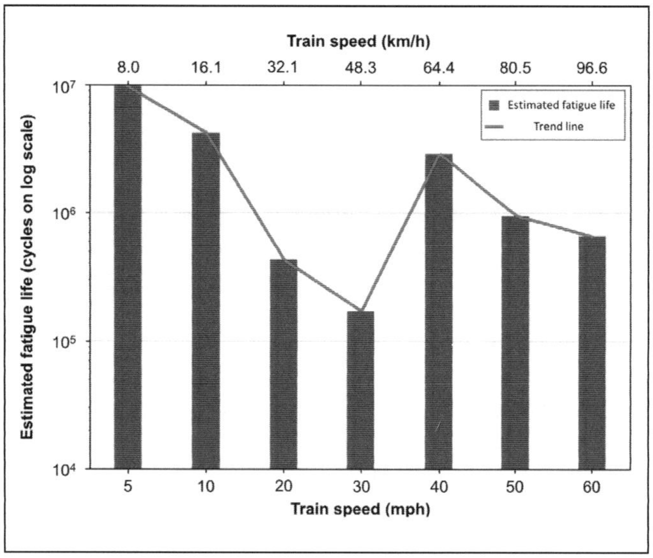

Figure 11. Estimated fatigue life at rail-end upper fillet at various operation speeds.

Conclusion

This paper presents results from detailed FE simulations of the contact force at the wheel–rail interface, the stress distribution around the rail-end bolt-hole, and rail-end upper fillet areas under moving wheel loadings. Seven different train speeds, varying from 5 mph (8.0 km/h) to 60 mph (96.6 km/h), were simulated and compared to investigate the relationship between the fatigue life and train speed. The following conclusions can be drawn from the results of this study:

- At a rail joint, the contact force at the wheel–rail interface does not change monotonically with the changing train speed. When train speed is reduced, the maximum contact force at the wheel–rail interface may not necessarily reduce.
- The non-monotonic relationship between the contact force at the wheel–rail interface and train speed was due to both the negative correlation of the rail height mismatch and the operation speed and the positive correlation of the dynamic load and the operation speed.
- When deciding to impose a temporary speed restriction, it should be understood that reducing train speed may not necessarily extend the fatigue life of a track with joints. If the operation speed is reduced inappropriately, the fatigue life of the rail joints could even be reduced.

Acknowledgments

This research was partially funded by WSP, under contract with New York City Transit Authority (NYCTA). Additional supporting funding was provided by National University Rail (NURail) Center, a USDOT-OST Tier 1 University Transportation Center. J. Riley Edwards has been supported in part by grants to the UIUC Rail Transportation and Engineering Center (RailTEC) from CN and Hanson Professional Services.

Author Contributions

The authors confirm contribution to the paper as follows: study conception and design: Yu Qian, Hao Yin; model design: Hao Yin, Yu Qian, Kaijun Zhu; data collection: Hao Yin; analysis

and interpretation of results: Hao Yin, Yu Qian, John Riley Edwards, and Kaijun Zhu; draft manuscript preparation: Hao Yin, Yu Qian, and John Riley Edwards.
All authors reviewed the results and approved the final version of the manuscript.

References

1. Mayville, R. A., R. G. Stringfellow, and S. Gallagher. *Development and Application of Rail Defect Fracture Models to Assess Remedial Actions.* Final Report No. DOT/FRA/ORD-93/33. U.S. Department of Transportation, Federal Railroad Administration, 1993.
2. Mayville, R. A., and R. G. Stringfellow. Numerical Analysis of a Railroad Bolt Hole Fracture Problem. *Theoretical and Applied Fracture Mechanics*, Vol. 24, No. 1, 1995, pp. 1–12.
3. Carolan, M. E., D. Y. Jeong, and A. B. Perlman. *Engineering Studies on Joint Bar Integrity, Part II: Finite Element Analysis.* American Society of Mechanical Engineers (ASME), Colorado Springs, Co., 2014, pp. 1–10.
4. National Transportation Safety Board. *Derailment and Subsequent Collision of Two Metro-North Passenger Trains.* Railroad Accident Brief 1409. National Transportation Safety Board, 2013.
5. Schlesinger, D. *Sources of Transportation Accident Information.* American Society of Mechanical Engineers (ASME), Columbia, S.C., 2016.
6. Smith, R. A. The Wheel-Rail Interface—Some Recent Accidents. *Fatigue & Fracture of Engineering Materials & Structures*, Vol. 26, No. 10, 2013, pp. 901–907.
7. Lazo, L. Metro Speed Restrictions Could Add to Commuters' Travel Time. *The Washington Post*, February 3, 2015. Retrieved from https://www.washingtonpost.com/news/dr-gridlock/wp/2015/02/03/metro-speed-restrictions-could-add-to-commuters-travel-time/?noredirect = on&utm_term = .44772747024
8. Zhu, K., J. R. Edwards, Y. Qian, and B. Andrawes. Finite Element Analysis of the Effects of Bolt Condition on Bolted Rail Joints Stresses. *Transportation Research Record: Journal of the Transportation Research Board*, 2016. 2545: 36–45.
9. Zhu, K., Y. Qian, J. R. Edwards, and B. Andrawes. Finite Element Analysis of Rail End Bolt-Hole and Fillet Stress on Bolted Rail Joints. *Transportation Research Record: Journal of the Transportation Research Board*, 2017. 2607: 33–42.
10. Soylemez, E., and K. Ciloglu. Influence of Track Variables and Product Design on Insulated Rail Joints. *Transportation Research Record: Journal of the Transportation Research Board*, 2016. 2545: 1–10.
11. Uzzal, R. U. A., W. Ahmed, and S. Rakheja. Dynamic Analysis of Railway Vehicle-Track Interactions Due to Wheel Flat with a Pitch-Plane Vehicle Model. *Journal of Mechanical Engineering*, Vol. 39, No. 2, 2008, pp. 86–94.
12. Zhang, Z., S. Wei, B. Andrawes, D. A. Kuchma, and J. R. Edwards. Numerical and Experimental Study on Dynamic Behaviour of Concrete Sleeper Track Caused by Wheel Flat. *International Journal of Rail Transportation*, Vol. 4, No. 1, 2016, pp. 1–19.
13. Yamaguchi, Y. *Tribology of Plastic Materials: Their Characteristics and Applications to Sliding Components.* Elsevier, Amsterdam, 1990.
14. Kernes, R. G., J. R. Edwards, M. S. Dersch, D. A. Lange, and C. P. Barkan. Investigation of the Dynamic Frictional Properties of a Concrete Crosstie Rail Seat and Pad and Its Effect on Rail Seat Deterioration (RSD). Presented at 91st Annual Meeting of the Transportation Research Board, Washington, D.C., 2011.
15. Wen, Z., X. Jin, and W. Zhang. Contact-Impact Stress Analysis of Rail Joint Region Using the Dynamic Finite Element Method. *Wear*, Vol. 258, No. 7, 2005, pp. 1301–1309.
16. Matake, T. An Explanation on Fatigue Limit Under Combined Stress. *Bulletin of JSME*, Vol. 20, No. 141, 1977, pp. 257–263.
17. Brown, M. W., and K. J. Miller. A Theory for Fatigue Failure under Multiaxial Stress-Strain Conditions. *Proceedings of the Institution of Mechanical Engineers*, Vol. 187, No. 1, 1973, pp. 745–755.
18. Zheng, Z. G., T. Sun, X. Y. Xu, S. Q. Pan, and S. Yuan. Numerical Simulation of Steel Wheel Dynamic Cornering Fatigue Test. *Engineering Failure Analysis*, Vol. 39, 2004, pp. 124–134.
19. Glinka, G., G. Wang, and A. Plumtree. Mean Stress Effects in Multiaxial Fatigue. *Fatigue & Fracture of Engineering Materials & Structures*, Vol. 18, No. 7–8, 1995, pp. 755–764.
20. Morrow, J. Fatigue Design Handbook. *Advances in Engineering*, Vol. 4, 1968, pp. 21–29.
21. Zhai, W. M. *Vehicle-Track Coupling Dynamics*. China Railway Press, Beijing, People's Republic of China, 2007.

The Standing Committee on Railway Maintenance (AR060) peer-reviewed this paper (18-03116).

The opinions expressed in this paper are solely those of the authors and do not represent the opinions of the funding agency.

TRR
JOURNAL OF THE TRANSPORTATION RESEARCH BOARD

Article

Transportation Research Record
2018, Vol. 2672(10) 96–105

DOI: 10.1177/0361198118776525
journals.sagepub.com/home/trr
SAGE

Effects of Pier Deformation on Train Operations within High-Speed Railway Ballastless Track–Bridge Systems

De Zhang[1], Junhua Xiao[1], and Xiao Zhang[1]

Abstract

The deformation of a bridge foundation (i.e. pier) for a ballastless track of a high-speed railway may cause additional irregularities within the track, thereby affecting train operation. By using a unit slab ballastless track bridge system as the research object, this study built a finite element model and a train–track dynamic interaction model. The additional rail deformation caused by the vertical or lateral deformation of a bridge pier was calculated by the finite element model, and then the effects on train operation due to the additional rail deformation were analyzed by the train–track dynamic model. It was found that the lateral deformation of a single pier should be of the most concern for the management and control of a high-speed railway. Specifically, when a pier suffered settlement and lateral deformation concurrently, the evaluation indices of train operation were primarily affected by the magnitude of the lateral deformation, and were only slightly affected by the settlement.

Bridge engineering accounts for a large proportion of Chinese high-speed railway lines, and provides a smooth and stable foundation for high-speed train operations. However, with the rapid development of high-speed railways and the large-scale construction of municipal infrastructure in China, a number of developing municipal facilities, such as trench excavations for new roads, tunnel construction for urban subways, and excavations of building foundation pits, are adjacent to, or pass under existing high-speed railway bridges (*1–3*), and consequently may cause settlement or lateral deformation of railway bridge foundations. The deformation of bridge foundations results in additional track irregularity, thereby affecting the running safety and riding comfort of high-speed trains.

Based on calculations of a train–bridge coupled dynamic model, the design code (*4*) stipulated limit values for the post-construction settlement of piers of simply supported beam bridges within high-speed railway systems. For example, the uniform settlement of each pier for a ballastless track should not exceed 20 mm, and the settlement difference between adjacent piers should not exceed 5 mm. However, the design code (*4*) did not stipulate any control limit for the lateral deformation of bridge piers, nor the relationship between such a control limit and the evaluation indices of running safety and riding comfort for train operation. So lack of index for lateral deformation of piers is not conducive to the management and control of bridge foundation deformation for high-speed railways.

Fortunately, in recent years, deformation of bridge foundations on high-speed railways and its effeccts on train operation has become a popular subject for research. Some investigations focus on the deformation of railway bridge foundations due to soil consolidation and the dynamic load of trains (*5*). In contrast, most research focuses on deformation of bridge foundations arising from the adjacent construction of municipal infrastructure, for example, piling of buildings, excavation of foundation pits, dewatering, and ground loading (*2, 3*). Further, some studies elucidate the additional track irregularity arising from the deformation of bridge foundations and the consequent dynamic interaction between trains, tracks, and bridges (*6–10*). For example, Chen et al. (*6*) deduced the mapping relationship between pier settlement and rail deformation for different slab ballastless track systems. Yau et al. (*7, 8*) and Ju (*9*) analyzed dynamic responses of trains and bridges due to non-uniform settlement of a single pier. Cao et al. (*10*)

[1]Key Laboratory of Road and Traffic Engineering of the Ministry of Education, Tongji University, Shanghai, China

Corresponding Author:
Address correspondence to Junhua Xiao: jhxiao@tongji.edu.cn

analyzed the deformation of a single pier within a ballast track bridge system arising from the excavation of adjacent foundation pits, as well as its impacts upon train operation.

However, there have been few investigations regarding the effects of lateral deformation of bridge foundations on train operation in ballastless track high-speed railway systems. In particular, research into the effects on train operation of the deformation of contiguous bridge foundations, and by a concurrence of lateral deformation and settlement of bridge foundations, is barely reported, although these phenomena are very common in practice.

By using a unit slab ballastless track system on a high-speed railway bridge as the research object, this study established a finite element model (including track structure, bridge superstructure, supports, and piers) and a train–track dynamic interaction model, and analyzed the additional rail deformation under various displacement patterns (e.g. the settlement or lateral deformation or both) and deformation magnitudes of bridge piers, and the corresponding responses of train operation. The research findings provide a useful reference for determining appropriate limit values used to control the deformation of bridge piers within ballastless track high-speed railway systems.

Calculation of Additional Deformation of Steel Rail Induced by Bridge Pier Deformation

It is an important prerequisite to determine the mapping relationship between bridge pier deformation and steel rail deformation. Chen et al. (*6*) assumed that the bridge and steel rail were of simply supported beam structures, the track slab was of a free beam structure, and linear springs were used for the fasteners and packing layer, as shown in Figure 1. Accordingly, Chen et al. (*6*) proposed an analytical equation for the mapping relationship between the settlement of a bridge pier and the deformation of steel rail for a unit slab ballastless track.

Using the unit slab ballastless track system on a bridge as the research object, this study determined the relationship between the vertical/lateral deformation of a bridge pier and rail deformation via finite element simulations, and compared the numerical simulation result of rail vertical deformation with the calculation result obtained by analytical equation (*6*), so as to verify the competence of the numerical model and parameters.

Finite Element Numerical Model

Using the finite element software ABAQUS, this paper developed a finite element model (including a unit slab ballastless track, bridge superstructure, support, and pier), and subsequently calculated the additional deformation of a steel rail arising from the deformation of a bridge pier.

A four-span simply supported beam bridge was selected to build a spatial finite element model. The standard length of each span was 32.6 m, the net width of the inner side of the bridge floor was 8.8 m, and the beam width was 12.0 m. The beam depth at the core of the bridge cross section was 3.078 m, the center of the support was 55 cm away from the beam end in the longitudinal direction, the transverse distance between the centers of two supports under the same box beam was 4.5 m, and the pier height was 20 m. The geometrical parameters of the simply supported box beam and bridge piers were available from the standard drawing for the pre-stressed concrete simply supported beam of the ballastless track (*4*). The railway route was double-straight-line, and the design running speed was 350 km/h. Figure 2 shows the cross section of the bridge structure.

In the finite element model, the steel rail was simulated by spatial beam elements, the track slab, base slab, box beam, and bridge pier were all simulated by solid elements, and linear spring elements were used for both the fastener system and packing layer.

The parameters of CRTS II standard slab track structure were employed, and the section parameters of 60 kg/m steel rail were applied (*4*). The strength grades of the concrete for track slab, base slab, box beam, and bridge pier were C60, C40, C55, and C30, respectively, and the elastic modulus of the mortar was 200 MPa.

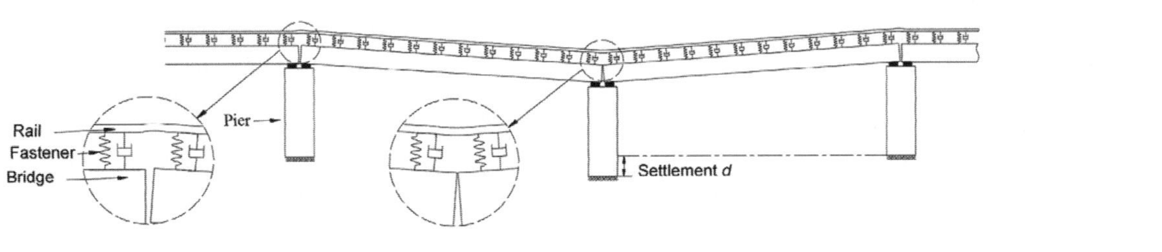

Figure 1. Rail deformation caused by pier settlement in a unit slab ballastless track bridge system, adapted from Chen et al. (6).

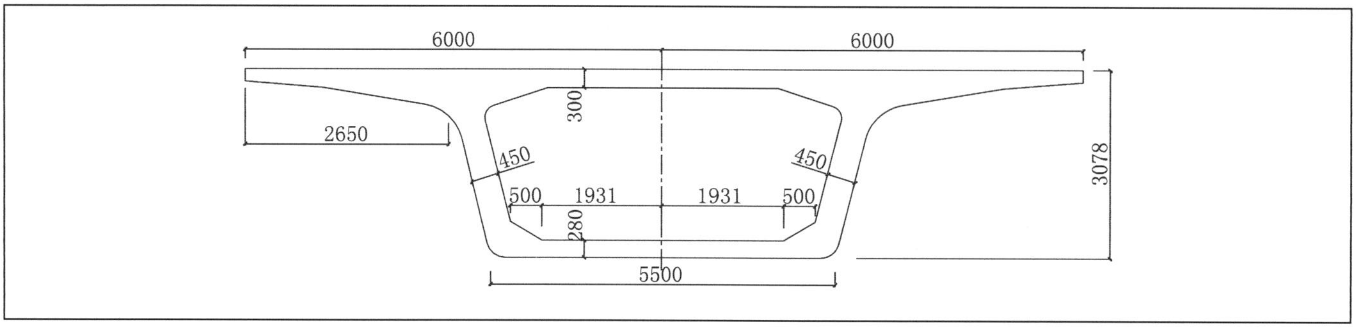

Figure 2. Mid-span section of a 32.6 m long simply supported beam bridge (unit: mm).

Table 1. Spring Stiffness of Fixed and Movable Supports in Different Directions

	GD	DX	ZX	HX
Vertical stiffness (N/m)	7.5×10^8	7.5×10^8	7.5×10^8	7.5×10^8
Lateral stiffness (N/m)	6.0×10^8	1.5×10^7	6.0×10^8	1.5×10^7
Longitudinal stiffness (N/m)	6.0×10^8	3.0×10^6	3.0×10^6	1.2×10^7
Vertical rotation (N·m^2)	1.0×10^{11}	1.0×10^{11}	1.0×10^{11}	1.0×10^{11}
Lateral rotation (N·m^2)	0	0	0	0
Torsional stiffness (N·m^2)	1.0×10^{11}	1.0×10^{11}	1.0×10^{11}	1.0×10^{11}

Each bridge support was a KTPZ8360 V-shaped pot-type rubber support, and the vertical/lateral/longitudinal spring stiffness was simulated according to the relationship between the vertical/horizontal bearing capacity and the allowable displacement. Table 1 lists the stiffness values of the fixed and movable supports in different directions. In Table 1, GD indicates a fixed support, HX indicates a lateral movable support, ZX indicates a longitudinal movable support, and DX indicates a multidirectional movable support.

The settlement or lateral deformation of a bridge pier was simulated by applying a forced displacement to the bottom node of the pier, as shown in Figure 3.

Calculations of Steel Rail Deformations

Assuming that the settlement of pier 3 (Figure 3) was 5 mm, the rail deformation was calculated by adopting the finite element model and the analytical method in Chen et al. (*6*), respectively. Figure 4*a* shows the compared calculation results. As shown in Figure 4*a*, the steel rail was deformed arising from the settlement of the bridge pier; specifically, at two ends of the deformation zone, the steel rail was slightly up-warped, and in the middle of the deformation zone the steel rail was shaped as a mild transition curve. According to Figure 4*a*, it can also be seen that the deformation curve obtained by finite element modeling was basically the same as that obtained by the analytical method. This shows that the finite element model and parameters were reasonable.

Additionally, assuming that piers 2 and 3 suffered a settlement of 5 mm at the same time, the calculated rail deformation was compared with that arising from the settlement of pier 3 alone, as shown in Figure 4*b*. It can be seen that the deformation curve of steel rail arising from concurrent settlement of two contiguous piers had a larger wavelength than that arising from the settlement of a single pier, and the former was smoother at both ends and in the middle of the deformation zone than the latter.

Calculation of Train–Track Dynamic Interaction under the Action of Foundation Deformation

Track Irregularities

Geometric irregularities of the track were generated by German low-interference high-speed track spectrum (*11*), and the track irregularities were superposed on the additional rail deformation arising from bridge pier deformation. Then, the track irregularities and additional rail deformation were used as inputs in the train–track calculation model.

In numerical simulations, the vertical/lateral deformation of piers was changed within a range of 5 mm to 20 mm, with an interval of 5mm. For a single-pier deformation, the deformation was applied to pier 3 (Figure 3). For double-pier deformations, the deformations were

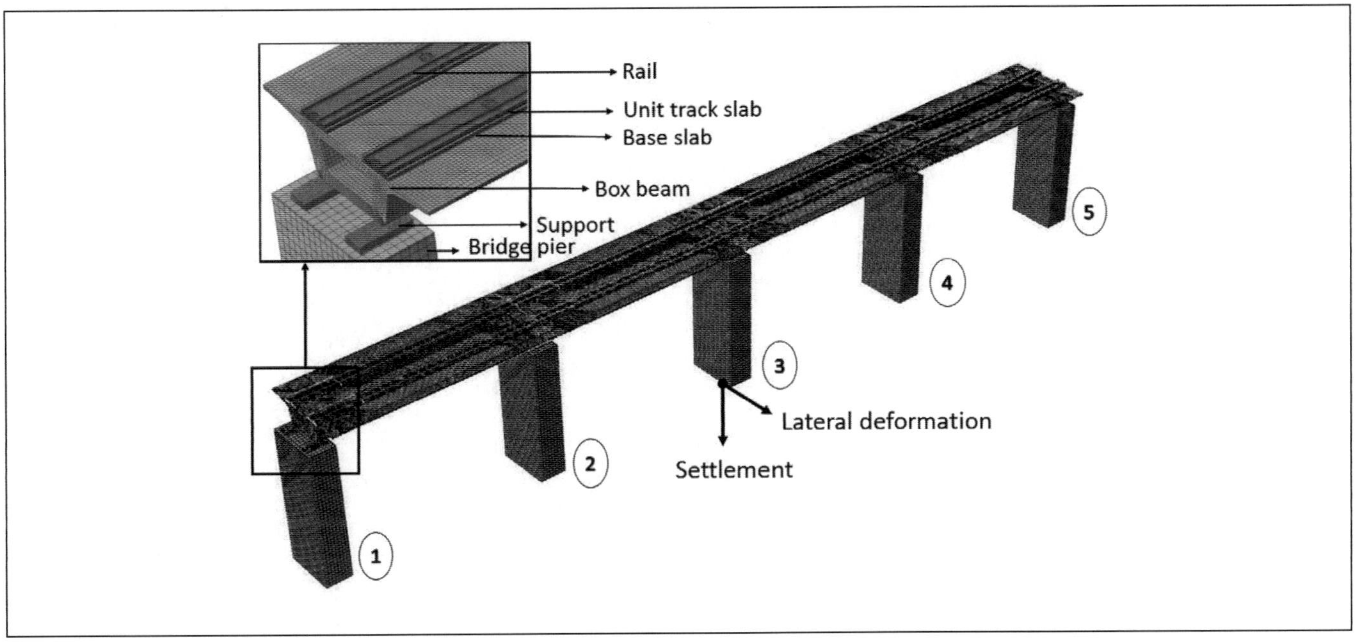

Figure 3. Finite element model for the ballastless track bridge system.

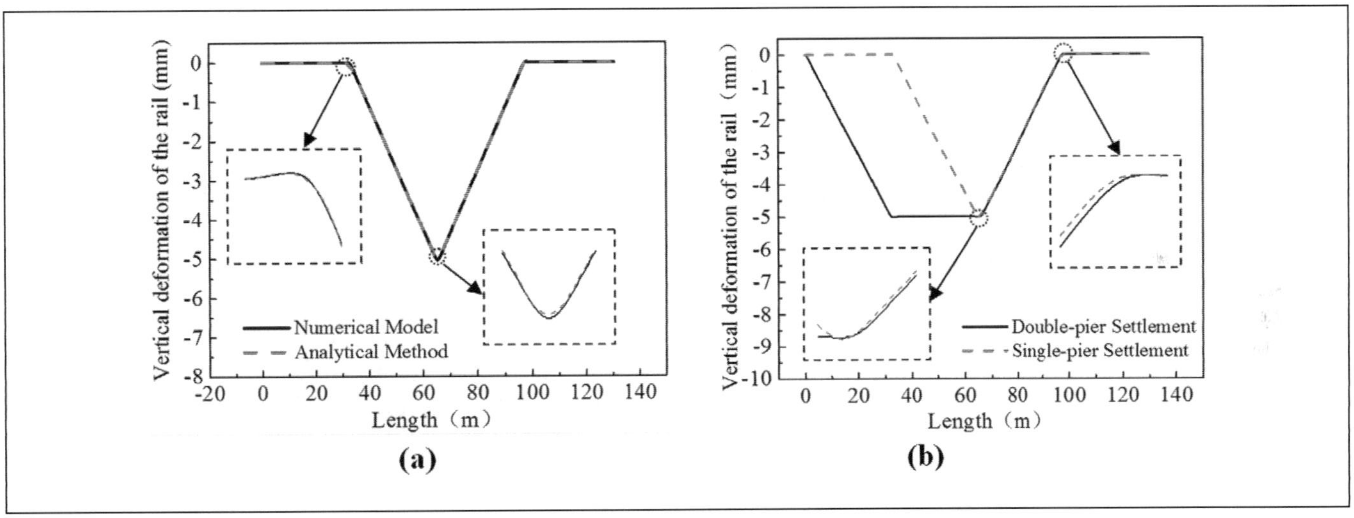

Figure 4. Calculation results of rail deformation arising from bridge pier settlement: (*a*) comparison between the results of numerical simulation and analytical method and (*b*) rail deformations due to single-pier settlement and double-pier settlements.

applied to piers 2 and 3 concurrently. Taking the single-pier deformation as an example, the curves of the additional rail deformation superposed on German track spectrum are shown in Figure 5.

Train–Track Dynamic Model

The train–track dynamic model was a combination of a train model and a track model. The relationship between the train and track was determined by a wheel-track relationship model, which assumed close contact in the vertical direction and Kalker creep-slippage in the horizontal direction herein (*11*). The train model was a multi-freedom vibration system that comprised one vehicle body, two bogies, four wheel sets, and a spring-damper suspension device. During vibration process, the elastic deformations of the vehicle, bogies, and wheel axes were not considered. Five degrees of freedom: yaw Y_c, side roll θ_c, head shake ϕ_c, heave Z_c, and nod φ_c, were considered for the vehicle. Five degrees of freedom: yaw Y_t, side roll θ_t, head shake ϕ_t, heave Z_t, and nod φ_t, were considered for a single bogie. Three degrees of freedom: yaw Y_w, side roll θ_w, and heave Z_w, were considered for a single wheel set. Therefore, the four-axis vehicle had a total of 27

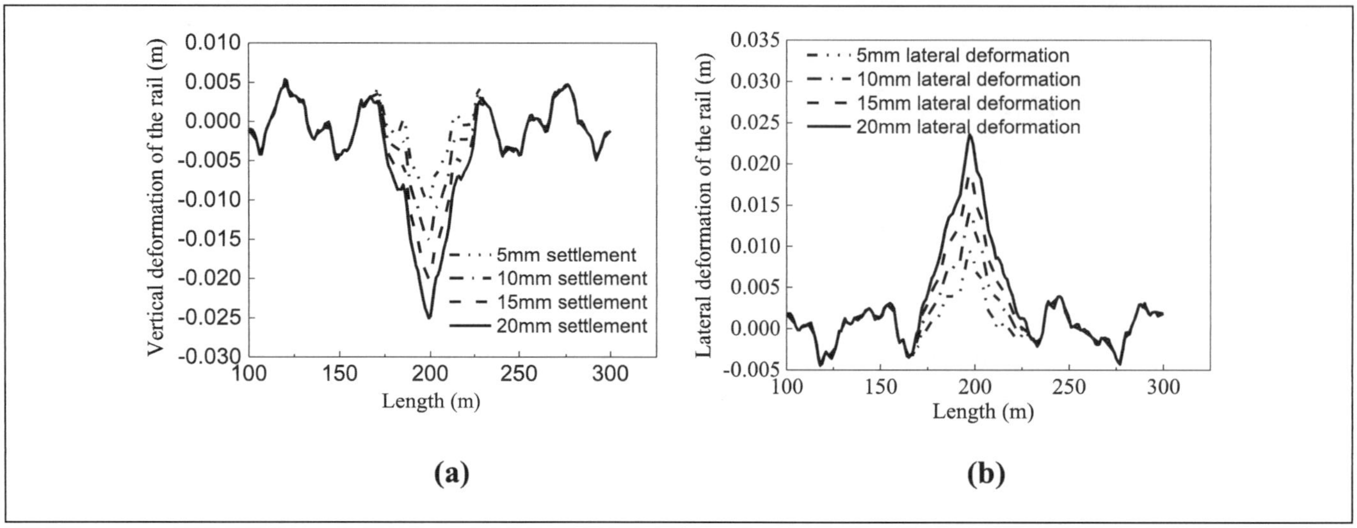

Figure 5. Deformation of steel rail arising from deformation of a single pier: (*a*) vertical deformation and (*b*) lateral deformation.

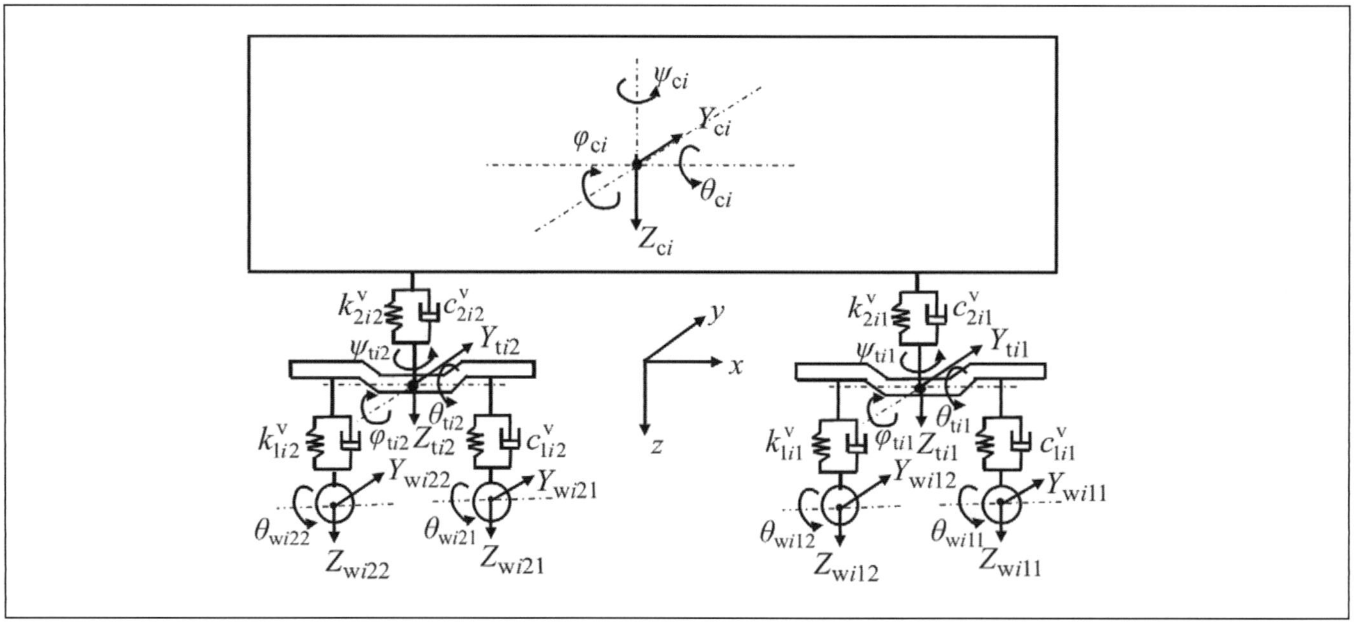

Figure 6. A 27-degree-of-freedom train model for a four-axle vehicle.

degrees of freedom. The vehicle was connected to two bogies via springs and dampers, and the bogies were connected to the corresponding wheel sets via springs and dampers, as shown in Figure 6.

Here, a Germany-made ICE3 high-speed train was selected as the research object, and the technical parameters for the train were described in Cao et al. (*10*) In order to consider the effect of train speed, the train speeds were set to 100 km/h, 250 km/h, and 350 km/h, respectively.

The steel rail was approximately considered as a Euler beam. The equations of the vehicle, the track, and the wheel-track relation were combined to constitute a large nonlinear dynamic equation set. The dynamic model is available and can be expressed as the following matrix:

$$[M]\{A\} + [C]\{V\} + [K]\{X\} = \{Q\} \qquad (1)$$

where $[M]$, $[C]$, and $[K]$ are the respective mass matrix, damp matrix, and stiffness matrix of the train–track space coupling dynamic system, $\{X\}$ is the general displacement vector of the system, $\{V\}$ is the velocity vector of the system, $\{A\}$ is the acceleration vector of the system, and $\{Q\}$ is the general load vector of the system and was

Table 2. Evaluation Indices for the Safety and Comfort of Train Operation

Item		Value
Safety indices	Derailment coefficient	$Q/P \leq 0.8$
	Rate of wheel load reduction	$\Delta P/\overline{P} \leq 0.6$
	Transverse wheelset force	$H \leq 0.85(10 + P_{st}/3)$
Comfort indices	Vertical vibration acceleration of the vehicle	0.13 g
	Transverse vibration acceleration of the vehicle	0.10 g

attained according to the wheel-track spatial contact relationship. The mass matrix, damp matrix, and stiffness matrix, as well as the expressions for the displacement vector, velocity vector, and acceleration vector were provided in Zhai (*11*).

When the train moves on the track, the position of each wheel is constantly changed. The dynamic equation of the space coupling system is a nonlinear differential equation set, which can be solved via the Newmark-β step-by-step integration method (*11*).

Analysis Results of Dynamic Calculations

By using a four-span standard simply supported beam bridge, this paper studied the effects on train operation of bridge pier deformation under the following circumstances: (1) a single pier and two contiguous piers respectively suffering vertical/lateral deformation; (2) a single pier suffering vertical and lateral deformations concurrently.

Limit Values Specified by the Design Code

Table 2 lists the evaluation indices for the safety and comfort of train operation specified by the design code (*4*).

Here, Q and P are lateral and vertical force acting on the flange of the rail climbing wheel respectively, $\overline{P}$ is average wheel weight, ΔP is half of the two wheel load difference, and P_{st} indicates the static axle load. The static axle loads of Germany-made ICE3 motor train and trailer are 156.96 kN and 143.22 kN, respectively. Therefore, the allowed transverse wheelset forces are 52.97 kN and 49.08 kN, respectively.

The derailment coefficient and rate of the wheel load reduction are derived from the Nadal formula for static or quasi-static conditions (i.e. transverse wheelset force $H = 0$). The regulatory requirements above are theoretically conservative but necessary for safety margin. Moreover, the limit of wheel load reduction rate can extend to 0.8 when the train speed exceeds 350 km/h, according to the explanation of the design code (*4*).

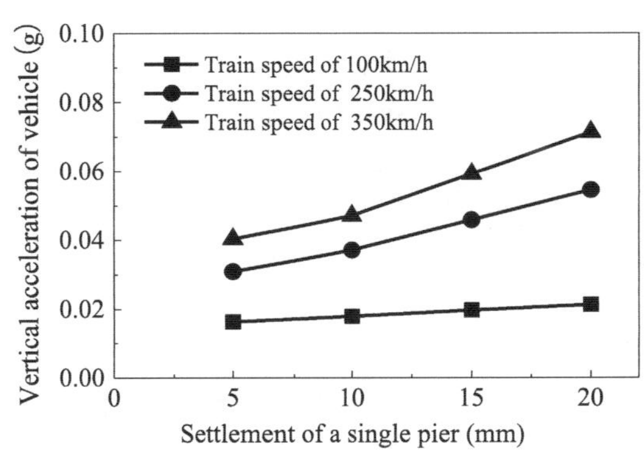

Figure 7. Relationship between the vertical acceleration of a vehicle and the settlement of a single pier.

Effects of Pier Deformation on Train Operation

The following sections analyze the effects of pier deformation on train operation. The analysis took into account three factors: train speed, the pattern of deformation of piers, and the magnitude of deformation of piers.

Mere Unidirectional Deformation of a Single Pier. Taking the relationship between the vehicle acceleration and the pier settlement as an example, Figure 7 shows the variations of the maximum vertical acceleration of a vehicle at various running speeds with the increase of the settlement of a single pier. It can be seen that, at the same speed, the maximum vertical acceleration of the vehicle had a linear positive correlation with the magnitude of the pier settlement. In addition, the higher the train speed, the more rapidly the maximum vertical acceleration of the vehicle increased.

However, both the results in Figure 7 and the calculation results from other evaluation indices for train operation show that when the settlement of a single pier was set within 20 mm, the safety indices did not exceed the specified limits, regardless of the train speed. In addition, the values of safety indices would not significantly

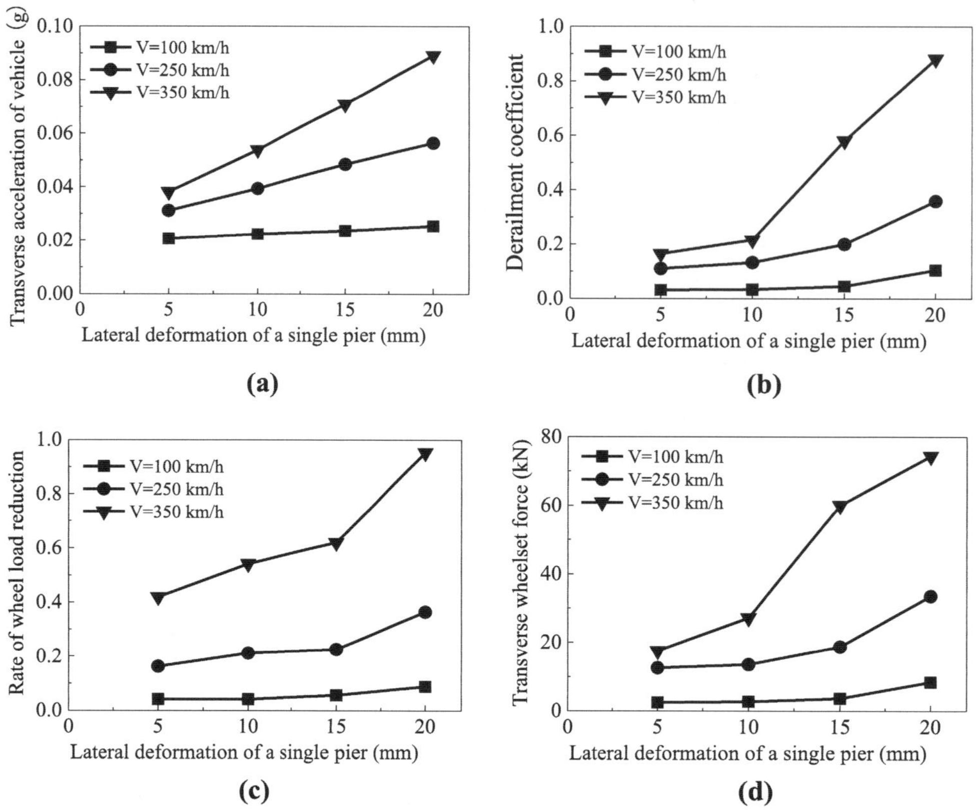

Figure 8. Effects of lateral deformation of a single pier on train operation: (*a*) transverse acceleration, (*b*) derailment coefficient, (*c*) rate of wheel load reduction, and (*d*) transverse wheelset force.

increase with increasing magnitude of pier settlement. The findings were similar to the conclusions set forth in Chen et al. (*6*) and Cao et al. (*10*) by their train–track dynamic models. Therefore, the calculation method, model, and parameters herein were appropriate. As the design code (*4*) stipulates the control standard of pier settlement according to the dynamic responses of train operation under various values of pier settlement, this study mainly discusses the change laws on dynamic responses of train operation arising from the lateral deformation of piers.

Figure 8 shows the curves depicting changes in the transverse acceleration, derailment coefficient, rate of wheel load reduction, and transverse wheelset force when a single pier was deformed laterally at varying magnitudes (5 mm, 10 mm, 15 mm, and 20 mm) and under varying train speeds (V = 100 km/h, 250 km/h, and 350 km/h).

Comparing Figure 8*a* with Figure 7, it can be seen that the lateral deformation of a single pier affected the vehicle acceleration more significantly than the pier settlement did. In addition, the values of the safety indices for train operation increased more drastically with increasing magnitude of pier lateral deformation, and were prone to exceed the specified limits under high train speeds. For example, when the train speed was 100 km/h, the values of safety indices varied slightly with pier lateral deformation. However, when the train speed was 350 km/h and the magnitude of pier lateral deformation reached 15 mm, both the rate of wheel load reduction and transverse wheelset force exceeded their specified limits. Actually, when the magnitude of a pier lateral deformation was 15 mm, the calculated deviation of track direction was approximately 4 mm, which just met the allowed maximum deviation (i.e. 4 mm) of the standard track direction for high-speed railways specified in the technical code (*12*). Therefore, the calculation results of the evaluation indices for train operation by train–track dynamic model herein were proven to be reliable.

Lateral Deformations of Two Contiguous Piers. In practice, multiple contiguous piers may be deformed concurrently. Figure 9 shows the curves representing the evaluation indices of train operation under the following circumstances: (a) train running at speeds (*V*) of 250 km/h and

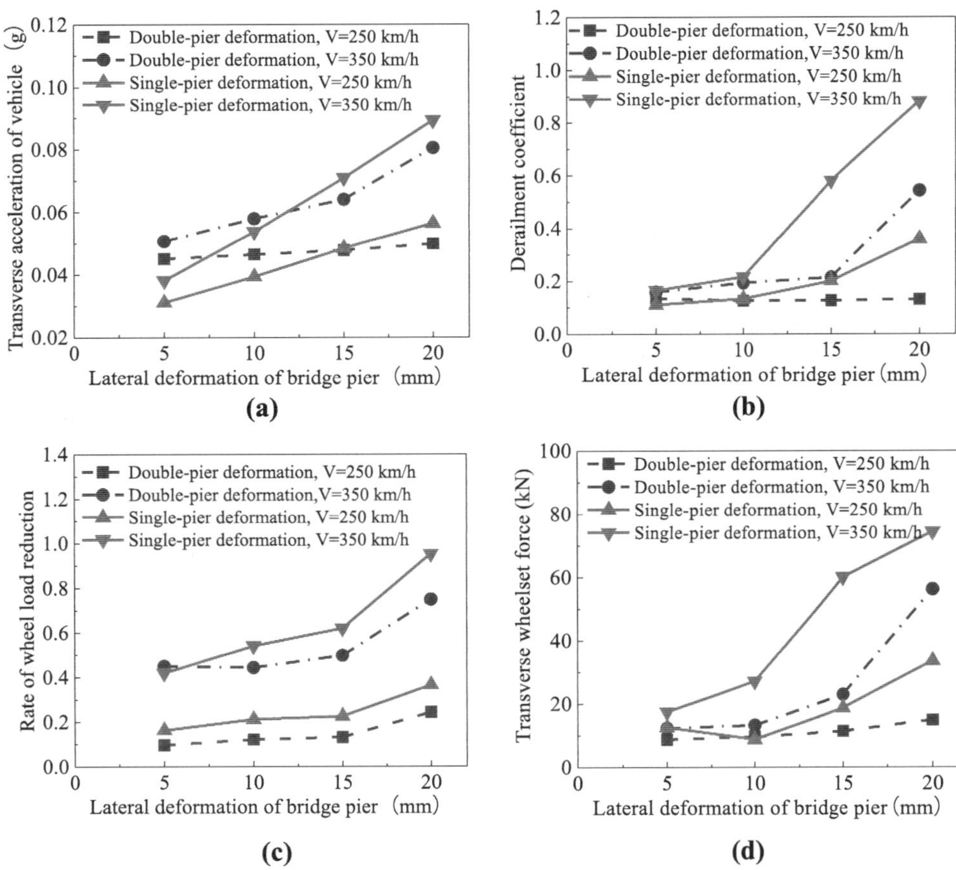

Figure 9. Effects of lateral deformation of two contiguous piers on train operation: (*a*) transverse acceleration, (*b*) derailment coefficient, (*c*) rate of wheel load reduction, and (*d*) transverse wheelset force.

350 km/h; (b) two contiguous piers laterally deforming to various magnitudes.

Overall, the values of the evaluation indices varied with the lateral deformations of two contiguous piers significantly less than with the lateral deformation of a single pier. This was mainly due to the fact that the curve of the additional rail deformation arising from deformation of a single pier had a small wavelength, and was very steep at both ends and in the middle of the deformation zone (Figure 4). After two contiguous piers suffered a lateral deformation of 15 mm, the values of safety indices began to increase rapidly.

Bidirectional Deformations of a Single Pier. The bridge piers often suffer vertical settlement and lateral deformation concurrently due to ground consolidation and the influence of adjacent constructions. Figure 10 shows the impacts on the safety indices for train operation when a single pier suffered a settlement of 5 mm and 20 mm (S means settlement and V means velocity of train) along with a superposition of lateral deformation.

As mentioned above, the settlement of a pier within a range of 0 to 20 mm did not substantially affect the evaluation indices of train operation; consequently, when a single pier suffered settlement and lateral deformation concurrently, the evaluation indices were primarily affected by the magnitude of the lateral deformation, and were slightly affected by the settlement. In particular, the effects of the pier settlement could essentially be ignored when the train speed was relatively low.

Conclusions

Using a multi-span simply supported railway bridge and a unit slab ballastless track system as the research subjects, this study developed a finite element model (including a ballastless track structure, bridge superstructure, support, and pier) and a train–track dynamic interaction model, and subsequently analyzed the effcts of vertical/lateral deformation of bridge piers on the operation of high-speed trains. In summary, this study drew the following conclusions:

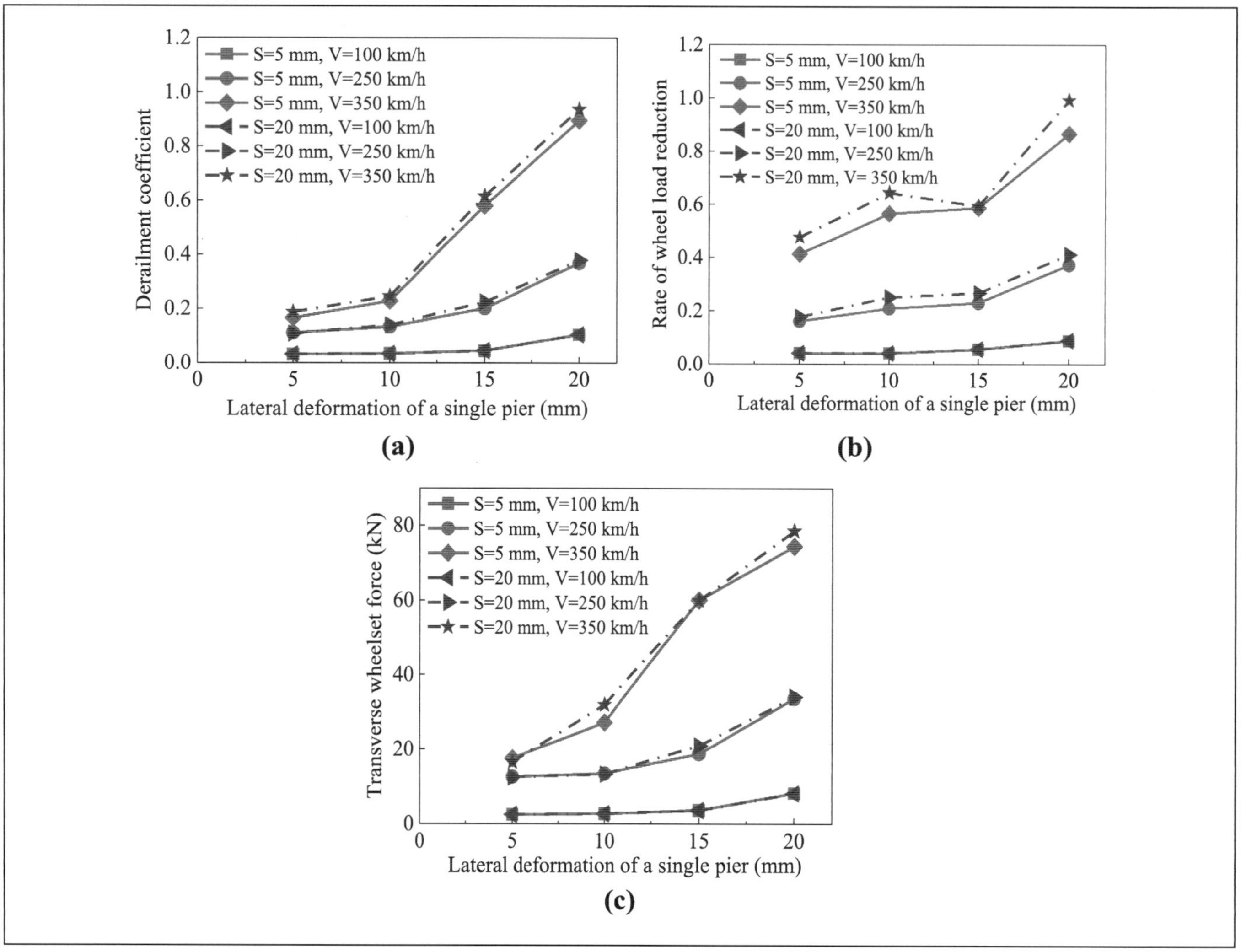

Figure 10. Effects on train operation of a combination of settlement and lateral deformation of a single pier: (*a*) derailment coefficient, (*b*) rate of wheel load reduction, and (*c*) transverse wheelset force.

1. The deformation of a single pier affected the safety of train operation more severely than the concurrent deformations of double or multiple contiguous piers did.
2. The lateral deformation of a bridge pier affected train operation more significantly than pier settlement. The values of safety and comfort evaluation indices for train operation did not exceed the specified limits when a single pier merely suffered a settlement within the extent of 20 mm. However, when a single pier suffered a lateral deformation of 15 mm, the values of the safety indices exceeded their specified limits for trains at high speeds.
3. When a pier suffered settlement and lateral deformation concurrently, the evaluation indices of train operation were primarily affected by the magnitude of the lateral deformation, and were only slightly affected by the settlement.

Therefore, the lateral deformation of a single pier should be of the most concern for the management and control of bridge foundation deformation for high-speed railways.

Author Contributions

The authors confirm contribution to the paper as follows: study conception and design: Junhua Xiao; De Zhang, data collection: De Zhang; analysis and interpretation of results: Junhua Xiao, De Zhang, Xiao Zhang; draft manuscript preparation: De Zhang.

References

1. Wang, X., L. Q. Wu, Y. Zhou, and Y. R. Wang. The Long-Term Settlement Deformation Automatic Monitoring System for the Chinese High-Speed Railway. *Shock and Vibration*, Vol. 2015, No. 2, 2015, pp. 1–12.
2. He, W., Y. L. Duan, L. W. Deng, and W. G. Zhou. Risk Assessment and Early-Warning System for High-Speed

Railway During the Construction and Operation of Underpass Bridges. *Journal of Performance of Constructed Facilities*, Vol. 30, No. 1, 2016, p. C4015003.
3. Peng, H., J. C. Wei, and B. Zhang. Settlement Control and Impact Analysis of the Subway Shield Under-Passing Beijing-Shanghai High Speed Railway Bridge. *Applied Mechanics and Materials*, Vol. 226–228, 2012, pp. 1655–1660.
4. National Railway Administration of China, Code for Design of High Speed Railway (TB10621-2014) (English version), China Railway Press, Beijing, China, 2015.
5. Yang, Q., W. M. Leng, S. Zhang, R. S. Nie, L. M. Wei, C. Y. Zhao, and W. Z. Liu. Long-Term Settlement Prediction of High-Speed Railway Bridge Pile Foundation. *Journal of Central South University*, Vol. 21, No. 6, 2014, pp. 2415–2424.
6. Chen, Z. W., W. M. Zhai, C. B. Cai, and Y. Sun. Safety Threshold of High-Speed Railway Pier Settlement Based on Train-Track-Bridge Dynamic Interaction. *Science China Technological Sciences*, Vol. 58, No. 2, 2015, pp. 202–210.
7. Yau, J. D. Response of a Train Moving on Multi-Span Railway Bridges Undergoing Ground Settlement. *Engineering Structures*, Vol. 31, No. 9, 2009, pp. 2115–2122.
8. Yau, J. D., and L. Fryba. Interaction Dynamics of a High-Speed Train Moving on Multi-Span Railway Bridges with Support Settlements. *Proceedings of ISMA (International Conference on Noise and Vibration Engineering)*, Leuven, Belgium, 2014, pp. 963–972.
9. Ju, S. H. 3D Analysis of High-Speed Trains Moving on Bridges With Foundation Settlements. *Archive of Applied Mechanics*, Vol. 83, No. 2, 2013, pp. 281–291.
10. Cao, Y. M., H. Xia, W. L. Lu, K. P. Wang, and R. Calcada. A Numerical Method to Predict the Riding Comfort Induced by Foundation Construction Close to a High-Speed-Line Bridge. *Proceedings of the Institution of Mechanical Engineers Part F-Journal of Rail and Rapid Transit*, Vol. 229, No. 5, 2015, pp. 553–564.
11. Zhai, W. M. *Vehicle-Track Coupled Dynamics*, 4th ed. Science Press, Ltd., Beijing, China, 2014.
12. Ministry of Railways of China, Line Maintenance Rules for Ballastless Track of High Speed Railway (TG/GW 115-2012), China Railway Press, Beijing, China, 2012.

The Standing Committee on Railroad Track Structure System Design (AR050) peer-reviewed this paper (18-01699).

Article

TRR
JOURNAL OF THE TRANSPORTATION RESEARCH BOARD

Transportation Research Record
2018, Vol. 2672(10) 106–115

DOI: 10.1177/0361198118790643
journals.sagepub.com/home/trr
SAGE

Evaluation of Ballast Behavior under Different Tie Support Conditions using Discrete Element Modeling

Wenting Hou[1], Bin Feng[1], Wei Li[2], and Erol Tutumluer[1]

Abstract

This paper presents findings of a railroad ballast study using the discrete element method (DEM) focused on mesoscale performance modeling of ballast layer under different tie support conditions. The simulation assembles ballast gradation that met the requirements of both American Railway Engineering and Maintenance-of-Way Association (AREMA) No. 3 and No. 4A specifications with polyhedral particle shapes created similar to the field-collected ballast samples. A full-track model was generated as a basic model, on which five different support conditions were studied in the DEM simulation. Static rail seat loads of 10 kips (44.5 kN) were applied until the DEM model became stable. The pressure distribution along the tie-ballast interface predicted by DEM simulations was in good agreement with previously published results backcalculated from laboratory testing. Static rail seat loads of 20 kips (89 kN) were then applied in the calibrated DEM model to evaluate in-track performance. Results from the validated full-track DEM simulations indicated that only a small portion of ballast particles participated in load distribution under static loading. Particles on the shoulders and particles in the areas with poor support conditions often experience no or very low contact forces. Load transfer mechanisms investigated through a contact force network varied greatly among different support conditions: lack of rail seat support, full support, and lack of center support had wider force distribution angles than the high center binding and severe center binding conditions. The severe center binding scenario was found to be the most critical support condition in terms of causing the highest tie-ballast contact pressure exceeding 30% of the AREMA allowable pressure.

In the U.S., most railway corridors are constructed as ballasted track. Ballasted track consists of two major parts, namely, superstructure and substructure. Superstructure includes rail, fastening system, and tie; while substructure includes ballast, subballast, and subgrade. Ballast is an essential layer of railroad track substructure which provides load distribution as well as drainage. Ballast layer under repeated train moving loads undergoes shakedown and permanent deformation, which eventually lead to ballast degradation and track settlement. As indicated, the excessive dynamic or impact wheel loading is often the primary cause for ballast related track geometry deterioration (*1–3*). Therefore, it is important to better understand deterioration mechanisms of the particulate nature of ballast behavior and the effects of its degradation modes on track condition assessment.

The discrete element method (DEM) has been proved as an effective method to simulate ballast layer deformation behavior under repeated train loading (*4–6*). Previous research studies on the DEM simulations of ballast behavior have focused on the particulate nature of ballast, for example, ballast particle size or gradation (*5*), ballast particle shape (*7*), degraded ballast (*8*), and geogrid reinforced ballast (*9*), and often studied ballast behavior from specific laboratory tests (*10*). Most of the studies listed above only considered full support interface between ballast and tie, which is certainly not representing the true individual particle point contact in-track support condition. McHenry et al (*11*) quantitatively characterized ballast-tie interface pressure distribution in both laboratory testing and in-track testing using matrix-based tactile surface sensors. Results confirmed the variability of support conditions (load distribution) considering contact area as well as peak pressure. They also

[1]Department of Civil and Environmental Engineering, University of Illinois at Urbana-Champaign, Urbana, IL
[2]College of Civil Engineering and Architecture, Zhejiang University, Hangzhou, China

Corresponding Author:
Address correspondence to Erol Tutumluer: tutumlue@illinois.edu

found that the location of measured peak pressures disagreed with the pressure distribution recommended in the American Railway Engineering and Maintenance-of-Way Association (AREMA) *Manual for Railway Engineering* (*12*).

In the past, few studies concentrated on ballast support conditions. Kaewunruen and Remennikov investigated the influence of three types of ballast support conditions to concrete tie flexural performance by using finite element modeling, including half side support, symmetric support, and non-symmetric support conditions (*13*). Results indicated that the bending moments were affected significantly by the uneven ballast stiffness distribution. It was found that the nominal bending moment at mid-span could be eight times larger than the standard design moment at the most critical support condition. Gao et al recently introduced a ballast support condition backcalculator, which was used to quantify ballast pressure distributions under ties (*14*). In their study, strain gages were installed on concrete ties to measure bending moments. Together with the measured rail seat loads, the ballast support condition backcalculator could estimate the in-track ballast support conditions beneath the tie. Figure 1 shows the laboratory experiment setup in their study. By varying the location of rubber pads underneath the crosstie, they simulated five different support conditions, including full support, lack of rail seat support, high center binding, lack of center support, and severe center binding.

Note that no previous literature can be found to focus on the ballast mesoscale behavior and investigate the ballast particle contact force network, such as all normal and shear forces between two contacting particles within the ballast layer, under different ballast support conditions. This paper intends to fill this gap in our understanding by conducting DEM simulations of particulate nature ballast behavior under loading of strain gage instrumented concrete ties with known support conditions in the laboratory.

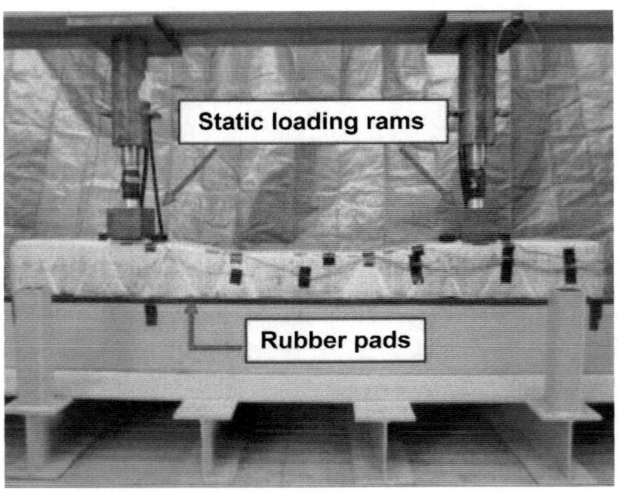

Figure 1. Laboratory experiment setup for simulating different support conditions (*14*).

Objective and Scope

The main objective of this paper is to investigate ballast-tie interaction under different tie support conditions and to better understand the stress distribution in ballast layer. The ultimate goal is to determine the most critical tie support scenarios leading to tie failures and track geometry deterioration and, accordingly, provide guidance to railway maintenance and rehabilitation activities to ensure better performance of ballasted track. To achieve these interconnected objectives, DEM modeling has been adopted in the scope to simulate the particulate nature ballast layer behavior under loading for five different support conditions. The five different support conditions were chosen based on a previous laboratory study conducted by Gao et al and are related to recorded field track conditions (*14*). The use of the same support conditions in DEM modeling allows a comparison of the predictions with the results from experiments, which essentially provides a methodology to calibrate the DEM model parameters. Next, the validated DEM model is utilized to produce more detailed mesoscale analyses of the ballast layer at current in-track loading level of 20 kips (315 kip gross rail load per car) under different support conditions, such as force chain and stress distribution.

Dem Simulations of Different Tie Support Conditions

Ballast Material Properties

In this study, polyhedral particles were employed to represent realistic shapes of ballast particles in the DEM simulations. A polyhedral 3-D DEM code, BLOKS3D, was used for the DEM simulations (*15*, *16*). The key particle morphological indices such as the Flat and Elongated (F&E) ratio and Angularity Index (in degrees) of field-collected ballast particles were determined using the Enhanced University of Illinois Aggregate Image Analyzer (E-UIAIA) (*17*). Figure 2 shows the imaging-based particle shape indices obtained by E-UIAIA. The gradation of ballast material adequately met both No. 3 and No. 4A gradations of AREMA. Table 1 lists the percent passing sieve sizes of the granite type ballast material used as the gradation input for the DEM simulations. The polyhedral particles were generated by BLOKS3D automatically based on the inputs of gradation and the shape properties quantified by E-UIAIA.

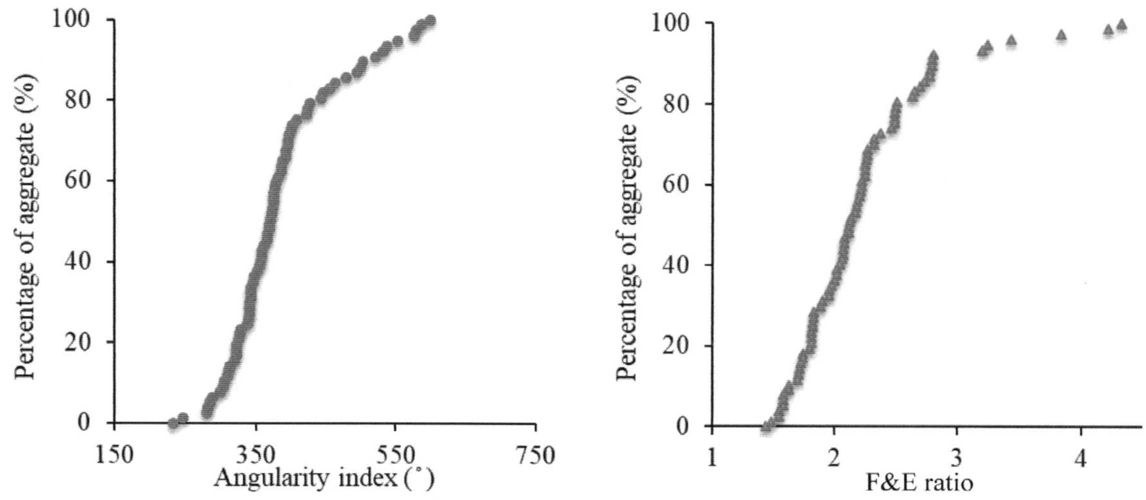

Figure 2. Imaging-based 3-D ballast particle shape properties quantified by E-UIAIA.

Table 1. Particle Size Distribution of the Ballast Material Used in DEM Simulations

Particle size (in.)		Percent passing (%)
2.5	63.5 mm	100
2	50.8 mm	95
1.5	38.1 mm	65
1	25.4 mm	10
0.5	12.7 mm	0

Simulation Approach

A full-track model was generated in BLOKS3D with the layout shown in Figure 3. A simple concrete tie representation with dimensions of $10\frac{3}{8} \times 9\frac{5}{8} \times 102$ in. ($0.2635 \times 0.244 \times 2.59$ m) was placed on top of a 14-in. (0.36-m) thick ballast layer. Full width of the track was modeled with 11.8-in. (0.3-m) wide ballast shoulders and 2:1 shoulder slope. The length in the train moving direction was chosen as 24 in. (0.61 m), which is approximately one-tie spacing. After the geometry of DEM simulation was determined, ballast particles were gravity dropped into the defined geometry. Around 11,000 particles were used in the simulation. Compaction of the ballast particles was accomplished by placing a non-deformable plate on top of the ballast layer and pushing downward until the density targeted in track was achieved. Once the basic ballast creation and compaction in DEM was done, the concrete tie was generated as a non-deformable master block in BLOKS3D and placed on top of the ballast layer. Note that as only a non-deformable master block is available in the code, the tie created is not able to bend as in reality, which may affect the tie-ballast pressure distribution to some extent. It was considered as acceptable in this study since the focus is the ballast layer performance, rather than crosstie bending flexural performance.

In the DEM simulation, some key model parameters were determined after trial and error and based on the previous knowledge from DEM simulations, such that the normal contact stiffness was adjusted to 20 MN/m; the shear contact stiffness was adjusted to 10 MN/m; global damping ratio was nil; contact damping ratio was adjusted to 0.40; and the surface friction angle related to the roughness of contacting aggregate particles was 31° (*4, 8, 10*).

Model Calibration with Laboratory Test Results

The purpose of model calibration is to determine the model parameters which produce the DEM simulation results that match with laboratory test results. The schematic profile view drawing of the five different support conditions simulated in the laboratory experimental setup are shown in Figure 4*a* which illustrates: 1) lack of rail seat support condition, representing the in-track condition caused by high impact wheel loads, as insufficient support under rail seats; 2) full support, representing an idealized track, condition; 3) lack of center support condition, representing the idealized track condition right after tamping, which provides adequate support under the rail seats and decreased support on the center of the crosstie; 4) high center binding; and finally, 5) severe center binding support, representing extreme cases in which

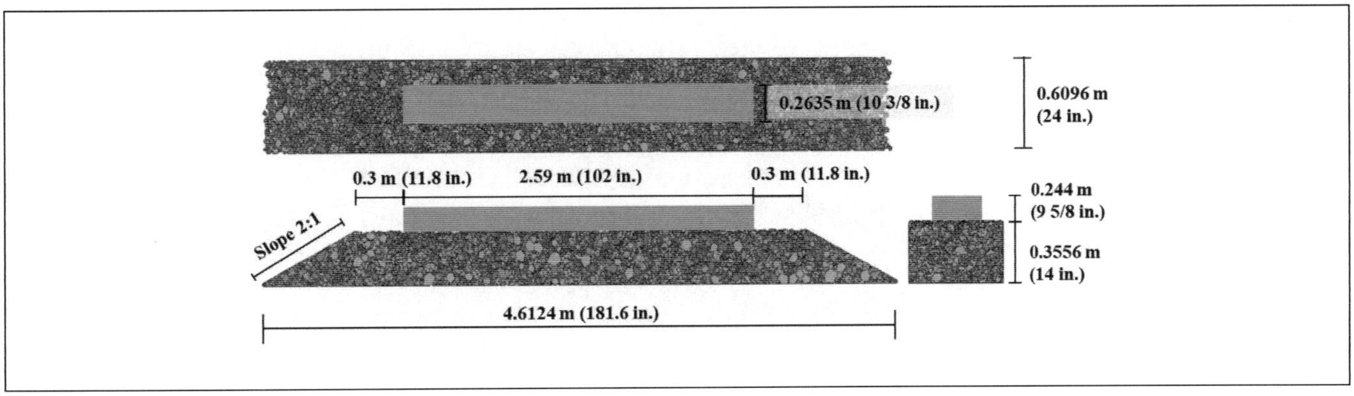

Figure 3. Basic one-tie ballast layer geometry established in DEM simulation.

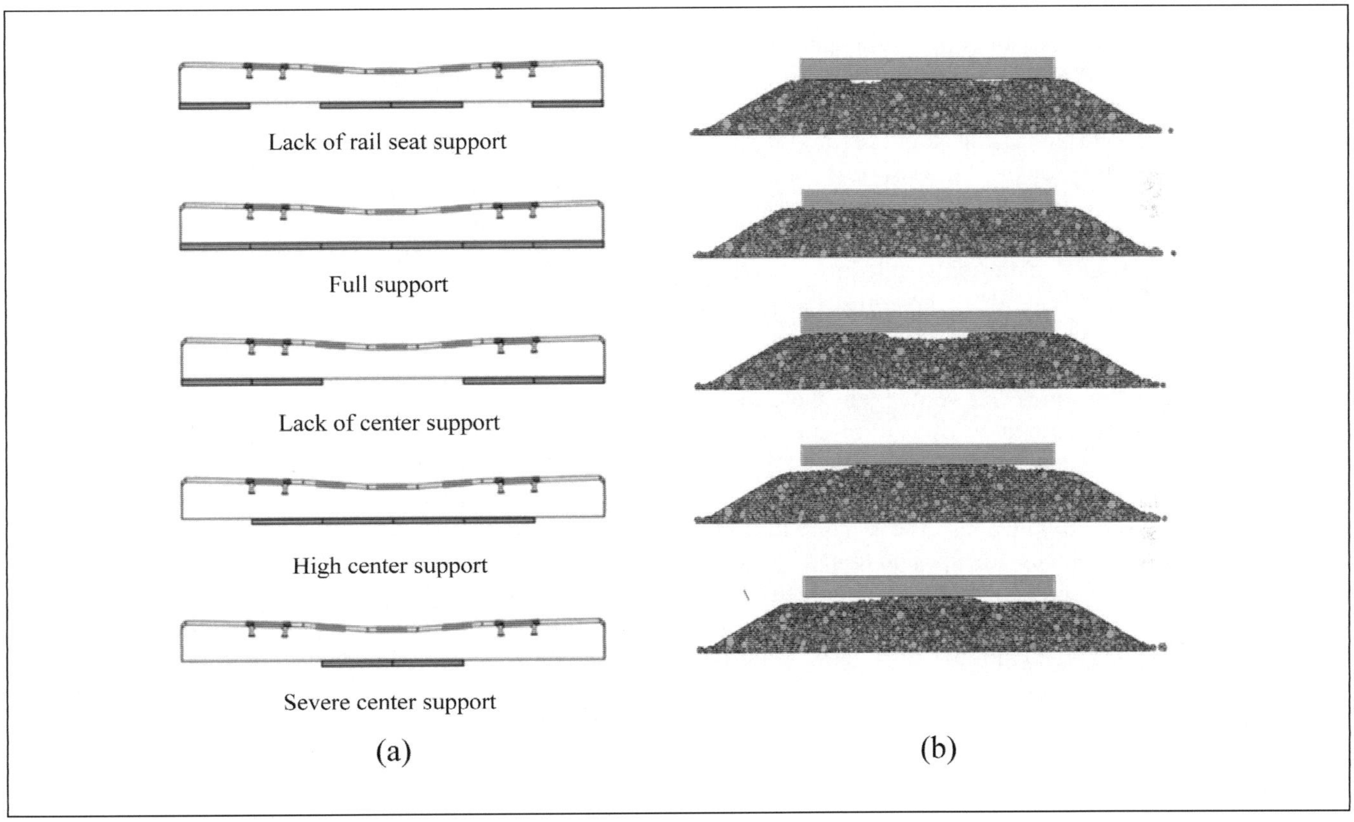

Figure 4. (*a*) Schematic profile view drawings of the five different support conditions in laboratory testing (*14*) and (*b*) Profile views of five different support conditions in DEM simulations.

the track support migrates toward the center of the crosstie.

A total of five strain gages were installed on the crosstie, with one at each rail seat, one at the center, and another located halfway between each rail seat and center as shown on top of the crosstie in Figure 4 (*18*). Rubber pads were used beneath the crosstie to simulate ballast support (*19*). Overall, the bottom of the concrete crosstie was divided into six equally spaced sections. By varying the locations of rubber pads, five different support conditions were simulated. A static load of 10 kips (44.5 kN) was applied to each rail seat. The bending moments and two static rail seat loads were used as inputs in the backcalculator program to calculate the tie-ballast reaction force distribution (*14*).

To simulate the five different support conditions in laboratory testing, five full-track models were generated based on the basic DEM model. Figure 4*b* shows the profile views of track models for different support conditions. For example, to create the severe center binding support condition, before placing the tie on top of the ballast layer, more ballast particles were dropped only at the center of existing ballast layer and compacted to target density. After the full-track models were formed, two static rail seat loads of 10 kips (44.5 kN) each were applied vertically until the ballast layer was stable under the external loading. The force exerted on each ballast particle was obtained from the DEM simulations. From the laboratory testing backcalculation results, the tie-ballast reaction force at each rubber pad (17 in./0.43 m) was consistent (*14*). To make a valid comparison, the external forces exerted on surface ballast particles at the six corresponding rubber pad locations were summed up separately to obtain tie-ballast reaction force distributions from the DEM simulations.

Figure 5 presents the tie-ballast reaction force distributions obtained from both the DEM simulations and the laboratory tests. For almost all support conditions simulated, the overall reaction force distribution trends as well as the magnitude in percentage from both the laboratory and DEM simulations were in good agreement. One inconsistent comparison was the full support condition (see Figure 5*b*), in which the laboratory results showed that the reaction force was very low in the center section and more concentrated below the rail seat sections, rather than more concentrated in the left half of the crosstie section from the DEM simulation. This is probably because of the following: 1) the discrete nature of DEM simulation may cause abrupt and sharp corner/edge on the left half side of the tie-ballast contact surface, which will lead to force concentration in the DEM simulation; 2) the rubber pad used in laboratory testing simulated an ideal uniform support condition, whereas in DEM simulation the particles that were generated randomly do not guarantee symmetrical or uniform stiffness; 3) the concentrated force may cause ballast particle breakage in reality and thus redistribute the forces evenly among particles, whereas in the current BLOKS3D code no particle breakage is considered and, accordingly, force concentration would not dissipate; and finally, 4) under the external forces applied on two rail seats, the edges and center of the tie will bend and lift up in the laboratory tests whereas the tie is non-deformable in the DEM simulations. Despite the inconsistent trends observed from full support condition, it can be noticed that the tie-ballast force at each observation point is still within a reasonable range. From Figure 5, it may be concluded that the DEM simulation results based on the current set of model parameters could reasonably match the laboratory test results. Therefore, the calibrated DEM model can be used to predict different in-track ballast support conditions.

Results and Discussion

In this section, mesoscale stress tensor formulas are derived. The formulas are adequate to calculate stresses in particles with rotation. To better study the in-track ballast behavior under different tie support conditions, the field-monitored rail seat load of 20 kips (89 kN) from the Northeast Corridor was applied as the static loading in the calibrated DEM model. The magnitude and distribution of particle vertical normal stress σ_{33} and particle contact force network were visualized and analyzed using the calibrated DEM model.

Calculation of Stress Tensor

In order to analyze the mesoscale structure of a ballast layer under external loading, mathematical formulas are required to properly calculate the stress tensor for a single particle. Roul and Schineer derived stress tensor formulas for particles without rotation as follows (*20*):

$$\langle \sigma_{ij} \rangle = \frac{1}{V} \sum_{p=1}^{n} \sum_{c=1}^{c^p} x_j^c F_i^c \qquad (1)$$

where

F^c = contact force acting at point c, and

x^c = corresponding branch vector from the center of the mass of the particle p to contact point c.

During discrete element modeling, relative rotation exists between polyhedral particles and creates inertia forces on the particle. In order to adjust Equation 1 for polyhedral particles with rotation, assuming the angular velocity and angular acceleration for a polyhedral particle P are ω^P and $\dot{\omega}^P$, forces caused by particle rotation can be expressed in the form of:

$$R_{ij}^P = \omega_k^P \omega_j^P J_{ik}^P - \omega_k^P \omega_k^P J_{ij}^P + \Phi_{jkl} \dot{\omega}^P J_{il}^P \qquad (2)$$

where

$J_{ij}^P = \oint \rho x_i r_j dV^P = \oint \rho r_i r_j dV^P$, and

Φ_{jkl} = permutation tensor.

Inserting Equation 2 in Equation 1, the stress tensor formula for dynamic conditions is obtained in Equation 3. Sigma is used on R_{ij}^P to sum all forces caused by rotation on particle P.

$$\langle \sigma_{ij} \rangle = \frac{1}{V} \left(\sum_{p=1}^{n} \sum_{c=1}^{c^p} x_j^c F_i^c + \sum_{P=1}^{n} R_{ij}^P \right) \qquad (3)$$

where

F^c and x^c are defined as the same as Equation 1, and

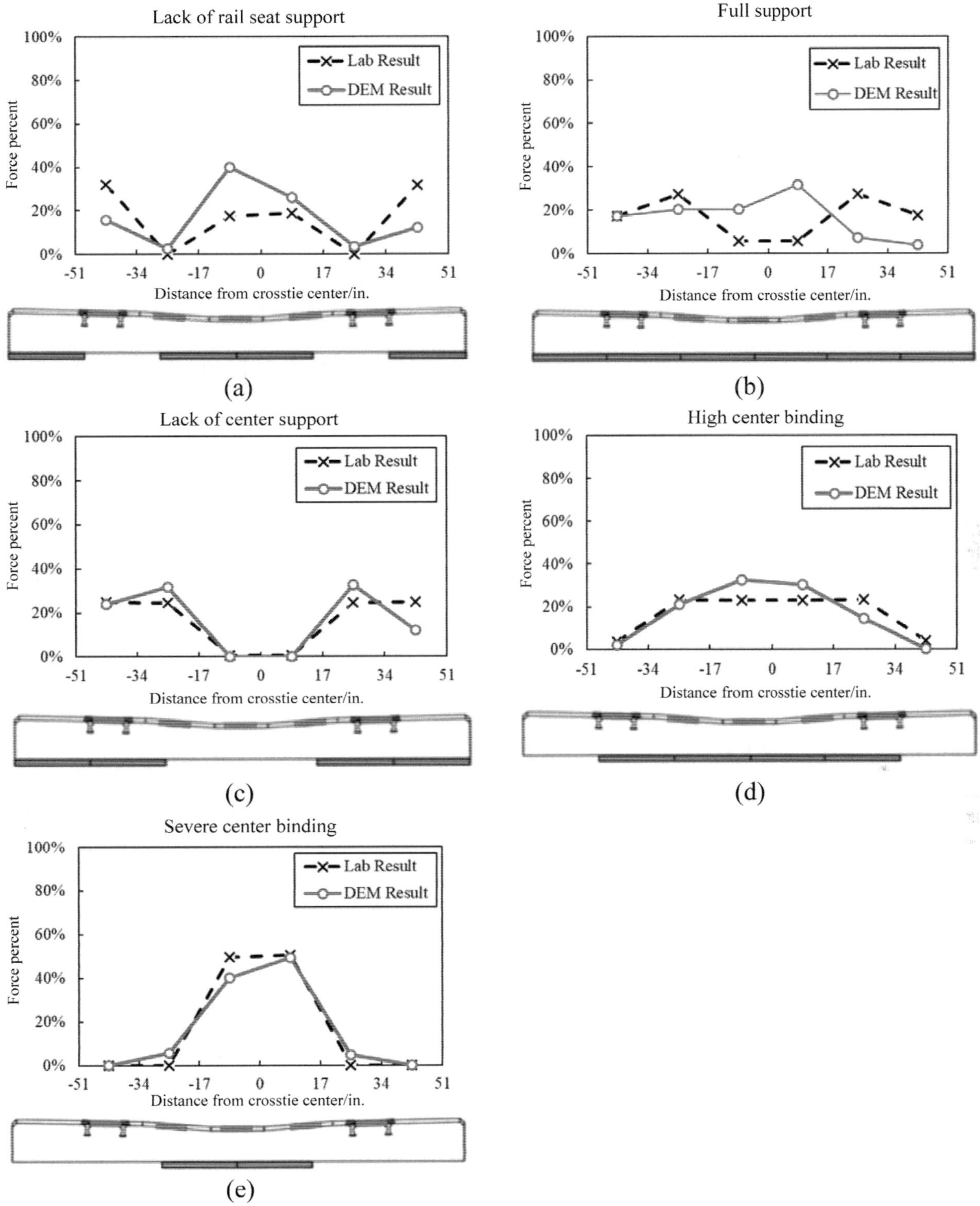

Figure 5. Comparisons of tie-ballast reaction forces backcalculated from laboratory testing to those predicted from DEM simulations: (*a*) lack of rail seat support, (*b*) full support, (*c*) lack of center support, (*d*) high center binding, and (e) severe center binding.

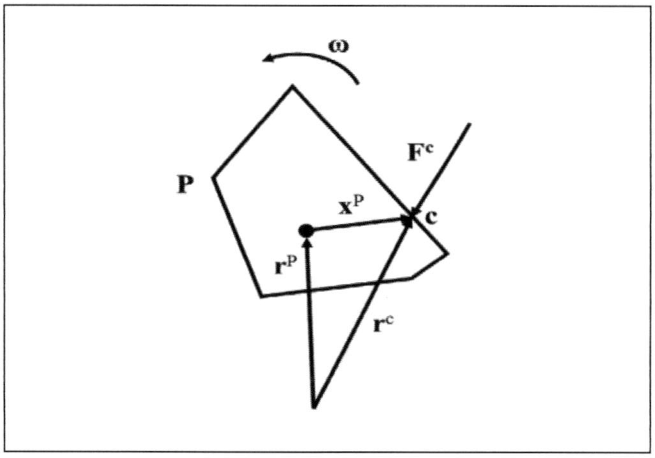

Figure 6. Schematic of random particle *P*.

R^P = force acting on particle P caused by rotation (Figure 6).

Vertical Normal Stress σ_{33} Distribution

Using Equation 3, the force distribution can be calculated as a 3×3 matrix indicating normal forces and shear forces in x, y, z directions obtained for every single particle. Indices 1, 2, 3 represent x, y, z directions, respectively. σ_{33} is the vertical normal stress in every ballast particle to react the external force applied on the crosstie. The magnitude and distribution of σ_{33} from the analysis can offer a better insight into the ballast layer performance in mesoscale.

Vertical normal stress σ_{33} magnitudes and distributions for the DEM model with five different support conditions were computed next. The visualized results are presented in Figure 7. Darker color in the figure indicates higher vertical normal stresses whereas lighter color indicates none or low vertical normal stresses.

From the visualized results, clear differences exist in the vertical normal stress distributions among the different support conditions. In the lack of rail seat support condition, few ballast particles under rail seat contribute to reacting the external force; high stress concentration was observed in the center area of tie and ballast layer. Such a high stress level would probably lead to higher settlement in the center and eventually develop a gap between tie and ballast layer. On the contrary, a better vertical normal stress distribution appears throughout the ballast layer in full support condition: particles with high stresses are fewer than the other four different support conditions. As for the lack of center support condition, several ballast particle zones with high normal stress concentrations exist with depth directly under rail seats where ballast particles are prone to crushing. Center binding conditions usually create higher vertical normal stresses in the middle and little load spreading with depth. To be more specific, in the high center binding condition, particles close to the middle of the crosstie experience very high vertical normal stresses; a high stress concentration is not seen on either end of the

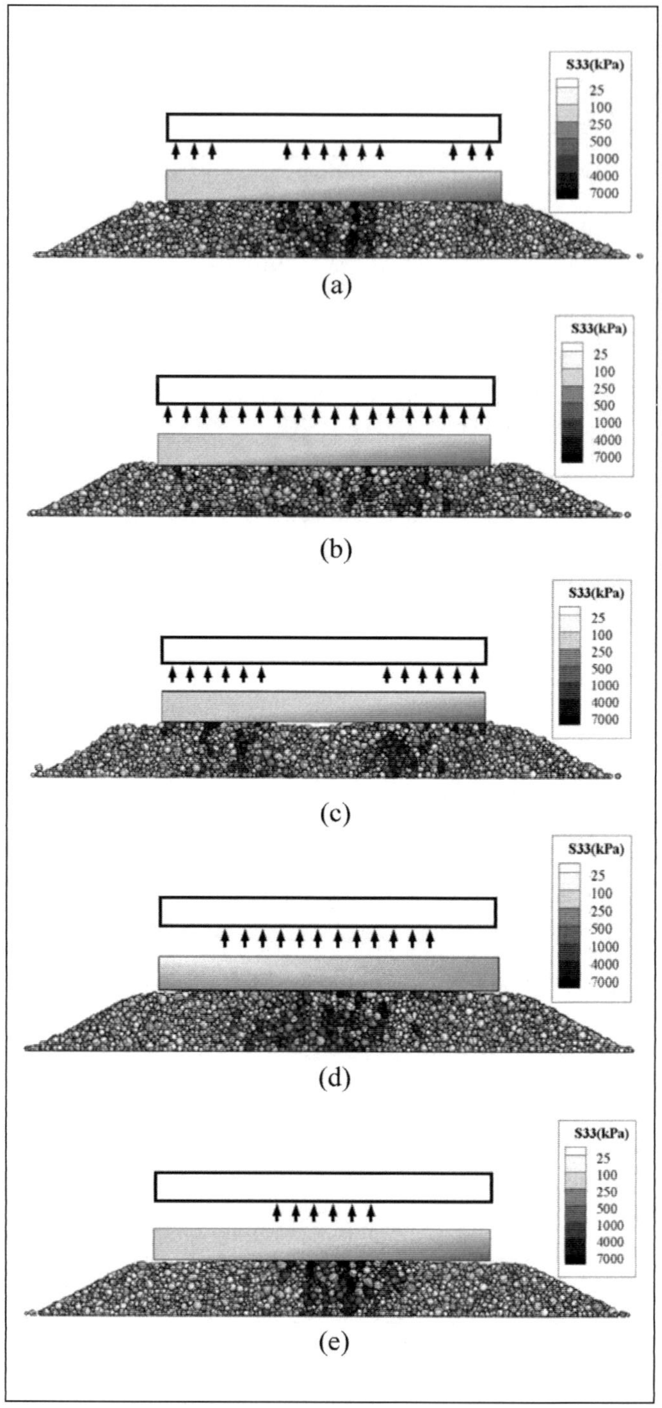

Figure 7. Vertical normal stress distributions of the ballast layers with five different support conditions: (*a*) lack of rail seat support, (*b*) full support, (*c*) lack of center support, (*d*) high center binding, and (*e*) severe center binding.

ballast cross section. Furthermore, in the severe center binding condition, high vertical normal stresses concentrate in the middle section of the ballast layer from top to bottom.

Besides the apparent differences, similarities are also observed among the presented five different support conditions. The ballast layer was constructed with about 11,000 particles in the DEM model; however, only a small portion of particles contributes to the reaction force. Particles on the shoulder and particles in the area with poor support condition often experience none or very low stresses under the external loading. Only a few particles near the tie-ballast contact surface tend to experience extremely high vertical normal stresses under the external loading. With such large stresses, particles have a high possibility of breaking down into smaller particles. After breakage, such high stresses will transmit to other particles and lead to further breakage until a lower magnitude contact stress network than can be tolerated by ballast particle rigidity is maintained. Note that more particle breakage and degradation will occur with the growth of external loading cycles to contribute to an increasing Fouling Index (FI), as defined by Selig and Waters, in the ballast layer (*1*). Higher FI values will eventually affect the drainage ability of the ballast layer and lead to faster ballast degradation and track deterioration. Among all the studied ballast support conditions, severe center binding seems to be the most critical scenario with reference to reaction force concentration and the ballast degradation potential.

Contact Force Network

Visualizing the contact force network among particles is an effective and widely used method to study the mesoscale structure of unbound granular materials (*21*, *22*). One of the advantages of using contact force network is that ballast layer can be visualized through the width with no concern about ballast particles in the front obstructing the view of the ballast particles in the back. Contact force networks for the five different support conditions are visualized in Figure 8, where the line thickness and darkness are both indicative of the higher magnitude contact forces.

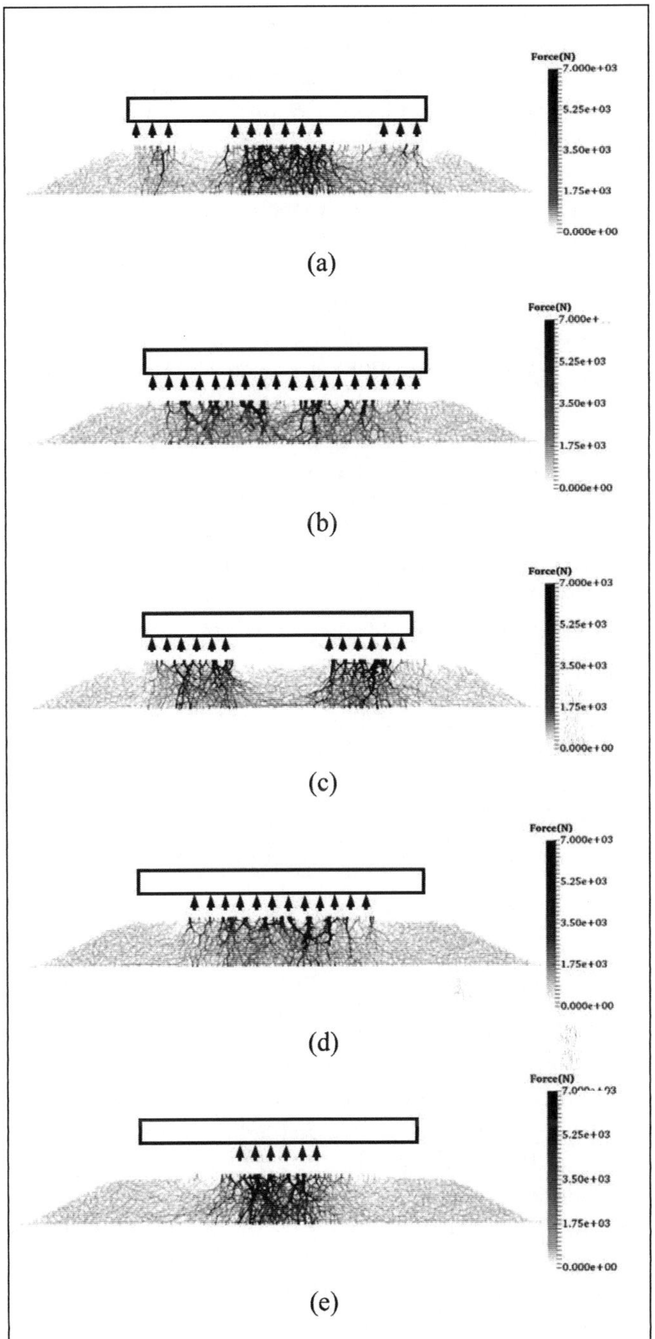

Figure 8. Contact force networks of the ballast layers with five different support conditions: (*a*) lack of rail seat support, (*b*) full support, (*c*) lack of center support, (*d*) high center binding, and (e) severe center binding.

Figure 8 clearly indicates that the contact force networks match well with their actual support conditions. Darker and thicker lines concentrate in the areas with a good support condition whereas lighter and thinner lines prevail in the areas with a poor support condition. One exception is the force network in full support condition, in which the contact force network seems a little concentrated in the left half side of the crosstie rather than evenly distributed along the crosstie as expected. A possible reason is that the crosstie first contacted with some ballast particles on the left when the loads were applied. Even though the tie-ballast contact surface appears flat in the model, because of its discrete nature, ballast particles are not easily distributed evenly across the contact surface. Similar findings were observed by McHenry et al that the pressure distribution between ballast and tie measured in the laboratory at the full support condition was not uniform but had point load concentrations (*11*). Force diffusion is obvious in the lack of rail seat support condition and the lack of center support condition, whereas it is not clear in the high center binding condition and the severe center binding condition. Force network results are in good agreement with the observations from the vertical normal stresses in the previous section: fewer dark network lines indicate a more concentrated reaction force. Large magnitudes of the reaction force exist and transmit only in a few particles. Ballast shoulders and areas with the poor support condition rarely have dark and thick lines. The poorer the support condition, the more the contact forces concentrate in a limited number of ballast particles, which may eventually lead to particle breakage. The maximum tie-ballast contact pressure was also recorded as 97, 70, 102, 77, and 114 psi, for support conditions from (a) to (e), respectively. As expected, the full support condition exhibits the smallest tie-ballast contact force. AREMA assumes an allowable ballast pressure under concrete crossties to be 85 psi (586 kPa) (*12*). Only the full support and high center binding conditions satisfy this requirement for the in-track load level. Note that the maximum tie-ballast contact pressures under lack of rail seat support and lack of center support conditions are similar in magnitude. Finally, the most critical condition is severe center binding, which exceeds the allowable contact pressure by 30%.

Conclusion

Five realistic full-track ballast simulation models, each representing one unique support condition commonly found in track, were developed and analyzed using the DEM with the consideration of realistic 3-D polyhedral ballast particles. The DEM models were based on the available test data obtained from previously conducted laboratory testing conditions. For the model calibration effort, the DEM simulation results of the tie-ballast contact force distributions matched well in both magnitude and overall trend with the backcalculated results obtained from the laboratory tests, which had used strain gages instrumented on the tested crosstie. The calibrated DEM model was then used to predict in-track performance under 20 kip rail seat loads. Stress tensor formulas were derived for calculating ballast particle stresses with relative rotation. Vertical normal stress magnitudes and distributions within ballast layer were successfully visualized. In addition, contact force networks showing the force chains in the ballast layer were successfully visualized for the different tie support conditions. The following conclusions can be drawn from this study:

1. Only a portion of ballast particles participated in load distribution when external forces were applied.
2. Good support condition indicates better force distribution throughout the ballast layer in terms of lower magnitude normal stresses and more evenly distributed force network.
3. Poor support condition leads to higher stress concentrations on a few particles, which would increase the breakage possibility and ballast fouling tendencies, leading to faster track deterioration.
4. DEM simulation results show high normal stresses on ballast layer under loading. Although no particle breakage was considered in the simulation, in reality such high normal stresses can lead to particle breakage.
5. DEM simulation results show that different support conditions influence ballast contact force distribution greatly. Lack of rail seat support, full support, and lack of center support have wider force distribution angles than high center binding and severe center binding conditions.
6. The severe center binding scenario seems to be the most critical support condition in terms of generating non-uniform contact force distributions and highest tie-ballast contact forces among the studied five support conditions that exceed the AREMA allowable pressure of 85 psi by 30%.

Acknowledgments

The support for this study was partly provided by the Amtrak Concrete Tie and Track Structure Improvement project undertaken in the Rail Transportation and Engineering Center (RailTEC) at the University of Illinois at Urbana-Champaign (UIUC). The authors would like to acknowledge J. Riley Edwards and the project team members for their useful advice at different stages of this research.

Author Contributions

The authors confirm contributions to the paper as follows. Study conception and design: Wenting Hou, Erol Tutumluer; data collection: Wenting Hou, Bin Feng; analysis and interpretation of results: Wenting Hou, Bin Feng, Wei Li; draft manuscript preparation: Wenting Hou, Bin Feng, Wei Li, Erol Tutumluer. All authors reviewed the results and approved the final version of the manuscript.

References

1. Selig, E. T., and J. M. Waters. *Track Geotechnology and Substructure Management*. Thomas Telford, London, 1994.

2. Li, D., and D. Davis. Transition of Railroad Bridge Approaches. *Journal of Geotechnical Engineering*, Vol. 131, No. 11, 2005, pp. 1392–1398.
3. Mishra, D., E. Tutumluer, T. D. Stark, J. P. Hyslip, S. M. Chrismer, and M. Tomas. Investigation of Differential Movement at Railroad Bridge Approaches through Geotechnical Instrumentation. *Journal of Zhejiang University Science A*, Vol. 13, No. 11, 2012, pp. 814–824.
4. Tutumluer, E., H. Huang, Y. M. A. Hashash, and J. Ghaboussi. Discrete Element Modeling of Railroad Ballast Settlement. In CD-ROM Proceedings of the *AREMA 2007 Annual Conference*, Chicago, IL, 2007.
5. Tutumluer, E., H. Huang, Y. M. A. Hashash, and J. Ghaboussi. AREMA Gradations Affecting Ballast Performance Using Discrete Element Modeling (DEM) Approach. In Conference Proceedings: *AREMA 2009 Annual Conference*, Chicago, IL, 2009, pp. 20–23.
6. Tutumluer, E., Y. Qian, Y. M. A. Hashash, J. Ghaboussi, and D. D. Davis. Field Validated Discrete Element Model for Railroad Ballast. In Proceedings of the *AREMA 2011 Annual Conference*, Minneapolis, MN, 2011, pp. 18–21.
7. Lu, M., and G. R. McDowell. The Importance of Modelling Ballast Particle Shape in the Discrete Element Method. *Granular Matter*, Vol. 9, No. 1–2, 2007, p. 69.
8. Huang, H., and E. Tutumluer. Discrete Element Modeling for Fouled Railroad Ballast. *Construction and Building Materials*, Vol. 25, No. 8, 2011, pp. 3306–3312.
9. Ngo, N. T., B. Indraratna, and C. Rujikiatkamjorn. DEM Simulation of the Behavior of Geogrid Stabilized Ballast Fouled with Coal. *Computers and Geotechnics*, Vol. 55, 2014, pp. 224–231.
10. Qian, Y., S. Lee, E. Tutumluer, Y. Hashash, D. Mishra, and J. Ghaboussi. Simulating Ballast Shear Strength from Large-Scale Triaxial Tests: Discrete Element Method. *Transportation Research Record: Journal of the Transportation Research Board*, 2013. 2374: 126–135.
11. McHenry, M. T., M. Brown, J. LoPresti, J. Rose, and R. Souleyrette. Use of Matrix-Based Tactile Surface Sensors to Assess Fine-Scale Ballast–Tie Interface Pressure Distribution in Railroad Track. *Transportation Research Record: Journal of the Transportation Research Board*, 2015. 2476: 23–31.
12. *Manual for Railway Engineering*. American Railway Engineering and Maintenance-of-Way Association, Lanham, Md., 2012.
13. Kaewunruen, S., and A. M. Remennikov. Influence of Ballast Conditions on Flexural Responses of Railway Concrete Sleepers. *Concrete in Australia: Journal of Concrete Institute of Australia*, Vol. 35, No. 4, 2009, pp. 57–62.
14. Gao, Z., H. E. Wolf, M. S. Dersch, Y. Qian, and J. R. Edwards. Field Measurements and Proposed Analysis of Concrete Crosstie Bending Moments. *AREMA 2016 Annual Conference*, Orlando, FL, 2016.
15. Ghaboussi, J., and R. Barbosa. Three-Dimensional Discrete Element Method for Granular Materials. *International Journal for Numerical and Analytical Methods in Geomechanics*, Vol. 14, No. 7, 1990, pp. 451–472.
16. Zhao, D., E. G. Nezami, Y. M. Hashash, and J. Ghaboussi. Three-Dimensional Discrete Element Simulation for Granular Materials. *Engineering Computations*, Vol. 23, No. 7, 2006, pp. 749–770.
17. Moaveni, M., S. Wang, J. M. Hart, E. Tutumluer, and N. Ahuja. Aggregate Size and Shape Evaluation Using Segmentation Techniques and Aggregate Image Processing Algorithms. *Transportation Research Record: Journal of the Transportation Research Board*, 2013. 2335: 50–59.
18. Wolf, H. E. *Flexural Behavior of Prestressed Concrete Mono Block Crossties*. Doctoral dissertation. University of Illinois at Urbana-Champaign, 2015.
19. César Bastos, J. Analysis of the Performance and Failure of Railroad Concrete Crossties with Various Track Support Conditions. Doctoral dissertation. University of Illinois at Urbana-Champaign, 2016.
20. Roul, P., and A. Schinner. Simulation Study on Micro and Macro Mechanical Behavior of Sand Piles. *Powder Technology*, Vol. 204, No. 1, 2010, pp. 113–123.
21. O'Sullivan, C. A., L. Cui, and S. C. O'Neill. Discrete Element Analysis of the Response of Granular Materials during Cyclic Loading. *Soils and Foundations*, Vol. 48, No. 4, 2008, pp. 511–530.
22. Sazzad, M. M., and K. Suzuki. Micromechanical Behavior of Granular Materials with Inherent Anisotropy under Cyclic Loading Using 2D DEM. *Granular Matter*, Vol. 12, No. 6, 2010, pp. 597–605.

The Standing Committee on Railroad Track Structure System Design (AR050) peer-reviewed this paper (18-04833).

Article

TRR
JOURNAL OF THE TRANSPORTATION RESEARCH BOARD

Transportation Research Record
2018, Vol. 2672(10) 116–124

DOI: 10.1177/0361198118782250
journals.sagepub.com/home/trr
SAGE

Laboratory Characterization of Structural Capacity of North American Heavy Haul Concrete Crossties

Josué C. Bastos[1], Alejandro Álvarez-Reyes[1], Marcus S. Dersch[1], J. Riley Edwards[1], and Christopher P. L. Barkan[1]

Abstract

Over the past few decades, the use of concrete crossties in North America has increased as a high-performance alternative to timber crossties, especially in heavy-haul freight and higher speed rail corridors. To accommodate heavier axle-loads and prevent center cracking, railroads and suppliers have consistently increased the bending moment thresholds that a crosstie must withstand, leading to stiffer elements that may be more prone to brittle cracking. This paper attempts to characterize the structural capacity of the crosstie at two critical design locations, the center and the rail seat cross sections. Concrete crossties were subjected to four-point bending tests at both the center and rail seat, while recording corresponding deflections using protocols adapted from the American Railway Engineering and Maintenance-of-Way Association (AREMA) Manual on Railway Engineering. Experimentation was performed on some of the most representative concrete crosstie designs that are currently installed in North America, with 70 individual crossties undergoing tests. The resulting load-deflection curves illustrate the variability of each crosstie's response to load and characterize the flexural performance of the crossties. Results showed that North American crosstie designs frequently have significant reserve capacity and high stiffness. The results also indicate that even severely abraded crossties can have enough flexural capacity to withstand typical field loading conditions.

Crossties are an essential component of ballasted railroad track, and the use of engineered concrete crossties has been increasing in North America on demanding heavy axle load (HAL) freight corridors, often considered as a longer-lasting, lower-maintenance alternative to wood crossties (*1–4*). Such crossties are concrete members that are commonly expected to withstand over 40 years of service (*5*), frequently subjected to severe conditions imposed by heavy axle loads, weather exposure, proximity to ground moisture, cyclic loading, and variable support conditions at the ballast interface. To fulfill these expectations, manufacturers have used high-strength concrete to manufacture monoblock crossties that are highly prestressed with large cross sections, as will be shown in this paper. However, these crossties are commonly overdesigned, having considerable reserve flexural capacity beyond the loading environment requirements as quantified in the field under typical HAL freight service loads (*6, 7*).

As new concrete crosstie designs are introduced in the North American railway industry, it is beneficial to quantify their flexural capacity and additional properties that can contribute to better-informed decisions for future designs and specifications. Current crossties tend to have previously unquantified reserve capacities. This was indicated by Wolf (*6*), who quantified in-track crossties' reserve capacity to range from 25% to 75% with respect to their cracking bending moment and similar findings were noted by Edwards et al. (*7*). Furthermore, more comprehensive tests will be necessary if the design of concrete crossties in North America follows the international trend of shifting from allowable stress to a limit state design approach (*8*). The work presented in this paper contributes to the characterization of the structural capacity of North American heavy haul concrete crossties. New specimens representative of common North American concrete crosstie designs were tested in flexure to obtain the stiffness, ultimate load capacity, and ultimate displacement of these crossties. This was done at the center and rail seat sections, the most common locations for structural analysis of concrete crossties (*6, 9*). Additionally, aged and abraded specimens were also tested for evaluation of the effect of mild and severe wear on the investigated properties.

In the following sections, a new center negative bending moment test is proposed, which is a modified version of the American Railway Engineering and Maintenance-of-Way Association (AREMA) center negative bending test protocol (Chapter 30, Article 4.9.1.6) (*9*). The crossties are supported

[1]RailTEC, University of Illinois at Urbana-Champaign, Urbana, IL

Corresponding Author:
Address correspondence to Josué C. Bastos: cesarba2@illinois.edu

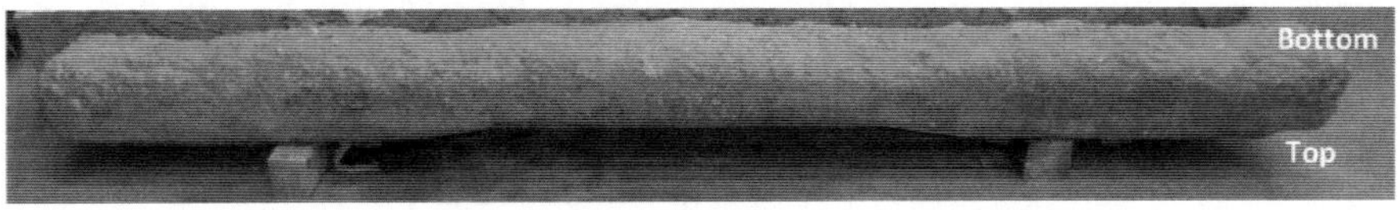

Figure 1. Severely bottom-abraded specimen (Design 7a).

by half-moon steel bars as opposed to rubber pads to accurately record displacements that are due to bending of the crossties (not from the compression of the rubber pads) and to better control the contact area and friction forces at the support interface with the crossties. In addition, the crossties are loaded to ultimate failure as opposed to a pre-determined load which identifies the presence, or lack-thereof, of a crack. Loading to ultimate failure allows for quantification of flexural reserve capacity and deformation ability before the ultimate state. The test output is the entire load-displacement curve, and the main points of interest are the crosstie displacements in the linearly elastic load range as well as the load and displacement at failure, defined as the peak load and the associated displacement achieved during testing. Similarly, a rail seat positive bending test is developed based on the equivalent AREMA recommended test protocol (Chapter 30, Article 4.9.1.8), with the load-displacement curve being the test output (*9*).

Typical Tests Background and Analysis

The most prevalent evaluative tests for prestressed concrete crossties in the North American railway industry are defined within the recommended practices of the AREMA Manual for Railway Engineering (*9*). Among these, the center negative bending moment test is of particular importance, as it aims to assess a crosstie's flexural strength at one of the two key locations where design bending moments are considered by AREMA. Moreover, it has been reported that center negative crosstie bending is among the five most critical track structure conditions with respect to the occurrence of accidents on concrete crosstie tracks in the U.S. (*10*).

Despite its conceptual importance, the current AREMA center negative bending moment test falls short of its potential benefits. The single test output is whether or not there are cracks originating in the tensile face of the crosstie that extend to the "outermost level of reinforcement or prestressing tendons" at the load required to produce the specified negative center design moment (*9*). Consequently, the crosstie is not loaded to an ultimate flexural failure, characterized by either rupture of prestressing material or crushing of concrete, reducing the concept of design failure to the simple presence of hairline cracks that go through the concrete cover on the tensile surface.

Another example of a standard test that could be improved to provide more information is the AREMA rail seat repeated-load test (Chapter 30, Article 4.9.1.5). This test protocol only addresses whether or not there was tendon slippage of more than 0.025 mm (0.001 inch), even though it eventually does load a crosstie rail seat to ultimate failure (*9*). An improvement to the protocol would require documentation of the amount of slippage and generation of the load-displacement curve for the crosstie. In general, the current "pass/fail" approach of AREMA recommended tests for the evaluation of concrete crosstie designs may be leading the industry to limit its ability to explore all the information that could be ascertained from these tests, inhibiting the development of new, more optimized, crosstie designs.

Experimentation Plan

Laboratory experiments assessed the stiffness, ultimate load, and ultimate displacement of concrete crossties subjected to a four-point bending test configuration both for center negative and rail seat positive bending moments. Eight crosstie designs (hereafter referred to as Designs 1 through 8) were subjected to center negative bending moment tests, and four were also subjected to rail seat positive bending moment tests (Designs 1, 2, 3, and 8). Designs 1, 2, 3, and 5 represent some of the most prevalent designs installed on U.S. HAL freight railroads or higher speed passenger lines including UP, BNSF, Amtrak, and CSX. Designs 4 and 6 are emerging designs that adopt differentiated characteristics, the former being pre-tensioned and the latter, post-tensioned. Design 7 represents a crosstie that is no longer manufactured, but is still present on many railroad lines. While the remaining designs were new and unused, all replicates of Design 7 had been removed from service for showing signs of deterioration, typically worn fastening system shoulders. One particular crosstie, however, presented severe bottom abrasion and was very dissimilar to the other Design 7 specimens, leading to the creation of its own category that will be referred to as Design 7a (Figure 1). Finally, Design 8 represents an additional crosstie that is no longer produced, and its specimens were removed from service for the specific purpose of being tested in this study.

The selected designs, although not identical, have similar dimensions, being expected to withstand similar loading conditions (i.e., revenue North American HAL freight and/or higher speed passenger service). The typical rail seat width at mid-height was 244 mm (9.6 in.), while the typical rail seat height was 226 mm (8.9 in.). At the center section, the typical mid-height width and height were 246 mm (9.7 in.) and

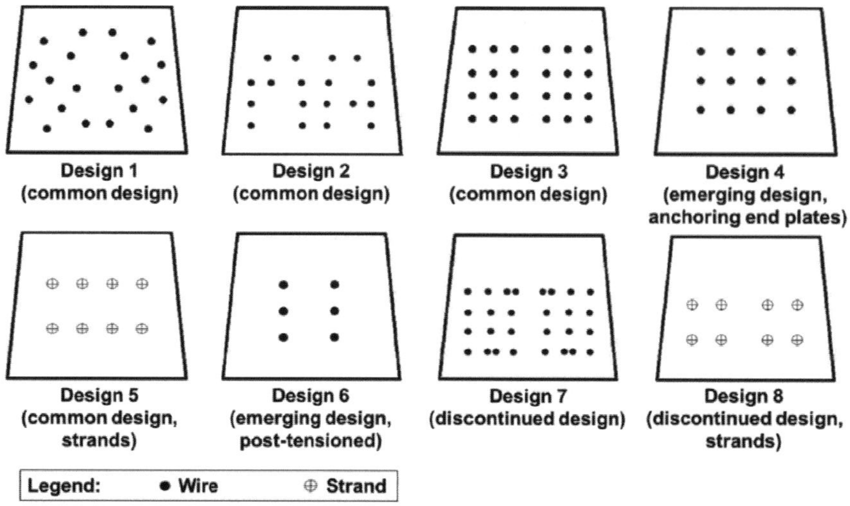

Figure 2. Schematic of prestressing arrangement for tested crosstie designs.

Table 1. Characteristics of Tested Crosstie Designs

		Crosstie design							
		1	2	3	4	5	6	7	8
Concrete compressive strength (lower bound)	ksi	8.2	7.3	7.1	7.7	12.4	–	9.1	9.7
	(MPa)	(56.8)	(50.2)	(49.3)	(52.9)	(85.5)	–	(62.6)	(66.9)
Ratio of steel in cross sectional area	Center	0.011	0.009	0.011	0.009	0.086	–	0.014	0.083
	Rail seat	0.008	0.007	0.009	0.007	0.075	–	0.011	0.061

180 mm (7.1 in.), respectively. Although most crossties were 2,591 mm long (102 in.), the crossties of Design 4 were 2,438 mm long (96 in.) and those of Design 7 were 2,515 mm long (99 in.). In the case of prestressing tendons, the only crossties that had strands were those of Designs 5 and 8, while all the others had wires. Figure 2 shows a schematic of the number and arrangement of prestressing elements for all designs. The ratio of reinforcement area per cross-sectional area is provided in Table 1.

In addition, concrete cylinders were cored from one replicate of each design to estimate the concrete strength of the tested specimens. All cores were drilled in a direction parallel to the casting direction (vertical) with a diameter of 92 mm (3.625 in.) and height of 184 mm (7.25 in.). The resulting height–diameter ratio of 2 is recommended by the ASTM C42/C42M-16 standard (*11*), which also allows for testing cores with embedded metals when impracticable to avoid. To minimize the effect of the presstressing force on the compressive strength of the concrete cylinders, all cores were drilled within the region where the prestressing force had not fully developed in the concrete. The drilling direction and diameter were based on conclusions of related research, but, since steel wires were present in these cylinders, the resulting concrete strengths are expected to be a lower bound of the strength when there was no steel reinforcement (*12–16*). The resulting concrete strength is reported in Table 1.

Replicates were performed to account for experimental and manufacturing variability, and a total of 70 crossties were tested, 54 of these in center negative and 16 in rail seat positive bending. Table 2 shows the number of replicates used for each application. The numbers in the "Total" column vary for several experimental reasons, and tests that deviated from the protocol were removed to ensure accurate results.

Center Negative Bending Moment Test

Individual concrete crossties were placed upside down in a loading frame where both rail seats were simply supported by half-moon steel bars spaced 1,524 mm (60 in.) apart (Figure 3). A vertical load was applied at the crosstie bottom at two locations 152.4 mm (6 in.) apart and symmetrically positioned about the crosstie center line. In addition to being similar to the AREMA center negative bending moment test configuration (Chapter 30, Article 4.9.1.6), the use of four-point bending is appropriate for assessing the crosstie flexural capacity because it applies a constant bending moment between the contact points of the loading head and eliminates shear forces in the same region.

Table 2. Replicates Used to Estimate Properties of Each Crosstie Design

Crosstie design		1	2	3	4	5	6	7	7a	8	Total
Center	Total number of specimens	9	8	9	1	7	7	6	1	6	54
	Ultimate load replicates	8	8	9	1	7	7	6	1	5	52
	Ultimate displacement replicates	8	4	9	1	7	7	6	1	4	47
	Stiffness replicates	9	4	9	1	7	7	6	1	6	50
Rail seat	Total number of specimens	4	4	5	–	–	–	–	–	3	16
	Ultimate load replicates	4	4	5	–	–	–	–	–	3	16
	Stiffness replicates	3	3	4	–	–	–	–	–	3	13

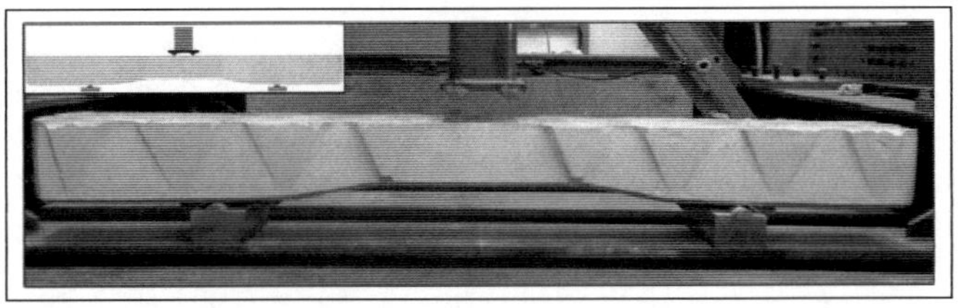

Figure 3. Four-point bending configuration for center section testing (crosstie is upside down when loaded).

To measure the crosstie center displacement, potentiometric linear transducers and one linear variable-differential transformer were employed. All devices were calibrated, and the loading frame displacements were monitored to ensure that boundary condition movements (if any) were considered for accurate computation of the crosstie deflection results. The load was applied at a constant rate of 22.24 kN/min (5 kips/min). Crossties that did not present a smooth bottom surface due to irregularities or indentation required application of white gypsum cement to provide a flat contact surface. Additionally, every crosstie was first subjected to three cycles of seating load ranging from 2.22 to 66.7 kN (0.5 to 15 kips) to ensure full contact between the loading head and the specimen during the ultimate test. This seating-procedure also reduced small surface irregularities that would chip off during the cycles, thus, producing a smoother displacement curve in the final test. After the seating load procedure was completed, the crosstie was loaded continuously from 2.22 kN (0.5 kips) to failure.

Once each crosstie was loaded to failure, the key characteristics considered to be of most interest were the ultimate load, ultimate displacement, and displacement at 33.4 kN (7.5 kips). This last data point was within the linearly elastic region of the load versus displacement curve, and it was used for calculating elastic region slope in units of force per displacement, a metric that will be referred to as "stiffness" for the remainder of this paper.

Rail Seat Positive Bending Moment Test

Four crosstie designs were also subjected to a rail seat positive bending moment test in a four-point bending configuration (Figure 4). Two half-moon steel bars were placed underneath the crosstie, each of them located 356 mm (14 in.) away from the rail seat vertical center line. On top of the rail seat, two additional half-moon steel bars were positioned 57 mm (2.25 in.) away from the rail seat vertical center line to apply the load. This configuration is similar to the recommended practices within the AREMA Manual on Railway Engineering (Chapter 30, Article 4.9.1.8), with the difference that steel bars were used in place of the prescribed rubber pads.

Figure 4. Four-point bending configuration for rail seat section testing.

Unlike the center negative bending test protocol, no seating load procedure was performed for most of the rail seat tests. Consequently, the rail seat load-displacement curves were not as smooth as the crosstie center ones due to the contact adjustment between loading head and crosstie. For this reason, the resulting displacements at ultimate condition are not analyzed in this paper. Nevertheless, it was possible to estimate the linear region stiffness as the mentioned contact adjustment was typically finalized within the elastic portion of the load-displacement curve. For rail seat tests, the stiffness values presented in this paper are the linear slope between the points at 267 kN (60 kips) and 356 kN (80 kips).

Results of Experimentation

Figure 5 displays the resulting load versus displacement curves of the center negative bending moment tests. Every line

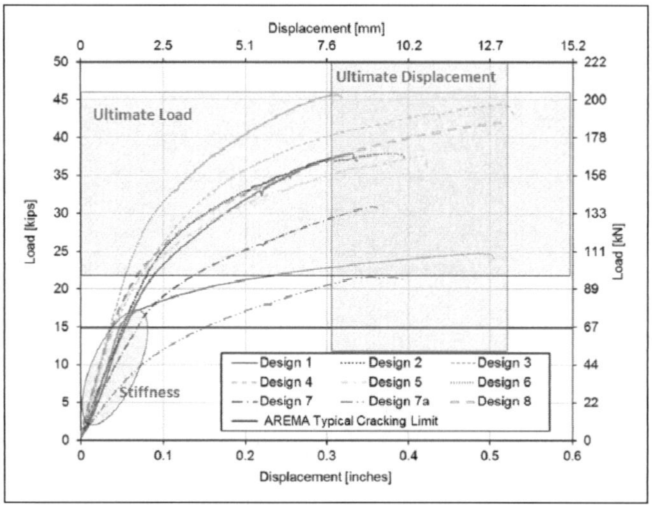

Figure 5. Load-displacement results for center negative bending tests—typical replicates.

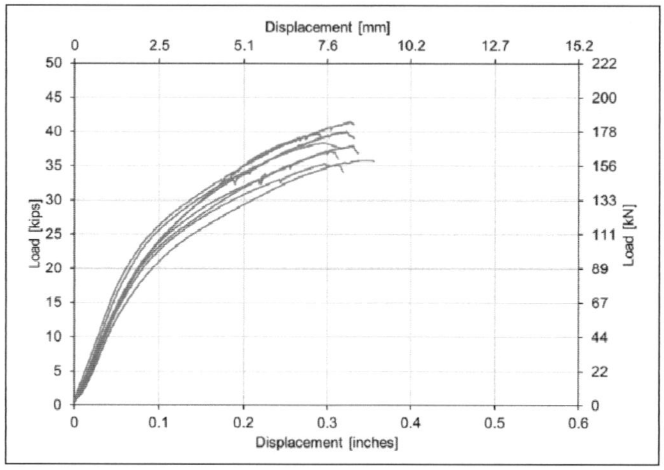

Figure 6. Load-displacement results for center negative bending tests—Design 1 replicates.

represents one typical replicate (i.e., closest to the average properties) of each crosstie design, allowing for a direct qualitative comparison between them. In this figure, three regions are highlighted corresponding to the previously mentioned variables of interest: ultimate load, ultimate displacement, and stiffness. The ultimate load is directly related to the crosstie's ultimate flexural capacity, indicating the crosstie's strength before there is crushing of the concrete in compression or failure of the prestressing steel in tension. The ultimate displacement indicates the capacity of the crosstie to deform and adjust itself to the ballast support conditions before a catastrophic failure takes place. Lastly, the stiffness is also a metric of the crosstie's displacement behavior, but it accounts primarily for elastic deflections in service (as opposed to the ultimate state deflection, which should not be reached under normal service conditions). In addition, a typical AREMA cracking limit (*6*) is presented for reference. The difference between the ultimate load and the AREMA cracking limit can be interpreted as the reserve flexural capacity of the crosstie.

In Figure 5, it is noticeable that all the properties vary significantly among the different designs. The stiffness, however, seem to have the least variability, which is an expected outcome given the similar geometries of the designs tested. It was observed in the experiments that cracking (i.e., cracking reaching the first layer of prestressing steel) occurs roughly at the end of the linear region, thus beyond the plotted AREMA limit for all designs. In addition, the unbonded post-tensioned crosstie design (Design 6) presents a dissimilar post-linear behavior than the other designs, showing the highest displacement rate after cracking. As expected, the severely abraded specimen (Design 7a) shows clear signs of deterioration, but it still presents approximately 70% of the load capacity and 60% of the stiffness of the healthier specimens of the same design (Design 7). This seems to indicate that bottom abrasion may not be a hazardous defective condition for designs with multiple rows of prestressing steel, even though this type of degradation has been reported to generate unsafe scenarios for designs with only two rows of prestressing steel (*17*).

As previously mentioned, replicate tests were conducted for most crosstie designs. Figure 6 illustrates this by showing test results for Design 1. It is evident that there is variability among replicates and a statistical treatment of the results is necessary. Therefore, a confidence interval for the means was estimated using Equation 1, which is derived from the Central Limit Theorem (*18*). In this process, the sample standard error was estimated from the mean square error of a completely randomized design in which crosstie model (design) was the single statistical effect considered.

$$D_{(1-\alpha)} = \frac{t_{\alpha/2,\,n-1} \times s}{\sqrt{n}} \quad (1)$$

where,

n: Number of observations (replicates).

$t_{\alpha/2,n-1}$: two-tailed t-value from t distribution for significance level α and $n-1$ degrees of freedom.

α: Significance level (0.1 adopted).

s: Sample standard error (square root of mean square error used).

$D_{(1-\alpha)}$: Detectable deviation of sample mean relative to the population mean.

A summary of the experimental results is presented in Table 3, where each design characteristic takes four rows: two for the mean result in different unit systems, and two for the 90% confidence margin also with different unit systems. For example, the mean ultimate load for crosstie Design 1 is 170.2 kN (38.3 kips), with 90% confidence that the error in

Table 3. Experimental Results and 90% Confidence Margin for Center Tests

		Crosstie design								
		1	2	3	4	5	6	7	7a	8
Ultimate load	kip	38.3	36.7	46.9	44.4	38.9	24.8	30.4	21.6	43.3
	(kN)	(170.2)	(163.4)	(208.6)	(197.4)	(173.0)	(110.3)	(135.3)	(96.1)	(192.5)
$D_{0.90}$	kip	1.5	1.5	1.4	–	1.6	1.6	1.8	–	1.8
	(kN)	(6.6)	(6.6)	(6.1)	–	(7.3)	(7.3)	(8.2)	–	(8.2)
Ultimate displacement	in	0.317	0.337	0.324	0.512	0.418	0.456	0.348	0.356	0.486
	(mm)	(8.1)	(8.6)	(8.2)	(13.0)	(10.6)	(11.6)	(8.8)	(9.0)	(12.3)
$D_{0.90}$	in	0.031	0.054	0.028	–	0.034	0.034	0.038	–	0.038
	(mm)	(0.8)	(1.4)	(0.7)	–	(0.9)	(0.9)	(1.0)	–	(1.0)
Stiffness	kip/in	269.7	312.7	358.5	372.1	387.1	415.3	234.3	138.4	334.6
	(kN/mm)	(47.2)	(54.8)	(62.8)	(65.2)	(67.8)	(72.7)	(41.0)	(24.2)	(58.6)
$D_{0.90}$	kip/in	36.8	69.9	36.8	–	43.6	43.6	48.863	–	48.863
	(kN/mm)	(6.4)	(12.2)	(6.4)	–	(7.6)	(7.6)	(8.6)	–	(8.6)

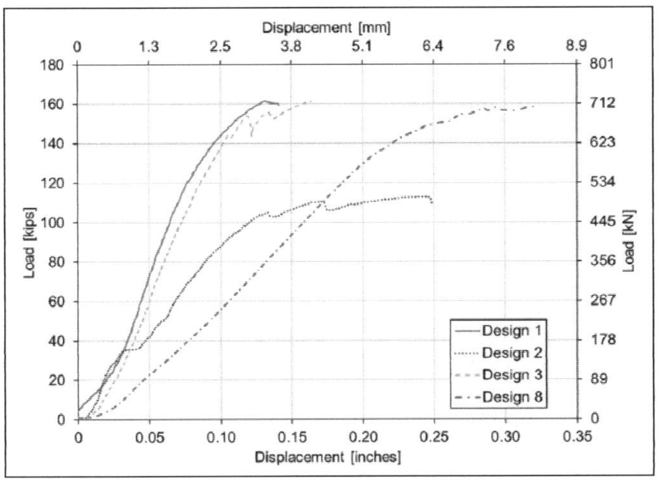

Figure 7. Load-displacement results for rail seat positive bending tests – typical replicates.

estimating the mean is not greater than plus or minus 6.6 kN (1.5 kips). Design 3 had the highest load capacity, while Design 4 presented the highest displacements at failure. In addition, the highest linear elastic displacements were associated with Design 1 (least stiff).

Based on the results presented in Table 3, the crosstie designs with highest displacements and load capacity were identified. Nevertheless, it is necessary to evaluate which characteristics are desirable. While it is generally accepted that high load capacity is a preferable characteristic, the desired level of displacement is not always clear for a system. While extreme displacements can be unsafe for train operations, zero displacement will result in excessively stiff systems that are subjected to higher impact loads and premature cracking in brittle components (*19*, *20*), perhaps even accelerating ballast deterioration. Considering that wood crossties have a modulus of elasticity that is three to five times lower than concrete crossties (*9*), the displacements in Table 3 could likely be increased without posing additional risk for railroad operation (even considering the shorter crosstie spacing in wood crosstie track). Previous research has shown that the bending of one concrete crosstie tested out of track leads to a small gauge-widening effect (approximately 3 mm (0.12 inch) for a typical freight crosstie design in the US) (*21*). When installed in track, the gauge widening effect due to crosstie bending should be even smaller given the contribution of adjacent crossties and the rail itself, which is an indication that crossties that are more ductile would have a negligible effect on railroad safety. In this regard, at least one such reduced-modulus concrete is being developed to "alleviate premature cracking due to high stress concentrations" (*22*). Therefore, while more research is needed to prescribe ideal stiffness values, having greater deflections seems to be a positive attribute in this context, so long as the minimum bending capacity requirements are met.

The load-displacement curves of the rail seat positive bending tests are shown in Figure 7, where one typical replicate is displayed per design. The roughness in the graphs is mostly explained by the lack of a seating load procedure for rail seat tests, yielding inaccurate absolute displacement results, as mentioned earlier. However, the lack of accurate ultimate-state-displacement data should not pose a problem for optimization of future designs since allowing for the rail seat cross sections to go beyond the elastic region and crack should be avoided to prevent an undesirable progressive increase of track stresses and deflections (*23*). In addition, rail seat cracks resulting from positive bending moments originate at the bottom of the crosstie, making it impossible to visually inspect such cracks. This is not the case for the crosstie center, as center cracks originate from the crosstie top – thus being more visible – and pose a smaller risk than at the rail seat (*21*, *23*, *24*).

Table 4 summarizes the rail seat test results in a manner similar to that of Table 3. Designs 1, 2, and 3 are possibly the most prevalent designs of concrete crossties with typical pretensioning wires used in the U.S. and Design 8 is recognized

Table 4. Experimental Results and 90% Confidence Margin for Rail Seat Tests

		Crosstie design			
		1	2	3	8
Ultimate load	kip	155.1	114.4	159.4	153.0
	(kN)	(690.0)	(508.9)	(708.9)	(680.6)
$D_{0.90}$	kip	10.5	10.5	8.5	15.1
	(kN)	(46.8)	(46.8)	(37.9)	(67.1)
Stiffness	kip/in	1937.2	924.3	2306.6	765.6
	(kN/mm)	(339.2)	(161.9)	(403.9)	(134.1)
$D_{0.90}$	kip/in	346.4	346.4	241.8	346.4
	(kN/mm)	(60.7)	(60.7)	(42.3)	(60.7)

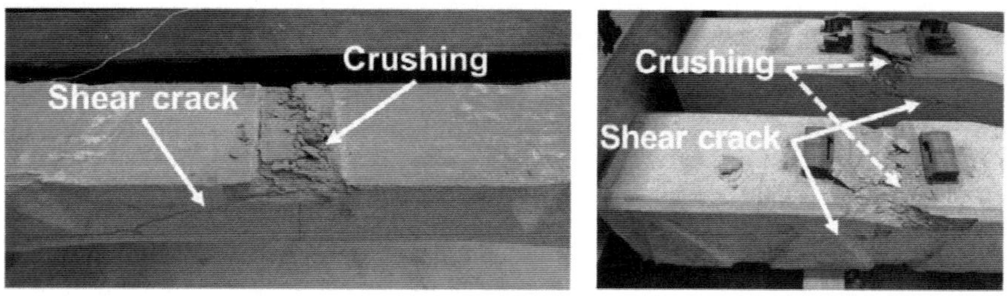

Figure 8. Failed crossties with crushed concrete and shear crack at (*a*) center and (*b*) rail seat.

as being a durable crosstie despite being discontinued. Given that each of these designs have proven performance records in heavy haul lines, it seems that Design 3 has the highest reserve capacity for both center and rail seat cross sections, while Design 2 seems to be a more economical alternative. Based on Tables 3 and 4, some of the most prevalent designs in the U.S. (Designs 1–3) have ultimate capacities within the ranges of 163.4 kN (36.7 kips) to 208.6 kN (46.9 kips) for the center negative bending moment test and 508.9 kN (114.4 kips) to 708.9 kN (159 kips) for the rail seat positive bending moment test. These values correspond to 55.9 kNm (495 kip-in) to 71.5 kNm (633 kip-in) center negative bending moments and 75.9 kNm (672 kip-in) to 105.8 kNm (936 kip-in) rail seat positive bending moments. It is worth noting that common in-service demands are lower than 22.7 kNm (201 kip-in) center negative bending moments and 33.9 kNm (300 kip-in) for rail seat positive bending moments, which are typical AREMA cracking limits (*25*).

It is important to highlight that most tested specimens failed in compression at the center and at the rail seat, with concrete crushing in the outmost fibers. Subsequently, with the exception of a few cases, shear failure also occurred. With the flexural failure happening first, it can be inferred that the experimental protocol succeeded in assessing the flexural strength of the crossties. In some cases, however, it was difficult to determine which failure mode happened first (i.e., shear or flexural failure), as they seem to happen simultaneously. In these cases, the ultimate load could indicate either the flexural or the shear capacity of the specimen depending on which the primary failure mode was. Nevertheless, the authors were comfortable in assuming that the ultimate load represented reaching the bending capacity of the specimen, as crushing of concrete still happened, even if only as a secondary failure mode. Figure 8 shows failed specimens to illustrate the presence of both crushing of concrete and cracking due to shear. The sudden crushing failure seems to be an indication that crossties are compression-controlled beams as defined by the ACI 318-14 code (*26*), which is generally undesirable because the low ductility (i.e., high stiffness) of these elements limits their ability to redistribute loads away from an overloaded section (*27*). This could be prevented by, for example, reducing the amount of tension reinforcement tendons to allow for less stiff beams.

Conclusions

This paper has proposed and implemented testing protocols adapted from AREMA recommendations for assessing the center negative and rail seat positive bending capacities by four-point loading with half-moon steel bars. Load-displacement curves are the test's output, with special attention given to the slope of the linear region to assess stiffness, the ultimate load to assess strength, and the ultimate displacement for the center section to assess the overall deformation capacity before failure. The primary benefits of the proposed test protocol with its revisions to the current AREMA recommended practices are listed below:

- provides opportunity to objectively compare various designs
- provides opportunity to objectively compare crossties of same design to assess deterioration or manufacturing variability
- provides quantifiable results (not simply a pass/fail test)
- associates underlying concept of failure to ultimate condition (as opposed to presence of cracks)
- provides a simple test protocol that can be easily implemented by the rail industry

This new testing approach can provide more details about each crosstie's flexural design characteristics and the possibility of better comparing various designs. Generally, HAL freight railroads are interested in high-toughness crossties that limit brittleness and are ductile enough to conform to the track support conditions without failing due to excessive deformation. Additionally, manufacturers can benefit from this methodology, as it could be used to demonstrate that current crossties tend to have previously unquantified reserve capacities, which could lead to more economical designs with less material. The test protocols proposed in this paper, therefore, have potential benefits for both current and future concrete crosstie design methods.

Structural properties of representative heavy haul concrete crosstie designs in North America were quantified for the rail seat and center sections by testing 70 individual crossties. Variability was observed among the flexural capacity of the designs tested, even if only considering the most prevalent wire-based pre-tensioned crossties. This indicates that some designs have greater reserve capacity while others are more economical. For these designs (Designs 1–3) the ultimate capacities were within the ranges:

- center negative bending moments: 55.9 kNm (495 kip-in) to 71.5 kNm (633 kip-in)
- rail seat positive bending moments: 75.9 kNm (672 kip-in) to 105.8 kNm (936 kip-in)

In addition, the failure mode observed in laboratory experimentation was of sudden crushing both at the center and at the rail seat. Finally, results on center bending of bottom-abraded crossties indicate that even severe abrasion throughout the entire cover may reduce the ultimate capacity of a crosstie to about 70% of its capacity with normal wear.

Acknowledgments

This research effort is funded by the Federal Railroad Administration (FRA), part of the United States Department of Transportation (U.S. DOT) and the National Railroad Passenger Corporation (Amtrak). Most laboratory experiments were performed by the Rail Transportation and Engineering Center (RailTEC) of the University of Illinois at Urbana-Champaign (UIUC) at the Research and Innovation Laboratory (RAIL) at the Harry Schnabel, Jr. Geotechnical Laboratory in Champaign, Illinois. Some experiments, however, were performed at UIUC's Nathan M. Newmark Civil Engineering Laboratory (NCEL). The authors also would like to acknowledge the following industry partners: Union Pacific Railroad; BNSF Railway; National Railway Passenger Corporation (Amtrak); Progress Rail Services, Inc.; GIC USA; Hanson Professional Services, Inc.; and CXT Concrete Ties, Inc., an LB Foster Company. J. Riley Edwards has been supported in part by the grants to the UIUC Rail Transportation and Engineering Center (RailTEC) from CN and Hanson Professional Services. Alejandro Álvarez-Reyes has been partially supported by the National Council of Science and Technology of Mexico (CONACYT).

Author Contributions

The authors confirm contribution to the paper as follows: study conception and design: Josué C. Bastos, Alejandro Álvarez-Reyes, Marcus S. Dersch, and J. Riley Edwards; data collection: Josué C. Bastos and Alejandro Álvarez-Reyes; analysis and interpretation of results: Josué C. Bastos, Alejandro Álvarez-Reyes, Marcus S. Dersch, and J. Riley Edwards; draft manuscript preparation: Josué C. Bastos, Marcus S. Dersch, J. Riley Edwards, Alejandro Álvarez-Reyes, and Christopher P.L. Barkan. All authors reviewed the results and approved the final version of the manuscript.

References

1. Harris, D., R. Lutch, T. Ahlborn, and P. Duong. Optimization of a Prestressed Concrete Railroad Crosstie for Heavy-Haul Applications. *Journal of Transportation Engineering*, Vol. 137, No. 11, 2011, pp. 815–822.
2. Zeman, J. C. *Hydraulic Mechanisms of Concrete-Tie Rail Seat Deterioration*. Master's thesis. University of Illinois at Urbana-Champaign, Urbana, IL, USA, 2010.
3. Zeta-Tech. *Assessment of Concrete Tie Life on US Freight Railroads*. Railway Tie Association, 2010, p. 12.
4. Weber, J. W. Concrete Crossties in the United States. *Prestressed Concrete Institute Journal*, Vol. 14, No. 1, 1969, pp. 46–57.
5. Remennikov, A. M., and S. Kaewunruen. Experimental Load Rating of Aged Railway Concrete Sleepers. *Engineering Structures*, Vol. 76, 2014, pp. 147–162.
6. Wolf, H. E. *Flexural Behavior of Prestressed Concrete Monoblock Crossties*. Master's thesis. University of Illinois at Urbana-Champaign, Urbana, IL, USA, 2015.
7. Edwards, J. R., Z. Gao, H. E. Wolf, M. S. Dersch, and Y. Qian. Quantification of Concrete Railway Sleeper Bending Moments Using Surface Strain Gauges. *Measurement*, Vol. 111, No. Supplement C, 2017, pp. 197–207.
8. Murray, M. Heavy Haul Sleeper Design – a Rational Cost-Saving Method. Presented at the IHHA 2015 Conference, Perth, WA, Australia, 2015.
9. AREMA. Chapter 30 – Ties. In *Manual for Railway Engineering*. The American Railway Engineering and Maintenance of Way Association, Landover, MD, USA.
10. Bastos, J. C., M. S. Dersch, and J. R. Edwards. Determination of Critical Track Conditions and Their Impact on the Performance

of Concrete Crossties and Fastening Systems. Presented at the AREMA 2015 Annual Conference, Minneapolis, MN, USA, 2015.

11. American Society for Testing and Materials. *C78/C78M – 16, Standard Test Method for Flexural Strength of Concrete (Using Simple Beam with Third-Point Loading)*. ASTM International, 2016.
12. Khoury, S., A. A.-H. Aliabdo, and A. Ghazy. Reliability of Core Test – Critical Assessment and Proposed New Approach. *Alexandria Engineering Journal*, Vol. 53, No. 1, 2014, pp. 169–184.
13. Yip, W. K., and C. T. Tam. Concrete Strength Evaluation through the Use of Small Diameter Cores. *Magazine of Concrete Research*, Vol. 40, No. 143, 1988, pp. 99–105.
14. Bartlett, F. M., and J. G. Macgregor. Effect of Core Diameter on Concrete Core Strengths. *Materials Journal*, Vol. 91, No. 5, 1994, pp. 460–470.
15. Bungey, J. H. Determining Concrete Strength by Using Small-Diameter Cores. *Magazine of Concrete Research*, Vol. 31, No. 107, 1979, pp. 91–98.
16. Gaynor, R. D. Effect of Horizontal Reinforcing Steel on the Strength of Molded Cylinders. *Journal Proceedings*, Vol. 62, No. 7, 1965, pp. 837–840.
17. Marquis, B., J. LeBlanc, H. Yu, and D. Jeong. *Volpe Report on CSX Train Derailment on MN Tracks*. National Transportation Safety Board, USA, 2014, p. 41.
18. Ott, R., and M. Longnecker. *An Introduction to Statistical Methods and Data Analysis*. Cengage Learning, 2008.
19. Gettu, R., Z. P. Bazant, and M. E. Karr. Fracture Properties and Brittleness of High-Strength Concrete. *ACI Materials Journal*, Vol. 87, No. 6, 1990, pp. 608–618.
20. Taherinezhad, J., M. Sofi, P. A. Mendis, and T. Ngo. A Review of Behavior of Prestressed Concrete Sleepers. *Advances in Rail Track Infrastructure Research and Practice (ARTIRP)-2013*, Vol. 13, No. 1, 2013, pp. 1–16.
21. Bastos, J. C. *Analysis of the Performance and Failure of Railroad Concrete Crossties with Various Track Support Conditions*. Master's Thesis. University of Illinois at Urbana-Champaign, Urbana, IL, USA, 2016.
22. Rizos, D. C. High-Strength Reduced-Modulus High Performance Concrete (HSRM-HPC) for Prestressed Concrete Tie Applications. Presented at the 2016 Joint Rail Conference (JRC), Columbia, SC, USA, 2016.
23. Domingo, L. M., C. Zamorano Martin, C. Palenzuela Aviles, and J. I. Real Herraiz. Analysis of the Influence of Cracked Sleepers under Static Loading on Ballasted Railway Tracks. *The Scientific World Journal*, Vol. 2014, 2014, p. 363547.
24. Dean, F. E. *The Effect of Service Loading on the Bending Strength of Concrete Ties*. FRA/TTC-81/11 Intrm Rpt. Battelle Columbus Laboratories, Federal Railroad Administration, and Transportation Systems Center, Pueblo, Colorado, USA, 1981.
25. Gao, Z., H. E. Wolf, M. S. Dersch, Y. Qian, and J. R. Edwards. Field Measurements and Proposed Analysis of Concrete Crosstie Bending Moments. Presented at the American Railway Engineering and Maintenance-of-Way Association Annual Conference, Orlando, Florida, 2016.
26. ACI Committee 318. *Building Code Requirements for Structural Concrete (ACI 318-14) and Commentary*. American Concrete Institute, 2014.
27. Wight, J. K. *Reinforced Concrete: Mechanics and Design*. Prentice Hall, Hoboken, NJ, USA, 2016.

The Standing Committee on Railway Maintenance (AR060) peer-reviewed this paper (18-05811).

The material in this paper represents the position of the authors and not necessarily that of FRA or Amtrak.

TRR
JOURNAL OF THE TRANSPORTATION RESEARCH BOARD

Article

Transportation Research Record
2018, Vol. 2672(10) 125–135

DOI: 10.1177/0361198118784379
journals.sagepub.com/home/trr
SAGE

Full-Scale Model Testing on Ballasted High-Speed Railway: Dynamic Responses and Accumulated Settlements

Wei Li[1], Xuecheng Bian[1], Xiang Duan[1], and Erol Tutumluer[2]

Abstract

High-speed trains generate much higher vibrations in track structures than conventional trains and intensive train passages (e.g., on the Beijing–Shanghai high-speed railway line where the train passage interval is less than 5 minutes) cause accumulated permanent settlement in the railway track substructures, which will decrease track performance and jeopardize the safety of trains. Since very few field measurements on ballasted high-speed railways are available in literature, this paper presents experimental results of vibration velocity, dynamic soil stress, and the accumulated settlement of a ballasted high-speed railway from a full-scale model testing facility with simulated trains moving loads at various speeds. A portion of a realistic ballasted railway consisting of track structure, ballast layer, subballast, embankment, and piled foundation was constructed in a larger box. An eight-actuator sequential loading system was used to generate equivalent vertical loadings on the track structure for simulating the dynamic excitations due to train movements. Dynamic stresses measured in the track substructure layers (ballast, subballast, and embankment) were found to be strongly dependent on train speeds especially for speeds higher than 144 km/h. It was found that both the vibration velocity and the dynamic soil stress were greatly amplified as the train speed increased to 300 km/h, and the ballast layer effectively reduced the vibrations transmitted from the track structure to underlying soil. The accumulated settlement of the substructure did not reach a stable state even after 100,000 moving train loads at a speed of 300 km/h.

High-speed railway networks are developing rapidly in China, with 20,000 km of track at the end of 2016. The length of ballasted high-speed track is growing with the extension of high-speed rail lines to western areas of China. Ballasted track has lower construction costs in contrast to ballastless slab track.

A ballasted railway generally consists of a track superstructure (rails, fasteners, crossties) and a geotechnical substructure, which commonly comprises compacted granular soils in the embankment and coarse aggregates used in the ballast and subballast layers. Because modern train speeds have increased to more than 300 km/h, high-speed railways are exposed to intensive traffic loadings at high speeds, which will cause excessive vibration as well as settlement to the substructure and eventually accelerate track deterioration (*1–4*). Thus, it is of utmost importance to gain a better understanding of the dynamic responses and settlement behavior of track substructures under train traffic loading.

Physical model tests are efficient tools for investigating track dynamic behavior trends under train loads. Anderson and Key (*5*) developed a two-layer ballast bed of 1 m length, 0.8 m width and 0.6 m height, and carried out a series of model tests to study deformation of the ballast layer under stationary cyclic loadings. Al Shaer et al. (*6*) established a 1:3 scale model of a ballasted track with three hydraulic jacks applying vertical loads directly to the crossties with a shift in loading time to simulate the passage of a train bogie. Momoya et al. (*7*) and Ishikawa et al. (*8*) reported test results using a 1:5 scale ballasted track with a small moving wheel at speeds of 600 mm/min and 700 mm/min, respectively. Experimental results revealed that the loading method has a considerable influence on the cyclic plastic deformation of railway ballast because settlement in a moving-wheel loading test is generally much larger than that in a fixed-point cyclic loading test. Most currently available model test facilities use small-scale models of railways and focus on the development of settlement of the railway under either cyclic or moving wheel loads at fairly low speeds. To evaluate the performance of railways under high-speed trains, moving train loads should be simulated in physical model tests. Besides, a full-scale model is preferable to represent realistic conditions on site.

Bian et al. (*9–12*) first developed a full-scale physical model (15 × 5 × 6 m) for ballastless high-speed railways and successfully verified its ability to simulate moving train loads. They then reproduced realistic dynamic responses of

[1]Department of Civil Engineering, Zhejiang University, Hangzhou, China
[2]Department of Civil and Environmental Engineering, University of Illinois at Urbana-Champaign, Urbana, IL

Corresponding Author:
Address correspondence to Xuecheng Bian: bianxc@zju.edu.cn

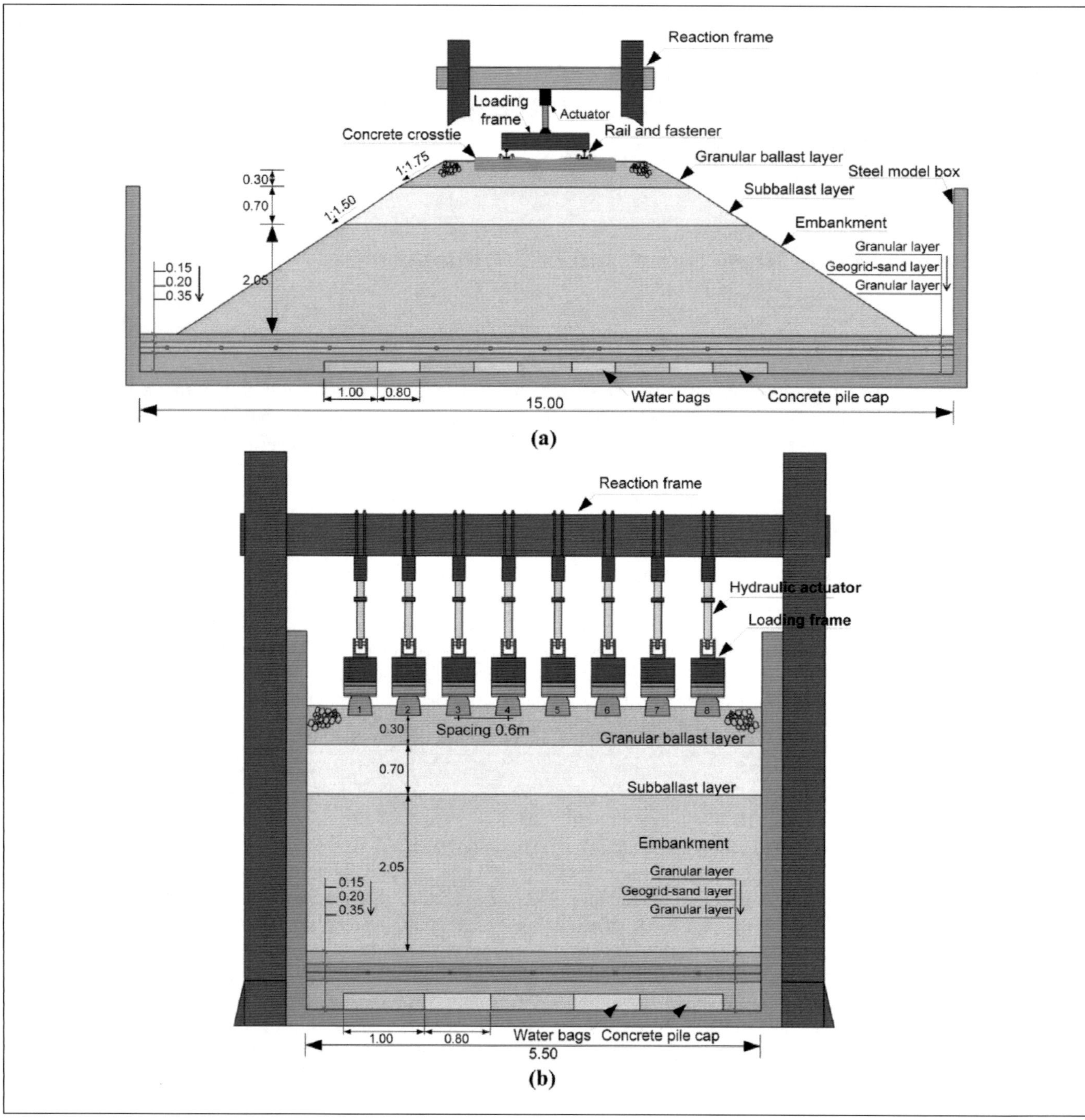

Figure 1. Geometry details of the physical model test for ballasted high-speed railway: (*a*) transversal; (*b*) longitudinal cross-section view (unit: meters).

on-site railways. In this full-scale test set-up, a ballasted railway track test section was recently built to investigate dynamic loading behavior of ballasted track substructure under simulated moving train loads. This paper first introduces this full-scale model testing facility with simulated moving train loads. Next, test results including vibration velocities, dynamic soil stresses, and accumulated settlements at various speeds are presented and discussed.

Full-Scale Physical Model of Ballasted Railways

Geometry and Construction of Full-Scale Physical Model

A full-scale physical model of a ballasted high-speed railway is shown in Figure 1, and was constructed following the

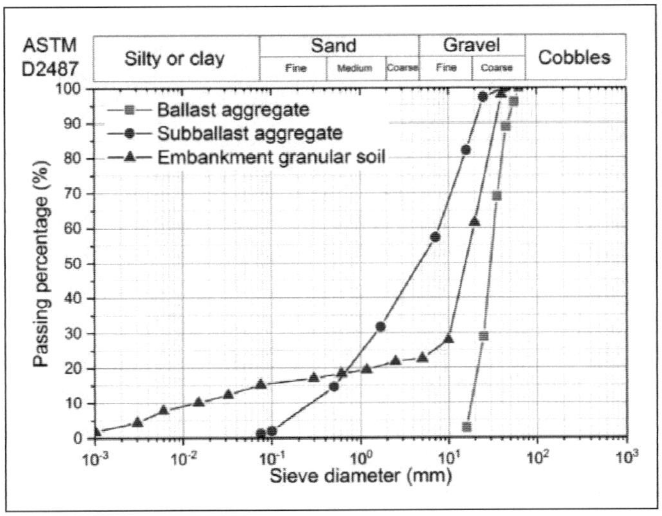

Figure 2. Grain size distributions of geomaterials in the tested ballasted railway.

Chinese high-speed railway design code in a rectangular steel box, which was 15.0 m long and 5.5 m wide. The numerical analyses proved that the dynamic responses of track structure and underlying soils would have least boundry effects with these sizes (*10*). Moreover, a 15 mm-thick smooth latex layer was spread on the inner walls of the steel box to minimize the boundary effects.

The track structure consisted of eight pairs of segmented rails and fasteners, eight IIIc-type concrete crossties, and the underlying ballast layer. Figure 1*b* shows that the ballast layer was 0.3 m thick underneath the crossties, while it was 0.485 m thick between two neighboring crossties. The substructure comprised, from the bottom up, a 0.7 m-thick piled foundation, a 2.0 m-thick embankment and a 0.7 m-thick subballast layer. The slopes of ballast layer and its underlying roadbed were 1:1.75 and 1:1.50 respectively.

Fifteen 0.2 m-thick C20 concrete pile caps were cast *in situ* with square dimensions of 1 m, and they were equidistantly distributed with 1.8 m spacing. PVC water bags were placed around the pile caps to model the subsoil, while the pile-soil differential settlement was controlled by discharging water from these bags (*13*). Above these pile caps, a 0.35 m-thick granular layer, a 0.20 m-thick geogrid-sand layer and a 0.15 m-thick granular layer were constructed to support the embankment. The embankment was constructed using a granular soil, while the subballast and the ballast layer were built with unbound aggregate materials.

The grain-size distributions of these three geomaterials are presented in Figure 2, and the properties of the aggregates and the granular soil are summarized in Table 1. A tamping machine was used to compact these three layers to the desired stiffness, which is usually represented by the modulus of subgrade reaction called K30. Static plate load tests using a circular plate with a diameter of 30 cm were conducted to determine the stiffness characteristics listed in Table 1.

For the track structure, eight 2.6 m-long IIIc-type concrete crossties were placed in the ballast layer with 0.60 m spacing. Eight pairs of elastic strip V-type fasteners were installed on the surface of the crossties using anchor bolts of 30 mm diameter, screwed at the positions of the insulated sleeves with a torque of 160 N•m. The elastic strip V-type fasteners consist of two T-shaped screw spikes, two W2 type clips and a rubber pad. Two CHN60-type rails were fixed to the crosstie using W2-type clips and the T-shaped screw spikes at a target torque of 160 N•m. Finally, the loading apparatus consisting of eight loading frames, eight hydraulic actuators, and a set of reaction frames weas installed on the physical model.

Sequential Loading System

Traffic loading, induced by train wheels moving on the rail surface, transfers first to the track structure via fasteners and then to the underlying track structures. In this process, the fasteners play an important role in transforming train loads moving on the rail surface into vertical time-varying loadings on the crossties at each fastener's position. The model testing took this advantage to simulate the dynamic impact of moving loads on the track structure by applying the equivalent dynamic loads at each fastener's position with actuators.

The equivalent dynamic loadings were determined by a previously verified rail-crosstie-foundation dynamic analysis model (*14*) as shown in Figure 3. The rails were simplified as Euler beams. The fasteners were modeled as discrete viscoelastic springs while the underlying geotechnical track substructure was modeled as distributed viscoelastic springs. From this simplified analytical model, the dynamic loading acting at the positions of the fasteners can be obtained for the train running along the rails. It is possible, therefore, to remove the rails in the model test and apply the same time varying loads directly onto the fasteners by servo-controlled hydraulic actuators but with a time shift to simulate a train's movement. As shown in Figure 3, for the Chinese CRH3-type train with a170-kN axle load and a 25-m carriage length, the magnitude of the M-shaped signal was 90 kN, and the phase lag between each actuator was 9°.

The loading apparatus consists of eight hydraulic actuators equipped with servo valves, load and displacement transducers, a computer-based control system, and a set of reaction frames. The fastener forces or other kinds of loads are input into the controller program of the control system. The controller program sends electronic command signals to the servo valve in each actuator, which then directs oil from power cylinder to drive the movement of the actuator. Figure 4 compares the output loads from the actuator with the input command loads, taking a load equivalent to a train moving at, for example, 36, 108, 216 and 300 km/h.

Table 1. Properties of Geomaterials Used in Full-Scale Physical Model Test

Material	Wet density (kg/m^3)	Water content (%)	Dry density (kg/m^3)	K30 (MPa/m)
Dry ballast aggregate	/	/	1740.1	216
Subballast aggregate	2005.8	6.62	1880.5	336
Embankment granular soil	1931.1	4.13	1854.5	227

Note: The minimum K30 values for ballast, subballast and embankment layers are 190 MPa/m, 190 MPa/m and 130 MPa/m respectively according to the Chinese code for design of high-speed railway.

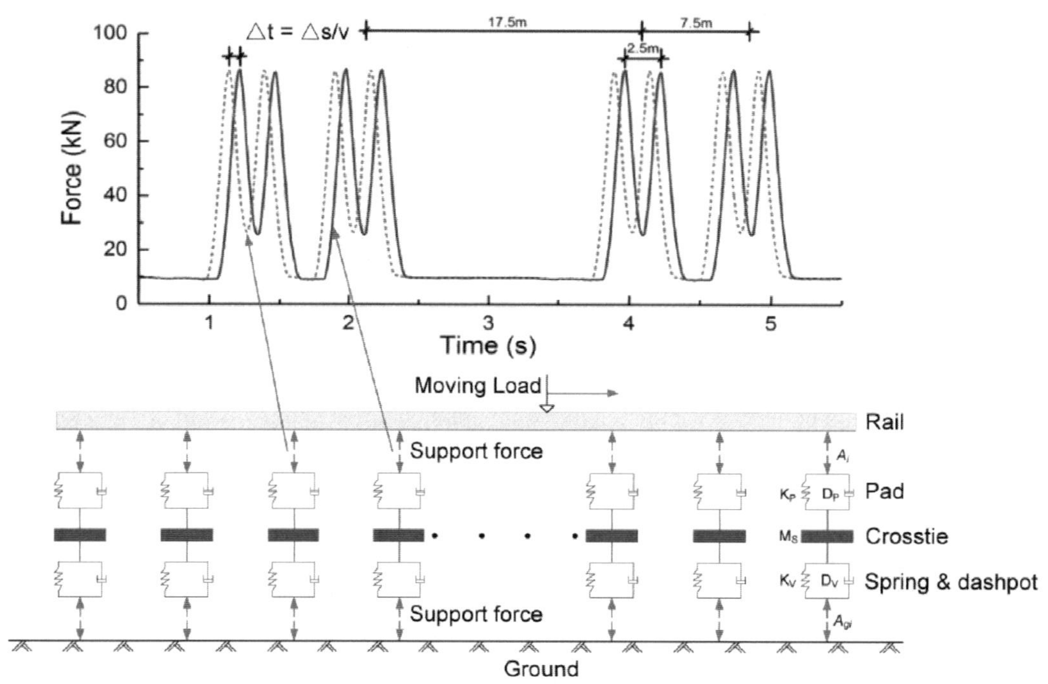

Figure 3. Dynamic loads in a sequential loading system to simulate moving train loads on track structure.

Instrumentation

Instrumentation consisted of 23 soil pressure sensors, 8 vibration velocity sensors and 12 settlement transducers, all installed in the physical model to record dynamic responses and permanent deformations. These transducers were calibrated in the laboratory before installation. As shown in Figure 5*a* and 5*c*, the vibration velocity sensors were mounted on the concrete crossties, the ballast layer, and the subballast layer. Figure 5*c* shows that soil pressure sensors were embedded in the ballast layer, the subballast layer, the embankment, and the piled foundation in the middle cross-section of the physical model. To record the accumulated settlement of each substructure layer, eight settlement plates, consisting of an LVDT, a settlement bar, a casing steel tube, and a settlement plate with square dimension of 0.2 m were installed on the subballast layer as well as the embankment, and furthermore, four LVDTs were mounted on the concrete crossties to record the settlement of the track, as shown in Figure 5*a* and 5*b*.

Test Results and Discussion

Since CRH3-type trains with the axle load of 170 kN are widely used in Chinese high-speed railways, this kind of train load was adopted in the following full-scale model tests. Dynamic response data of the ballasted track, for example, vibration velocity, dynamic stress, and so forth were monitored under nine simulated train speeds ranging from 36 to 300 km/h. Besides, up to 100,000 simulated train loads at the speed of 144, 216, and 300 km/h were applied to the track structure to record the accumulated settlements of the substructure.

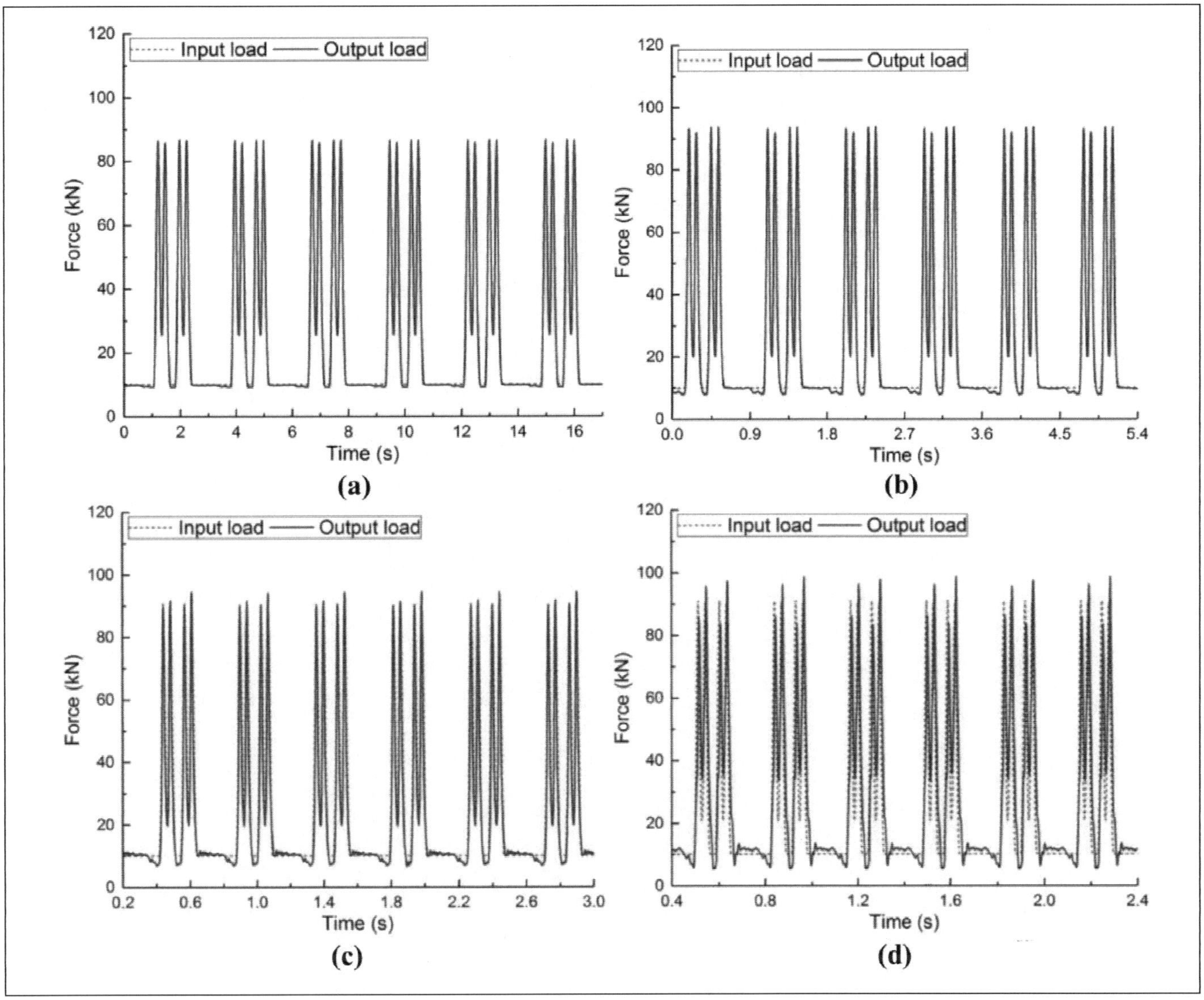

Figure 4. Comparisons of input and output loading curves of the test apparatus. Train speeds are (*a*) 36 km/h, (*b*) 108 km/h, (*c*) 216 km/h, and (*d*) 300 km/h.

Vibrations of the Ballasted Track Structure and the Railway Substructure

Time histories of vibration velocities at the concrete crosstie and ballast surface measured in the model tests are presented in Figure 6 for three different train speeds. The positive sign indicates the downward vibrations. The locations of the train axles can be identified easily from the peaks and troughs in the dynamic responses. In addition, higher speed train operation induces stronger vibrations than lower speed operation, and vibrations attenuate dramatically from the concrete crosstie to the ballast layer. The ballast consists of large unbound aggregate particles that typically move independently in the assembly and interact only at contact points. Such particle-scale behaviors could hardly be captured by the conventional vibration sensors. "SmartRocks" (*15*) is an innovative instrument for recording particle–particle contacts in realistic railway ballast layers and offers an insight into the particle–scale interactions under a moving train load. "SmartRocks" can be adopted in the next phase of the full-scale testing to record particle–scale vibration behaviors.

Figure 7 shows the vibration velocity amplitudes at the ballast and underlying subballast in the transversal direction from the track center due to moving train loads at three different speeds. The vibration velocity amplitudes peak at the ballast near the fasteners, and they are 20.58, 18.39, and 8.12 mm/s at the train speeds of 300, 252, and 108 km/h, which decrease by 25.7%, 27.8% and 39.9%, respectively, at the ballast near the crosstie end and further decrease by 12.7%, 18.49% and 24.9%, respectively, at the ballast center. A higher train speed

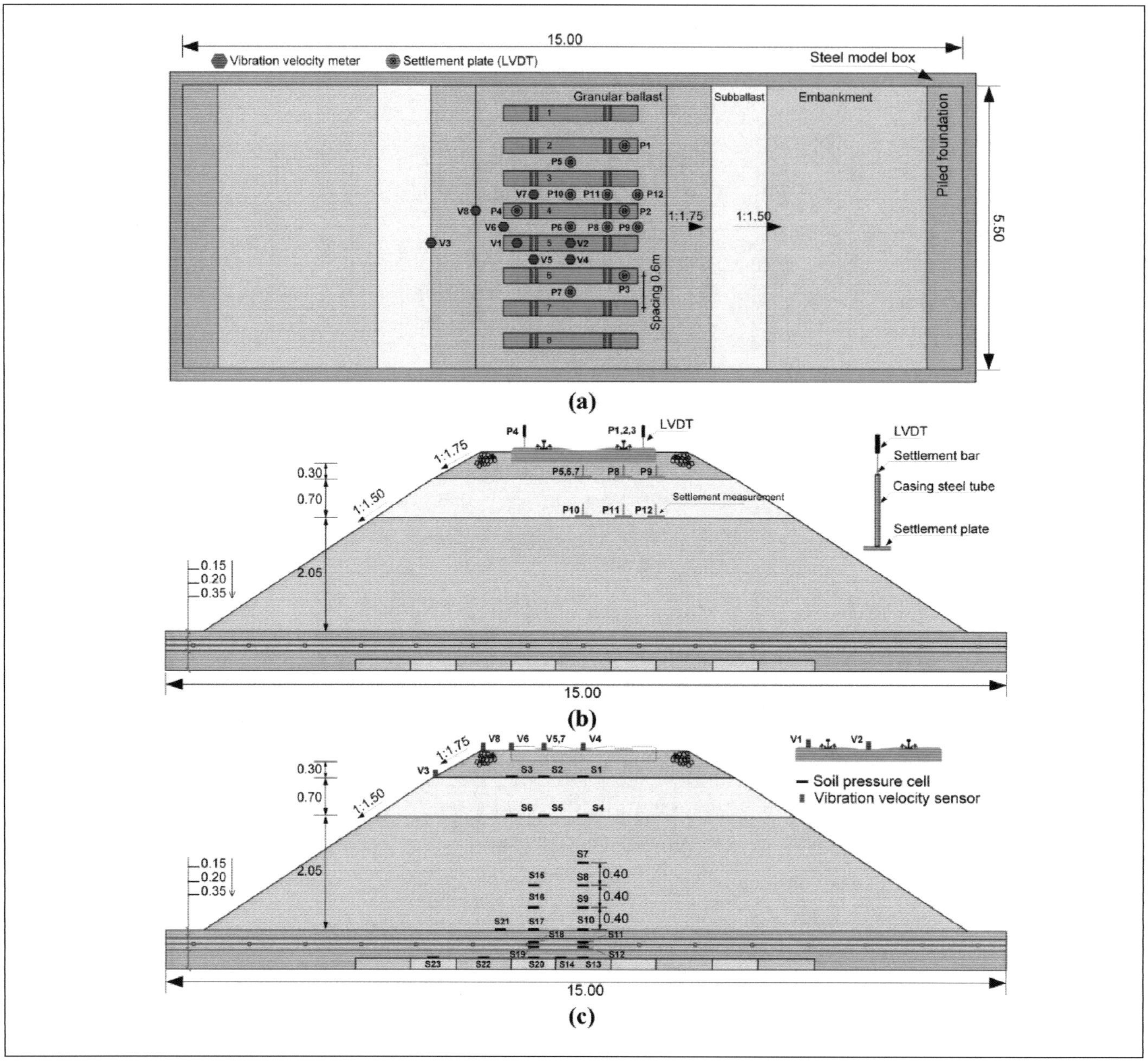

Figure 5. Layout of the installed sensors: (*a*) top view of the velocity sensors and the settlement plates; (*b*) cross-section view of the settlement plates; (*c*) cross-section view of the velocity meters and soil pressure sensors (unit: meters).

results in slower attenuations of the vibration velocities. The amplitude of vibration velocity at the ballast near the fasteners reaches 20.58 mm/s when the train speed increases to 300 km/h. Such strong vibrations at the ballast near the fasteners will jeopardize the stability of the unbound materials. With the propagation of the vibrations, the vibration velocity amplitudes decrease at a higher rate from the ballast to the subballast. The ballast under the concrete crossties has a considerable reduction effect on vibration propagation.

The dependence of the track's vibration velocity on train speed is illustrated in Figure 8. The vibration velocities of the track structure and the underlying aggregates present an almost monotonically increasing tendency with the train speed. Vibration responses at the concrete crosstie have higher increasing rates than those on the underlying aggregates (ballast and subballast layers). Such phenomena may be attributed to the different vibration characteristics of track structure and aggregates under train loads moving at high speeds. The vibration velocities of the concrete crossties (V1 and V2) grow slowly when train speed is above 216 km/h. It is noted that vibration velocities at the shoulders of ballast and subballast (V3 and V8) increase considerably as the train

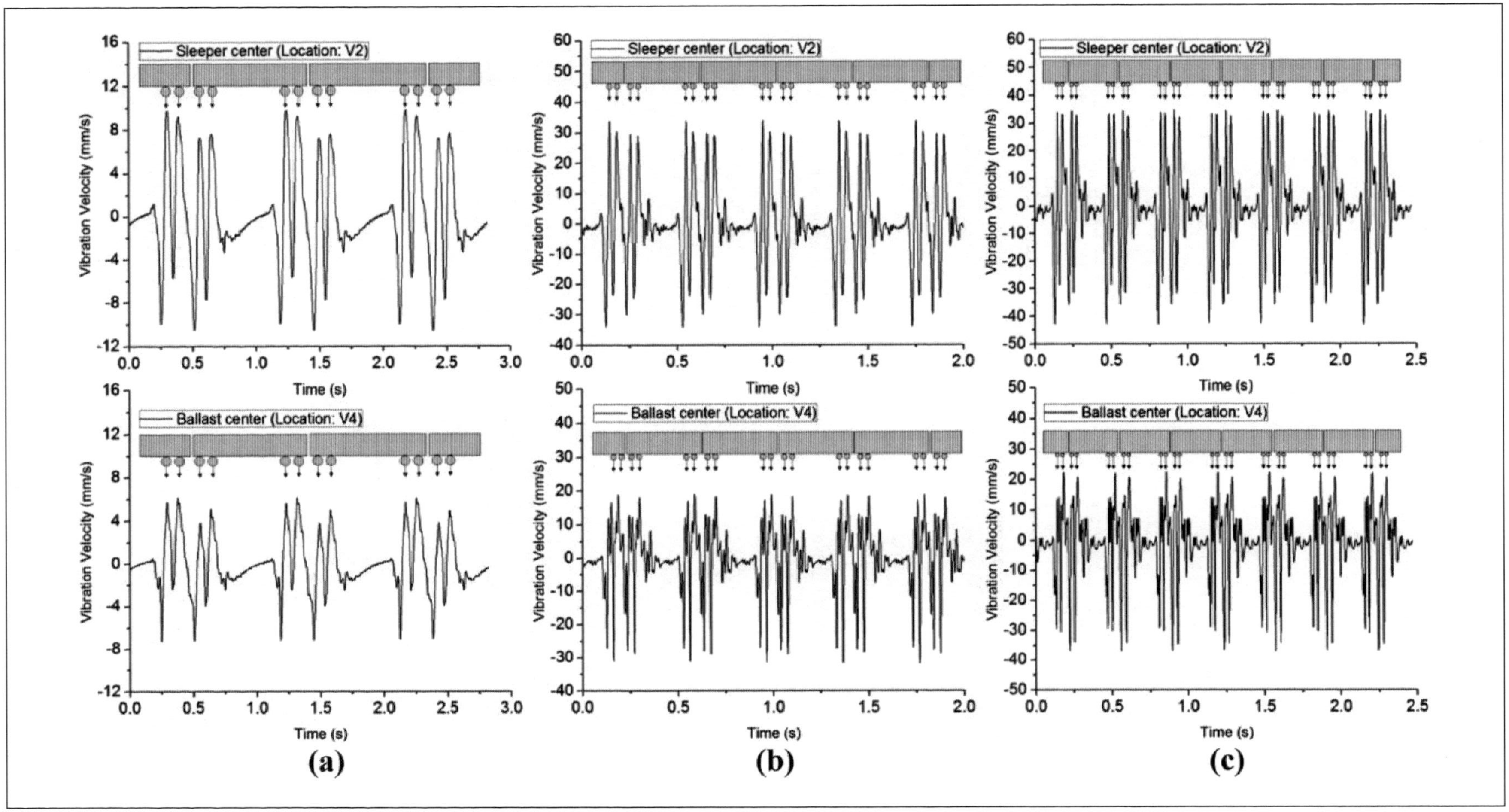

Figure 6. Time histories of vibration velocities at concrete crosstie and ballast surface (axle load = 170 kN): (*a*) train speed = 108 km/h; (*b*) train speed = 252 km/h; (*c*) train speed = 300 km/h.

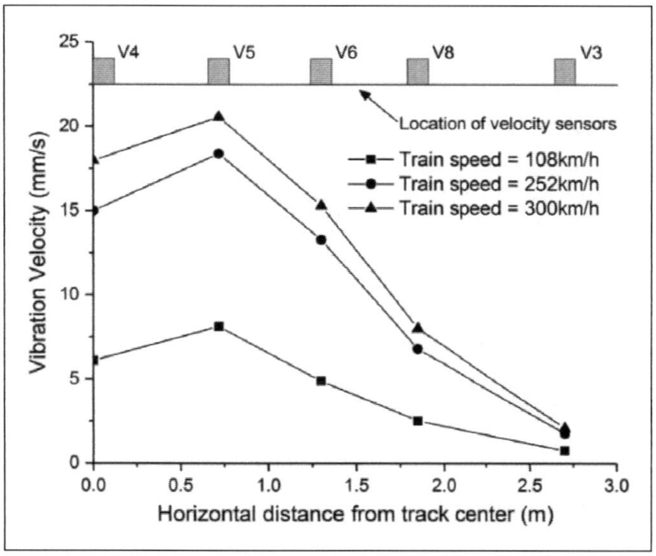

Figure 7. Distribution of vibration velocity in the horizontal direction from the track center (axle load = 170 kN).

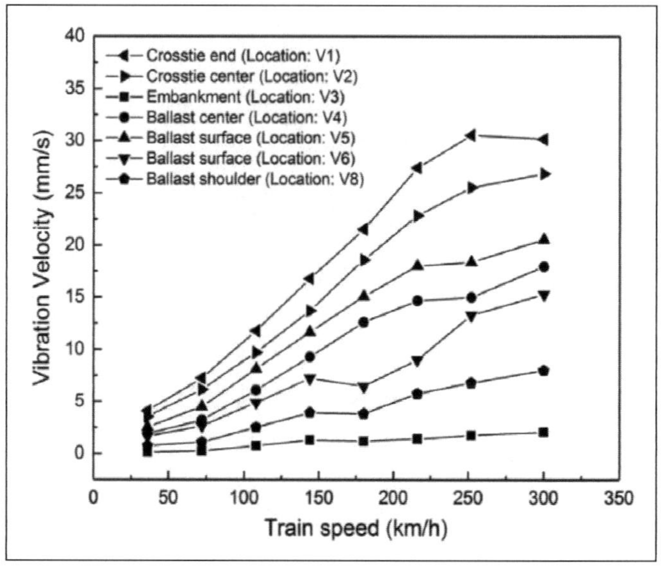

Figure 8. Relationship between the vibration velocity and the train speed based on the model test (axle load = 170 kN).

speed is raised to 300 km/h, but their amplitudes do not show a pronounced increasing trend as other sensors (such as V4, V5, and V6). The factor leading to this phenomenon is that the granular material at the shoulder is relatively loose due to lack of confining pressure. This will cause significant vibration reduction when waves propagate from the track to the ballast shoulder and finally to the subballast shoulder.

Distribution of Dynamic Soil Stresses in Track Substructure

Figure 9 shows the time histories of vertical dynamic soil stresses at different depths from the ballast surface. The variation of each dynamic soil stress exhibits the shape of a letter "M", which resembles the profile of train loads applied on the fasteners. With soil depth increasing, the vertical dynamic soil stresses caused by the train loads decrease from 157.26 kPa at the subballast surface to 14.49 kPa at 1.0 m depth below ballast surface at a train speed of 108 km/h.

The distribution patterns of contact pressure between the crosstie and the roadbed in the ballasted railways have been widely studied (*16*); they are known to affect the horizontal stress distribution pattern at various depths beneath the crosstie. Figure 10 shows the transverse distribution of the train-induced stress at various depths under the crosstie at three different speeds. The dynamic stress under the concrete crosstie is not even in the transverse direction. It is the lowest under the rail and increases slightly toward the crosstie center, while it peaks under the crosstie edge. This is not consistent with the distribution patterns of contact pressure described in (*16*), it is attributed to the discrete nature of ballast aggregates and the compacting operation which pose uneven support conditions to concrete crossties. In this case, a higher percentage of axle load is borne by ballast particles near the crosstie edge.

Figure 11 shows the test results of dynamic stress amplitudes at various locations beneath the track under moving train loads at different speeds. In this figure, the amplitudes of dynamic soil stresses at places near the embankment surface (e.g., location S4, S5, and S6) are not very sensitive to the train speed, but they show clear ascendant tendencies with the increase of the train speed. While for the places at the subballast (e.g., locations S1, S2, and S3), the situation becomes different. The dynamic stress amplitudes generally exhibit high ascendant tendencies with some fluctuations when the train speed is lower than 144 km/h, but these values increase quickly when the train speed is in the range from 144 to about 300 km/h. This dynamic amplification of stresses in the subballast and embankment have also been observed in previous field measurements (*17*).

Accumulated Settlement of Track Structure under Moving Train Loads

Figure 12 presents the accumulated settlement within the substructure layers graphed with the number of moving train loads for the speeds 144, 216, and 300 km/h. The number of moving train loads is up to 100,000 for each train speed considered. It is noted that it is is easier for the accumulated settlement to reach a stable state under repeated train moving loads when train speed is below 216 km/h. While the accumulated settlement increases faster at 300 km/h, it does not reach a stable condition even after 100,000 moving train loads. Increasing the train speed leads to a higher stress level with higher loading

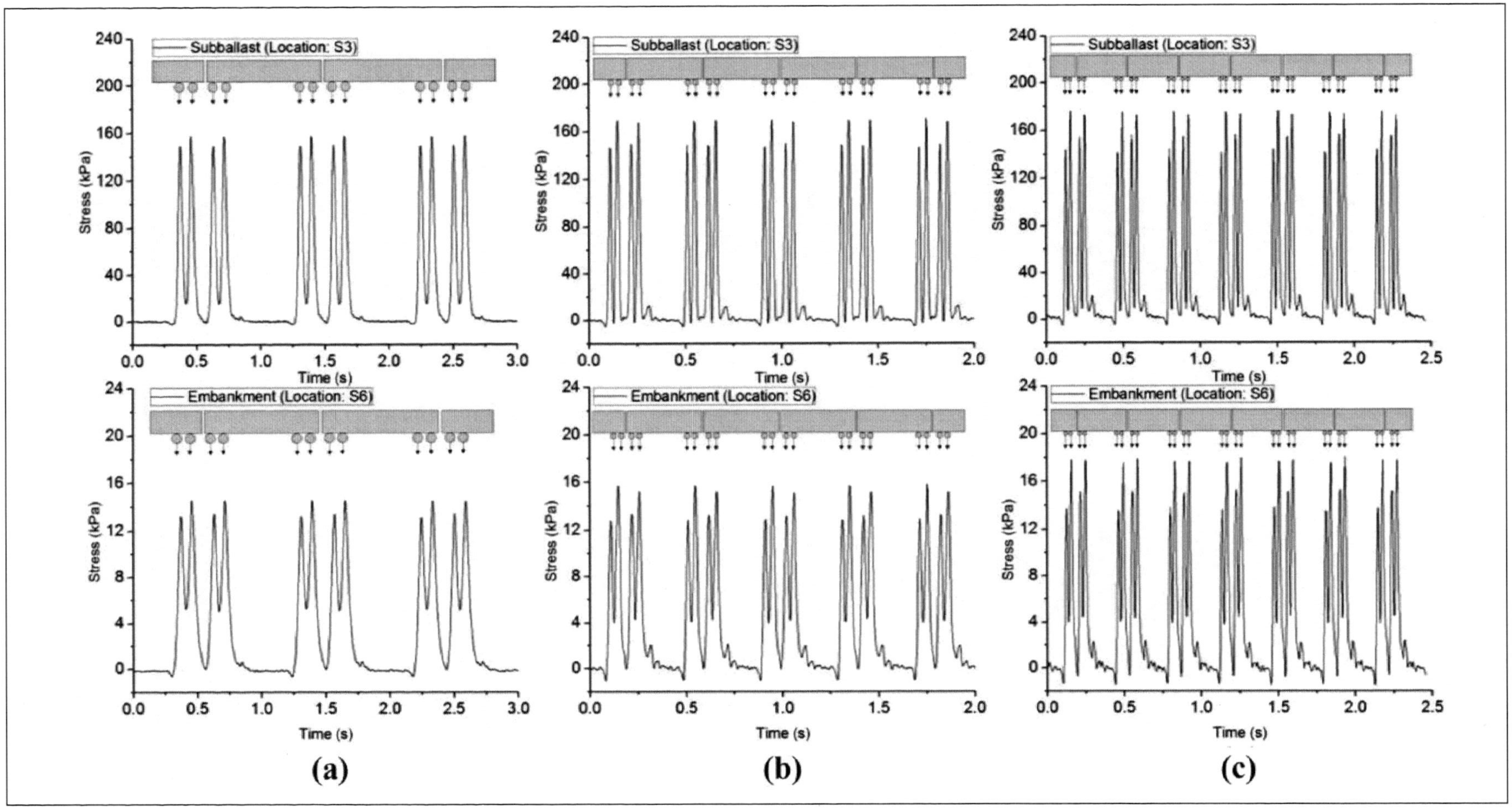

Figure 9. Time histories of dynamic soil stresses measured at various locations: (*a*) train speed = 108 km/h; (b) train speed = 252 km/h; (*c*) train speed = 300 km/h.

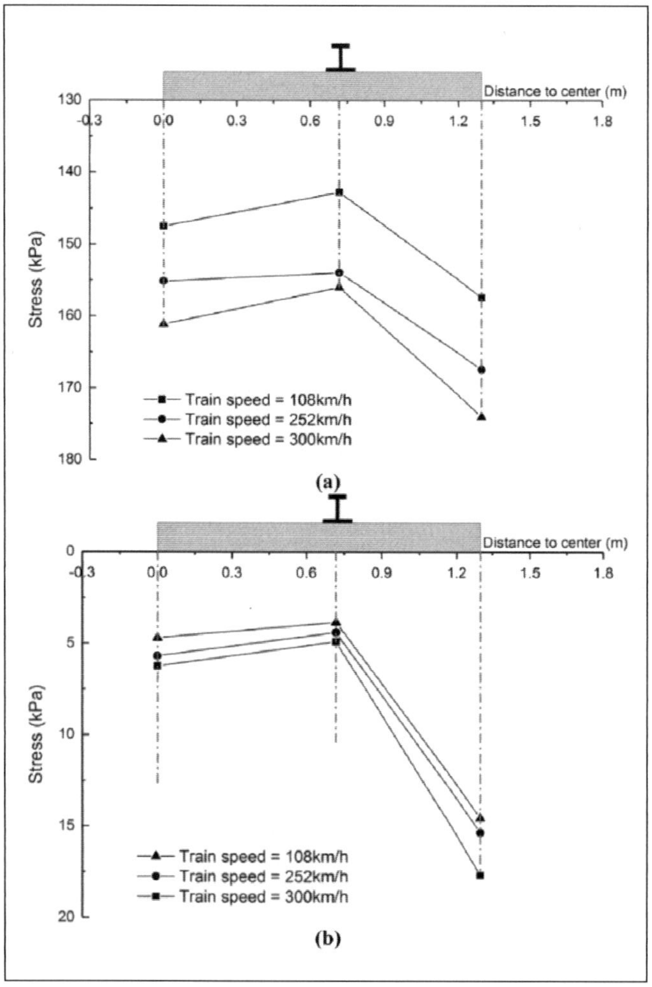

Figure 10. Stress distribution under track structure in the depth of (*a*) 0.3 m and (*b*) 1.0 m from concrete crosstie bottom.

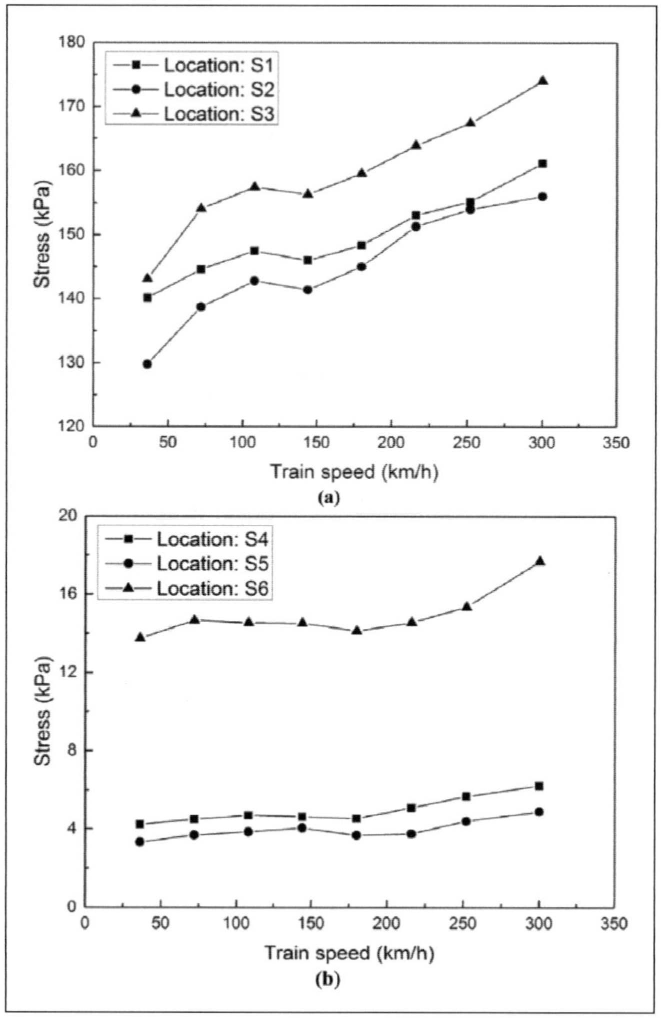

Figure 11. Relationship between dynamic soil stresses and train speeds at the depth of (*a*) 0.3 m and (*b*) 1.0 m from concrete crosstie bottom.

frequencies distributed on each substructure layer, as shown in Figure 11. This indicates that the loading frequency has a profound effect on the accumulated settlement within substructure layers comprising compacted granular materials with a higher stress level. This trend has also been observed in laboratory triaxial tests on ballast aggregates (*18*).

Conclusion

This paper has presented an experimental study conducted in a full-scale model testing facility with the objective of investigating the dynamic response and settlement behavior of a ballasted track under simulated moving train loads. From the test results, the following conclusions can be drawn:

1) Vibration velocities of the track structure and the underlying substructure presented a monotonically ascending tendency with the train speed. The ballast layer played a key role in reducing vibrations transmitted from the track structure to underlying soils.
2) Dynamic stresses measured in the track substructure layers (ballast, subballast, and embankment) were found to be strongly dependent on train speed especially when the train speed rises above 144 km/h. Dynamic soil stresses measured on the surfaces of subballast and embankment were distributed unevenly beneath the crosstie in transverse direction, and in this case, the stress level was much higher near the crosstie edge.
3) Increasing train speeds lead to additional settlement development. It was easier for the accumulated settlement to reach a stable state under repeated moving train loads when the train speed was below 216 km/h. As the train speed increased to 300 km/h, a stable state could not be reached even after 100,000 moving train loads.

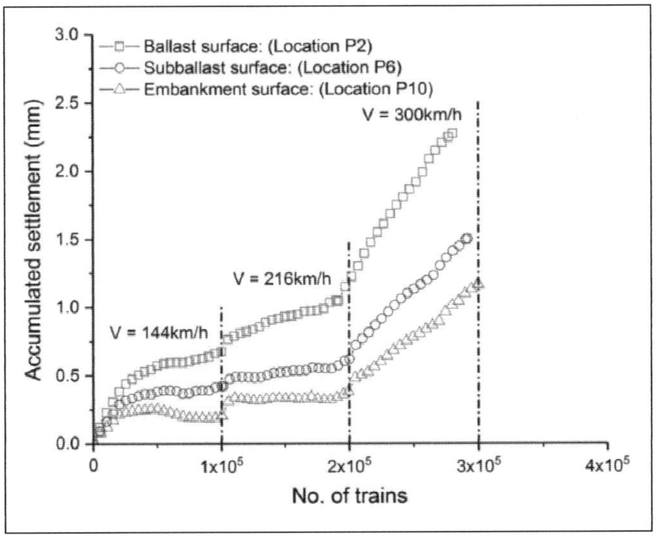

Figure 12. Accumulated settlement measured at different substructure surface layers under three train speeds.

Acknowledgments

Financial support from the National Natural Science Foundation of China (Grant No.51678524), Advanced Newton Fellowship of Royal Society of UK (NA140060) and Distinguished Young Scientist Program of Zhejiang Province (LR16E080002) are gratefully acknowledged.

Author Contributions

The authors confirm contribution to the paper as follows: study conception and design: Xuecheng Bian; data collection: Wei Li, Xiang Duan; analysis and interpretation of results: Wei Li, Xuecheng Bian, Erol Tutumluer; draft manuscript preparation: Wei Li, Xuecheng Bian, Erol Tutumluer. All authors reviewed the results and approved the final version of the manuscript.

References

1. Bian, X. C., Y. M. Chen, and T. Hu. Numerical Simulation of High-Speed Train Induced Ground Vibrations Using 2.5 D Finite Element Approach. *Science in China Series G: Physics Mechanics and Astronomy*, Vol. 51, No. 6, 2008, pp. 632–650.
2. Huang, H., and S. Chrismer. Discrete Element Modeling of Ballast Settlement Under Trains Moving at "Critical Speeds". *Construction and Building Materials*, Vol. 38, 2013, pp. 994–1000.
3. Tutumluer, E., Y. Qian, Y. M. A. Hashash, J. Ghaboussi, and D. D. Davis. Discrete Element Modelling of Ballasted Track Deformation Behaviour. *International Journal of Rail Transportation*, Vol. 1, No. 1–2, 2013, pp. 57–73.
4. Bian, X., Y. Gao, H. Huang, and Y. Xie. Hanging Tie Study Using the "Moving Deflection Spectrum". *Proceedings of the Institution of Mechanical Engineers Part F-Journal of Rail and Rapid Transit*, Vol. 231, No. 8, 2017, pp. 982–990.
5. Anderson, W. F., and A. J. Key. Model Testing of Two-Layer Railway Track Ballast. *Journal of Geotechnical and Geoenvironmental Engineering*, Vol. 126, No. 4, 2000, pp. 317–323.
6. Al Shaer, A., D. Duhamel, K. Sab, G. Foret, and L. Schmitt. Experimental Settlement and Dynamic Behavior of a Portion of Ballasted Railway Track Under High Speed Trains. *Journal of Sound and Vibration*, Vol. 316, No. 1–5, 2008, pp. 211–233.
7. Momoya, Y., E. Sekine, and F. Tatsuoka. Deformation Characteristics of Railway Roadbed and Subgrade Under Moving-Wheel Load. *Soils and Foundations*, Vol. 45, No. 4, 2005, pp. 99–118.
8. Ishikawa, T., E. Sekine, and S. Miura. Cyclic Deformation of Granular Material Subjected to Moving-Wheel Loads. *Canadian Geotechnical Journal*, Vol. 48, No. 5, 2011, pp. 691–703.
9. Bian, X. C., H. G. Jiang, Y. M. Chen, J. Q. Jiang, and J. Han. A Full-Scale Physical Model Test Apparatus for Investigating the Dynamic Performance of the Slab Track System of a High-Speed Railway. *Proceedings of the Institution of Mechanical Engineers Part F-Journal of Rail and Rapid Transit*, Vol. 230, No. 2, 2016, pp. 554–571.
10. Bian, X. C., H. G. Jiang, C. Cheng, Y. M. Chen, R. P. Chen, and J. Q. Jiang. Full-Scale Model Testing on a Ballastless High-Speed Railway Under Simulated Train Moving Loads. *Soil Dynamics and Earthquake Engineering*, Vol. 66, 2014, pp. 368–384.
11. Jiang, H. G., X. C. Bian, Y. M. Chen, and J. Han. Impact of Water Level Rise on the Behaviors of Railway Track Structure and Substructure Full-Scale Experimental Investigation. *Transportation Research Record: Journal of the Transportation Research Board*, 2015. 2476: 15–22.
12. Jiang, H. G., X. C. Bian, C. Cheng, Y. M. Chen, and R. P. Chen. Simulating Train Moving Loads in Physical Model Testing of Railway Infrastructure and Its Numerical Calibration. *Acta Geotechnica*, Vol. 11, No. 2, 2016, pp. 231–242.
13. Chen, R. P., Y. W. Wang, X. W. Ye, X. C. Bian, and X. P. Dong. Tensile Force of Geogrids Embedded in Pile-Supported Reinforced Embankment: A Full-Scale Experimental Study. *Geotextiles and Geomembranes*, Vol. 44, No. 2, 2016, pp. 157–169.
14. Takemiya, H., and X. C. Bian. Substructure Simulation of Inhomogeneous Track and Layered Ground Dynamic Interaction Under Train Passage. *Journal of Engineering Mechanics*, Vol. 131, No. 7, 2005, pp. 699–711.
15. Liu, S., H. Huang, T. Qiu, and L. Gao. Comparison of Laboratory Testing Using SmartRock and Discrete Element Modeling of Ballast Particle Movement. *Journal of Materials in Civil Engineering*, Vol. 29, No. 3, 2017, p. D6016001.
16. Sadeghi, J., and P. Barati. Evaluation of Conventional Methods in Analysis and Design of Railway Track System. *International Journal of Civil Engineering*, Vol. 8, No. 1, 2010, pp. 44–55.
17. Kempfert, H. G., and Y. Hu. Measured Dynamic Loading of Railway Underground. *Proc., 11th Pan-American Conference on Soil Mechanics and Geotechnical Engineering*, International Society for Soil Mechanics, Foz do Iguaçu, Brazil, 1999, pp. 843–847.
18. Sun, Q. D., B. Indraratna, and S. Nimbalkar. Deformation and Degradation Mechanisms of Railway Ballast under High Frequency Cyclic Loading. *Journal of Geotechnical and Geoenvironmental Engineering*, Vol. 142, No. 1, 2016, p. 04015056.

The Standing Committee on Railway Maintenance (AR060) peer-reviewed this paper (18-03555).

TRR
JOURNAL OF THE TRANSPORTATION RESEARCH BOARD

Article

Transportation Research Record
2018, Vol. 2672(10) 136–145

DOI: 10.1177/0361198118793500
journals.sagepub.com/home/trr
SAGE

Quantification of Loading Environment and Flexural Demand of Prestressed Concrete Crossties under Shared Corridor Operating Conditions

Ricardo J. Quirós-Orozco[1], J. Riley Edwards[1], Yu Qian[2], and Marcus S. Dersch[1]

Abstract

Optimization of the design of railway infrastructure and its components requires a comprehensive understanding of the loading demands that are expected. Currently, many design guidelines for track components use historical wheel loads and calculate bending moments based on broad assumptions. However, tools are available to accurately quantify and characterize the variability each load has on a particular component. This is particularly important in shared use rail corridors where higher speed passenger services operate on the same infrastructure as heavy axle load (HAL) freight trains. Each traffic type generates unique demand variabilities, and these should be incorporated into a holistic approach to optimized track design. Therefore, researchers at the University of Illinois at Urbana-Champaign (UIUC) are conducting field research aimed at the characterization of field conditions on Amtrak's Northeast Corridor through the use of wheel impact load detector (WILD) data and concrete crosstie surface strain gauges. Results from this experimentation show high variability of loads resulting from varied types of train operations and significant differences in impact load ratios. Finally, laboratory measured flexural capacity for the concrete crossties showed a conservative design with a potential margin of improvement in reduction of residual capacity (i.e., factor of safety).

Amtrak's Northeast Corridor (NEC) is the most densely traveled rail corridor in the United States, with almost 11.7 million annual passengers traveling on Amtrak services and another 250 million annual commuter rail passengers distributed among eight operators (*1*). In addition, the corridor is traversed by approximately 50 freight trains per day from two Class I freight railroads. As such, the corridor's infrastructure experiences a variety of loading conditions ranging from heavy axle load (HAL) freight operations to high-speed passenger train services. Historically, the track structure and components such as crossties have been designed through a process based to a large extent on practical experience (*2*) or have assumed simplistic static loading cases. Optimization of track components for highly variable loads and speed requires an in-depth understanding of the effect of each service.

Before this study, research efforts had been undertaken to better understand the loading environment and improve design methodologies on HAL freight and intercity passenger rail corridors (*3, 4*). These efforts were accomplished, in part, through the analysis of data from wayside systems used for monitoring the performance of rolling stock such as the wheel impact load detector (WILD) (*5*). On a component level, surface strain gauge instrumentation has been used successfully in the quantification of flexural demands on concrete crossties (*2, 6*). In the case of Amtrak's NEC, field instrumentation dates back to 1983 when one of the first WILD sites was deployed to address the causes of transverse rail seat cracks in concrete crossties (*7*). The use of both WILD sites and instrumented crossties allows for a comprehensive characterization of the demand to which the track and its components are subjected. These data, when

[1]Rail Transportation and Engineering Center (RailTEC), Department of Civil and Environmental Engineering (CEE), University of Illinois at Urbana-Champaign (UIUC), Urbana, IL
[2]Department of Civil and Environmental Engineering, The University of South Carolina, Columbia, SC

Corresponding Author:
Address correspondence to Ricardo J. Quirós-Orozco: rjq2@illinois.edu

evaluated in conjunction with supporting laboratory experimental data facilitate the evaluation of the efficiency of a design to support the actual field loading demands. This analysis was performed as part of a project devoted to improvement of concrete crosstie specifications for Amtrak's NEC.

Overview of Field Instrumentation

WILD

A WILD is an electronic data collection device designed to measure and isolate vertical and lateral wheel forces with the use of either rail mounted strain gauges or accelerometers (*8, 9*). Although its primary objective is to evaluate the performance of the rolling stock and measure the impact forces caused by out-of-round or otherwise damaged wheels (*9, 10*), it has also proven to be a practical mechanism for producing reliable wheel load data that can serve rail infrastructure researchers and practitioners (*2*).

A common strain gauge–based WILD site is over 15 m (50 ft) in length, with a series of strain gauges microwelded to the neutral axis of the rail's web. They quantify the wheel load by a direct mathematical or a calibrated relationship between strain and force (*9*). Instrumentation is divided into several ballast cribs at various intervals to capture a single wheel's rotation up to five times, recording peak impact and average forces at a data collection rate of up to 30 kHz (*7, 11*). Electronic signal processors housed in a wayside enclosure analyze the data using an algorithm which isolates wheel tread irregularities and computes both nominal and peak load values (*12*). Rail infrastructure owners commonly define loading thresholds according to their operational procedures which facilitate the delivery of alerts when limits are exceeded (*8*). A WILD located adjacent to the surface strain–gauged crossties on Amtrak's NEC can provide information related to nominal load, dynamic load, peak load as well of peak ratios lateral load, car type, lateral load, hunting index, and speed. Further processing of WILD data was used to create a loading input database on which the present analysis is based.

Concrete Crosstie Surface Strain Gauges

Researchers in the Rail Transportation and Engineering Center (RailTEC) at the University of Illinois at Urbana-Champaign (UIUC) have previously used surface mounted strain gauges to measure bending moments experienced by concrete crossties under revenue service HAL freight trains (*13*) and in rail transit applications (*14*). In the case of Amtrak's NEC, temporary instrumentation of concrete crossties was performed in 1983 to identify causes of premature cracks (*7*) and again in 2014 to investigate a later generation of crosstie design and performance questions (*15*).

Field experimentation for this project was conducted in Edgewood, Maryland. Researchers instrumented a total of seven crossties on Track 2 (which handles primarily northbound traffic) near an operative WILD that provided wheel loads for the project. The sensors deployed allowed for the quantification of bending strains at discrete locations along the length of the crosstie caused by train loading (Figure 1). Surface strain gauges were oriented longitudinally along the chamfer near the top surface of the crosstie. Four crossties were instrumented with three gauges: under the two rail seats and one in the center of the crosstie. A fifth crosstie was instrumented with two additional gauges located halfway between each rail seat and center. Finally, two external crossties were instrumented with a single center strain

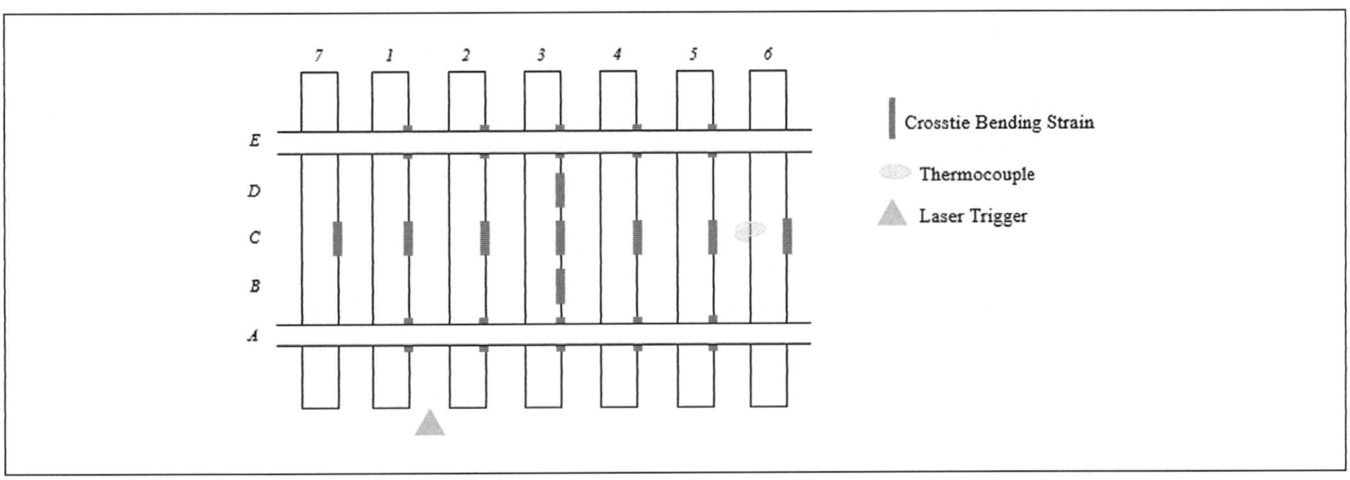

Figure 1. Plan view of instrumented crosstie section on Amtrak's NEC in Edgewood, Maryland.

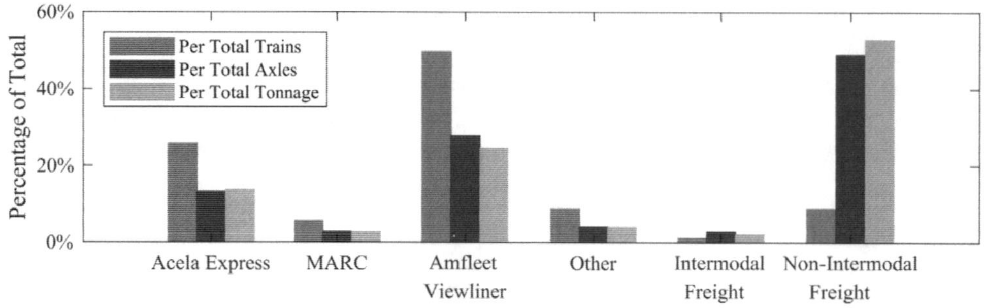

Figure 2. Distribution of services on the Amtrak NEC Edgewood, MD field site by total number of trains, axles, and gross tonnage.

gauge to capture additional bending data at the crosstie center given its critical nature as a design region. Independently of strain gauge instrumentation, thermocouples were deployed to capture ambient, top, and bottom of crosstie temperature. Collection of data using the aforementioned forms of instrumentation was automatically initiated through the use of a laser monitoring the track. Autonomous operation of the site allowed for uninterrupted data collection for weeks at a time.

Concrete crosstie calibration factors were determined through laboratory experimentation to relate the measured strains to a known bending moment, through a process that was detailed by Edwards et al. (*6*). In summary, these parameters were found by instrumenting several crossties of the same design and vintage as those installed in track using the same strain gauge layout as was used in the field. A calibrated load was applied to each crosstie at its rail seats in a configuration adapted from the design validation tests presented in Chapter 30, Section 4.9 of the American Railway Engineering and Maintenance-of-way Association (AREMA) *Manual for Railway Engineering* (*16*).

Wheel Load Analysis

Nominal and Peak Loading (using WILD Data)

Data were collected and processed from the Amtrak NEC WILD site located in Edgewood, Maryland from 1 January to 30 June 2017. These data are an aggregate of the information available for the two tracks at the WILD site. As such, an average of 100 trains per day were recorded for a total of 18,117 trains, 1,127,422 axles and 19.28 million gross tons (MGT) during the analysis period. Passenger train services at this location vary widely and include Amtrak's Acela Express and multiple intercity and regional rail services using Viewliner and Amfleet rolling stock. Commuter rail service is operated by the Maryland Area Regional Commuter (MARC) Agency with a mixture of different passenger coaches and locomotives.

In terms of total number of trains, traffic is dominated by Amtrak's intercity services, followed by Acela Express and MARC commuter trains (Figure 2). Freight transport is the dominant traffic type when measured by both total axles and tonnage. The "Other" category relates to maintenance or inspection equipment and trains in which the automatic equipment identification (AEI) tag was not properly read as the vehicle passed by the WILD site.

Descriptive statistical analysis of the wheel load data allows for the comparison of the variability of the rolling stock currently in use along this section of Amtrak's NEC. Results for nominal loads are presented in Figure 3 as percentage exceeding curves for each train type classification. On further analysis of both tracks, it was found that the distribution of passenger services is quite similar, whereas loaded freight traffic mostly used Track 3 and empty freight trains typically operated on Track 2.

For passenger trains, the variation between the weight of cars and their respective power units can be seen from the bimodal nature of the data (i.e., sharp change of slope of the data) around the 20 kips (89 kN). The commuter rail service operated by MARC also shows the presence of two types of locomotives with a significant disparity in their static wheel loads, as evidenced by the gradual slope of the curve. For freight services, the disparity is particularly noticeable between empty and loaded freight cars within the non-intermodal category. Meanwhile, for the intermodal cars, one can observe a higher variability in loads because of the varied nature of their payloads.

Figure 4 shows the same data once dynamic and impact factor effects are considered. The dynamic load is the highest load that is captured by the WILD for a given wheel as it passes through the series of instrumented cribs (i.e., the maximum reading from all of the strain-gauged cribs). As expected, there is an increase in the magnitude of loads in all train type categories. Although the differences between passenger coaches and power units are still visible, the change is more gradual, indicating a relationship of the impact factor with the nominal load of each wheel. This effect is particularly

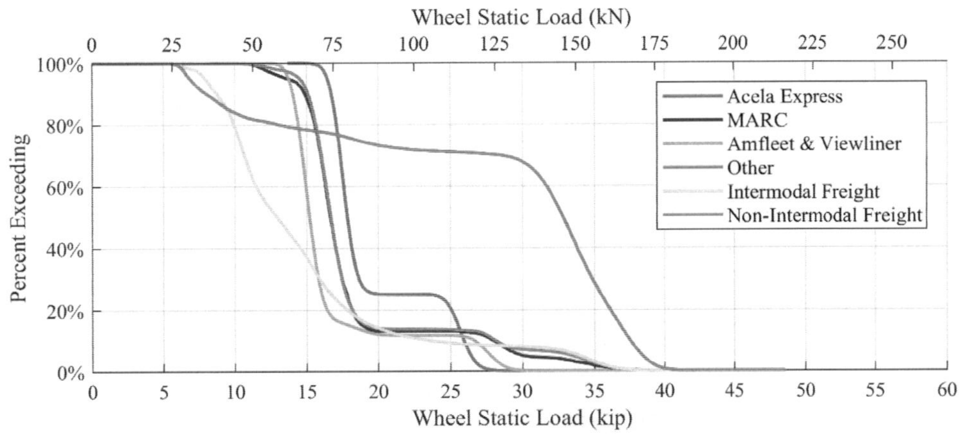

Figure 3. Nominal wheel load distribution on the Edgewood, MD field site by car type.

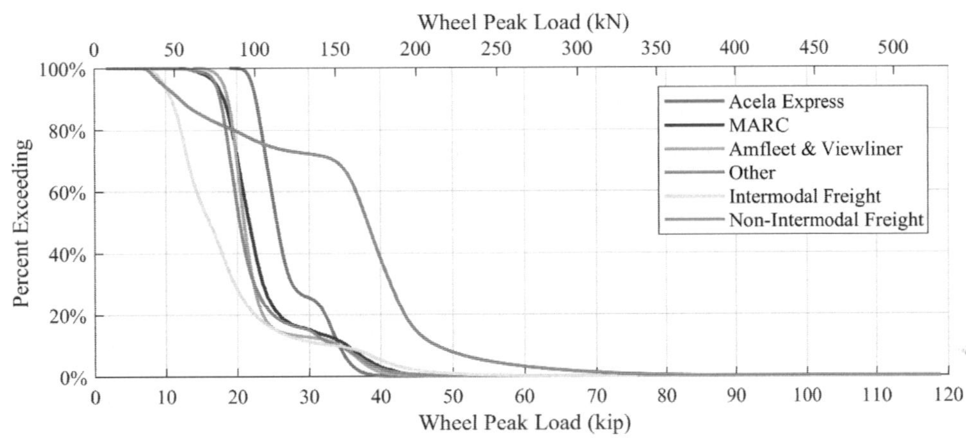

Figure 4. Peak wheel load distribution on the Edgewood, MD field site by car type.

noticeable for the HAL freight traffic, a possible indication that the impact factors for the empty cars are considerably higher than the loaded vehicles. Amfleet and Viewliner coaches also seem to have a higher average impact factor than MARC trainsets as the relationship between nominal and dynamic loads between both categories is reversed.

Impact Factory Variability

The relationship between nominal and dynamic loads for each wheel's passage is shown in Figure 5, in which black lines represent specific impact factor values. Most of the data (98.3% of all wheel loads) are located under the line representing an impact factor of 3 (200% over the nominal load), which is the design value proposed by AREMA for concrete crosstie design (*16*). Out of the 1.7% of wheels exceeding this threshold, the majority of wheels are from unloaded freight cars, with an average peak load of 27.9 kips (124 kN). This finding reinforces the assumption that 1) empty freight vehicles typically experience higher impact factors than loaded cars and 2) the total magnitude of these loads is unlikely to be damaging to the infrastructure, even with a high impact factor.

Flexural Bending Moments on Concrete Crossties

Flexural Bending Moments Results

Concrete crosstie bending moment data represent a subset of the total traffic passing Edgewood, MD on Track 2 during several multiweek time periods during the late

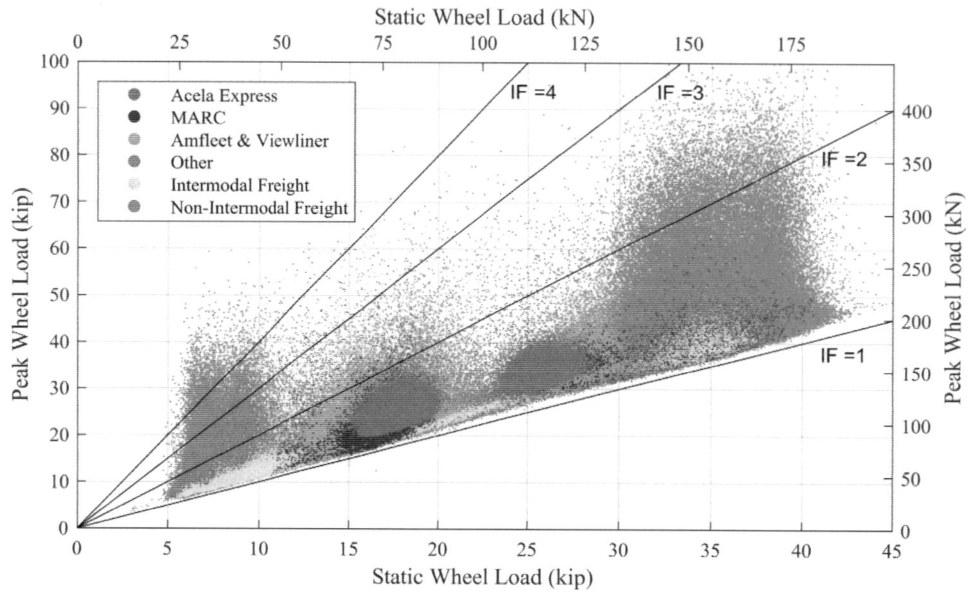

Figure 5. Relationship between static and peak wheel loads by vehicle classification.

part of 2016 and early 2017. In total, 4,612 trains were processed from 13 December 2016 to 17 May 2017. This dataset includes 4.5 out of the 7.6 MGT of traffic previously discussed with respect to WILD data.

Data processing techniques were required to identify peak data accurately. The most common noise encountered in strain gauges is related to typical alternate current (AC) interference at a frequency of 60 Hz (*17*). However, because of the presence of the electrified catenary on the NEC operating at 12 kV 25 Hz (*18*), interference in the strain gauge signal was identified at 25 Hz through visual analysis of the frequency spectrum. Filtering techniques based on the Chebyshev Type II band stop filter were used to remove the interference at the 25 Hz frequency and its harmonics.

Each dataset was compared with its respective WILD site data to obtain a classification of the rolling stock, quantify the number of axles and speed, and refine the processing algorithms. Descriptive statistics were prepared for bending moments the crossties were subjected to, based on the type of train traffic. Figures 6 and 7 show the results for the center and rail seat section, respectively.

As was expected, higher bending moments were recorded for loaded non-intermodal trains, reaching a maximum of 149 and 275 kips·in. [16.8 and 31.1 kN·m] for center negative and rail seat positive regions, respectively. It is important to note that these values have not reached the specified capacity of the crossties which are defined as 208 and 306 kips·in. [23.5 and 34.6 kN·m] for center negative and rail seat positive regions, respectively. Furthermore, laboratory quantified performance of the current crosstie design revealed actual first crack capacities of 347 and 487 kips·in. [39.2 and 55.0 kN·m] for center negative and rail seat positive regions, respectively. Center positive bending moments are likely to be related to temperature curling effects and associated changes in the support condition as well as relatively low magnitude vertical loads (*19*).

A comparison of the field bending moment data with the WILD site loads was performed to qualitatively assess the relationship between the variables. In general, there was a steady increase of bending moments as wheel load increased, as expected. However, it was also noted that center bending moments are not as sensitive to wheel load increases as rail seat moments are. This behavior was previously mapped to crossties with sufficient support under their rail seats (*20*), and good support under the rail seats is a plausible scenario considering the traffic characteristics at this location. A more detailed analysis of the support conditions for the site is presented in subsequent sections.

Detailed analysis of the top 10% of bending moments is presented in Table 1 as they reflect potential design level values. It is interesting to note the significant jump in the values from the two sources of data from the 99.5% to 100% percentiles in both flexural bending moment data and wheel peak loads.

Crosstie Support Condition Evaluation

Evaluation of ballast existing support conditions was performed through the use of a numerical back calculator

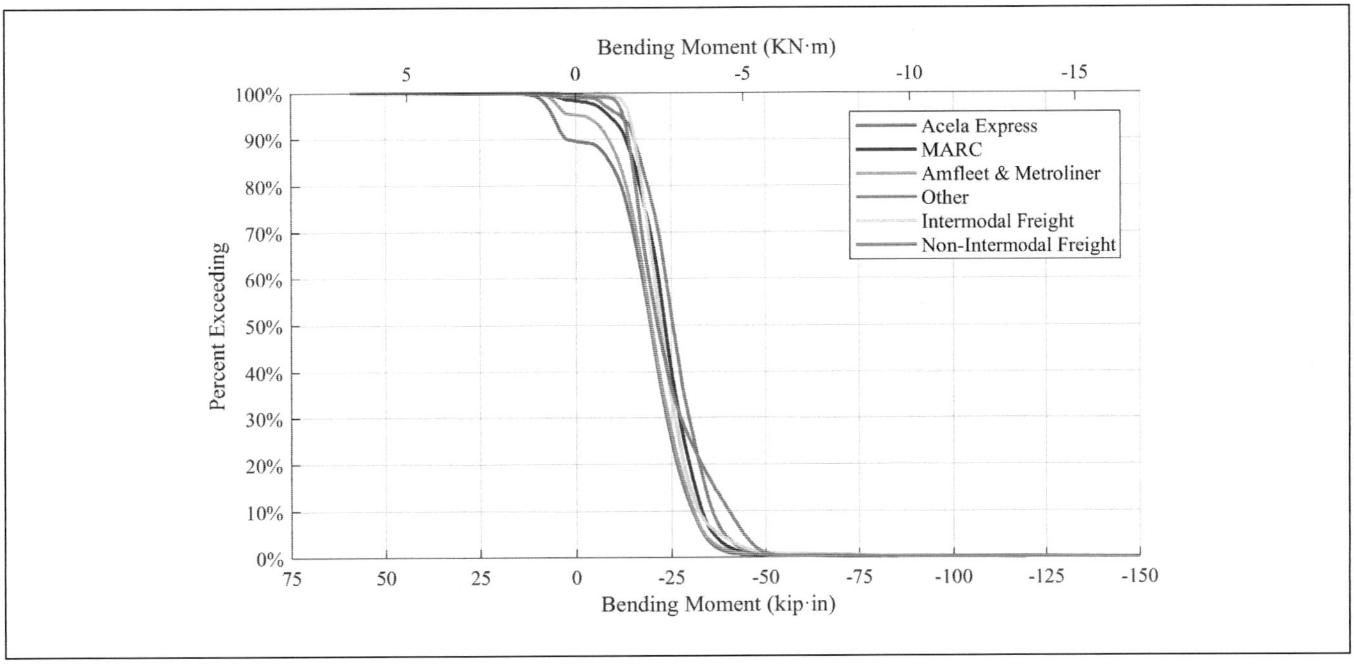

Figure 6. Flexural bending moments of prestressed concrete ties on the center region by train type.

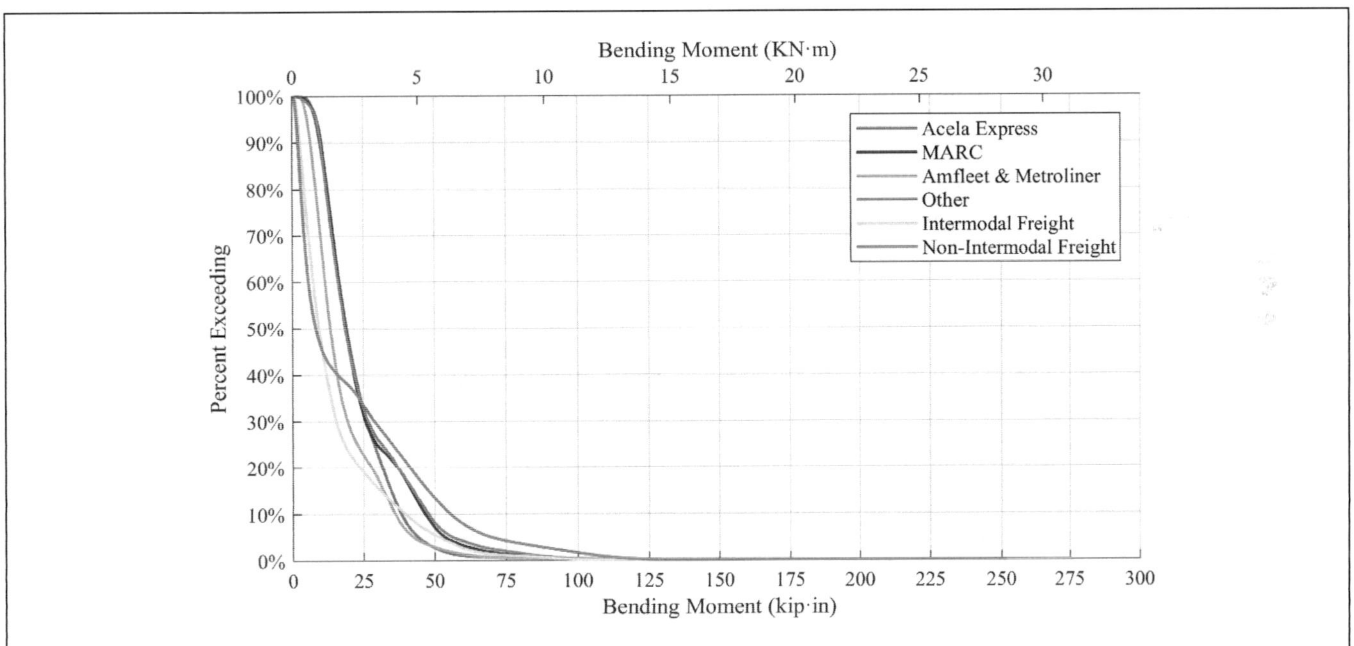

Figure 7. Flexural bending moments of prestressed concrete ties on rail seat region by train type.

developed by Gao et al. (*21*). This calculator uses bending moment profiles from instrumented concrete crossties and approximated rail seat loads using WILD data as inputs (*21*). Understanding the existing support condition on an instrumented site is a fundamental element in the analysis of data and prestressed concrete crossties as the bending moment is highly sensitive to variations of the support condition (*20*). The computational algorithm divides the crosstie into six bins and optimizes the percentage of total ballast reaction in each bin required to re-create the bending moment collected with the use of surface strain gauges in the field.

Table 1. Top Ten Percentiles of Peak Wheel Load and Bending Moment Data per Car Classification for the Edgewood, MD Field Site

Classification	Measured data	90%	95%	98%	99%	99.5%	100%
Acela	Wheel load (kips)[a]	33.9	35.2	36.2	37.5	38.6	50.5
	C- moments (kips·in.)[b]	−31	−34	−36	−40	−44	−111
	RS + moments (kips·in.)[b]	38	44	51	60	80	232
MARC	Wheel load (kips)[a]	31.9	38.5	41.1	43.3	44.9	57.4
	C- moments (kips·in.)[b]	−33	−36	−39	−44	−48	−119
	RS + moments (kips·in.)[b]	47	54	64	80	90	211
Amfleet and Viewliner	Wheel load (kips)[a]	35.4	39.2	41.3	43.4	45.2	68.8
	C- moments (kips·in.)[b]	−31	−34	−37	−42	−50	−112
	RS + moments (kips·in.)[b]	36	43	52	64	75	276
Other	Wheel load (kips)[a]	32.5	37.2	40.3	43.8	46.8	65.3
	C- moments (kips·in.)[b]	−36	−39	−42	−46	−48	−85
	RS + moments (kips·in.)[b]	48	56	70	86	95	160
Intermodal freight	Wheel load (kips)[a]	31.8	40.9	44.2	47.9	50.5	73.2
	C- moments (kips·in.)[b]	−32	−38	−43	−50	−73	−82
	RS + moments (kips·in.)[b]	40	52	61	77	90	125
Non-intermodal freight	Wheel load (kips)[a]	43.1	47.2	52.7	62.6	69.2	119.5
	C- moments (kip·in)[b]	−40	−44	−47	−50	−52	−150
	RS + moments (kip·in)[b]	56	70	91	107	115	252

[a] 1 kip=4.45 kN.
[b] 1 kip·in=0.113 kN·m.

A subset of data from the aforementioned bending moment dataset was used in the analysis back calculator. Particular focus was placed on selecting train passes of similar load levels and negative temperature differentials to capture worst-case conditions for center negative bending moments. Upward curl in concrete crossties has the potential for inducing high ballast reaction in the center of the element and therefore maximizing the effect of load on the center negative bending moment (*13*). For that reason, the analysis was limited to selected non-intermodal freight locomotives on the night of 23 April 2017. Negative temperature differentials recorded on site ranged from −4.4°F to −5.4°F (−2.5°C to −3°C). Locomotive wheels showed both the highest loads and the least variation and therefore represent a stable condition on which to perform the analysis. Figure 8 presents the results of the ballast pressure distribution of six locomotives passing the Edgewood, MD field site. Each line represents the average of the conditions calculated for each of the six axles of the individual locomotives. Even with the effect of upward temperature curling, there is a high concentration of support under the rail seats. It is possible to conclude that the dominant support condition at the site is one of a recently tamped track, with adequate support of rail seats. This aligns with the expected results as presented in previous sections.

Although existing support conditions at the field site align with those expected for well-maintained track, certain sections of the NEC might experience different support conditions because of differences in maintenance frequencies or because of track irregularities such as transition zones (e.g., bridge approaches). Using the same computational algorithm developed by the support condition back calculator (*21*), fixed hypothetical support conditions adapted from Bastos (*20*), and the aforementioned wheel load data, it is possible to generate simulations of expected bending moments for different support scenarios.

Results from this analysis are presented in Figure 9 for the center region of the crosstie. Also included in Figure 9 are the aggregated field measured data from Figure 6, specified and laboratory measured first crack flexural capacity of the current crosstie design. Depending on the support case, simulated values exceed current design first crack and even ultimate capacity, especially those related to localized support under the center of the crosstie (i.e., moderate center binding). Although the analytical model does not consider potential dynamic changes in the support condition caused by the interaction between the tie deformed shape and ballast, the data presented show bending moments that are feasible based on current NEC wheel loading conditions. If the track is maintained to the current level that was observed in the field, it is not expected that bending moments should exceed current specification values, which is probably a strong indicator of favorable long-term crosstie performance.

Results for the rail seat region are presented in Figure 10 and provide a different outlook. The existing support conditions generate the highest potential bending moment. In contrast with the center region, it is not expected that changes in the support condition will

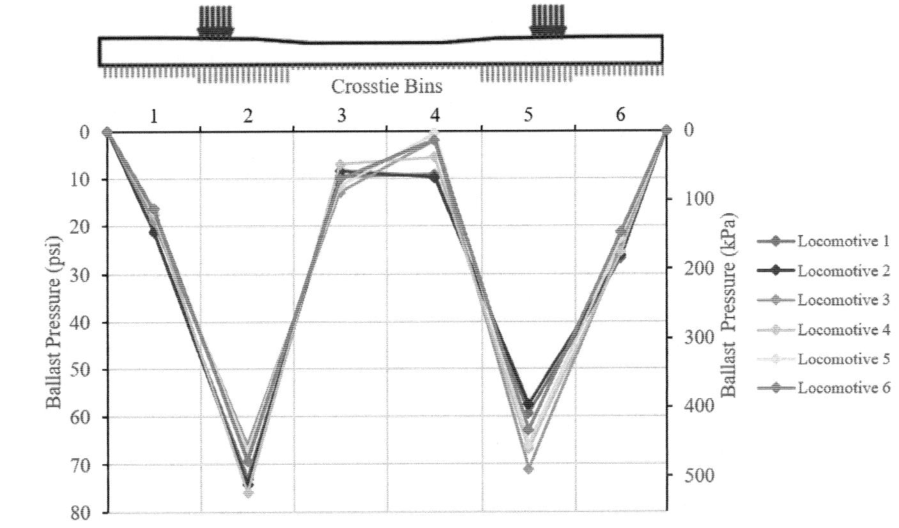

Figure 8. Average ballast pressure distributions for six separate freight locomotives on April 23, 2017.

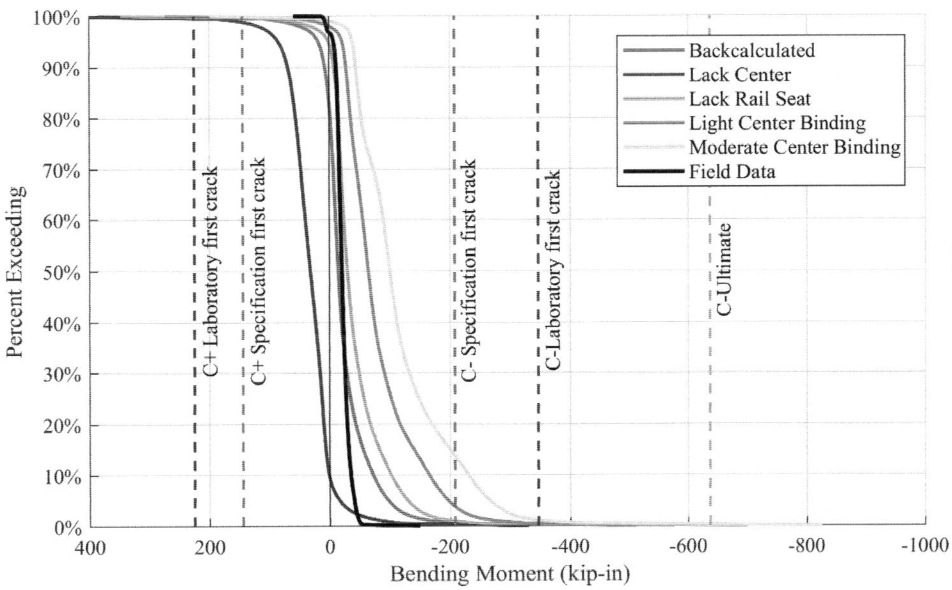

Figure 9. Extrapolated bending flexural bending moments of prestressed concrete ties on the center region associated with different support conditions based on field loading data.

generate higher flexural values at the rail seat. Considering that the current specification value of 306 kips·in. (34.6 kN·m) is surpassed neither by field measured data nor by analytical scenarios, it can be concluded that the current design provides a strong rail seat section that can be further optimized by potentially reducing the specification value of the rail seat section.

Conclusion

Optimization of the track structure and its components requires a comprehensive understanding of the effect and variability of loads. Field instrumentation such as WILD sites and instrumented crossties proved to be a valuable resource for addressing these questions through loading and bending moment quantification. From the field

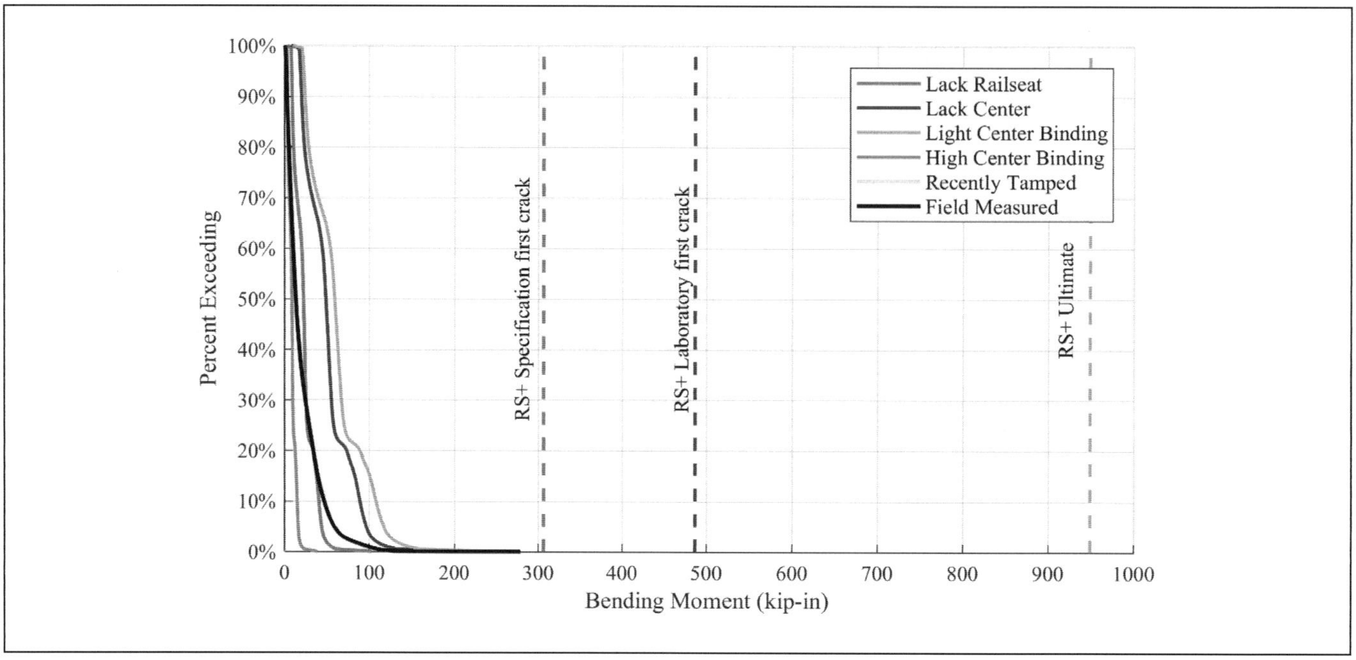

Figure 10. Extrapolated bending flexural bending moments of prestressed concrete ties on the rail seat region associated with different support conditions based on field loading data.

experimentation conducted on Amtrak's NEC at Edgewood, MD, and supporting laboratory experimentation several conclusions can be drawn:

- The loading environment of the NEC shows high variability of load levels between equipment types and operators.
- The top 1% of peak wheel loads show extreme values, which was also observed in the distribution of bending moments.
- Crosstie center bending moments are less sensitive to load increases than rail seat bending moments, implying adequate support conditions under the rail seats at this field location.
- Considering measured field values, neither specification nor laboratory flexural capacities of the current crossties are exceeded by current operations.
- Back calculation of existing support conditions of the Edgewood, MD site confirmed the previous observations and showed sufficient support under the rail seats even when the crosstie is subjected to negative temperature differentials (upward curl).
- There is potential for high flexural demands in the center region of the crosstie which exceed current specification values and actual measured capacity in localized parts of the NEC.
- Infrastructure owner experience with the performance of the current design on those more demanding sections of the NEC will govern the necessity of increasing the center negative specification value to align with potential flexural demands.
- For the rail seat region of the crosstie, there is no expected scenario in which flexural specification value can be surpassed; this indicates room for design optimization to levels of either the actual field demand or currently capacity specified by the owner.
- Maintaining track condition to the level quantified in the field will ensure bending moments, not in exceedance of current specification values, which will probably be a strong indicator of good long-term performance.
- In the near term, future research for the project should focus on the incremental optimization of the crosstie, based on the existing design to align recorded demand to actual element capacity.
- Longer term research objectives should focus on the development of a new crosstie design considering recent and innovative trends in the industry that could optimize the use of resources and provide flexural capacities in alignment with the actual flexural demand in the field.

Use of these data for track structure and component performance could influence future designs that are better suited to the operational characteristics of Amtrak's NEC. Additionally, they provide a basis for the future implementation of a reliability-based design approach, in alignment with the vision for the mechanistic design of track systems.

Acknowledgments

This research effort is funded by The National Railroad Passenger Corporation (Amtrak). Additional support was provided by National University Rail (NURail) Center, a U.S. DOT-OST Tier 1 University Transportation Center. The authors would like to thank Mike Hudson of LB Foster Inc., Salient Systems for his assistance. J. Riley Edwards has been supported in part by grants to RailTEC from CN and Hanson Professional Services, Inc.

Author Contributions

The authors confirm contribution to the paper as follows: study conception and design: J. Riley Edwards, Marcus S. Dersch, Yu Qian, Ricardo J. Quirós-Orozco; data collection: Marcus S. Dersch, Yu Qian, Ricardo J. Quirós-Orozco; analysis and interpretation of results: Ricardo J. Quirós-Orozco, J. Riley Edwards, Yu Qian; draft manuscript preparation: Ricardo J. Quirós-Orozco J. Riley Edwards. All authors reviewed the results and approved the final version of the manuscript.

References

1. Amtrak. *The Amtrak Vision for the Northeast Corridor 2012 Update Report*. Amtrak, Washington, D.C., 2012, p. 42.
2. Van Dyk, B. *Characterization of the Loading Environment for Shared-Use Railway Superstructure in North America*. MS thesis. University of Illinois at Urbana-Champaign, Champaign, IL, 2015.
3. Van Dyk, B. J., C. T. Rapp, M. S. Dersch, J. R. Edwards, C. J. Ruppert, Jr., and C. P. L. Barkan. Evaluation of Existing Loading Environment in North America for Improved Concrete Sleepers and Fastening Systems. Presented at the 2013 World Congress of Railway Research, Sydney, Australia, 2013.
4. Edwards, J. R., M. S. Dersch, and R. G. Kernes. *Improved Concrete Crosstie and Fastening Systems for US High Speed Passenger Rail and Joint Corridors*. RailTEC, University of Illinois at Urbana-Champaign, Champaign, IL, 2017.
5. Remennikov, A., M. Murray, and S. Kaewunruen. Dynamic Design Guidelines for Prestressed Concrete Sleepers. *Proc., 20th Australasian Conference on the Mechanics of Structures and Materials*, Queensland, Australia, 2008, pp. 743–748.
6. Edwards, J. R., Z. Gao, H. E. Wolf, M. S. Dersch, and Y. Qian. Quantification of Concrete Railway Sleeper Bending Moments Using Surface Strain Gauges. *Measurement*, Vol. 111, No. Supplement C, 2017, pp. 197–207. https://doi.org/10.1016/j.measurement.2017.07.029.
7. Harrison, H. D., and J. M. Tuten. Wheel Load Measurement. WO1986003582, 1984.
8. LB Foster Salient Systems. *Wheel Impact Load Detector Product Sheet*. http://www.lbfoster-salientsystems.com/pdf/LBF_Wild_Product_sheet_LOWRES.pdf. Accessed July 10, 2017.
9. Barke, D., and W. K. Chiu. Structural Health Monitoring in the Railway Industry: A Review. *Structural Health Monitoring*, Vol. 4, No. 1, 2005, pp. 81–93. https://doi.org/10.1177/1475921705049764.
10. Harrison, H. D., L. R. Cheng, and W. GeMeiner. Tracking the Performance of Heavy Axle Load Vehicles in Revenue Service. Presented at the ASME 2006 International Mechanical Engineering Congress and Exposition, Chicago, IL, 2006.
11. Canadian National. *Wheel Impact Load Detectors: The Early History on CN*. St. Louis, MO, 2011.
12. Stratman, B., Y. Liu, and S. Mahadevan. Structural Health Monitoring of Railroad Wheels Using Wheel Impact Load Detectors. *Journal of Failure Analysis and Prevention*, Vol. 7, No. 3, 2007, pp. 218–225. https://doi.org/10.1007/s11668-007-9043-3.
13. Wolf, H. E. *Flexural Behavior of Prestressed Concrete Monoblock Crossties*. Master's thesis. University of Illinois at Urbana-Champaign, Champaign, IL, 2015.
14. Edwards, J. R., A. Cook, M. Dersch, and Y. Qian. Quantifying Rail Transit Concrete Monoblock Sleeper Field Bending Moments. *Journal of Transportation Engineering: Part A (Systems)*, 2017.
15. Mayville, R. A., L. Jian, and M. Sherman. *Performance Evaluation of Concrete Railroad Ties on the Northeast Corridor*. Publication DOT/FRA/RPD-14/03. Federal Railroad Administration, Waltham, MA, 2014, p. 171.
16. American Railway Engineering and Maintenance-of-Way Association. Chapter 30, Ties. In *Manual for Railway Engineering*. The American Railway Engineering and Maintenance of Way Association, Landover, MD, 2017.
17. National Instruments. *Making Accurate Strain Measurements*. http://www.ni.com/white-paper/6186/en/. Accessed July 11, 2017.
18. Fisher, R. B. Introduction of Static Frequency Converters on SEPTA's 25 Hz Commuter Rail System. *Proc., ASME/IEEE Joint Conference on Railroads,* Chicago, Ill., IEEE, New York, NY, 1990.
19. Wolf, H. E., Y. Qian, J. R. Edwards, M. S. Dersch, and D. A. Lange. Temperature-Induced Curl Behavior of Prestressed Concrete and Its Effect on Railroad Crossties. *Construction and Building Materials*, Vol. 115, 2016, pp. 319–326. https://doi.org/10.1016/j.conbuildmat.2016.04.039.
20. César Bastos, J. *Analysis of the Performance and Failure of Railroad Concrete Crossties with Various Track Support Conditions*. Master's thesis. University of Illinois at Urbana-Champaign, Champaign, IL, 2016.
21. Gao, Z., H. E. Wolf, M. S. Dersch, Y. Qian, and J. R. Edwards. Non-Destructive Estimation of Concrete Crosstie Support Conditions Using Field Bending Moments. Presented at 96th Annual Meeting of the Transportation Research Board, Washington, D.C., 2017.

The Standing Committee on Rail Transit Infrastructure (AR055) peer-reviewed this paper (18-02775).

The material in this paper represents the position of the authors and not necessarily that of Amtrak.

Article

TRR
JOURNAL OF THE TRANSPORTATION RESEARCH BOARD

Transportation Research Record
2018, Vol. 2672(10) 146–155

DOI: 10.1177/0361198118775872
journals.sagepub.com/home/trr
SAGE

Application of Granular Material Pressure Cells to Measure Railroad Track Tie/Ballast Interfacial Pressures

Jerry G. Rose[1], David B. Clarke[2], Qinglie Liu[3], and Travis J. Watts[1]

Abstract

This paper describes the development of a method to measure railroad track tie/ballast interfacial pressures using pressure cells specially designed for granular materials. Repeat measurements were taken during a several-month period on a Norfolk Southern Corporation high-tonnage mainline. The research employed new wood crossties routed so pressure cells could be recessed within the ties. Thus, the active surfaces of the pressure cells were flush with the tie bottoms. Cabling was run through a recess to the tie end. This greatly reduced the likelihood of damage to the instrumentation during track surfacing and lining activity. The ties were installed such that multiple cells were directly under consecutive rail seats of one rail. Several ties also had cells either at the center or the rail seat of the opposite rail. The researchers expended considerable effort to provide consistent ballast conditions for the instrumented ties and adjacent, undisturbed (transition) ties. Norfolk Southern crews surfaced and tamped through and on either side of the test section. This, plus consolidation through normal accruing train traffic, resulted in consistent measurements through the section. The paper presents ballast pressure magnitudes and distributions and discusses results, including the effects of variable ballast support, wheel loadings, and impact loadings. Typical vertical ballast pressure measurements directly under the rail seat, with compacted ballast and minimal impact forces ranged from 20 to 30 psi (140 to 210 kPa) under the heaviest common revenue wheel loadings.

The magnitudes and relative distributions of typical pressures induced by heavily loaded revenue freight cars and locomotives at the tie-ballast interface have been difficult to quantify and assess. The American Railway Engineering and Maintenance-of-Way Association (AREMA) design methodology assumes that only the outer thirds of the wood tie conveys rail loads to the ballast. The center-third of the tie is typically not tamped during tie installation and subsequent track surfacing. Thus, the ballast is not tightly consolidated in this area. As the tie is loaded through the rails, the center-third of the tie is not initially in contact with the ballast, therefore no pressure distribution occurs.

Figure 1 contains the calculations for wood tie/ballast pressure determination as contained in the AREMA *Manual for Railway Engineering* (*1*). This procedure is largely based on the original Talbot design practice. The default calculations are for a typical 100 ton (91 tonne) capacity car assuming 40% of the axle load is carried by the tie directly under the axle and further assuming an impact factor of 50% for a 50 mph (80 km/h) velocity.

The calculated average tie/ballast dynamic interfacial pressure is 65 psi (450 kPa) under the outer two-thirds of the tie. The analysis makes no allowance for variations in the transmitted pressure along the effective tie length. If the center third of the tie is in equal bearing, the average tie-ballast interface calculated pressure decreases to 43 psi (300 kPa). However, this scenario has not been measured or confirmed from direct in-track tie-ballast interfacial pressure measurements.

Previous Tie/Ballast Pressure Testing

Considerable effort has been expended at developing a consistent, reliable, and accurate procedure to measure pressures at the tie/ballast interface directly below the rail seat. This is the location considered to have the highest ballast pressures. The following sections briefly describe initial tests, with the bulk of the paper addressing the most recent series of in-track tests at Mascot, TN.

[1]Department of Civil Engineering, University of Kentucky, Lexington, KY
[2]Center for Transportation Research, University of Tennessee, Knoxville, TN
[3]School of Civil Engineering and Architecture, East China Jiaotong University, Nanchang, China

Corresponding Author:
Address correspondence to Jerry G. Rose: jerry.rose@uky.edu

$$P_{Dyn} = P_S + \theta P_S$$

P_{Dyn} = Dynamic Wheel Load, lbf

P_S = Static Wheel Load, lbf

θ = Impact Factor

$$\theta = \frac{33 \times V}{D_w \times 100}$$

V = Velocity, mph

D_w = Wheel Diameter, in.

$$P_a = \frac{2P_{Dyn}}{\left(\frac{2}{3}\right) bL} \times \%$$

P_a = Tie/Ballast Pressure, psi

b = Width of Tie, in.

L = Length of Tie, in.

% = Percent of Wheel Load carried by Tie directly under the load, typically 40%

For a static 33,000 lbf wheel load, 33-in. diameter wheel, 50 mph velocity, 9 in. wide and 8 1/2 ft long wood tie, the calculated tie/ballast interfacial pressure is 65 psi (450 kPa).

Figure 1. Typical calculations for tie/ballast contact pressures as contained in the AREMA *Manual for Railway Engineering* (*1*).

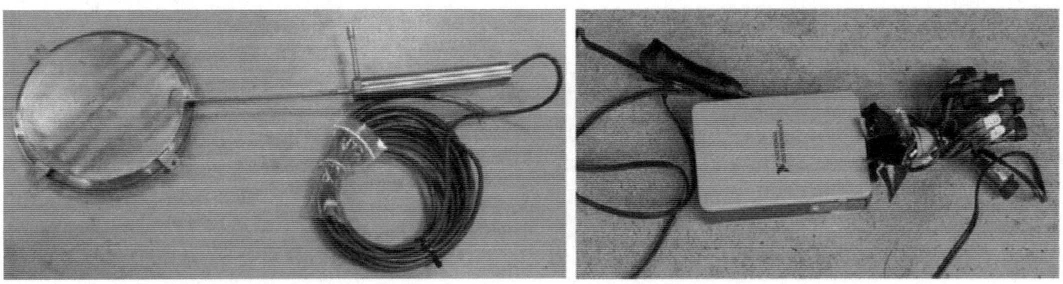

Figure 2. Pressure cell (left) and data logger module (right).

Instrumentation

All tests employed Geokon Model 3515 granular materials pressure cells. These are designed to measure dynamic pressure changes in aggregate of similar size and grading to railroad ballast. A cell consists of two circular 8 in.- (200 mm-) diameter stainless steel plates welded together around the periphery, separated by a small gap (void) filled with hydraulic fluid. The pressure cells have an active area of 50.3 $in.^2$ (324 cm^2). Applied pressure squeezes the two plates together, creating fluid pressure in the cell. The two plates are sufficiently thick so they do not deflect locally under the point loads from surrounding large aggregate particles.

A pressure transducer installed in the steel cell housing transforms the fluid pressure to a current signal. The measured pressure is the average pressure on the active area. The pressure range of the pressure transducer is 0 to 360 psi (0 to 2.5 MPa).

A National Instruments Model NI 9203 C Series Current Input Module was used for data acquisition. The 12Vdc module has eight analog −20 mA to 20 mA current input channels with a maximum 200 kHz sample rate. Figure 2 shows a pressure cell and the data logger module.

A user-friendly program for data collection was developed using LabVIEW. The program can change units from mA to MPa and psi synchronously and show the pressure magnitude trace in real time during a test. The experiment sampling rate was 2,000 samples/s.

Laboratory Prototype Tests

To develop and validate in-service test procedures, the researchers initially conducted laboratory tests of pressure cells installed in a prototype trackbed section consisting of a rail, wood tie, ballast, and subballast. This simulated a typical trackbed.

Figure 3. TTI Railroad test site (left) and Flatrock test site (right).

The pressure cell was installed in the tie section directly below the rail seat. Using a testing machine, the team applied controlled and known loadings to the rail/tie section. The cell measured the average pressure at the tie-ballast interface.

The research determined that the cell measurements were accurate and consistent. Cell measurements correlated perfectly with calculated tie/ballast interface pressures based on applied loads (*2*, *3*). Measurement repeatability within and between individual cells was excellent for loadings encompassing typical maximum tie/ballast pressure magnitudes. The test data indicates that the cells accurately reflect pressure under laboratory conditions where the applied load can be measured and controlled.

Transition to In-Track Installations

The next step was to transition to in-track testing. In-service track differs from the laboratory tests in that a tie carries only a portion of the imposed wheel load. The rail is continuous and distributes the wheel load to several consecutive ties

For in-track tests, the team placed pressure cells at the ballast interface below the rail seat for six consecutive ties. The experiment protocol was to maintain ballast in an undisturbed, compact condition when the cells were installed. However, this proved to be difficult in practice and ballast was disturbed.

Initial tests were conducted at two separate sites. A summary of these tests follows.

TTI Railroad at Paris, KY. The first site was on a TTI Railroad track connecting CSX Transportation's nearby mainline with the TTI Paris, KY yard. Maximum operating speed over the track was 10 mph (15 km/h). The lead track has 132 lb jointed rail affixed with cut spikes to timber crossties. The trackbed was a tightly compacted, highly fouled mixture of ballast, degraded ballast, coal dust, and soil. The pressure cells were positioned at the ballast interface below the rail seat of in-situ ties, with the transducer housing and electrical cable in the crib area. Figure 3 shows the TTI track site.

Test results varied as a function of how tightly the particular cell was in contact with the bottom of the tie. Cells in direct contact with a tie recorded reasonably to extremely high pressure levels. However, cells that were positioned lower, thereby providing a "gap" or void between the tie and the cell, recorded pressures much lower, since the tie "bridged over" the cell.

Due to the extremely low tonnage and speeds and the fouled ballast, the cells were basically fixed in position so it was difficult to determine if the measured pressures were typical of those developed in an undisturbed track. The cells must be positioned properly so that the applied pressures reflect undisturbed track.

Norfolk Southern at Flatrock, KY. The second site was at Flatrock, KY, on a very heavy tonnage Norfolk Southern (NS) Corporation mainline freight track. This was a concrete tie track positioned on clean thick ballast with asphalt underlayment support adjacent to a wheel impact load detector (WILD) installation. The maximum train speed was 45 mph (72 km/h). Figure 3 shows the Flatrock test site.

As with the Paris test, the cells were inserted at the tie/ballast interface under six consecutive ties of one rail. However, the disturbed ballast was reasonably loose in the vicinity of the cells, due to the effects of inserting the cells, so the ties initially "bridged across" the cells. Little to no force (pressure) from the axles passed through the cells. The team installed thin metal shims between the ties and cells so the ties would be in direct contact with the compacted ballast.

After multiple phases of shimming, each following 2 weeks of train traffic, higher pressure magnitudes were initially obtained. The typical tie/ballast pressures obtained under locomotives or loaded freight cars varied significantly between cells (*4*). Ballast was disturbed (loosened) during the installation of the cells. There is no feasible method to adequately consolidate/compact the disturbed ballast without destroying the exposed instrumentation.

In-Track Test Equipment and Data Acquisition Practices at Mascot Site

The site subsequently elected to evaluate the applicability of recessing the cells in the underside of wood ties to accurately and consistently measure the pressures at the tie/ballast

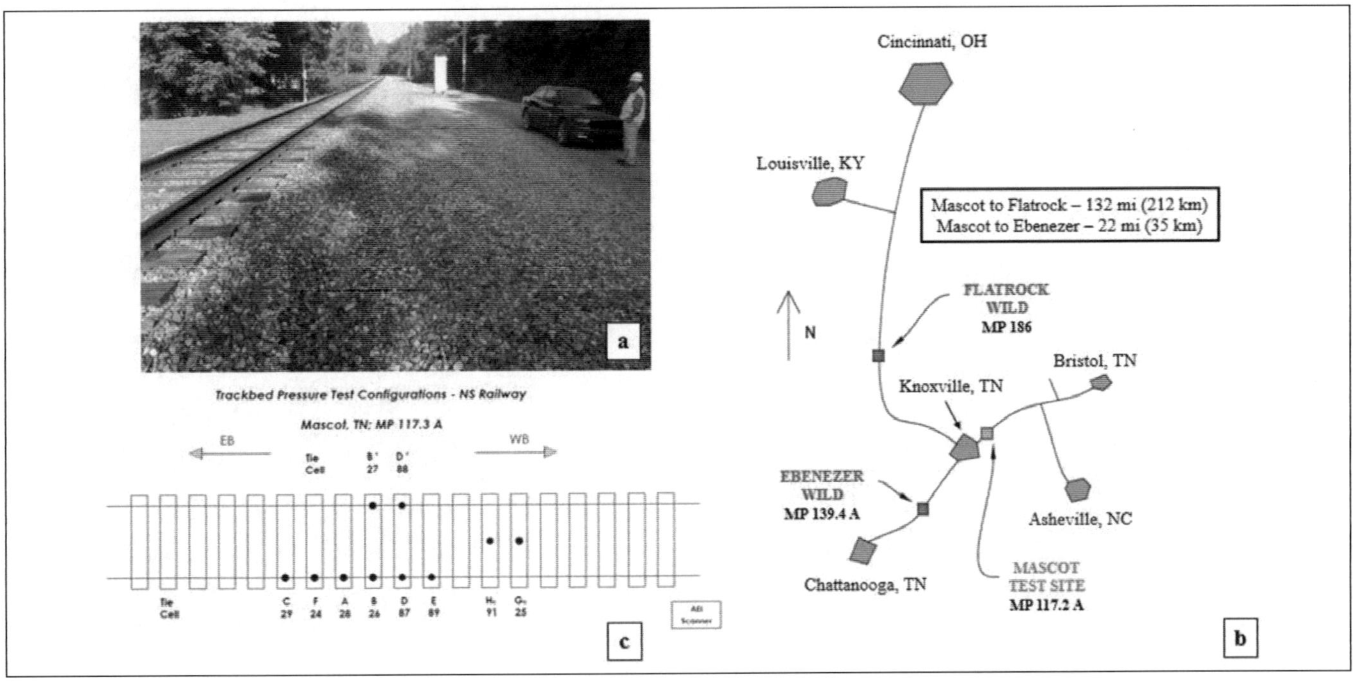

Figure 4. (*a*) View of the test site, (*b*) schematic of the NS through mainlines in the area, and (*c*) detailed schematic for the test site.

interface on an NS mainline at Mascot, TN. The following sections describe recent activities associated with this test site and provide the major aspects of this paper.

The Mascot test site is located on a mainline track with 136 RE continuous welded rail secured with cut spike fasteners to wood ties. Ties are positioned on 20 in.- (500 mm)- centers and each tie is box anchored. The track support consists of standard NS mainline granite ballast on a well-seasoned roadbed. There are no indications of mud or fouling. NS personnel report that the area has a long record of stable roadbed/trackbed requiring minimal track maintenance. The test area was timbered and surfaced in November 2015.

The site is on a horizontal tangent with a 0.25% vertical grade eastbound ascending. The track annually carries 37 million gross tons (33.6 million gross tonnes) of traffic, with a maximum train speed of 45 mph (72 km/h).

All east–west bound trains passing through Knoxville traverse the test section. A wayside automatic equipment identification (ADI) reader adjacent to the test site documents passing train consists. In addition, through trains pass over WILD sites west of Knoxville at either Ebenezer, TN, or Flatrock, KY. Data from these installations permits subsequent comparisons of the tie/ballast pressures versus axle loads.

Figure 4 contains a view of the test site, a schematic of the through NS mainlines in the Knoxville area, and a detailed schematic of the test site showing the locations of the instrumented ties.

Instrumented Tie Installation Procedures

In order to obtain representative and consistent pressure data (*4*, *5*), the tie bottoms were recessed to precisely conform to the shape of the cells. The recesses provided space for the active circular area of the pressure cell, the transducer housing, and the instrument cable. This operation was performed in a machine shop using a computer-controlled router. Compliance with shop environmental requirements required copper-naphthenate-treated ties.

The routing was precisely 1 in. (25 mm) deep. The active cell area was flush with the tie bottom to ensure representative ballast contact. The cell was attached to the tie using either screws or corner braces. Thin metal plates protected the transducer housing and instrument cable. A waterproof plastic box beyond the end of the tie contained the coiled cable length when not in use. With all of the instrumentation protected, railroad personnel could raise/surface the track using normal procedures.

Cells were placed directly under one or both rails or the center of the tie. Figure 5 shows the three routing patterns, the routing process, and a cell partially installed within a tie prior to inserting the tie in the track.

NS provided the equipment and personnel to install the five instrumented ties. The crew took extreme care to avoid damage to the instrumented ties during the installation process. Figure 6 illustrates the handling and placement of the instrumented ties.

The instrumented ties were immediately tamped and the initial testing procedure followed. Figure 7 shows the tamping

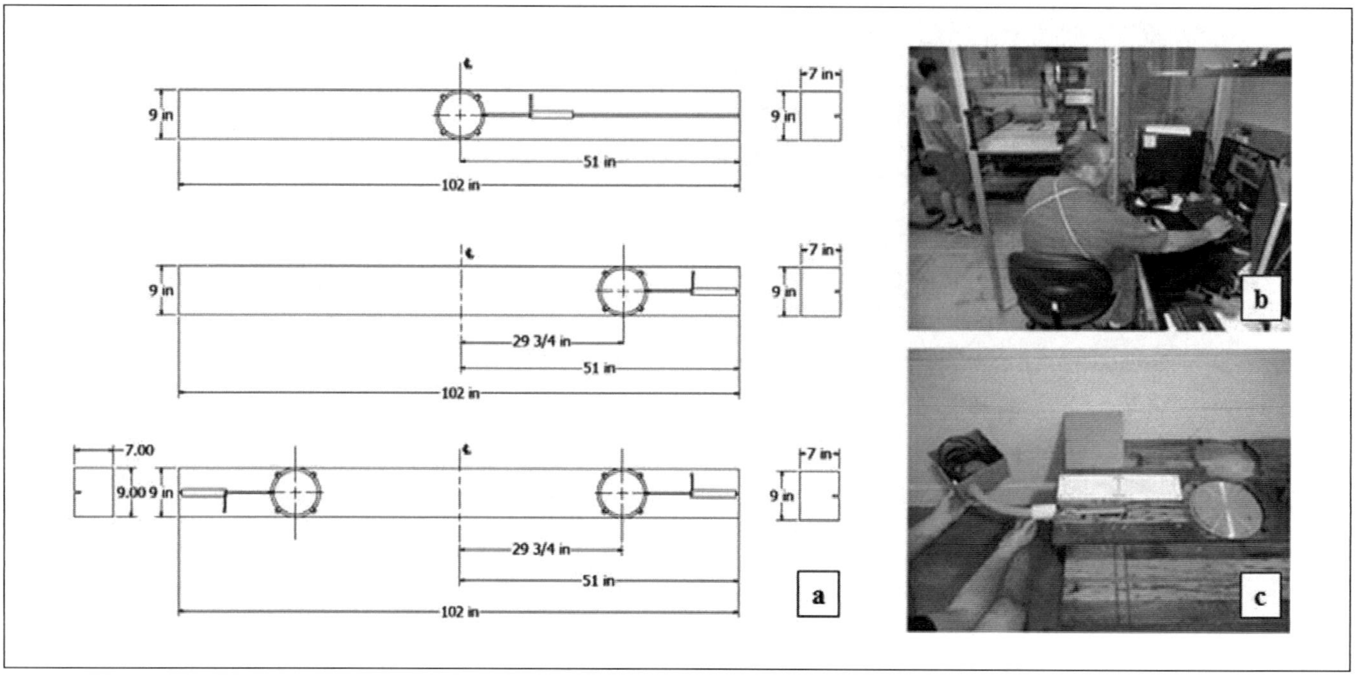

Figure 5. (*a*) The three routing patterns for under the rails and in the center of the track, (*b*) the routing process, and (*c*) a partially inserted cell and housing.

Figure 6. Inserting instrumented ties in the track.

process and the instrumented ties that were tamped. Initially only the instrumented ties were tamped (shaded gray), as shown in (c); subsequently after the initial test series the approach ties were also tamped (shaded gray), as shown in (d).

Test Procedure

Figure 8 contains a schematic of the data acquisition equipment and the wayside test equipment. The pressures exciting the six individual cells were simultaneously

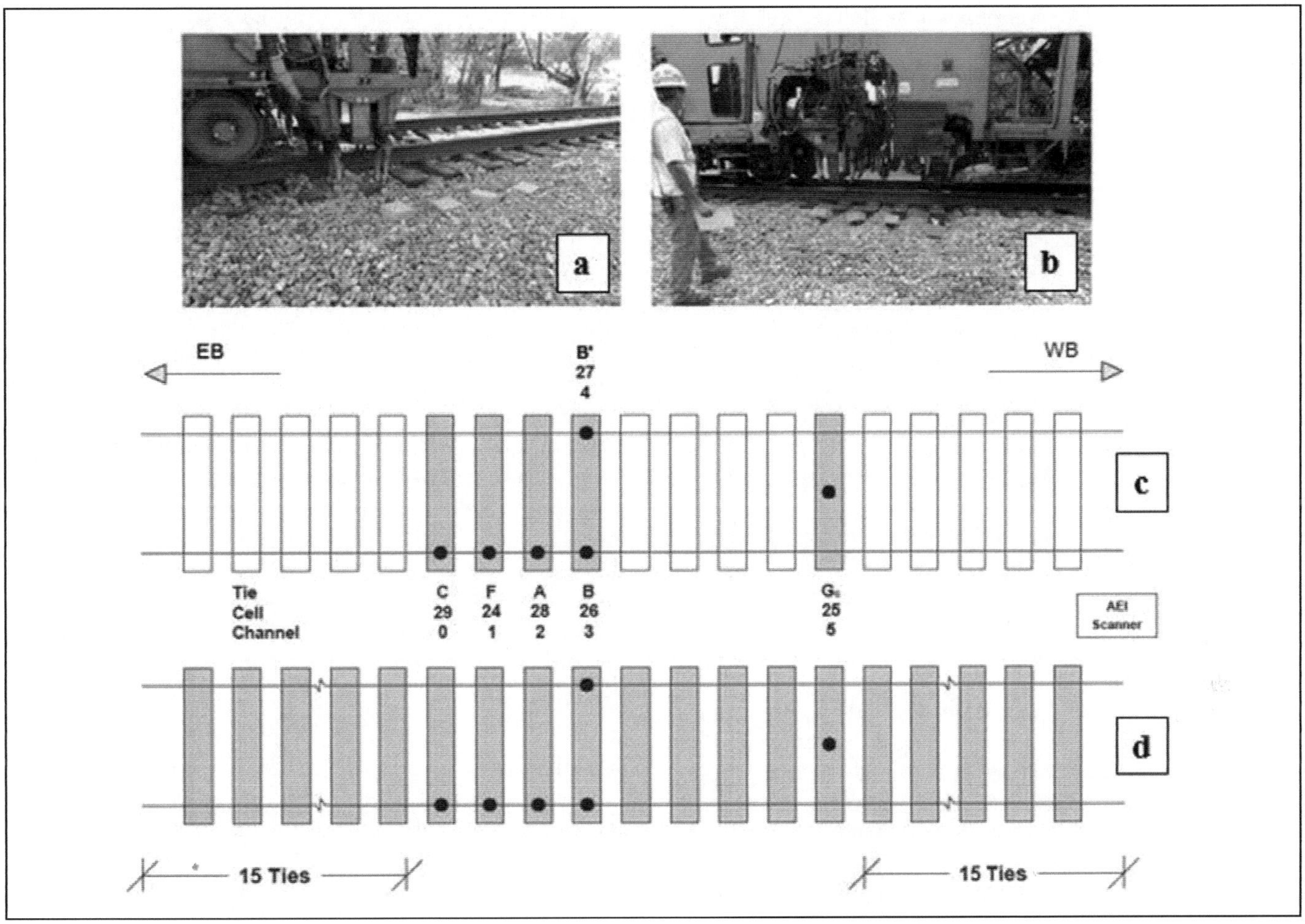

Figure 7. Track tamping process (a & b) and the tamped instrumented ties shown in shaded gray (c & d). The upper sketch (c) shows the instrumented ties (shaded) that were tamped for the initial two test series. After the initial two test series the 15 ties on each approach plus the ties in the instrumented area were tamped as shown in the lower sketch (d).

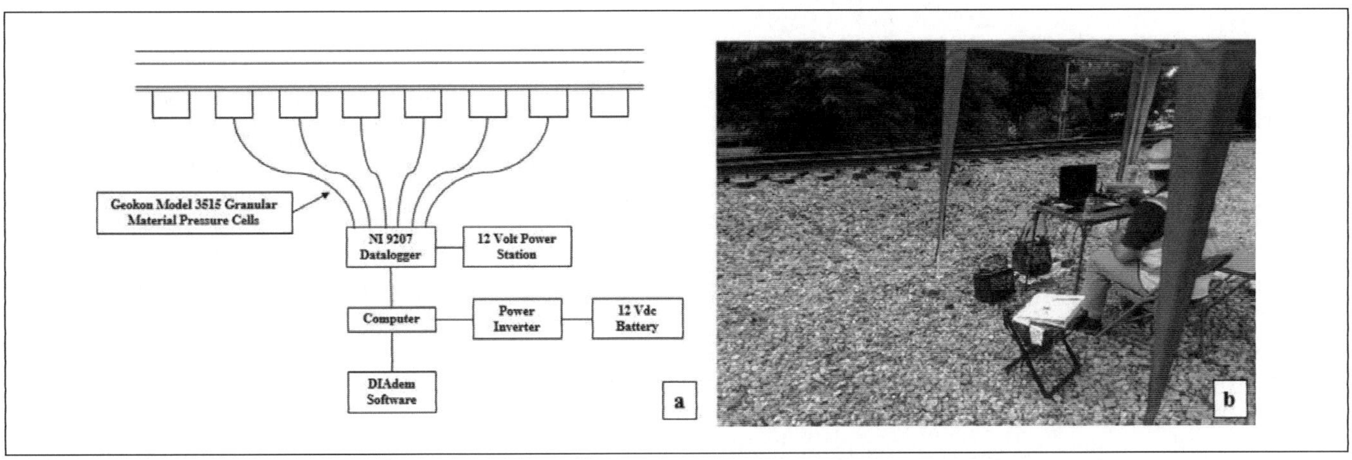

Figure 8. (a) Schematic of data acquisition equipment and (b) wayside measuring equipment.

recorded in real time sequences (sampling rate of 2,000 samples/s). The following information was obtained for each test train:

- Train Number
- Lead Locomotive Number
- Time
- Number of Locomotives
- Direction Of Travel
- Speed Of Train
- Tonnage Of Train
- Type Of Train
- Number of Axles
- Gallons of Fuel
- Length Of Train
- Number Of Cars

Table 1. Chronology for Tamping and Testing the Instrumented Ties

Mascot, TN, site	
Date	
2016	
September 26	Installed 6 Sensors (5 Ties) *Single Tamped 5 Ties with Sensors* Test – (variable and low)
October 12	Test – (variable and low) *Single Tamped 39 Ties (15 + 9 + 15)*
October 26	Test – (more consistent)
November 7	Test – (more consistent)
November 15 and 16	*Double Tamped 39 Ties with Double Insertion*
November 28	Test – (more consistent)
December 15	Test – (more consistent)
2017	
April 13	Test – (little change, except centerline of tie)
April 27 and 28	Test – (little change, except centerline of tie)

Table 1 provides the chronologies for the tamping and testing sequences. For the initial two series of tests only the five instrumented ties were tamped as per the layout in Figure 7. Additional tamping included nine ties in the instrumented area and 15 ties on each approach. Later in the testing sequence, a more thorough double-tamping procedure was performed to insure that the instrumented and approach ties were uniformly tamped so that the contact pressures would be equivalent for all ties. Since that time, four more series of tests have been conducted, extending the test program to 7 months.

Mascot Data Presentation

Nine series of pressure measurements have been conducted. Figure 9 contains typical pressure traces for a mixed freight train after all tamping was complete. The bottom trace is for the cell installed in the center of the track. All other cells were located under the rail seat as per Figure 7. The train's consist was six 6-axle locomotives, shown by the 12 uniform peaks at the left of the traces, and 20 trailing freight cars (six loads and 14 empties). Subsequent peaks of similar magnitude reflect axles of loaded cars.

Figure 10 shows the locomotives axles using an expanded time scale. Each individual peak is a wheel load, with six loads for each locomotive. Individual wheel loads for the locomotives are approximately 33,000 lbf (148 kN).

Table 2 provides average cell pressure readings for the nine test series. For simplicity of data presentation, only the locomotive pressure data is shown. The typical wheel loading from the six-axle locomotives is approximately 33, 000 lbf (148 kN) so direct comparisons can be made from cell-to-cell and date-to-date. Pressure readings for the five cells directly under the rail are shown.

The pressures were lower for the first two series of tests due to the ballast compaction being disturbed during insertion of the five instrumented ties. The instrumented ties slightly "bridged over" the inadequately consolidated ballast. The pressures averaged just 20 psi (140 kPa), excluding the September 26 and October 12 series of tests.

Figure 11 graphically presents the average pressure data in Table 2, along with the range of pressure measurements, excluding data for the center-of-tie pressure cell.

Figure 12 summarizes the average pressure readings over time for the cell installed in the tie center. The pressures remained very low for the initial 2 months of traffic. As the tonnage accumulated during the 7-month period, pressures at the tie center increased to values near those recorded under the rail seat. The initial low pressures appear related to the lack of initial compaction/tamping of ballast under the tie middle-third. Ballast is not in firm contact with the tie bottom. However, as the ballast compacts with accumulating tonnage, the tie settles and more load is transferred in the middle-third. Thus, the tie center cell records higher pressure readings.

Conclusions

A group of six wood ties containing granular material pressure cells imbedded within recesses in the undersurface of the ties were successfully installed on an NS mainline track at Mascot, TN. A series of nine trackbed tie/ballast interfacial pressure measurements were taken of typical revenue trains over a 7-month period.

Ballast in the vicinity of the inserted instrumented recessed ties was disturbed (loosened) much less during the installation of the ties than previous attempts to excavate under existing solid ties to provide space for the cells positioned under the ties.

Imbedding the active area of the cells, transducer housing, and instrument cable within the recesses in the underside of the routed ties leaves the crib areas between the ties void of the cell instrumentation so that the instrumented ties can be adequately tamped to consolidate/compact the partially disturbed ballast.

Initial testing revealed that the properties of the track and pressure distribution within the support are highly dependent on the relative consolidation/denseness of the ballast. Initial pressure measurements were considered marginally representative of typical pressure distributions for a well-consolidated, ballast supported trackbed.

It is highly desirable to thoroughly tamp approach (undisturbed) existing ties in addition to instrumented ties to homogenize the compaction of the ballast in the vicinity of the ties.

Within cell pressure variations, for repeated train passages, are less variable than between cell variations for the same trains.

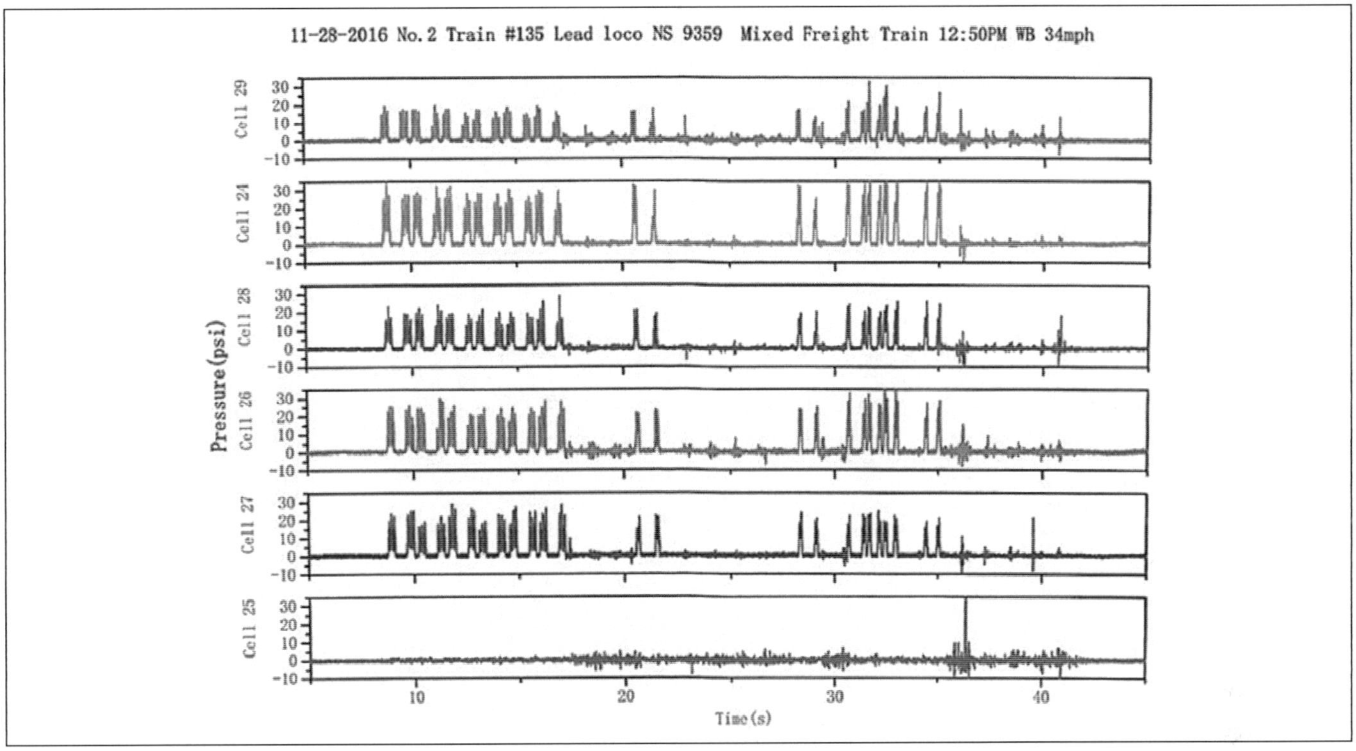

Figure 9. Pressure traces for a typical mixed freight train. The 12 peaks on the left represent the tie/ballast pressures recorded during the passage of the six locomotives, the remaining peaks represent six loaded freight cars. The upper five traces are for cells located under the rail, the bottom one is for a cell located in the center of the track.

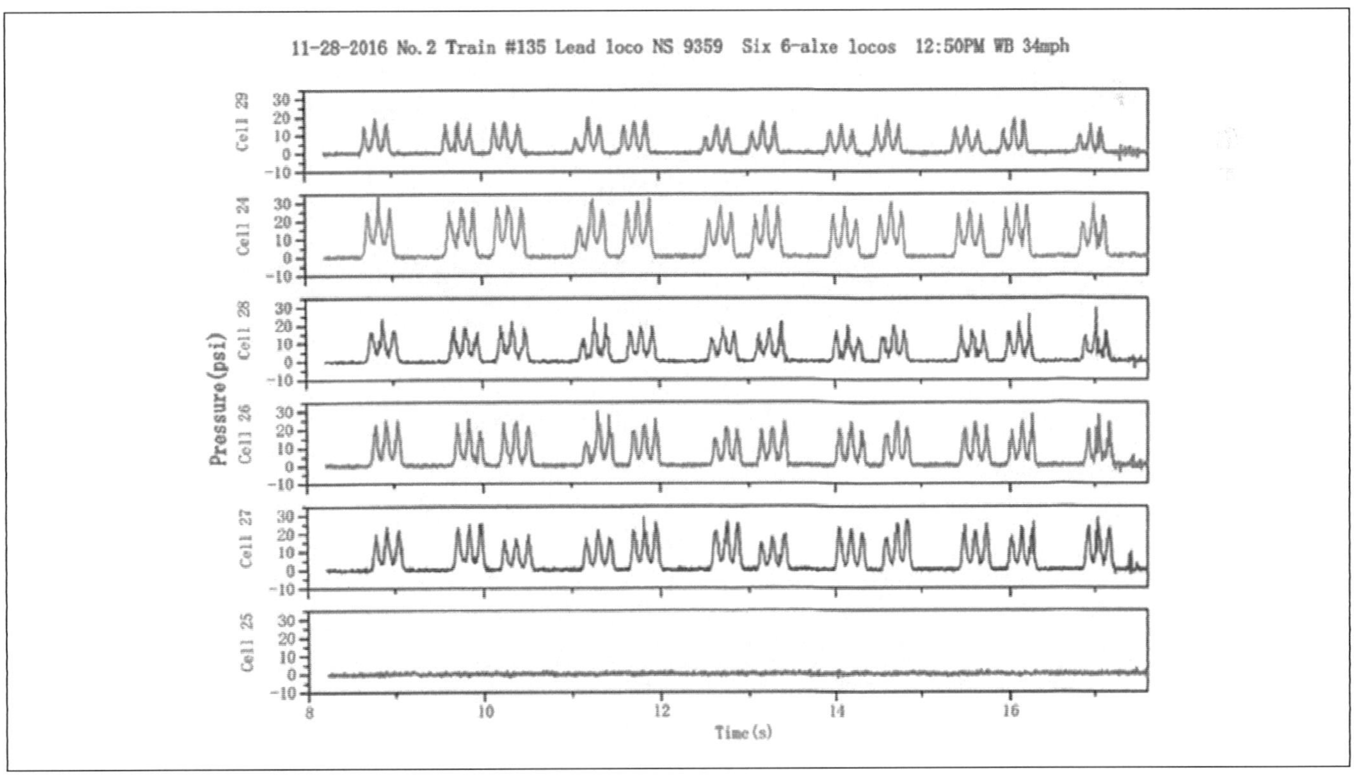

Figure 10. Expanded time scale for the six locomotives shown in Figure 9.

Table 2. Average Rail Seat Cell Pressure, Locomotives, by Cell and Test Date

Cell/date	29	24	28	26	27	Average
9/26[1]	21.14	23.20	9.21	22.72	12.30	17.71
10/12[2]	6.62	19.32	6.23	12.81	5.97	10.19
10/26	12.39	26.07	13.39	28.45	19.93	20.05
11/7[2]	15.08	29.25	15.59	25.30	17.40	20.52
11/28	15.54	27.53	20.31	24.63	22.89	22.18
12/15[3]	9.26	26.94	14.33	21.10	18.05	17.94
4/13	15.10	23.20	15.40	23.20	24.70	20.32
4/27	19.30	29.70	15.70	21.30	22.00	21.60
4/28	15.30	24.00	12.30	19.00	19.00	17.92
Average	14.41	25.47	13.61	22.06	18.03	18.71[4]

[1]Single tamping performed on instrumented ties.
[2]Double tamping performed on instrumented and approach ties.
[3]12/15 data – very cold temperatures.
[4]Average is 20.08 psi excluding the 9/26 and 10/12 test series.

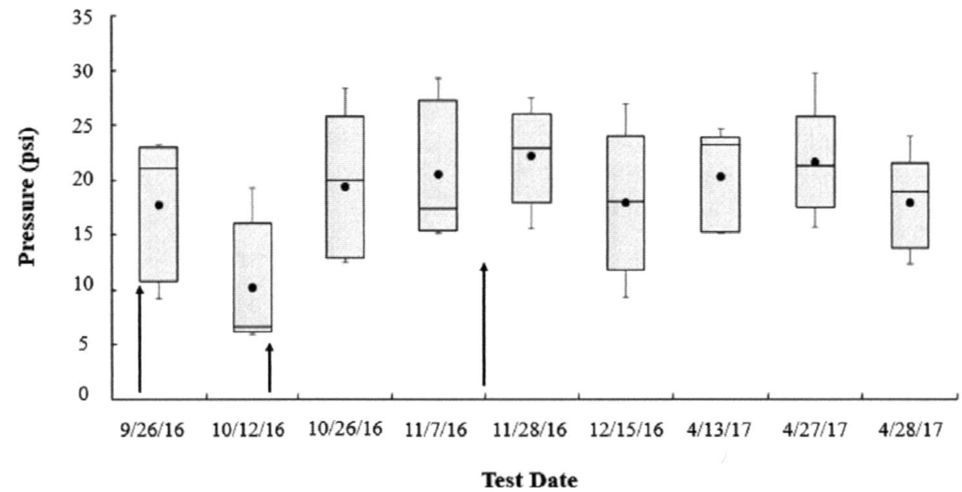

Figure 11. Graphical representation of the average pressures from Table 2 and the range of values. Arrows indicate dates of tamping.

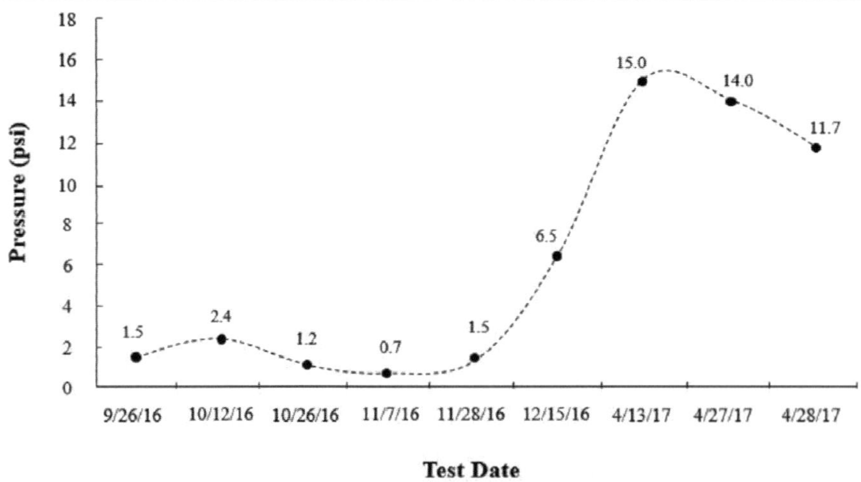

Figure 12. Average pressure readings for the cell installed in the center of the track for the nine test dates.

Imbedding the cells within the ties and securing the cells to the ties prevents the cells from settling in the ballast over time to develop gaps and "bridging," however the ballast must be adequately and uniformly tamped to realize this advantage.

Disregarding the initial two series of measurements, during which the ballast did not appear to be adequately and uniformly compacted, the average tie/ballast interfacial pressure under the rail seat for six-axle locomotives is 20 psi (140 kPa). The data is very consistent, with the maximum to minimum range by date averaging 18 to 22 psi (125 to 150 kPa) during the measurements for over 20 trains during the 6-month period.

The measured pressures are considerably lower than typically assumed for a high-tonnage ballasted timber tie track.

Pressures at the tie/ballast interface in the center of the track are initially very low. This is due to the center of the track not being routinely tamped during the installation of ties and surfacing track. However, with the passage of trains the track settles and the ballast in the center of the track readily contacts the ties providing the media for pressure transferal from the tie to the ballast.

The ballast in the vicinity of the instrumented and approach ties should be uniformly consolidated/tamped to achieve equal vertical support for the ties assuring an equalized track modulus throughout the test area.

It is desirable for the test area to accumulate several months of normal train traffic and tonnage to further homogenize the trackbed support prior to drawing specific conclusions relative to the results of a testing program.

Acknowledgments

This research was primarily funded by the National University Rail Center (NURail), a U.S. DOT OST Tier 1 University Transportation Center and the Economic and National Science Foundation of China (51368021). Graduate Assistant Macy L. Purcell assisted significantly during the early phases of the research. Also appreciated are the financial assistance, cooperation, and significant contributions of Norfolk Southern Corporation with the track site installations activities and continued monitoring and logistical aspects.

Author Contributions

The authors confirm contribution to the paper as follows: study conception and design: J.G.R., D.B.C., Q.L.; data collection: T.J.W., Q.L.; analysis and interpretation of results: T.J.W., J.G.R., D.B.C., Q.L.; draft manuscript preparation: J.G.R., D.B.C., T.J.W. All authors reviewed the results and approved the final version of the manuscript.

References

1. AREMA. *Manual for Railway Engineering*. The American Railway Engineering and Maintenance-of-Way Association, Lanham, M.D., 2017.
2. Rose, J. G., M. L. Purcell, and Q. Liu. *Suitability of Earth and Granular Materials Pressure Cells for Measuring Railway Trackbed Tie/Ballast Interfacial Pressures. Proc., 2016 Joint Rail Conference*, Paper JRC 2016-5735, American Society of Mechanical Engineers, New York, N.Y., 2016. 8 p.
3. Rose, J.G., D. Clarke, Q. Liu, and T. Watts. Development of a Laboratory Test Method for Measuring Trackbed Pressures at the Tie/Ballast Interface. Presented at 97th Annual Meeting of the Transportation Research Board, Washington, D.C., 2018.
4. Liu, Q., X. Lei, J. G. Rose, and M. L. Purcell. *Pressure Measurements at the Tie-Ballast Interface in Railroad Tracks using Granular Material Pressure Cells. Proc., 2017 Joint Rail Conference*, Paper JRC 2017-2219, American Society of Mechanical Engineers, New York, N.Y., 2017, 27 p.
5. Rose, J. G. Measuring the Influence of Wheel Flats on Ballast using Pressure Plates. Presented at 98th Annual Meeting of the Transportation Research Board, Washington, D.C., 2017.

The Standing Committee on Railroad Track Structure System Design (AR050) peer-reviewed this paper (18-00593).

Article

TRR
JOURNAL OF THE TRANSPORTATION RESEARCH BOARD

Transportation Research Record
2018, Vol. 2672(10) 156–165

DOI: 10.1177/0361198118810780
journals.sagepub.com/home/trr
SAGE

Quantitative Analysis of the Derailment Characteristics of Loaded and Empty Unit Trains

Weixi Li[1], Geordie S. Roscoe[1], Zhipeng Zhang[2], M. Rapik Saat[3], and Christopher P. L. Barkan[1]

Abstract

Operation of unit trains has grown substantially over the past half century owing to their ability to provide economic and efficient transportation of bulk products. Although various aspects of train safety have been studied, there has been limited research examining the effect of train loading conditions on derailment occurrence, cause and severity. An algorithm was developed to identify derailments of loaded and empty unit trains on mainlines and sidings recorded in the Federal Railroad Administration database. A dataset of these accidents for the 15-year period of 2001–2015 was developed and analyzed. The frequency of derailments for both loaded and empty unit trains declined by more than 50%. The average number of cars derailed per accident fluctuated for both loading conditions, but showed no particular trend. Approximately five times more loaded unit train derailments were recorded in the database than empty unit trains, but in the absence of specific unit train traffic data, inferences about rates are not possible. Loaded unit trains were more than four times heavier than empty unit trains and loaded train derailments tended to involve more cars than empty trains. The distribution of derailment causes differed for loaded and empty unit trains. Loaded trains most frequently derailed because of broken rails and welds, while the leading cause of empty train derailments was obstructions, in particular severe weather. Over 90% of the derailments of loaded and empty unit trains considered in this study occurred on mainline tracks, and the distribution of causes differed between mainline and siding tracks.

Railroads play a critical role in the transportation and economic prosperity of North America. Train safety has improved considerably over the past decade. This trend continues; in 2016 the derailment rate was the lowest it has been since the Federal Railroad Administration (FRA) began recording data. Nevertheless, with the large volume of traffic, accidents still occur. Derailments are the most common type of train accident recorded in the FRA's Rail Equipment Accident/Incident (REA) database, comprising of almost 70% of these accidents in the 15-year period from 2001 to 2015. Freight train derailments, especially those involving hazardous materials, have the potential to cause serious damage if there is a release of these materials. These types of accidents have received increased attention from the rail industry and the government in recent years owing to the expansion in the transportation of flammable liquids and several high-profile derailments involving these products.

Unit trains are a specific type of rail service in which an entire train transports a single commodity from one origin to one destination. Unit trains increase the efficiency of railroad freight transportation by reducing the operating expenses, using bulk loading, improving asset utilization, reducing transit time, and in general creating economies of scale (*1–3*). Historically, unit trains transported coal and certain other bulk commodities (*1*). More recently, flammable liquid tank cars have begun traveling in unit train like movements. For the purposes of this paper, "unit trains" will refer to fully loaded or empty trains having train type prefixes designating them as unit trains. These were determined from the "train number" column in the REA database; however, a more precise definition of unit trains can be found in Starr (*1*).

[1]Department of Civil and Environmental Engineering, Rail Transportation and Engineering Center, University of Illinois at Urbana-Champaign, Urbana, IL
[2]Department of Civil and Environmental Engineering, Rutgers University, Piscataway, NJ
[3]Association of American Railroads, SW, Washington, D.C.

Corresponding Author:
Address correspondence to Weixi Li: weixili2@illinois.edu

Considering the loading condition, unit trains are either fully loaded or empty. Most previous research on unit trains has focused on operational and economic questions, such as their productivity and profitability (*1–5*).

Previous research on train operating safety has included analyses of derailment frequency and consequences based on train speed (*6*) and derailment causes (*7–9*), but relatively little attention has been given to the effect of the loading condition. Liu et al. developed a zero-truncated negative binomial regression model for derailment severity that factors in the loading condition (*10*); however, the authors are unaware of any prior research specifically focused on the relationship between unit train loading condition, and derailment occurrence and cause. In this paper, the frequency and the severity of freight train derailments were analyzed based on different train loading conditions, and the most common derailment causes for each unit train loading condition were investigated.

Research Objective

The objective was to identify and quantify the effect of the loading condition on freight train derailments, and to compare their causes for loaded and empty unit trains. To achieve this the following steps were taken:

- Develop a methodology to identify loaded and empty unit trains recorded in the REA database;
- Build a database for derailments of loaded and empty unit trains;
- Analyze the resulting dataset to quantify the relationship between train loading condition and derailment frequency and severity;
- Evaluate the top derailment causes with regard to their frequency and average severity.

Methodology

Previous studies have used monetary damage and number of cars derailed (*7*) to assess the severity of train derailments. In this paper, derailment severity is defined as the average number of cars derailed in an accident and the frequency of derailments is defined as the number from a particular cause.

Data Source

The FRA compiles train accident data based on reports submitted by railroads operating in the United States. The train derailment data used in this study were from the REA database. The REA database provides detailed accident information, including operational factors, environmental factors, train characteristics, damage conditions, and other information useful for understanding the circumstances and causes of accidents. Railroads are required to submit reports to the REA database for all accidents that exceed a monetary threshold for damage and loss to infrastructure and equipment (*11*). This reporting threshold is periodically adjusted to account for inflation, rising from $6,600 in 2001–$10,500 in 2014 (*12*). Derailment accidents for Class 1 railroads over the period from 2001 to 2015 were used for the analysis in this paper.

Classification Method

The REA data were used to develop a dataset that included all Class I railroad derailments of freight trains operating on Class I owned mainline or siding tracks. A total of 6,047 such derailments were recorded for the 15-year period studied. To avoid including local trains derailments of trains with less than 30 cars were excluded, leaving 5,395 for the analysis. Approximately 75% of unit trains subsequently identified in the dataset were greater than 50 cars in length.

An algorithm was developed to identify loaded and empty unit trains in the REA dataset (Figure 1). The number of empty cars, the number of loaded cars, and the number of locomotives are recorded in the REA database. Using these fields, the length of a train with regard to the total number of cars and locomotives could be calculated. The majority of unit trains transport various non-regulated commodities; however, those transporting hazardous materials generally have "buffer" cars, as required by federal regulation. Buffer cars are placed between occupied locomotives and cars transporting hazardous materials (*13*). Buffer cars can be either empty, or loaded with an inert material. To account for the possible presence of buffer cars, a train was classified as loaded if 95% or more of its cars were loaded, or empty if 95% or more of its cars were empty. These percentages were calculated by dividing the number of loaded or empty cars by the total number of cars in the train. As the buffer car loading condition is independent of the loading condition of the rest of the train, the 95% criterion was used for both loaded and empty unit trains rather than a 100% criterion.

After obtaining all loaded and empty unit train derailments, the remaining derailments were filtered based on their train type. Train type information was obtained using online resources to interpret and classify the "train number" field in the REA database (*14, 15*). This was done to eliminate derailments that were not bulk unit trains, such as: manifest trains, intermodal trains, local trains, and work trains. Of the remainder, 1,536 derailments were classified as loaded unit trains, 303 were classified as empty unit trains, and 4,180 were classified as "other" trains (Figure 1).

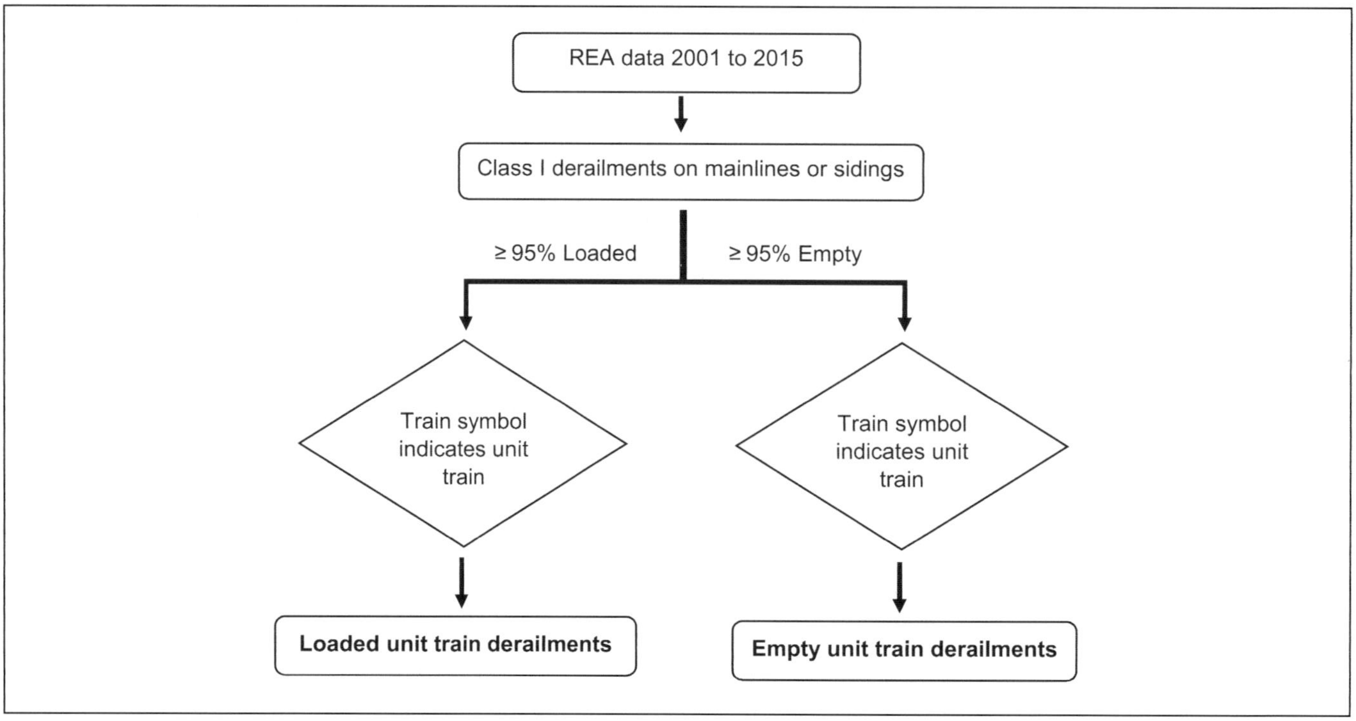

Figure 1. Flowchart for classifying loaded and empty unit trains.

Table 1. Summary Statistics of Derailments by Loading Condition

Loading condition	Number of accidents	Tons (1,000s)	Average train length	Average speed	Average number of cars derailed	Average POD	Average NPOD
Other	4,180	7.1	77.9	22.5	8.3	11.4	45.0%
Loaded	1,536	14.2	106.9	25.1	11.5	54.4	51.0%
Empty	303	3.0	106.8	24.8	8.9	41.8	40.2%
p-value	–	<0.001	0.945	0.786	0.001	<0.001	<0.001

Note: POD = position-of-derailment; NPOD = normalized position-of-derailment.

Loading-Condition Specific Derailment Analysis

There were about five times more records of loaded trains than empty trains in the dataset. Several pertinent characteristics of these derailments were summarized (Table 1), including the tonnage of the train, train length, speed at derailment, number of cars derailed, position-of-derailment (POD), and normalized position-of-derailment (NPOD), in which NPOD is the POD normalized by train length (*16*). Student's *t*-tests were used to assess the statistical significance of the differences in these characteristics for loaded and empty unit trains. The characteristics of other derailments were also included for comparison. Not surprisingly, the weight of loaded unit trains was considerably heavier than empty trains, differing by more than four fold. The average derailment speed of loaded and empty unit trains did not significantly differ (25.1 and 24.8 mph respectively), nor did the average train length (106.9 and 106.8 respectively). Empty unit trains derailed an average of 8.9 cars per derailment, while loaded train derailments averaged 11.5 cars or about 30% more, which was also significant. Considering that derailment speed and train length did not differ significantly, but total train weight did, the higher average number of cars derailed is consistent with Liu et al.'s (*10*) results on the effect of the loading factor. They found that derailment severity depends on the derailment speed, residual length, and loading factor. In addition to derailing more cars, the POD and NPOD for loaded unit trains was significantly farther back in the train compared to empty unit trains. This outcome could be caused by a difference in the derailment cause distributions for loaded and empty trains (*7*, *17*).

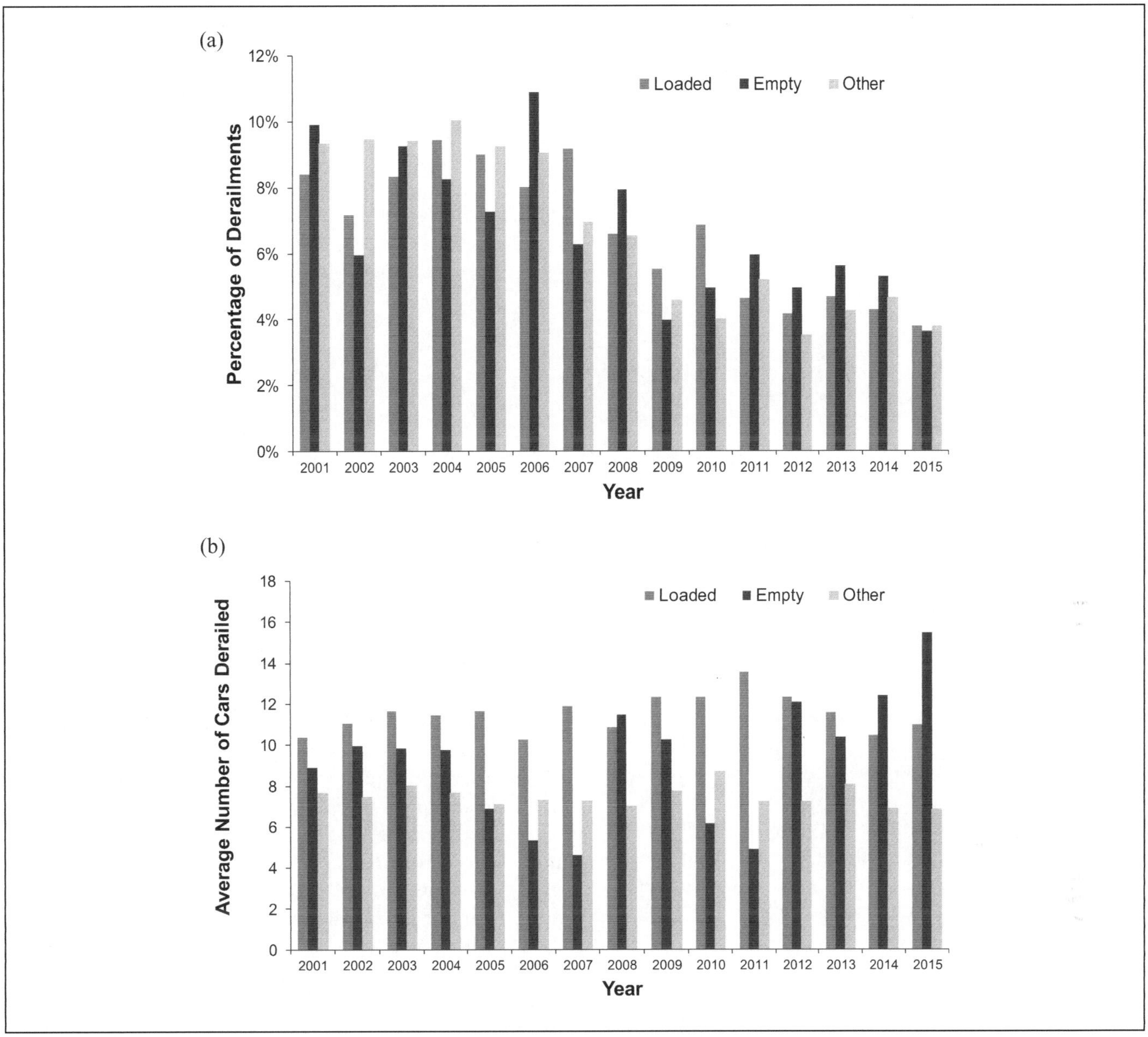

Figure 2. Derailment: (*a*) frequency and (*b*) severity by year, 2001–2015.

Derailment Frequency and Severity Trend

The trend in the frequency and severity of derailments was investigated (Figure 2). The number of derailments in each of the three categories of train type differed substantially so the data were normalized and percentages used to facilitate comparison. The derailment frequency for loaded unit trains, empty unit trains, and other trains declined by about 55%, 63%, and 60% respectively over the 15-year period (Figure 2*a*). This is consistent with the trend for all derailments, which declined by 58% over the 15-year study period. The derailment severity of empty unit trains was generally less than that of loaded unit trains (Table 1), but fluctuated widely from year to year (Figure 2*b*). For example, in 2015, derailment severity for empty unit trains was higher than that of loaded unit trains; however, this was because of a single incident in Iowa in which a tornado derailed 87 cars in an empty unit train. As the sample size for empty unit trains was relatively small, extreme incidents such as this one sometimes shifted the average for a given year. Although derailment severity for other trains is less than that of unit trains, it exhibited the same fluctuating trend.

Although extreme accidents can influence the average derailment severity, they are uncommon. To understand the distribution of derailment frequency and severity, the number of derailments was plotted against the number

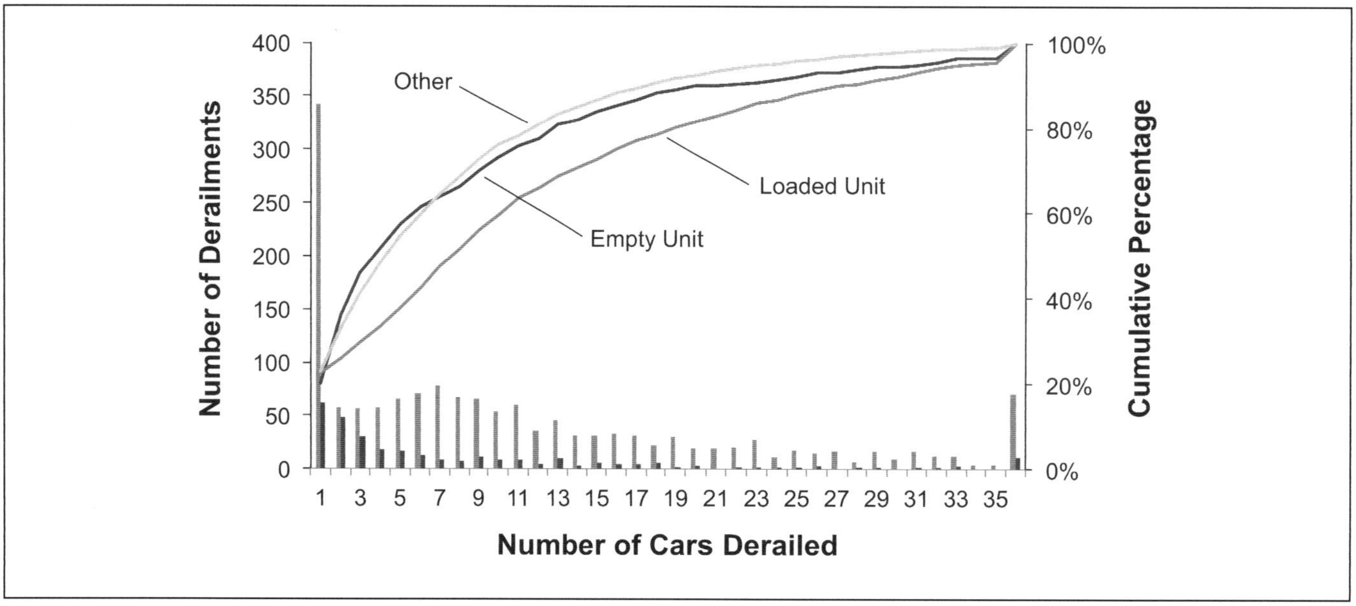

Figure 3. Distribution of derailment severity for loaded and empty unit, and other trains.

of cars derailed per accident (Figure 3). The cumulative percentage of cars derailed was also used to compare the distributions of the three train loading conditions. The cumulative curve for loaded unit trains was well to the right of the curve for empty unit trains, consistent with the finding that empty unit trains derail fewer cars than loaded unit trains. Interestingly, the cumulative curve for "other" trains was similar to the curve for empty unit train derailments. These other trains were on average about 2.4 times heavier than empty unit trains, but they still weighed only half as much on average compared to loaded unit trains. Other trains were also on average about 27% shorter than unit trains (Table 1).

Causal Analysis

ADL Accident Cause Group Comparison. The FRA provides a detailed list of accident causes and associated cause codes for railroads to use when reporting accidents to the REA database (*11*, *18*). FRA organizes the cause codes into a hierarchical structure of related groups of causes. Arthur D. Little (ADL) Inc. worked with the Association of American Railroads (AAR) in the early 1990s to develop a variant of the FRA grouping based on input from railroad engineering and mechanical experts (*19*). The objective of the ADL grouping was to better link causes that could be addressed through similar or related preventative measures. Each FRA cause code maps to a unique ADL cause group. The first step in the causal analysis was to identify the top ten ADL cause groups for the two loading conditions and rank them by the number of derailments (Table 2). The cause groups shared by both are shown in bold.

The top ten cause groups for the two loading conditions were plotted on a frequency versus severity graph (Figure 4). The graph is divided into four quadrants by the average frequency and average severity of the top ten derailment cause groups. The most frequent and severe causes fall in the upper right quadrant; cause groups in this quadrant have both above-average severity and above-average frequency (*7*, *20*, *21*). The top ten derailment cause groups for loaded and empty unit trains have different distributions (Table 2, Figure 4). For loaded unit trains, broken rails or welds was the leading cause group for both frequency and severity. This cause group accounted for about 20% of loaded unit train derailments with an average of about 15 cars derailing per accident. Broken rails or welds was the second leading cause of empty unit train derailments; however, obstructions accounted for the largest percentage of empty unit train derailments at 16.5% and also had the highest number of cars derailed with 18 cars on average. The obstructions cause group includes extreme environmental conditions, which will be further discussed below. Cause groups shared between the two loading conditions included broken rails or welds, track geometry excluding wide gauge, and buckled tracks.

As mentioned above, there was a five-fold difference in the number of derailment accidents for loaded versus empty unit trains. To facilitate comparison, the number of derailments owing to each cause group was normalized by the total number of derailments for each loading condition to calculate a percentage (Figure 5). The lines dividing the quadrants are the averages for severity and percentage of derailments for the top ten derailment cause groups for both loading conditions combined. The

Table 2. Frequency and Severity of the Top 10 Cause Groups for (*a*) Loaded and (*b*) Empty Unit Train Derailments

(*a*) Loaded unit trains

Rank	ADL cause group	Number of derailments	Percentage	Average number of cars derailed
1	**Broken rails or welds**	288	18.8%	14.7
2	Broken wheels (car)	175	11.4%	8.3
3	Other axle/journal defects (car)	127	8.3%	8.9
4	Bearing failure (car)	122	7.9%	6.7
5	**Buckled track**	93	6.1%	15.4
6	**Track geometry (excl. wide gauge)**	80	5.2%	9.2
7	Wide gauge	74	4.8%	10.5
8	Roadbed defects	44	2.9%	13.7
9	Turnout defects - switches	41	2.7%	5.4
10	Other Rail and Joint Defects	36	2.3%	24.9

(*b*) Empty unit trains

Rank	ADL cause group	Number of derailments	Percentage	Average number of cars derailed
1	Obstructions	50	16.5%	17.8
2	**Broken rails or welds**	31	10.2%	15.5
3	**Track geometry (excl. wide gauge)**	25	8.3%	6.4
4	Other wheel defects (car)	24	7.9%	2.8
5	**Buckled track**	15	5.0%	12.5
6	Lading problems	13	4.3%	3.3
7	Other brake defect (car)	10	3.3%	1.8
8	All other car defects	10	3.3%	3.9
9	Train Handling (excl. Brakes)	9	3.0%	5.6
10	Non-traffic, weather causes	8	2.6%	4.9

Note: Cause groups found in both unit train loading conditions shown in bold. ADL = Arthur D. Little Ltd.

labels for the three cause groups shared by both loading conditions – broken rails or welds, buckled track, and track geometry (excl. wide gauge) – are highlighted in yellow and shown in bold italic. Figure 5 enables comparison of the relative frequency and severity of derailment cause groups under the two loading conditions. For example, derailments caused by track geometry excluding wide gauge resulted in derailments with similar severity for both loading conditions; however, this cause group accounted for a greater percentage of empty unit train derailments.

To further understand the top causes for loaded and empty unit train derailments, they were analyzed in more detail. Specifically, loaded unit train derailments caused by broken rails or welds, and empty unit train derailments caused by obstructions were broken down by their individual FRA cause codes (Table 3). 66% of loaded unit train derailments caused by broken rails or welds were caused by just two FRA cause codes: detail fracture and transverse/compound fissure. In comparison, about 70% of empty unit train derailments caused by "obstructions" were owing to extreme environmental conditions, and most of these were caused by just two FRA cause codes as well: extreme wind velocity and tornado. Tornados were also the cause of the most severe accidents, derailing 52 cars on average in each of the five events recorded.

Top Ten Causes on Mainline and Siding Tracks. The data used in this study included derailments on both mainline and siding tracks. There were 1,426 (93%) derailments of loaded unit trains on mainlines and 110 (7%) on sidings. For empty trains, 275 (91%) occurred on mainline tracks, and 28 (9%) were on sidings. A question of interest was whether these two types of track differed in their distribution of derailment causes. For example, sidings might have a higher percentage of switch-related derailments owing to the greater relative frequency of turnouts on siding tracks compared to mainlines. Conversely, the higher operating speeds characteristic of mainline tracks might result in a higher percentage of derailments owing to equipment causes (*22*). There were insufficient empty unit train derailments for a meaningful comparison of their distributions and therefore the comparison of mainline and siding track accident causes was limited to loaded unit trains only. The top ten cause groups for these trains were ranked by number of derailments on each type of track (Table 4).

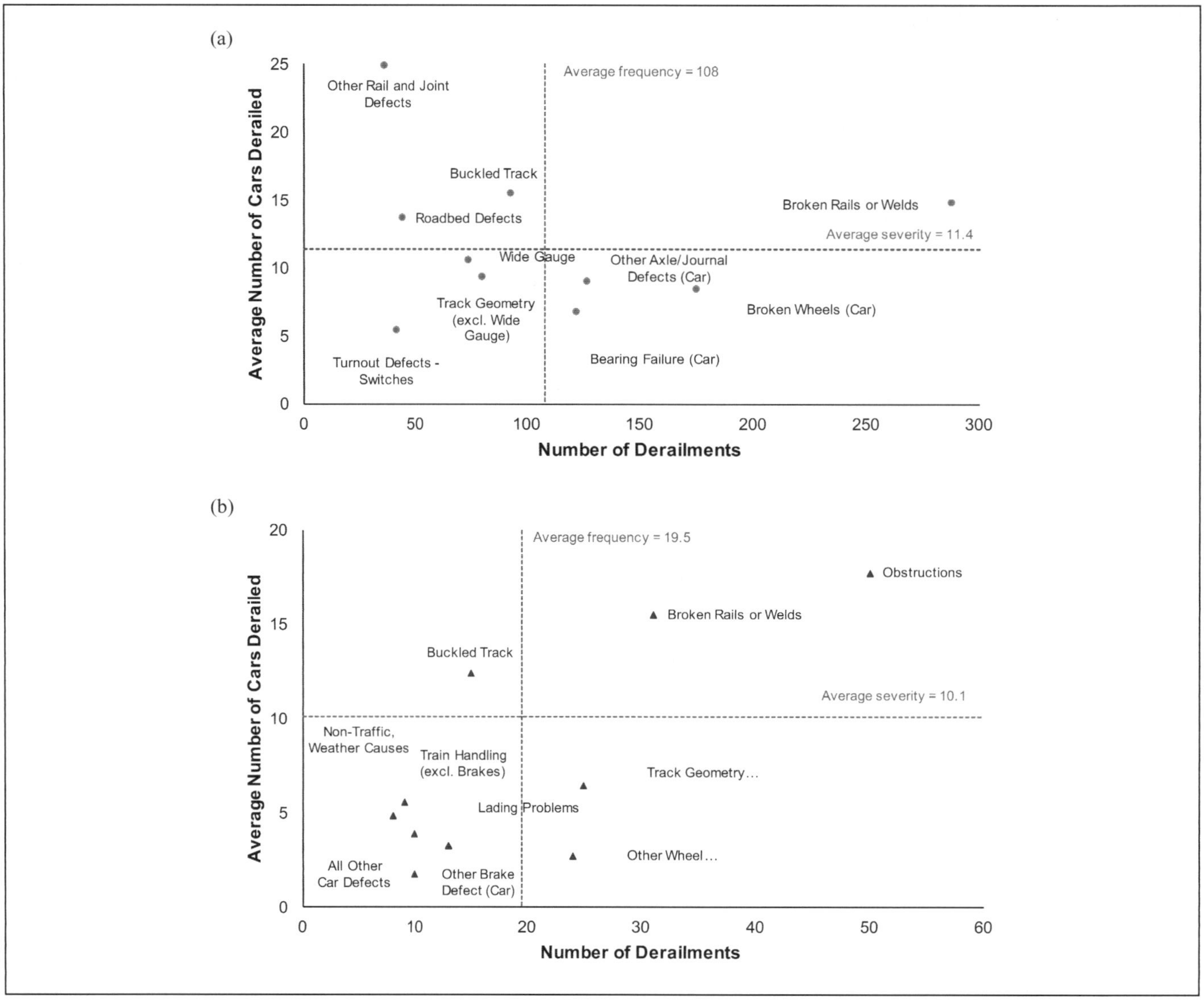

Figure 4. Frequency versus severity graphs for the top 10 cause groups for (*a*) Loaded and (*b*) Empty unit train derailments.

Among the top ten, broken rails or welds, wide gauge, track geometry excluding wide gauge, and buckled track were common to both mainline and siding tracks (indicated in bold in Table 4). As suggested above, three of the top ten cause groups for derailments on siding tracks were switch related, accounting for about 20% of the total, whereas none of the top ten mainline cause groups were switch related. Also, as suggested, four of the top ten cause groups on mainline tracks were equipment related, accounting for 32% of the total, while on sidings only one cause group was equipment related accounting for less than 3%.

Discussion and Future Research

The differing frequency of loaded versus empty unit train derailments recorded in the FRA REA database might suggest a difference in the derailment rate for the two loading conditions, or it may simply indicate that fully loaded unit trains accrue more mileage, and therefore more exposure to potential derailment circumstances than empty unit trains. The latter is unlikely to be the case for "shuttle" trains that travel back and forth as a unit between the same origin and destination for loading and unloading. However, there are other types of unit-train-like operations in which cars travel together while loaded, but after they reach their destination and are unloaded, the empty cars may not remain together, or even return to the same origin. Overall, railcars travel about 40% more car-miles while loaded than empty (*23*); however, this difference by itself is not enough to account for the possible difference in exposure mentioned above. Consequently, specific traffic data for loaded and empty unit trains are needed to address the effect of the loading

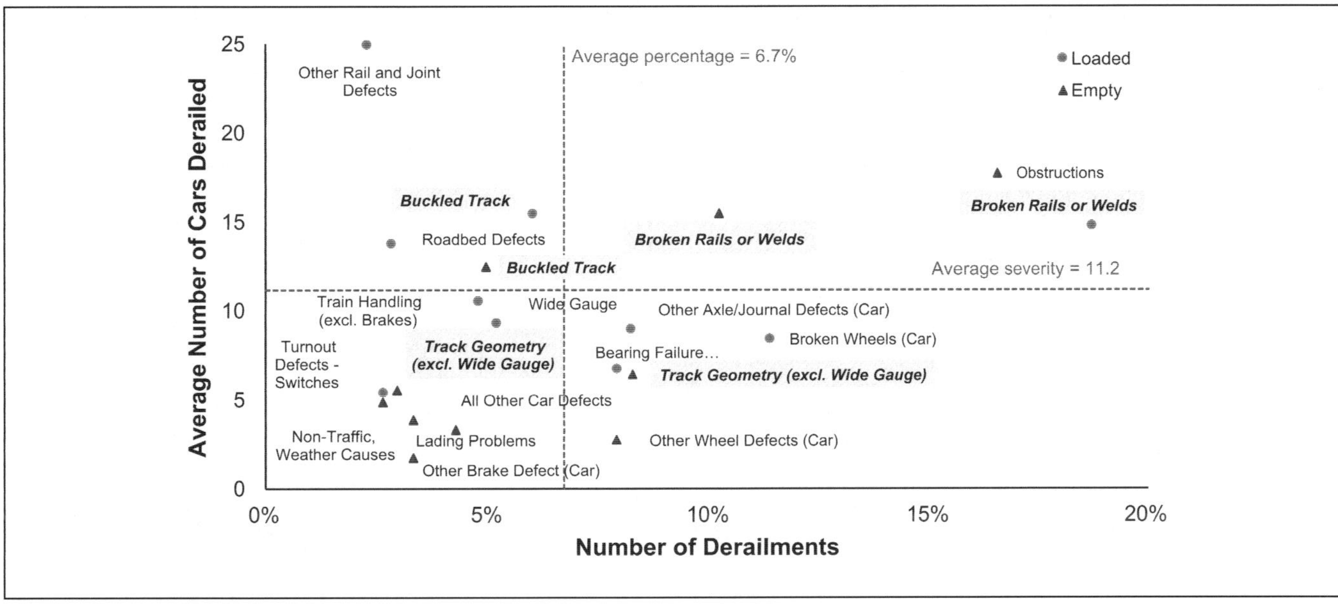

Figure 5. Relative frequency versus severity for the top 10 cause groups for loaded and empty unit train derailments.

Table 3. Frequency and Severity of the Top FRA Cause Codes for (*a*) Loaded and (*b*) Empty Unit Train Derailments

(*a*) Loaded Unit Train Derailments Caused by Broken Rails or Welds

Rank	FRA cause	Number of derailments	Percentage	Average number of cars derailed
1	Detail fracture from shelling or head check	95	33.0%	16.4
1	Transverse/compound fissure	95	33.0%	14.2
3	Vertical split head	28	9.7%	12.1
4	Weld (field)	24	8.3%	15.9
5	Head and web separation (outside joint bar limits)	21	7.3%	10.1
6	Base	14	4.9%	17.3
7	Engine burn fracture	4	1.4%	13.8
8	Horizontal split head	4	1.4%	14.3
9	Piped rail	2	0.7%	9.5
10	Weld (plant)	1	0.3%	23.0

(*b*) Empty Unit Train Derailments Caused by Obstructions

Rank	FRA cause	Number of derailments	Percentage	Average number of cars derailed
1	Extreme environmental condition – extreme wind velocity	27	54.0%	19.9
2	Snow, ice, mud, gravel, coal, sand, and so forth, on track	11	22.0%	3.0
3	Extreme environmental condition – tornado	5	10.0%	52.0
4	Object or equipment on or fouling track (other than above)	4	8.0%	6.3
5	Extreme environmental condition – flood	2	4.0%	10.0
6	Other extreme environmental conditions	1	2.0%	15.0

Note: FRA = Federal Railroad Administration.

condition on derailment rate. It is plausible that some derailment causes are influenced by the greater weight of a loaded rail vehicle. Such causes could include failure of certain railcar components or elements of the track structure it is traveling over. In general, these component failures would be expected to occur regardless of a car's placement in a unit train. FRA regulations, AAR interchange standards, American Railway Engineering and

Table 4. Frequency and Severity of the Top Ten Derailment Cause Groups for Loaded Unit Train Derailments on (*a*) Mainline Track and (*b*) Siding Track

(*a*) Mainline Track

Rank	ADL cause group	Number of derailments	Percentage	Average number of cars derailed
1	**Broken rails or welds**	262	18.4%	15.2
2	Broken wheels (car)	174	12.2%	8.3
3	Other axle/journal defects (car)	126	8.8%	8.9
4	Bearing failure (car)	121	8.5%	6.8
5	**Buckled track**	90	6.3%	15.5
6	**Track geometry (excl. wide gauge)**	72	5.0%	9.7
7	**Wide gauge**	53	3.7%	11.6
8	Roadbed defects	42	2.9%	13.9
9	Coupler defects (car)	36	2.5%	7.4
10	Other rail and joint defects	34	2.4%	25.9

(*b*) Siding Track

Rank	ADL cause group	Number of derailments	Percentage	Average number of cars derailed
1	**Broken rails or welds**	26	23.6%	9.8
2	**Wide gauge**	21	19.1%	7.8
3	Turnout defects - switches	12	10.9%	5.1
4	**Track geometry (excl. wide gauge)**	8	7.3%	4.8
5	Switching rules	6	5.5%	4.2
6	Use of switches	4	3.6%	3.5
7	All other car defects	3	2.7%	5.3
8	**Buckled track**	3	2.7%	11.3
9	Joint bar defects	3	2.7%	7.7
10	Miscellaneous track and structure defects	2	1.8%	6.0

Note: Cause groups found in both types of track shown in bold. ADL = Arthur D. Little Ltd.

Maintenance-of-Way Association recommended practices, and the track design, maintenance, and inspection standards of individual railroads are all intended to prevent such failures. It is hoped that the results presented here will provide insight into the failure modes associated with loaded and empty unit train operation and inform research on the most effective means of preventing these failures and managing the risks.

Conclusions

The use of unit trains has expanded substantially over the past half century owing to the substantial improvements in economies of scale, equipment utilization, operating efficiencies, and transit times that they offer when transporting bulk commodities such as coal and grain. Over the past decade and a half, growth in domestic production of flammable liquids has also led to expanded rail transport of these products in unit trains. This has led to increased interest in the factors affecting the safety performance of unit trains. Limited previous research has considered the relationship between the loading condition of unit trains and associated derailment causes. In this study, a methodology was developed to classify loaded and empty unit trains using the FRA REA database. The resultant dataset enabled analysis of several questions that had not previously been investigated and provides new insights about the factors affecting unit train safety. Over the 15-year period, the frequency of derailments for both loading conditions declined more than 50% while the average derailment severity for both loading conditions fluctuated, but showed no particular trend. Approximately five times more loaded unit train derailments were recorded in the REA database than empty unit trains, but in the absence of specific unit train traffic data, inferences about differences in derailment rates are not possible. Loaded unit trains were more than four times heavier than empty unit trains and loaded train derailments tended to involve more cars than empty trains. The distribution of derailment causes also differed, with broken rails or welds being the leading cause group for loaded unit trains, and obstructions, including extreme environmental conditions such as high winds, being the leading cause group for empty unit trains. Over 90% of the derailments of loaded and empty unit trains considered in this study occurred on mainline tracks, and the distribution of causes differed between mainline and siding tracks.

Acknowledgments

This research was supported by the National University Rail (NURail) Center, a U.S. DOT OST-R University Transportation Center, and the Association of American Railroads. Special thanks to Brandon Wang, Samantha Chadwick, Chen-Yu Lin, Manuel Martin Ramos and Tyler Dick at the University of Illinois at Urbana-Champaign and Xiang Liu at Rutgers University. Their helpful input and assistance greatly improved this research. The views expressed in this paper do not necessarily reflect those of the Association of American Railroads.

References

1. Starr, J. T. *Evolution of the Unit Train: 1960-1969*. Committee on Geographical Studies, Research Papers. University of Chicago, 1976.
2. Grimes, G. A. Planning for Unitrain Operations On Missouri Pacific. *Proc., Transportation Research Forum*, University of California, 1981, pp. 121–126.
3. Grimes, G. A. Terminal Facility Operations and Unitrain Planning. *Proc., Transportation Research Forum*, University of California, 1982, pp. 176–184.
4. Cacchiani, V., A. Caprara, and P. Toth. A Fast Heuristic Algorithm for the Train Unit Assignment Problem. *Transportation Modelling, Optimization and Systems*, Dagstuhl, Germany, 2012, pp. 1–9.
5. Clow, B., and W. Wilson. *Financial an Operating Performance of Cooperative Unit-Train Shipper in North Dakota*. Agricultural Economics Report 23114. North Dakota State University, Department of Agribusiness and Applied Economics, 1998.
6. Yang, T. H., W. P. Manos, and B. Johnstone. Dynamic Analysis of Train Derailments. In *Rail Transportation, The American Society of Mechanical Engineers*, New York, 1973, p. 8.
7. Barkan, C. P. L., C. T. Dick, and R. Anderson. Railroad Derailment Factors Affecting Hazardous Materials Transportation Risk. *Transportation Research Record: Journal of the Transportation Research Board*, 2003. 1825: 64–74.
8. Schafer, D. H., and C. P. L. Barkan. Relationship between Train Length and Accident causes and Rates. *Transportation Research Board Record: Journal of the Transportation Research Board*, 2008. 2043: 73–82.
9. Liu, X., M. R. Saat, and C. P. L. Barkan. Analysis of Causes of Major Train Derailment and Their Effect on Accident Rates. *Transportation Research Record: Journal of the Transportation Research Board*, 2012. 2289: 154–163.
10. Liu, X., M. R. Saat, X. Qin, and C. P. L. Barkan. Analysis of U.S. Freight-train Derailment Severity Using Zero-truncated Negative Binomial Regression and Quantile Regression. Accident Analysis and Prevention. *Accident Analysis and Prevention*, Vol. 59, 2013, pp. 87–93.
11. Federal Railroad Administration. *Guide for Preparing Accident/Incident Reports*. U.S. DOT Federal Railroad Administration, Washington, D.C., 2011.
12. Federal Railroad Administration. *Railroad Equipment Accident/Incident Reporting Threshold, 2012*. www.fra.dot.gov/eLib/details/L03622. Accessed November 7, 2011.
13. Federal Railroad Administration. *Safe Placement of Train Cars, 2005*. www.fra.dot.gov/eLib/Details/L03467. Accessed November 7, 2011.
14. *RailroadfanWiki*. railroadfan.com/wiki/index.php/Main_Page. Accessed November 7, 2011.
15. *Train Symbols*. sites.google.com/site/trainsymbols/. Accessed November 7, 2011.
16. Anderson, R. T., and C. P. L. Barkan. Derailment Probability Analyses and Modeling of Mainline Freight Trains. *Proceedings of the 8th International Heavy Haul Conference* (Jorge, C., ed.), Rio de Janeiro, Brazil, 14–16 June 2005, International Heavy Haul Association, Inc., Virginia Beach, VA, pp. 491–497.
17. Liu, X. Statistical Temporal Analysis of Freight-Train Derailment Rates in the United States: 2000 to 2012. *Transportation Research Record: Journal of the Transportation Research Board*, 2015. 1615: 119–125.
18. Federal Railroad Administration. *Train Accident Cause Codes*. http://safetydata.fra.dot.gov/officeofsafety/publicsite/downloads/appendixC-TrainaccidentCauseCodes.aspx?State=0. Accessed November 7, 2011.
19. Arthur, D. Little, Inc. *Risk Assessment for the Transportation of Hazardous Materials by Rail, Supplementary Report: Railroad Accident Rate and Risk Reduction Option Effectiveness Analysis and Data*, 2nd Rev. Arthur D. Little, Inc., Cambridge, MA, 1996.
20. Dick, C. T., C. P. L. Barkan, E. Chapman, and M. P. Stehly. A Multivariate Statistical Model for Predicting the occurrence and Location of Broken Rails. *Transportation Research Record: Journal of the Transportation Research Board*, 2003. 1825: 48–55.
21. Liu, X., M. R. Saat, and C. P. L. Barkan. Analysis of Major Derailment Causes on Heavy Haul Railways in the United States. *Proceedings of the 10th International Heavy Haul Association Conference*, New Delhi, India, 4–6 February 2013. Indian Railways, SI.
22. Liu, X., M. R. Saat, and C. P. L. Barkan. 2012. Analysis of Causes of Major Train Derailment and Their Effect on Accident Rates. *Transportation Research Record: Journal of the Transportation Research Board*, 2012. 2289: 154–163.
23. Association of American Railroads. *Analysis of Class I Railroads*. Association of American Railroads, Washington, D.C., 2015.

The Standing Committee on Railroad Operational Safety (AT030) peer-reviewed this paper (18-05260).

Article

TRR
JOURNAL OF THE TRANSPORTATION RESEARCH BOARD

Transportation Research Record
2018, Vol. 2672(10) 166–175

DOI: 10.1177/0361198118768532
journals.sagepub.com/home/trr
SAGE

The Life and Death of North American Rail Freight Electrification

John G. Allen[1] and Gregory L. Newmark[2]

Abstract

Although completely dieselized today aside from certain commuter and intercity passenger routes, U.S. railroads were world leaders in electrification in the early 20th century. The Pennsylvania Railroad and the Milwaukee Road had the most extensive electrifications, but several other railroads electrified largely for freight service. This paper explores the decisions to electrify freight railroads in the U.S., Canada, and Mexico (largely for short tunnels where steam locomotives were not practicable, mountain grades, and busy traffic districts), and why electrifications were discontinued (underpowered installations, aging electric infrastructure, and changes in ownership that made electrification geographically obsolete). Energy shortfalls and price spikes since the 1970s have provoked interest in electrification from freight railroads, but this interest has subsided whenever fuel prices decline. Although it is possible that environmental considerations may lead to electrification in some contexts, as long as fossil fuel prices remain low, electrification is unlikely to play a major role on North American railroads.

The early 21st century finds North American freight railroads using diesel-electric locomotives almost entirely. Experiments have been conducted with gensets (diesel-electric locomotives with multiple engines that can be turned on and off individually to save energy), hybrid locomotives with batteries for temporary energy storage and reuse, and liquefied natural gas, which some conjecture may be more desirable than diesel due to the low price of natural gas brought about through hydraulic fracturing.

One or more of these technologies may yet gain wider acceptance, but thus far no technology shift has occurred in motive power. To inform the consideration of future rail freight power, this paper recounts the history of a once-widespread mode that has received some renewed interest since the 1970s: electric traction.

The capital costs of electrification, particularly of substations, power distribution, electrical contact systems, and electric locomotives, often deterred railroads facing other capital investment needs (*1*). Yet in the early 20th century, the U.S. was a world leader in railroad electrification (*2*). "By 1931, when American railroads were operating nearly 5,000 electrified track-miles, U.S. electrification represented nearly 20 percent of the world total and far more than any other country" (*3*, p. 126). This included freight, intercity passenger, and commuter service alike, and, on two railroads, electric operation of all three train types (*4,5*). But today, no freight operation that is part of North America's general railroad network remains electrified.

Figure 1 shows the maximum extent of significant electrifications on North America's main line railroads installed primarily for freight or otherwise having a significant freight component. Table 1 shows basic data for these operations.

This paper explores the decisions to electrify freight railroads, the experience of electrified lines, and the abandonment of this traction technology – the life and death of North American freight rail electrification – to understand this important chapter in railroading history and glean lessons for future rail development. This study considers the 16 most significant rail freight electrifications in the U.S., Canada, and Mexico shown in Figure 1 and described in Table 1, plus one intercity passenger rail electrification that was similar in its rationale, technology, and history.

Not included are:

- Short operations between coal mines and electric power plants, such as the Black Mesa & Lake Powell or the Navajo Mine Railroad. As these properties electrified because the power generation process made electricity readily available, their experiences

[1]Independent Transportation Consultant, Chicago, IL
[2]Department of Landscape Architecture and Regional & Community Planning, Kansas State University, Manhattan, KS

Corresponding Author:
Address correspondence to John G. Allen: jgatransit@2200.com

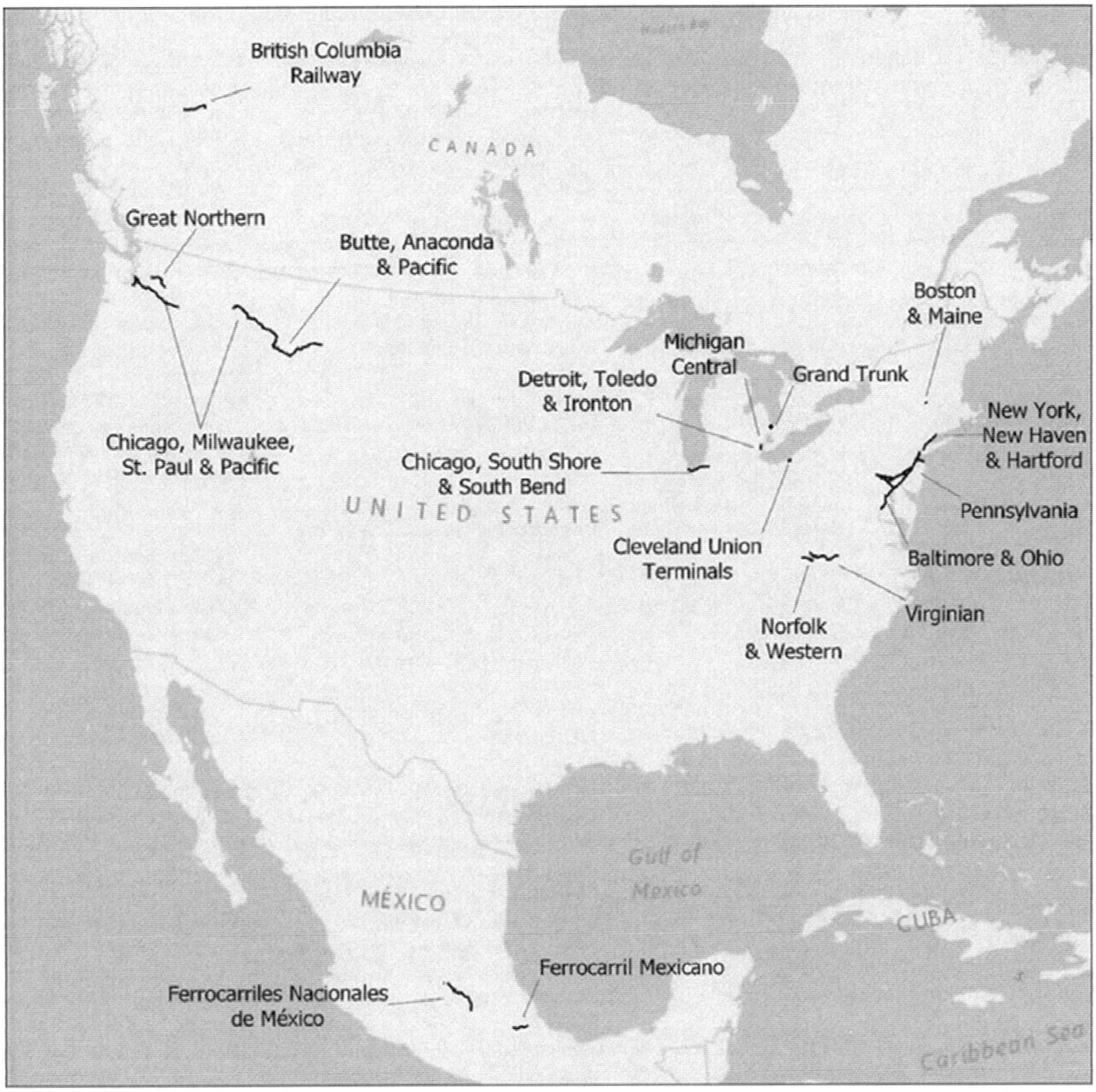

Figure 1. Location of significant electrifications of main line railroads involving freight in the U.S., Canada, and Mexico.

lack broader applicability to common-carrier railroads.

- The Niagara Junction in Niagara Falls, N.Y., because easily-available hydroelectricity made electrification unusually advantageous. This switching railroad had no line-haul function.
- The New York Central and Illinois Central, which electrified almost entirely for suburban and intercity passenger trains. Their electrified freight service was minimal and very short-distance.
- Electric interurban railways (*6*), which electrified for passenger service and only incidentally had electrified freight service. However, one such line serving a major industrial zone is included because its electrified freight service survived into the 1970s.

Why Railroads Electrified

To appreciate the appeal of electrification for freight railroads during the steam era in which circumstances justified the large capital investment, it is necessary to understand the benefits of electric (and, in the diesel era, diesel-electric) traction. One researcher, discussing the massive post–World War II change from steam to diesel-electric locomotives, explains that

> Diesel-electric locomotives have been the universal choice of North American railroads (and of railways on many other continents) because of the characteristics of their transmissions. Where a mechanical linkage forces a direct relationship between engine speed and track speed, an electric transmission allows for maximum power output at *any* speed, even a dead stop. … In practice this means that a diesel-electric produces maximum

Table 1. Basic Information on Selected North American Rail Freight Electrifications

Railroad	Distance	Voltage	Years	Electrification rationale	Why electrification ended
Boston & Maine Hoosac Tunnel	8 mi (13 km)	11 kV AC, 25 Hz	1911–1946	Operation through 4 ¾ mi (7.6 km) Berkshire Mountain tunnel	Diesel locomotives reduced fumes in tunnel to manageable level
New York, New Haven & Hartford (NH)	107 mi (172 km)	11 kV AC, 25 Hz	1907–1969[a]	Electric passenger, suburban and freight operation between New York and New Haven	Conversion to diesel freight reduced power draw on aging, unreliable electrification
Pennsylvania Railroad; subsequently Penn Central, then Conrail	656 mi[b] (1,056 km)	11 kV AC, 25 Hz	1916–1981[c]	General efficiency of operation between northern New Jersey, Harrisburg, Pa., and Washington, D.C. for passenger, freight and commuter trains	Northeast Corridor passed to Amtrak, which raised fees for freights; Conrail responded by shifting trains to other, non-electrified-lines
Baltimore & Ohio (B&O) Baltimore tunnel	3.6 mi (5.8 km)	650 V DC, third rail[d]	1895–1952	Steam traction unworkable in tunnels under central Baltimore	Diesel locomotives reduced fumes in tunnel to manageable level
Virginian Railway (VGN) merged into Norfolk & Western (N&W) in 1959	134.5 mi (216.4 km)	11 kV AC, 25 Hz	1925–1962	Increased throughput for heavy-haul freight through mountains	Merger with N&W resulted in directional running one way on Virginian, other way on N&W line, making electrification difficult to operate
Norfolk & Western (N&W) original electrification	55.9 mi (89.9 km)	11 kV AC, 25 Hz	1915–1950	Increased throughput for heavy-haul freight through mountains	Installation wore out prematurely; railroad replaced part of electrified segment with lower-grade non-electrified-alignment
Cleveland Union Terminals (Cleveland, Ohio)	17.0 mi (27.4 km)	3 kV DC, overhead wire	1930–1953	Steam traction not suited for station below street level; also, office buildings located above tracks as part of station complex	Diesel locomotives reduced fumes in station vicinity to manageable level
Michigan Central (New York Central subsidiary)	4.5 mi (7.2 km)	650 V DC, third rail	1910–1952	Operation through tunnel under Detroit River	Diesel locomotives reduced fumes in tunnel to manageable level
Grand Trunk (Canadian National subsidiary)	4.2 mi (6.8 km)	3.3 kV AC, 25 Hz	1908–1958	Operation through tunnel under St. Clair River	Diesel locomotives reduced fumes in tunnel to manageable level
Detroit, Toledo & Ironton (DT&I)	16.0 mi (25.7 km)	22 kV AC, 25 Hz	1926–1929	Development of high-voltage AC for eventual use on main line between Detroit area and southern Ohio	Railroad sold in 1929; new management not interested in electrification
Chicago, South Shore & South Bend (CSS&SB)	75 mi (121 km)	1.5 V DC, overhead wire	1926–1981[e]	Electrified as passenger-carrying interurban railway; economies of scale with electrified freight operation	Cheaper to replace with diesels as installation aged and traffic shifted toward heavy unit coal trains
Butte, Anaconda & Pacific (BA&P)	37.4 mi (60.2 km)	2.4 kV DC, overhead wire	1913–1967	General economy of operation for Rocky Mountain mining railroad	Cheaper to replace with diesels as traffic declined and installation aged
Chicago, Milwaukee, St. Paul & Pacific (Milwaukee Road)	663.4 mi (1,067.4 km)	3 kV DC, overhead wire	1915–1974[f] 1919–1972[g]	General economy of operation in two mountainous segments, also concern about forest fires with steam locomotives	Installation became physically worn out, cheaper to replace with diesels than to renew electrification
Great Northern (GN), New Cascade Tunnel and approaches	73.5 mi (117.3 km)[h]	11 kV AC, 25 Hz[i]	1927–1956	Increased throughput and general economy of operation through tunnel and associated mountainous segment	Cheaper to operate with diesels than to continue change of motive power
British Columbia Railway (BC Rail)	81.0 mi (130.3 km)	50 kV AC, 60 Hz	1984–2000	New line with two tunnels serving coal mines; concern with diesels stalling out in tunnel moving heavy-haul freight upgrade	Electrification was technically sound, but became uneconomic as coal mines became exhausted
Ferrocarril Mexicano	64 mi (103 km)	3 kV DC, overhead wire	1926–circa 1974	General economy of operation in mountainous territory	Installation became physically worn out, easier to operate with diesels
Ferrocarriles Nacionales de México (FNM)	159 mi (256 km)	25 kV AC, 60 Hz	1994–circa 1996	General economy of operation on busy segment of railroad	Railroad privatized at time when double-stack container service emerging, wires dismantled to provide higher clearances

Note: Data from reference (*5*). Later 20th century short line coal-hauling operations omitted. AC = Alternating current; DC = Direct current; Hz = Hertz; V = Volts.
[a]Electric freight on former NH discontinued in 1969; electrification remains in service for commuter rail and intercity passenger trains.
[b]Includes lines electrified partly, primarily, and in some cases exclusively for suburban and through passenger service.
[c]Some electrification remains in service for commuter rail and intercity passenger trains.
[d]Third rail mounted above and to one side, 1895–1902; repositioned to ground level in 1902.
[e]CSS&SB's predecessor originally electrified at 6.6 kV AC, 25 Hz in 1909; electrified freight service began in 1916 at that voltage, but was poorly developed relative to the situation after 1926.
[f]Rocky Mountain Division.
[g]Coast Division.
[h]Replaced a 1909, 4.0-mi (6.4-km) electrification through the first Cascade Tunnel.
[i]Original 1909 electrification used an unusual three-phase, two-wire system at 6.6 kV AC, 25 Hz.

tractive effort at zero speed—exactly what a railroad needs to start a heavy train (*7*, p. 411).

This advantage did not apply for steam (or to diesel locomotives with mechanical or hydraulic transmissions). Until diesel-electric locomotives became widespread in freight service after World War II, electrification was the only way for railroads to obtain this significant operating benefit. Steam locomotives commonly needed to be changed at division points, unlike electrics (or diesels).

Only electric motors provide maximum torque from a standstill, and may safely exceed their rated output for brief periods of time without overheating. Diesel-electric locomotives, despite their flexibility, lack one advantage of electrics: the ability—given adequately-powered motors and sufficient electricity—to draw as much remotely-generated power as needed to start and move a train.

Electrification was also advantageous, when justified by its capital costs, to different railroads for different reasons.

Tunnels

First, some roads installed short tunnel electrifications where steam operation was not practicable, as with the Baltimore & Ohio (B&O) beneath downtown Baltimore, Maryland (*8*), the Boston & Maine's Hoosac Tunnel in the Berkshire Mountains of Massachusetts, or under rivers, as with the Michigan Central and the Grand Trunk Western, crossing the Detroit and St. Clair Rivers, respectively.

Grades

Second, railroads sought to facilitate operation on lines with mountain grades. Thus, Norfolk & Western (N&W) and the Virginian Railway (VGN) sought to move long, heavy coal trains more efficiently through long mountainous segments than was practicable with steam locomotives. The Chicago, Milwaukee, St. Paul & Pacific (Milwaukee Road) electrified part of its transcontinental route in two discontinuous segments: one in the Rockies and the other in the Cascade Mountains. In addition to the increased operating efficiency through mountainous terrain, the railroad needed to avoid starting forest fires, particularly in federally owned forests that it traversed.

Intensive Traffic

Third, railroads were concerned with overall efficiency and economy. Unlike the other electrifications analyzed here, freight was not the dominant consideration for the New York, New Haven & Hartford (NH) or the Pennsylvania Railroad (PRR), which also had intensive commuter and busy intercity passenger services at levels too frequent for steam to be optimal. But given electrification for passenger and commuter trains, it made sense during the steam era (and beyond) to operate freights with purpose-designed electric locomotives.

Train Handling

The ability of electric locomotives to start trains more reliably and accelerate faster than steam, particularly for long, heavy trains, was another advantage that in some cases, such as the N&W, helped persuade management of the benefits of electrification.

Multiple Factors

These considerations were not mutually exclusive. For instance, the Great Northern (GN) and British Columbia Railway (BC Rail) were concerned with both grades and tunnels when they electrified. Similarly, although PRR was primarily concerned with overall efficiency, there were operational and safety benefits to electric running through that road's Baltimore tunnels, as had been the case earlier for competitor B&O.

For most of these electrifications, freight was the dominant consideration. However, commuter and intercity passenger trains were important for the NH and PRR electrifications, and passenger service was a lesser but still significant consideration for the Milwaukee Road. Whatever the operating needs of these railroads were, electrification met them—for a time.

Technology Development

Most of these electrifications addressed immediate needs, but one railroad was the exception that proved the rule. The Detroit, Toledo & Ironton (DT&I) was bought by automaker Henry Ford in 1921 (personally, not as a corporate subsidiary). DT&I was important for Ford's supply chain, linking southern Ohio (adjacent to coal mines) with the main Detroit-area manufacturing plant. Far from meeting immediate operating needs, the DT&I electrification was undertaken for reasons of technology development.

In 1923 DT&I announced plans to electrify, starting with a 16-mi segment in southeastern Michigan adjacent to Ford's River Rouge factory. Although the installation never went beyond this initial segment, DT&I envisioned electrifying its entire main line. Furthermore, a link was proposed from southern Ohio to West Virginia, connecting with VGN, which had its own electrification (*9*).

Perhaps presciently, a Westinghouse Electric manager wrote in 1923 that "at present the limit in the United States for the alternating current system is 11,000 volts, but there is no reason why this should not be increased as demands are made for heavier drafts of power" (*10*, p. 292). DT&I electrified at 22,000 Volts (22 kV) alternating current (AC), 25 Hertz (Hz), apparently the highest voltage on any railroad in the world at the time. Although DT&I still used 25 Hz (as lower-frequency AC was then necessary for modulating the speed of electric motors), it anticipated subsequent European development of 25 kV AC electrification at commercial frequency (*11*).

Why Railroads De-Electrified

There were three basic reasons for de-electrifying: diesel locomotives being safer to operate than steam through tunnels, technical limitations, changes in ownership, and maintenance costs.

Diesel in Short Tunnels

These electrifications, implemented to avoid the safety hazards of running steam locomotives through tunnels, were too short to be practical once diesel locomotives replaced steam. With proper ventilation, these tunnels could be safely operated with diesels, at least at the moderate traffic densities prevailing on these segments. Electrification, however, remains important where frequent commuter trains use tunnels, as in New York.

The Cleveland Union Terminals electrification in northern Ohio resembled the shorter tunnel freight electrifications in its technology and history, although it served passenger rather than freight trains and traversed no tunnels. This line was electrified in 1930 for amenity purposes, so that smoke would not pollute the intercity passenger station or office buildings above the tracks. Cleveland's installation was decommissioned in 1953, once the railroads using the station had replaced steam with diesel locomotives (*12*).

Technical Limitations

Several electrifications were ultimately abandoned because they were underpowered, or otherwise aged to the point of becoming liabilities, even though they were considered important when opened in the early 20th century.

Insufficient Power. The NH electrification was North America's first main line AC installation (*13*), but it was inadequately powered for the railroad's eventual needs (which for NH was a financial rather than a technological problem). These problems delayed and limited the use of electric locomotives for freight. NH retired its aging electric freight locomotives and replaced them with diesels in 1959.

However, after reexamining the unavoidable minimum costs of its electrification, NH bought 11 modern freight locomotives from the VGN that were surplus. These entered service in 1963 (*14*). But electrified freight became unsustainable as the power plant (inadequate from the outset) aged and the traction power supply became more brittle. Supplementary power tie-ins from utilities in New York City and Connecticut were both expensive and insufficient.

On December 31, 1968, NH was absorbed into Penn Central (PC), itself formed by merging the New York Central and the Pennsylvania Railroad. PC management ended electric freight service east of New York by mid-January 1969, and the ex-VGN locomotives spent the rest of their service lives on ex-PRR lines (*15*). Power shortages continued until Metro-North Railroad, which inherited the New York–New Haven, Connecticut segment in 1983, overhauled the electrification in 1986 with 12.5 kV AC at commercial frequency drawn directly from the power grid—17 years after electric freight operation had been discontinued.

Other installations used direct current (DC), which subsequently showed itself inadequate for long, heavy freights at any voltage. Among the best-known American electrifications were the Milwaukee Road's two technically identical but separate electrified districts in the Rocky and Cascade Mountains. There was a combined 663 route-miles of electrified track, separated by a 212-mi gap that was never bridged. When the Milwaukee Road electrified in the 1910s, it was not yet understood that even at the high pressure of 3 kV, direct current (DC) might not be adequate for freight service. This, however, became very clear with the longer, heavier trains of the post–World War II era.

To reduce the risk of substation failures or burnouts of electrical wires, longer freights were routinely operated with electric and diesel locomotives on the same train (*16*). Although installing two sets of pickup wires (as has been done on Soviet Railways in 3 kV territory) might have prevented some wire burnouts when trains were drawing power, ultimately the obsolescence and inadequacy of the installation made themselves felt. The financially troubled railroad could not afford the huge capital cost of reelectrification to a higher AC voltage, or even of reinforcing its existing installation for higher levels of power draw. Electric operation through the Cascades ended in 1972 and through the Rockies in 1974 (*17*).

The Ferrocarril Mexicano used the same system as the Milwaukee Road on a mountainous segment between Mexico City and Veracruz, Mexico. Faced with the same issues, successor Ferrocarriles Nacionales de México (FNM) discontinued electric operation by 1974.

Aging Electric Infrastructure. N&W electrified its main coal-carrying line in West Virginia in 1915 at 11 kV AC, then seen as the state-of-the-art voltage, to provide greater operating capability on sustained grades without adding track. The railroad built its own power plant, as rural utilities did not then have the electric capacity the railroad needed. But the installation aged prematurely, and by the end of World War II, modern steam locomotives had become more power-efficient. In 1950, N&W opened a new, non-electrified-segment with lower grades and a new tunnel, replacing the electrics with steam until the dieselization process took effect (*18*, pp. 87–88).

Electric interurban railway Chicago, South Shore & South Bend (CSS&SB) was modernized and reelectrified in 1926 for compatibility with the Illinois Central's suburban electrifications, which CSS&SB used to reach downtown Chicago (*19*). Given its electric status, electric freight locomotives were the obvious choice at the time. Indeed, freight operations in highly industrialized northwest Indiana ensured CSS&SB's survival. But as the railroad focused on unit coal trains starting in the 1970s, with their heavier power draw, and as the locomotives aged and spare parts became unavailable, dieselization made more sense. The last electric-powered freight train ran in 1981 (*20*). Commuter service is still electrified.

Butte, Anaconda & Pacific (BA&P), a Montana copper-mining railroad, electrified largely to more efficiently handle traffic between mines and smelters, using 2.4 kV DC. Although BA&P bought new electric locomotives as late as 1957, its installation was aging and traffic was declining as the mines became increasingly exhausted. A combination of these factors brought electrification to an end in 1967 (*21*).

Changes in Ownership

Other installations, when decommissioned, were fully capable of meeting the demands placed on them, but the fixed nature of electrification made them vulnerable to traffic shifts, particularly those caused by changes in ownership. All these railroads used AC, which if designed properly, provided sufficient power for decades.

The largest electrified freight operation to be dieselized after certain main lines changed hands was the Pennsylvania Railroad's huge installation (large parts of which survive in passenger and commuter service). Starting with a Philadelphia suburban segment in 1915, by the late 1930s it encompassed a large area between New York, Harrisburg, and Washington, D.C. (*22*). PRR apparently influenced decisions by VGN and GN to select 11 kV AC, 25 Hz.

During World War II, PRR's electrification contributed significantly to the war effort by moving freight and personnel reliably. Reexamining the future of electrification in 1958, PRR decided to reinvest in the future of electrified freight and ordered new locomotives in 1959 to replace life-expired units (*23*). Even in bankruptcy in the early 1970s, successor PC continued to operate electric freight locomotives on ex-PRR lines.

The creation of Conrail in 1976 led to a revival of rail freight in Northeastern U.S., but spelled the end of electrified freight operations. Concurrently with Conrail's creation, national passenger operator Amtrak took title to electrified ex-PRR lines between New York and Washington, and between Philadelphia and Harrisburg, used primarily by passenger and commuter trains. Although Conrail had trackage rights on Amtrak, the latter raised its fees to reflect what it believed was the high cost of maintaining tracks to meet the needs of heavy freight as well as fast passenger trains. Conrail responded by upgrading its own nonelectrified ex-Lehigh Valley and Reading Co. lines, where it shifted most of its traffic between northern New Jersey and Harrisburg. Conrail's last electric freights ran in 1982 (*24*).

Although N&W abruptly ended electric operations in 1950, rival VGN's 1925 installation remained efficient and in good condition. VGN bought four new electric locomotives in 1948, and 12 more in 1956. But matters changed drastically in 1958 when N&W acquired Virginian, and sought operating economies as it merged the two railroads in the coal country of Virginia and West Virginia. N&W adopted directional running, operating loaded trains eastbound over VGN, which had lower grades. Rather than shuttle electric locomotives against the current of traffic on VGN, or extend electrification so that it served both lines, N&W decommissioned the VGN installation in 1962 (*18*).

On the DT&I – purchased and electrified by automaker Henry Ford – electric operation began in 1926 (fed from the same power source as the River Rouge factory, where lights dimmed when a locomotive was accelerating). Ford, with his deep pockets, seemed unconcerned that the energy costs of the initial 16-mi electrification were much higher than for steam locomotives, perhaps because contracts with electric utilities might have reduced these costs for a longer electrification. But the Interstate Commerce Commission, uneasy with Ford's maverick role in the railroad industry, ordered a halt to further electrification work shortly after construction started on a further extension. Under anti-trust pressure, Ford sold DT&I in 1929 to a holding company controlled by PRR. Despite parent PRR's large electrification, the new management lacked Ford's interest in electrification, and the locomotives were scrapped (*9*).

The government-owned FNM operated a series of trunk lines between Mexico City and points north on busy routes to northeastern Mexico and the U.S. Seeking greater operating efficiencies, FNM procured electric locomotives in 1982 and built a priority freight corridor between Mexico City and Querétaro with a modern 25 kV AC, 60 Hz electrification. Electric service began in 1994, but lasted just 2 years. Rail freight concessioning, a process started in 1996, led to increased investment in modern diesel-electric locomotives and full adoption of modern operating practices used on U.S. and Canadian railroads. The wires were soon dismantled to accommodate double-stack container trains (*25–27*).

Maintenance Costs

The GN installation began in 1909 as a short electrification through its original Cascade Tunnel in 1909 using an unusual and cumbersome three-phase, twin-wire version of 11 kV AC (most AC electrifications on railroads involve only one phase). But the tunnel and the steep alignment, even with electrification, proved inadequate for the railroad's operating needs. GN therefore built a new 8-mi Cascade Tunnel, with gentler grades and a lower summit, along with a new and longer single-phase

11 kV electrification that opened in 1927 (*28*). These changes made the GN electrification, at 73.5 mi, comparable with the N&W and VGN (at 55.9 mi and 134.5 mi, respectively).

Unlike the better-known Milwaukee Road electrification, GN's installation provided sufficient power. During the steam era, electric haulage through the tunnel was a necessity, but with dieselization, the costs of maintaining electric locomotives, substations, and wires exceeded the benefits. GN ended electric operations in 1956 after making ventilation improvements in the tunnel. Some of GN's locomotives found a second life in freight service on the PRR (*29*).

In the Canadian Rockies, BC Rail electrified its 81-mi Tumbler Ridge Subdivision in 1984. This branch, in an isolated part of the province in the Rocky Mountains, served two coal mines. With two long tunnels involving grades, BC Rail was concerned that diesel locomotives would stall out from a lack of fresh air while straining their engines to maintain speed, and decided on electrification. BC Rail selected 50 kV AC, 60 Hz, largely because at this very high voltage, the railway would need only one substation (*30–32*).

Technologically, BC Rail's electrification had a long future ahead of it, but it closed after only 17 years when one of the two coal mines became exhausted. The railway decided that the remaining mine's production did not warrant the fixed cost of electrification.

Analysis

North American railroads converted almost completely from steam to diesel within about a dozen years from the late 1940s to the late 1950s. It is not that railroads failed to appreciate the advantages of electric traction; instead, this appreciation came in the form of the diesel-electric locomotive, which is essentially an electric locomotive powered by an on-board diesel engine. The flexibility of the diesel-electric has spared railroads the expense of erecting lineside power supplies, even though this comes at the cost of maintaining both diesel engines and electric motors aboard locomotives. For all North American freight railroads, the cost savings from fleet standardization and abandonment of costly electrification infrastructure more than outweighs the marginally better throughput (up to about 15%) that electrification potentially offers.

With the exception of Henry Ford's DT&I, considerations other than immediate operating needs did not enter into decisions to electrify freight railroads. As the case studies show, all other freight railroads electrified because they believed that doing so would make their operations more efficient, or, in some cases, feasible in the first place.

Technical Standards

One factor said to have retarded the growth of electrification was the lack of a technical standard. This was most evident in the so-called battle of the currents, between proponents of DC (led by Thomas Edison and continued by General Electric), and AC advocates (led by George Westinghouse and continued by Westinghouse Electric). Even within the DC camp, there were several alternatives. Third-rail systems tended to use between 600 and 750 V, and different overhead-wire systems used 1.2, 1.5, 2.4 and 3 kV. Although DC remains reasonably well-suited for commuter rail under some circumstances, the amounts of current required for heavy freight trains ultimately exposed its weaknesses, as the Milwaukee Road's experience showed.

On the AC side, early work with 6.6-kV current indicated that a higher voltage was preferable. NH's 1907 installation pioneered the use of 11 kV AC. The state of electric motor technology in the early 20th century was such that lower AC frequencies were needed to provide the level of speed control that railroads required. Therefore, 11 kV AC installations in the U.S. used 25 Hz. This voltage, used on longer electrifications (NH, PRR, N&W, Virginian, GN) other than the Milwaukee Road, paralleled European development of 15 kV AC at 16.67 Hz (still used on national railways in Germany, Switzerland, Austria, Sweden, and Norway).

Yet 11 kV AC never emerged as an industry standard. The DC vs. AC issue remained unsettled until the 1950s, when 25 kV AC at commercial frequency (60 Hz in North America and 50 Hz in most other parts of the world) became the universally accepted norm for major new railway electrifications, with certain heavy-haul operations (such as BC Rail) choosing 50 kV. But by then, electrification was in retreat on U.S. railroads, as the diesel-electric locomotive combined the tractive-force benefits of electric traction with the diesel's ability to travel anywhere.

In that sense, electric traction triumphed over steam on North American railroads, but with a self-contained (diesel) generation source aboard the locomotive rather than with external electricity sources.

Even if 11 kV AC had emerged as a standard, resulting in somewhat more electrification, it is difficult to see how this alone would have safeguarded electrification against other factors leading to its removal. Neither has the emergence of 25 kV as the modern industry standard (along with its heavy-haul cousin 50 kV) led to the electrification of any lines other than BC Rail's mining branch and Mexico's freight corridor.

Energy Prices and Interest in Electrification

The 1973–74 energy shortage produced interest in electrification on the part of major freight railroads, including Illinois Central Gulf (*33*), Union Pacific, and Canadian Pacific (*34*). The 1979 energy shortfall led Conrail to study electrification between Harrisburg and Conway Yard west of Pittsburgh, and even settled on 25 kV AC as the preferred voltage despite its existing 11 kV AC, 25 Hz installation further east. But a cost–benefit analysis suggested that there were more productive uses for Conrail's investment capital, and there matters ended (*35*).

Nevertheless, when energy prices were high, interest in rail freight electrification rose. In 1977 the Transportation Research Board published a volume devoted to electrification from a freight railroad perspective (*36*). Even after railroads lost interest in electrification when oil prices fell in the 1980s, other parties continued to investigate the economics and operations of rail freight electrification (*37*). There was renewed interest in electrification in the early 2000s when energy prices spiked (*38*), but no electrification resulted. Then, hydraulic fracturing ushered in a new period of inexpensive oil and natural gas.

There appears to be a fundamental mismatch between the high capital cost, the long lead time for implementation, and the sometimes-volatile nature of energy supplies and prices. The time frame within which energy price spikes have run their course has been shorter than the lead time needed for such a major investment as electrification (particularly in today's procedure-conscious age).

Aging Technology

Even if some of the technologically superior electrifications had survived, experience on Amtrak's Northeast Corridor (NEC) and electrified commuter railroads suggests that freight railroads would have encountered the same difficulties that surviving installations have experienced with maintaining adequate power for the needs of more, longer, and faster Amtrak and commuter trains (*39*).

The ex-PRR portion of the NEC, with the voltage raised from 11 to 12.5 kV but still at 25 Hz, has come under strain as the strong acceleration and high speeds of modern trains raises levels of power draw. Electric freight operation, had it continued, would have placed additional stress on the ex-PRR installation. Similarly, the 11 kV installations on the VGN and the GN would have felt the strain from the rising demands of modern traffic levels had they survived.

Technological Unfamiliarity

Another factor was the loss of momentum among North American railroads and suppliers. As one observer noted in 1984,

> a big part of the problem of making profitable sense out of electrification stems from an American gap in the technology. Our technological heritage … is keyed to the Pennsylvania Railroad's electrification … In the 40 years since then, U.S. technology has waned. European and Far Eastern nations are now leaders (*40*).

Those roads that have electrified have drawn on European experience, where electrification (albeit largely passenger-driven) is widespread and expanding. BC Rail relied heavily on Swedish experiencc (*30*), and Amtrak's electrification between New Haven and Boston draws on British practice (*41*).

Other Factors

There is a contradiction between increased tonnage per train and the relatively infrequent nature of freights compared with commuter trains and (on the NEC) intercity passenger trains. Unless the operating circumstances clearly require electrification, as with BC Rail, the cost of providing traction power needs to be weighed against how often trains will operate. Additionally, issues of critical mass and high initial investment make electrification a difficult proposition for railroads:

> Operating officials prefer a single, ubiquitous, standardized pool of motorized power that can go anywhere and pull anything. Engineering officials do not want to bother with catenary and substation maintenance. Financial officers do not want the added debt … to fund electrification (*42*, p. 47).

Ultimately, rail freight electrifications became vulnerable to technological deterioration (NH, N&W, Milwaukee Road) or changes in traffic and operating patterns (PRR, VGN, BC Rail, FNM). Technological improvements, driven by environmental legislation, continue to lower emissions from diesel locomotives. But fossil fuels are unlikely to remain inexpensive indefinitely. At some point, supplies may tighten due to international events, environmental impacts of hydraulic fracturing, eventual exhaustion of reserves, or even shortages of water (a key component of today's fuel extraction). More localized electrification might conceivably result from air quality concerns, e.g., in southern California (*43*), although these issues may also be manageable through improvements in diesel engine emissions. The environmental balance of any possible electrification would, of course, depend on the original electric power sources.

Recommendations

The most recent cost–benefit analysis of rail freight electrification in the U.S. context appears to have been published in 1983, proposing electrification of 10,000–29,000 route-miles (*44,45*). Much has changed since then. Hydraulic fracturing has driven the price of fossil fuels down (at least for the immediate foreseeable future), and advances in diesel engine technology have reduced emissions greatly. These factors do not favor electrification.

However, freight railroads are busier today, and there is greater concern about fossil fuel emissions. Furthermore, electrified lines would not necessarily rely on any one original energy source. There may also be some complementarity with wind and solar energy, especially in rural areas. Thus, a new analysis may be appropriate.

Railroads face many investment needs, one of the most pressing of which is positive train control, mandated by the U.S. Congress. The emphasis on short-term financial performance makes it difficult for companies in any industry to justify long-term investments. Therefore, public–private partnerships might be needed, such as that between the U.S. Reconstruction Finance Corporation and PRR enabling electrification between Wilmington, Delaware, and Washington, D.C., to proceed in the 1930s. Some of the benefits of electrification accrue more to society than to the railroads. This would suggest a cost-sharing role in rail freight electrification for those state and federal levels of government reflecting these benefits. At the same time, railroads seeking the operating benefits of electrification should be expected to participate in the costs.

Future Directions

The early 21st century finds electrification for rail freight essentially nonexistent in North America, the world leader a century earlier. Inexpensive fossil fuel, brought about by hydraulic fracturing, has minimized the likelihood of freight railroads investing in electrification, which requires large upfront costs before operating savings can be obtained. The low energy prices of the late 2010s emerge against a background of increased concern about greenhouse gas emissions generally and emissions from mobile sources in particular, although great progress continues to be made in reducing diesel emissions (*46*).

Electrification advocates between the 1910s and 1940s were concerned about the different voltages and power delivery systems then in use, and pled for standardization. Settling on a basic specification has served the North American railroad industry excellently in other regards. Adoption of high-voltage, industrial-frequency AC, such as 25 kV AC, 60 Hz, would enable railroads to use a more robust electrification technology than any available in the first half of the 20th century.

Nor is it clear that any future installations would be modeled on those of the late 20th century. In 2011, during an oil price spike, one large railroad was considering a

> lineside storage and energy transfer system where helpers are used in train operations. Discontinuous electrification, with overhead line installed on short sections of steeply-graded route, would enable downhill trains to "export" regenerated dynamic braking energy for use by other trains heading uphill, for lineside storage or for sale to a third party (*47*, p. 48).

Whatever motive power technologies North American freight railroads pursue in the future will, in all likelihood, be profoundly influenced by today's almost-universal use of diesel-electric traction. To the extent that there is a motive power debate within the industry, for the time being it involves liquefied natural gas (LNG) rather than electrification (*48*).

Electrification may yet have a role to play, particularly in emissions-conscious jurisdictions such as in California. Similarly, technological breakthroughs in battery technology may well make electrification more desirable, particularly in specialized contexts such as helper districts. But barring a large cost differential in favor of electricity as opposed to diesel fuel and natural gas, or a future policy shift away from fossil fuels in favor of other energy sources, that role is unlikely to be a large one.

Future development of battery technology could conceivably result in battery-powered locomotives. If issues of cost and charging time can be resolved, electric operation may take a new technological form.

More immediately, the life and death of freight electrification on North America's railroads may be relevant for the debate over diesel vs. LNG as a locomotive fuel. Even though LNG may be less expensive in operating terms, the savings may not be enough to justify the necessary capital costs of fueling and maintenance facilities for LNG locomotives. If LNG does not revolutionize North American freight railroading, the dead hand of rail freight electrification may help explain why not.

Acknowledgments

The authors are grateful to John P. Aurelius for his insights on electrification technology, to Jay Guarneri for preparing the map in Figure 1, and to the reviewers for their helpful ideas.

References

1. Birch, L. W. Are the Overhead Distribution Costs Retarding Railroad Electrification? *Transactions of the American Institute of Electrical Engineers*, Vol. 67, No. 1, 1948, pp. 243–257.
2. Lothes, S. Wired Up: The Future of Electrification on America's Freight Railroads. *Trains*, November 2009, pp. 24–31.
3. Middleton, W. D. *From Bullets to BART*. Central Electric Railfans' Association, Chicago, Illinois, 1989.
4. Condit, C. W. Railroad Electrification in the United States. *Proceedings of the IEEE*, Vol. 64, No. 9, 1976, pp. 1350–1360.
5. Middleton, W. D. *When the Steam Railroads Electrified*. Indiana University Press, Bloomington, 2001.
6. Hilton, G. W., and J. F. Due. *The Electric Interurban Railways in America*. Stanford University Press, Palo Alto, California, 1960.
7. Resor, R. R. Innovation and Competition in Locomotive Manufacturing, 1950–2000. *Proc., 36th Annual Meeting of the Transportation Research Forum*, Daytona Beach, Florida, Vol. 2, 1994.
8. Condit, C. W. The Pioneer Stage of Railroad Electrification. *Transactions of the American Philosophical Society*, Vol. 67, No. 7, 1977, pp. 1–45.
9. Trostel, S. D. *Henry Ford: When I Ran the Railroads*. Cam-Tech Publishing, Fletcher, Ohio, 1988.
10. Manson, A. J. *Railroad Electrification and the Electric Locomotive*. Simmons-Boardman, New York, 1923.
11. Duffy, M. C. *Electric Railways 1880–1990*. Institution of Electrical Engineers, London, 2003.
12. Grasso, J. *Operation CUT, The First 30 Years*. RAE Publishing, Flanders, New Jersey, 2003.
13. Swanberg, J. W. *New Haven Power 1838–1958*. Alvin F. Staufer Books, Medina, Ohio, 1995.

14. Pinkepank, J. A. Why NH Re-Electrified. *Trains*, August 1964, pp. 20–26.
15. Frattasio, M. NH Locomotive Integration, Part 15: GE EF-5 Electrics. *The Post* (Penn Central Historical Society), No. 3, 2013, pp. 26–31.
16. Holley, N. T. *The Milwaukee Electrics: An Inside Look at Locomotives and Railroading*. N.J. International, Hicksville, New York, 1987.
17. Clark, R. A. The Milwaukee Electrification - A Proud Era Passes. Special supplement to *The Milwaukee Road Magazine*, July-August 1973.
18. Cooper, M. Y. *Norfolk & Western Electrics*. Norfolk & Western Historical Society, Forest, Virginia, 2001.
19. Middleton, W. D. *South Shore: The Last Interurban*. Indiana University Press, Bloomington, Indiana, 1999.
20. Kaplan, D. R. *Duneland Electric: South Shore Line in Transition*. PTJ Publishing, Homewood, Illinois, 1984.
21. Mutschler, C. V. *Wired for Success: The Butte, Anaconda & Pacific Railway, 1892–1985*. Washington State University Press, Pullman, 2002.
22. Bezilla, M. *Electric Traction on the Pennsylvania Railroad, 1895–1968*. Pennsylvania State University Press, University Park, 1980.
23. Garner, T. Last of PRR's Freight Motors - The E44. *The East Wind* (New England Chapter, Pennsylvania Railroad Technical & Historical Society), Winter 2010, pp. 18–23.
24. Graham-White, S. EMD's Freight Electrics. *Diesel Age*, October 2007, pp. 48–54.
25. Ferrocarril Suburbano de La Zona Metropolitana Del Valle de México. *Mirada Ferroviaria* (digital magazine), August 2009, pp. 57–67.
26. Gorostiza, F. J. *Los Ferrocarriles En La Revolución Mexicana*. Siglo XXI Editores, Mexico City, Mexico, 2010.
27. Campos, J. Lessons from Railway Reforms in Brazil and Mexico. *Transport Policy*, Vol. 8, 2001, pp. 85–95.
28. Marshall, E. Great Northern Electrification. *Railway Age*, November 14, 1925, pp. 899–901.
29. Pennypacker, B. *Cross Continent Electrics*. N.J. International, Hicksville, New York, 1994.
30. Olson, P. E. Electrification of the Tumbler Ridge Branch Line in British Columbia, Canada. *Transportation Research Record: Journal of the Transportation Research Board*, 1985. 1023: 11–18.
31. Sanders, D. *Northern Light - A Portrait of BC Rail*. White River Productions, Kansas City, Missouri, 2010.
32. Fisher, G. T. *Powering Freight Railways for the Environment and Profit: Electrification–Why and How*. Presented at the Transport Canada Rail Conference, Toronto, Ontario, 2008.
33. 50 kV for Illinois Central Gulf? *Modern Railways*, May 1974, pp. 199–200.
34. The Quest for Higher Voltage: A Brief Look at Developments Around the World. *Modern Railways*, May 1974, pp. 201–203.
35. Marchinchin, J. R. Conrail's Electric Dreams. *Trains*, October 2013, pp. 40–49.
36. *Railroad Electrification: The Issues*. Transportation Research Board of the National Academies, Washington, D.C., 1977.
37. Cooper, H. B., Jr., and C. A. Webb. Energy Considerations for Railroad Electrification in the United States. *Transportation Research Record: Journal of the Transportation Research Board*, 1981. 802: 41–49.
38. Time to Revisit Electrification? *Railway Age*, September 2008, pp. 14–16.
39. Allen, J. G., J. P. Aurelius, and J. Black. Electric Power Supply for Commuter Rail: Are Railroads Keeping Up? *Transportation Research Record: Journal of the Transportation Research Board*, 2011. 2219: 88–96.
40. Robertson, W. E. *Some Pros and Cons of Electrification*. Submitted in Partial Fulfillment of the Requirements for Membership in the American Society of Transportation and Logistics, Pittsburgh, Pennsylvania, 1984.
41. Kramer, J. Developments in Catenary Design and Maintenance. *Railway Track & Structures*, March 1994, pp. 34–37.
42. Tennyson, E. L. Economic Synergism in Railroad Electrification. *Transportation Research Record: Journal of the Transportation Research Board*, 1984. 953: 47–51.
43. Cogswell, R. U. Los Angeles Basin Railway Electrification Study. *Transportation Research Record: Journal of the Transportation Research Board*, 1994. 1470: 10–21.
44. Spenny, C. H., and G. B. Mott. *A Model for Evaluating the Costs and Benefits of Railroad Electrification*. Publication SS-742-U-12-12. Transportation Systems Center, Cambridge, Massachusetts, 1983.
45. Spenny, C. H., and G. B. Mott. Cost/Benefit Evaluation of Electrification of a US Rail Network. *Transportation Research Record: Journal of the Transportation Research Board*, 1983. 939: 35–44.
46. Jackson, C. Emissions Regulations Tightened Again. *Railway Gazette International*, February 2011, pp. 35–37.
47. Jackson, C. Electrifying Times at Norfolk Southern. *Railway Gazette International*, September 2011, pp. 45–48.
48. Frailey, F. W. Bigger than the Switch from Steam. *Trains*, June 2013, pp. 14–15.

The Standing Committee on Freight Rail Transportation (AT045) peer-reviewed this paper (18-02855).

Article

TRR
JOURNAL OF THE TRANSPORTATION RESEARCH BOARD

Transportation Research Record
2018, Vol. 2672(10) 176–187

DOI: 10.1177/0361198118794532
journals.sagepub.com/home/trr
SAGE

Accident-Cause-Specific Risk Analysis of Rail Transport of Hazardous Materials

Xiang Liu[1], Tejashree Turla[1], and Zhipeng Zhang[1]

Abstract

Rail plays a key role in the transportation of hazardous materials (hazmat). Improving railroad hazmat transportation safety is a high priority for both industry and government. Many severe railroad hazmat release incidents occur because of train accidents. The Federal Railroad Administration identifies over 300 accident causes, including infrastructure defects, rolling stock failures, human factors, and other causes. Understanding how hazmat transportation risk varies with accident cause is a key step in identifying, developing, evaluating, and prioritizing cost-justified accident prevention strategies, thereby mitigating hazmat transportation risk. The objective of this paper is to develop an integrated, generalized risk analysis methodology that can estimate accident-cause-specific hazmat transportation risk, accounting for various train and track characteristics, such as train length, speed, point of derailment, the number and placement of tank cars in a train, tank car safety design, and population density along rail lines. Using the two major causes of accidents on freight railroads—broken rails and track geometry defects—as an example, this paper demonstrates a step-by-step analytical procedure and decision support tool to assess how accident frequency, severity, and hazmat transportation risk vary by accident cause. The research method can be adapted for risk analysis at corridor- or network-level accounting for other accident causes.

Rail transports over two million carloads of hazardous materials (hazmat) in the United States annually. Unlike a truck trailer that carries a single hazmat car, a train can carry multiple hazmat cars (e.g., 50 to 120 flammable liquid cars) with greater transportation efficiency. Although over 99.99% of rail hazmat shipments are safe, a train accident may result in the derailment and release of multiple tank cars.

There are three principal strategies to reduce hazmat transportation risk, which are the 1) prevention of tank car derailment in the first place; 2) mitigation of release probability by tank car safety improvement; or 3) mitigation of release consequences. Prior research has largely focused on the latter two strategies, and there has been relatively less work with respect to reducing tank car derailment probability. One risk management strategy in this category is reducing train derailment likelihood by preventing major accident causes. The objective of this paper is to develop an accident-cause-specific railroad hazmat transportation risk analysis model that is adaptable to a variety of infrastructure-related, train-related, and operational factors. The model outputs the amount of risk associated with a specific accident cause. The model can assist decision makers in the evaluation of the effect of a particular accident cause on hazmat transportation risk.

Literature Review

Prior research related to train derailment risk has covered many aspects of the situation including the wheel–rail interaction (*1*), derailment causal analysis (*2*, *3*) and hazmat transportation risk (*4–12*). These studies have focused on various risk mitigation strategies, related to infrastructure (*4*, *7*), rolling stock (*13*), tank car safety enhancement (*13–16*), train makeup (*17*, *18*), or mitigation of release consequence (*19*). Each of these risk reduction strategies focuses on at least one event in the causal chain of an accident-caused release incident (Table 1).

Train Accident Occurrence

Many severe hazmat release incidents are caused by train accidents, particularly train derailments. Derailments account for over 72% of all types of accidents on freight railroads (*28*). The Federal Railroad Administration (FRA) of the U.S. Department of Transportation (U.S.

[1]Department of Civil and Environmental Engineering, Rutgers, The State University of New Jersey, New Brunswick, NJ

Corresponding Author:
Address correspondence to Xiang Liu: xiang.liu@rutgers.edu

Table 1. Risk Management Strategies and Risk Factors

Risk management strategies	Risk factors
• Train accident occurrence (*7, 13, 20–24*)	• Infrastructure condition • Equipment condition • Human factors • Traffic exposure, etc.
• Number of cars derailed (*25, 26*)	• Speed • Accident cause • Point of derailment • Train length, etc.
• Number of hazmat cars derailed (*17, 18, 27*)	• Number of hazmat cars in the train • Train length • Placement of hazmat cars in the train, etc.
• Number of hazmat cars releasing contents (*14–16, 24*)	• Accident speed • Hazardous materials car safety design, etc.
• Release consequences (*10, 19*)	• Chemical property • Population density • Spill size • Environment, etc.

DOT) identifies around 389 distinct accident causes (*29*) related to infrastructure, rolling stock, human factors, and other factors (*24*). Prior research found that over 70% of freight train mainline derailments were caused by either infrastructure defects or rolling stock failures (*24*). The hazmat risk reduction strategies include the prevention of track defects (*7*), equipment condition monitoring to reduce in-service failures (*30*), or the use of more advanced train control technologies to reduce human error (*31*).

Number of Cars Derailed

The total number of cars derailed depends on accident cause, speed, train length, and point of derailment (*25, 27*). The number of cars derailed (a proxy indicator of accident severity) is an important consideration.

Number of Hazmat Cars Derailed

The total number of hazmat cars derailed is related to train length, number of hazmat cars and non-hazmat cars in a train, and their placement (*27, 32*). Possible strategies for reducing the probability of tank car derailment include reducing the speed of the train to reduce the total number of vehicles derailed (*8*) and the placement of tank cars in the positions that are less likely to derail (*17, 18*).

Number of Hazmat Cars Releasing Contents

Not all derailed or damaged tank cars release their contents. A tank car accident database has been developed by the Railway Supply Institute (RSI) and the Association of American Railroads (AAR). Using this database, it has been found that the tank car release probability can be reduced by using more robust tank car designs (*14–16*). In addition, reducing train speed can reduce the accident impact on the tank car, therefore decreasing release probability (*8*).

Release Consequences

The consequences of a release can be measured with different metrics, such as property damage, environmental impact, traffic delay, or the affected population. Geographical information systems (GIS) can be used for consequence analysis when integrated with other databases such as census and rail network data (*9*). The use of a lower-hazard chemical, rerouting of hazmat traffic to avoid populated areas, or improved emergency response and evacuation have been identified as potential strategies to mitigate release consequences, thereby reducing the risk (*10*).

Knowledge Gaps

Although the prior research recognized the importance of evaluating hazmat transportation risk on the causal level, it exclusively used empirical data analyzing approaches for this purpose. This paper aims to build a probabilistic risk analysis methodology that can estimate accident-cause-specific hazmat transportation risk using statistical approaches. Ultimately, this paper aims to address how railroad hazmat transportation risk varies by major causes given specific track and train characteristics. This information can further support the identification and prioritization of alternative risk mitigation strategies, particularly accident cause prevention techniques.

Data Sources

The Rail Equipment Accident/Incident Report (REAIR) form is used by railroads to report all accidents that exceed a monetary threshold of damages to infrastructure and rolling stock. FRA compiles these reports into the Rail Equipment Accident (REA) database, which records accident type, consist type, track type, accident cause, accident consequence, and other information. In this paper, the FRA REA database is used to calculate the frequency of freight train derailments on Class I railroad main tracks. In addition, the FRA Operational Database is used to calculate traffic volume, in terms of train-miles. Based on these two databases, it is possible to calculate accident-cause-specific derailment rate (number of derailments normalized by traffic exposure) and derailment severity (e.g., the number of cars derailed per derailment). Furthermore, the conditional probability of release (CPR) for a derailed tank car, which reflects its safety performance, is obtained based on published statistics from the AAR and RSI.

Methodology

The research methodology portrays the process of an accident-caused release incident. Each event in this process is subject to specified influencing factors. A probabilistic risk analysis (PRA) model is developed that integrates a set of probabilistic processes to calculate the probability of a release incident (Figure 1). This paper focuses on accident-caused release, without accounting for releases caused by thermal tearing. The latter will be addressed through a separate study in the next step.

Train Derailment Rate Calculation

Train derailments account for the majority of accident-caused release incidents, and thus this paper focuses on mainline derailments. The derailment rate is defined as the number of derailments normalized by traffic exposure:

$$Z = \frac{Y}{M} \tag{1}$$

where Y = number of train derailments, and M = traffic exposure.

Tank Car Derailment Probability

Position-dependent tank car derailment probability is related to its position in a train, the point of derailment (the first car in a sequence of cars derailed per accident), and the number of cars derailed. In this paper, "cars" refers to all types of vehicles in a train (locomotives as well as empty and loaded railcars) unless stated otherwise (*6*, *33*). Equation 2 presents a probabilistic model to

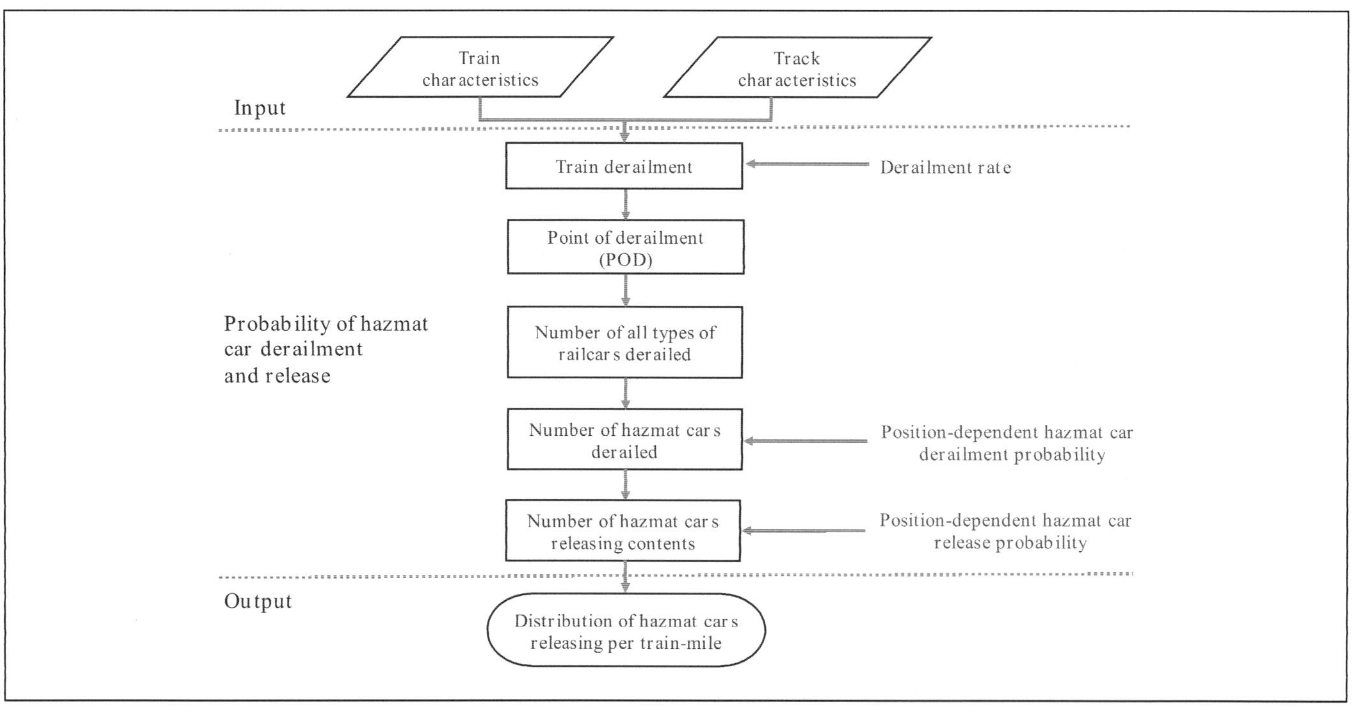

Figure 1. Accident-cause-specific hazmat release risk analysis framework.

estimate the derailment probability of a tank car at a given position of a train on a specific track segment.

$$PD_i(j) = \sum_{k=1}^{j}\{POD_i(k) \times \sum_{x=j-k+1}^{L-k+1} PN_i(x)\} \quad (2)$$

where

$PD_i(j)$ = probability of derailment for a vehicle at the jth position of a train on the ith segment,

$POD_i(k)$ = point of derailment probability for the kth position of a train on the ith segment,

$PN_i(x)$ = probability of derailing x vehicles in a train accident on the ith segment, and

L = train length (total number of vehicles in the train, including locomotives).

Release Probability of Derailed Tank Car

Tank car safety performance is reflected by its release probability once being derailed or damaged. This probability is referred to as the conditional probability of release (*6*), denoted by $CPR_i(j)$. Assuming that the releases of tank cars are independent, the probability that at least one hazmat tank car releases, denoted by $P_i(X_R|TD)$, is equal to one minus the total probability that none of the derailed tank cars release contents:

$$P_i(X_R|TD) = \left\{1 - \prod_j [1 - PD_i(j) \times CPR_i(j)]\right\} \quad (3)$$

where

$P_i(X_R|TD)$ = probability of a release incident after a train is derailed on the ith segment, and

$CPR_i(j)$ = conditional probability of release (CPR) of a derailed tank car.

Hazmat Transportation Risk

In the transport of hazardous materials, segment-level risk is generally defined as the multiplication of the probability of a release by the consequence of a release. Route risk is the summation of segment risk.

$$R = \sum_{i=1}^{N} P_i C_i \quad (4)$$

where

R = hazardous material transportation risk (e.g., expected release consequence),

P_i = probability of a release incident on the ith track segment,

C_i = consequence of a release (e.g., affected population) on the ith segment, and

N = number of segments.

Segment-specific release probability (P_i) per train shipment can be estimated by

$$P_i = P_i(TD) \times P_i(X_R|TD) \approx (Z_i L_i) \times P_i(X_R|TD) \quad (5)$$

where

$P_i(TD)$ = probability of a train derailment when traversing the ith segment,

$P_i(X_R|TD)$ = probability of a release incident after a hazmat train is derailed on the ith segment,

Z_i = train derailment rate on the ith segment, and

L_i = segment mileage.

Parameter Estimations in Risk Modeling

Freight Train Derailment Rate

This paper focuses on freight train derailments on Class I railroad mainlines, excluding other types of accidents (e.g., collisions or grade crossing accidents). Table 2 shows that the number of derailments and number of cars derailed vary with accident cause. This paper focuses on major cause groups. For illustration, this paper concentrates on the two most common freight train derailment causes, which are broken rails and track geometry defects, respectively. The top two causes are used as

Table 2. Major Causes of Class I Mainline Freight Train Derailments, 2000 to 2016

Cause group	Description	All types of train derailments	Hazmat train derailments	Total number of hazmat cars derailed
08T	Broken rails or welds	896	288	795
04T	Track geometry defects	444	193	223
10E	Bearing failure (car)	367	149	181
12E	Broken wheels (car)	332	85	164
09H	Train handling (excluding brakes)	288	136	112
01M	Obstructions	259	87	73
05T	Buckled track	236	70	210
03T	Wide gauge	234	63	152
04M	Track–train interaction	201	81	115
11E	Other axle or journal defects	190	34	37
	All causes (including the causes not listed in this table)	6,229	2,272	3,611

examples because they collectively account for around 20% of all freight train derailments on Class I mainlines (*24*). However, the generic PRA methodology developed here can be adapted to other causes as well.

The train derailment rate is affected by several factors, such as FRA track class, method of operation, and annual traffic density (*11*). This analysis focuses on FRA track class, which has been identified as a key influencing factor (*11, 25, 34, 35*). There are five principal track classes commonly used by U.S. freight railroads, ranging from Class 1, with the lowest maximum allowable freight train speed (10 mph), to Class 5, with the highest (80 mph in signaled territory). These classes include specifications for track structure, geometry, inspection frequency, and method of inspection, with more stringent requirements for higher track classes. The FRA standards represent minimum requirements; in fact, railroads can maintain various sections of their infrastructure to standards that exceed the minimum required by the FRA. This study does not further delineate accident statistics according to additional factors; this is to avoid a small sample size that will cause statistical sampling biases (*6*). Future research can account for additional factors affecting the train derailment rate. Track-class-specific train-mile distribution (Track Class 1—0.8%; Class 2—3.3%; Class 3—11.1%; Class 4—54.7%; Class 5—30.0%) is presented in a previous study using traffic data from Class I railroads (*26*).

The derailment rates caused by broken rails and track geometry defects by FRA track class were calculated (Figure 2). Future research may consider alternative traffic exposure metrics, such as car-miles or ton-miles on each track class, when the information is available. Figure 2 shows that the FRA track class has an inverse relationship with the train derailment rate for both broken rails and track geometry defects. This is probably

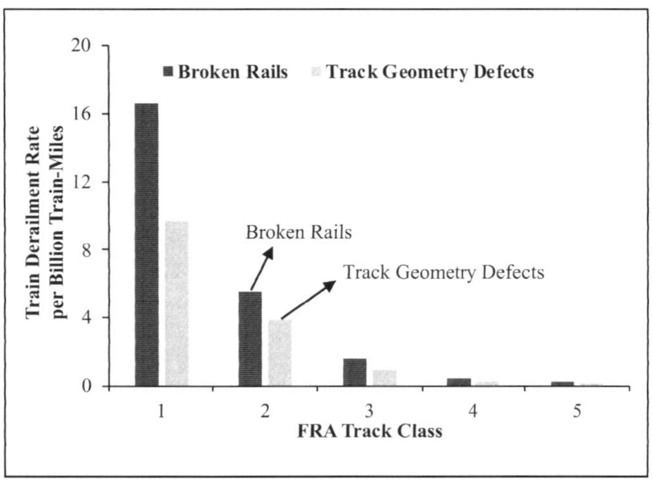

Figure 2. Accident-cause-specific train derailment rate by FRA track class.

attributable to better infrastructure conditions and more frequent inspection and maintenance on these high track classes, thereby reducing accident probability. Moreover, on each track class, broken rails result in a higher freight train derailment rate than track geometry defects. The derailment rate heterogeneity by accident cause and FRA track class is considered in the study's risk model.

For the Class 1 mainline freight train derailment rate from 2000 to 2016, the average annual decline rates were 10.6% and 8.7% for broken rails and track geometry defects, respectively (Figure 3). Assuming that this trend continues, these percentages can be used as temporal adjustment factors to estimate the derailment rate for a future year. A similar adjustment methodology has been used in the literature (*36*). In this study, let the reference year be 2008, as its derailment rate is closest to the 17-year average. Other analysts can also use this approach to update derailment rate statistics using future data.

Number of Cars Derailed

The model assumes that the number of cars derailed (both loaded and empty railcars, including locomotives) follows a truncated geometric distribution (*25, 26*). In Equations 6 and 7, PN(x) represents the probability of derailing a certain number of cars in a derailment, and x is the number of cars derailed or damaged. The probability of each car derailment, P, is affected by derailment speed and residual train length.

$$\mathrm{PN}(x) = \frac{P(1-P)^{x-1}}{1-(1-P)^{RL}} \tag{6}$$

$$P = \frac{e^z}{1+e^z} \tag{7}$$

For train derailment severity analysis, the response variable is the total number of cars derailed per accident. A truncated geometric (TG) regression model was developed to account for derailment speed, residual train length (RL $= L - \mathrm{POD} + 1$), proportion of loaded cars (LO), and whether the POD is loaded [I(POD)] as the explanatory variables. The fitted model is based on data covering Class I mainline freight train derailments between 2000 and 2016 for broken rails and track geometry defects (Equations 8, 9). The modeling details can be found in (*25, 26*).

$$\begin{aligned} z(\text{broken rails}) &= \text{logit}(P) = \log\left(\frac{P}{1-P}\right) \\ &= -0.891 - 0.387 \times \mathrm{IPOD} - 0.085 \times \mathrm{DS}^2 \end{aligned} \tag{8}$$

$$\begin{aligned} z(\text{track geometry defects}) &= \text{logit}(P) = \log\left(\frac{P}{1-P}\right) \\ &= (2.406 + 1.242 \times \mathrm{DS} - 0.343 \times \mathrm{LO} - 0.401 \times \mathrm{DS}^2) \end{aligned} \tag{9}$$

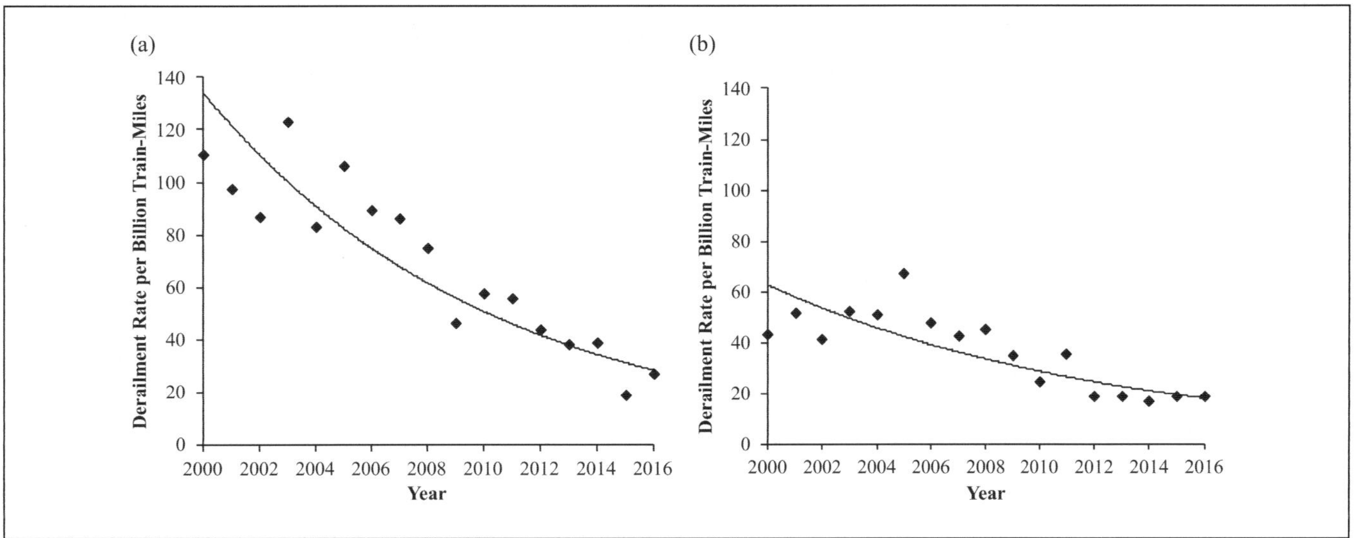

Figure 3. Class I mainline freight train derailment rates from 2000 to 2016 for (*a*) broken rails and (*b*) track geometry defects.

where

DS = logarithmic derailment speed (the speed is in mph),

I(POD) = whether the POD is a loaded car (1 indicates that the POD is loaded, 0 otherwise), and

LO = proportion of loaded cars to total number of cars (between 0 and 1, where 0 denotes an entirely empty train and 1 denotes a fully loaded train).

Point of Derailment

The normalized point of derailment (NPOD) is a metric accounting for train length heterogeneity. It is equal to the POD divided by train length (between 0 and 1). Beta distribution was found to be the best fit for FRA-reportable derailment data between 2000 and 2016 on mainlines. Given a train length L, the probability that the POD is at the kth position, POD(k), can be estimated using

$$\mathrm{POD}(k) = F\left(\frac{k}{L}\right) - F\left(\frac{k-1}{L}\right) \tag{10}$$

where

POD(k) = POD probability at the kth position of a train,

$F()$ = cumulative density distribution of the fitted NPOD distribution, and

L = train length (total number of cars in a train).

Figure 4 shows the cumulative POD distribution for derailments caused by broken rails and track geometry defects. It shows that the POD in a broken-rail-caused derailment is more likely to occur near the front of the train, whereas PODs caused by track geometry defects are relatively more uniformly distributed. This finding is consistent with a prior study based on older datasets (*6*).

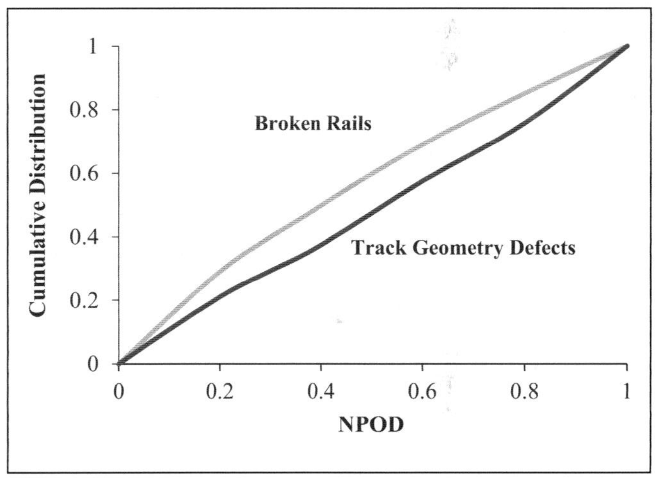

Figure 4. Fitted NPOD distributions by major accident causes.

Decision Support Tool

As part of this research, an automatic risk decision support tool is being developed (Figure 5). This tool consists of built-in formulas (presented above) required for calculating the cause-specific release probability of hazmat tank cars on Class I freight railroad mainlines. The tool is flexible enough to account for various risk factors (such as derailment speed, residual length, position of hazmat car, and so forth). A numerical example will be provided below to demonstrate the practical use of the tool for route-specific risk assessment.

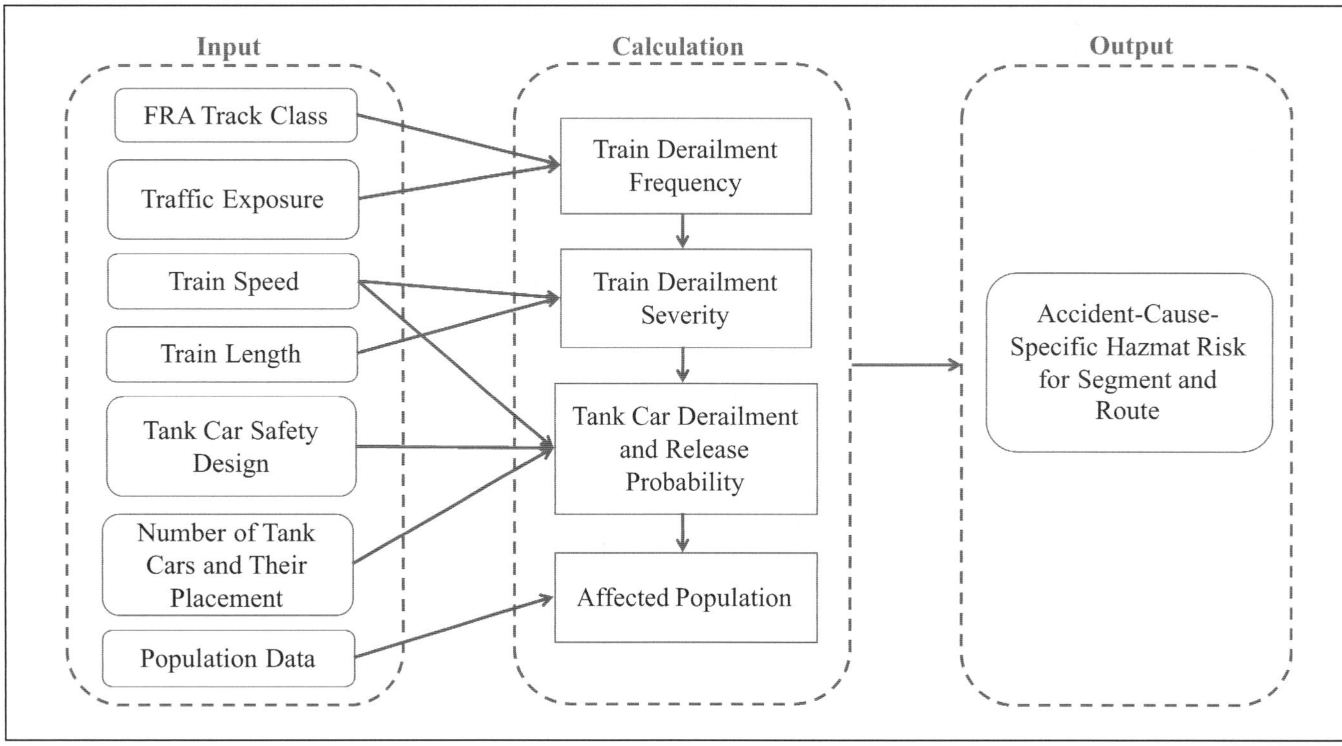

Figure 5. Implementation of decision support tool.

This computer tool has two practical advantages. First, it can automate sophisticated risk analysis processes based on user-defined inputs. Second, it can be further developed and used to compare alternative risk reduction strategies on specific routes or networks. Under various "what if" scenarios, the tool can generate different risk profiles, thereby providing guidance for data-driven risk management.

Numerical Example

Position-Dependent Derailment Probability

In this section, a numerical example is presented to explain the practical use of the methodology described above. For illustrative convenience, the paper focuses on broken rails and track geometry defects. The methodology can be adapted to other causes. In this paper, 100 vehicles per train are assumed, including locomotives, hazmat, and non-hazmat railcars, but the train length can be altered as desired. The position-dependent vehicle derailment probability can be estimated using Equation 1. Accident-cause-specific vehicle derailment probabilities depending on position are presented below for comparison. Given the train length (in this example, 100), the probability of derailment is estimated to vary with speed. The selected speeds (10, 25, 40, 60, and 80 mph) represent the maximum allowable speeds by FRA Track Classes 1 to 5, in signaled track territory.

For a given position of a tank car in a train, the higher the speed, the more likely that the car will be derailed (Figure 6). For example, consider a tank car located at the 40th position of a train, in which its derailment probability is 0.13 if this train is derailed at 25 mph, because of a broken rail. By comparison, the derailment probability is estimated to be 0.079 in a track-geometry-caused derailment at 25 mph. At 60 mph, the derailment probabilities increase to 0.318 and 0.123 for each cause, respectively. Moreover, it can be seen that, all else being equal, the vehicle derailment probability for broken rails is higher than for track geometry defects, because, on an average, broken rails tend to derail more cars than track geometry defects (*20*, *23*).

Position-Dependent Release Probability per Train-Mile

Table 3 shows derailment and release probabilities per train-mile for both accident causes, by FRA track class. It is assumed that all tank cars conform to DOT 117 standards, with the estimated conditional probability of release being 0.029 (*36*). It is also conservatively assumed that the accident speed is the maximum track speed for each track class. It is assumed that 10 hazmat cars are placed in the "worst case" scenario, in which these cars are located in positions that are more likely to derail. A sensitivity analysis of the effect of tank car position will be shown later.

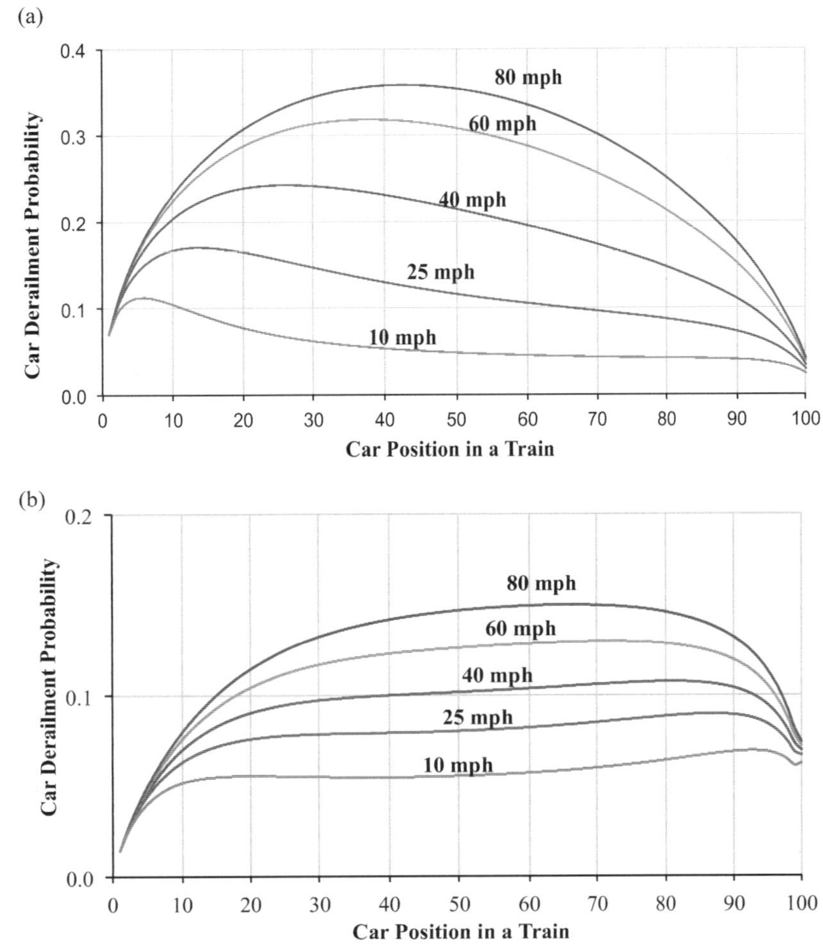

Figure 6. Position-dependent car derailment by accident cause for (*a*) broken rails and (*b*) track geometry defects.

Table 3. Train Derailment and Release Probability by Accident Cause

FRA Track class	Train derailment rate per train-mile	Probability of at least one hazmat car release in a train derailment	Release probability per train-mile
Broken rails			
1	166.44E−10	0.03074	51.17E−11
2	55.32E−10	0.03656	20.22E−11
3	16.25E−10	0.06178	10.04E−11
4	4.73E−10	0.08345	3.95E−11
5	2.04E−10	0.09363	1.91E−11
Track geometric defects			
1	96.64E−10	0.01702	16.44E−11
2	38.72E−10	0.02396	9.28E−11
3	9.29E−10	0.02970	2.76E−11
4	2.04E−10	0.03613	0.74E−11
5	0.39E−10	0.04147	0.16E−11

Table 3 shows that a higher FRA track class has a lower release probability per train-mile for each cause. The higher the track class, the lower the derailment rate caused by either broken rails or track geometry defects (as shown in Figure 2). FRA track class has a dual effect on derailment rate and tank car release probability. On one hand, a higher track class has a lower train derailment rate caused by broken rails or track geometry failures. On the other hand, a higher track class has a higher maximum operating speed, thus perhaps increasing the probability of tank car derailment and release (*36*). Because the decline in train derailment rate more than offsets the increase of tank car derailment and release probabilities, the net result is that higher track classes are associated with lower hazmat release probabilities per train-mile. Using broken rails as an example, the hazmat release probability per train-mile on Track Class 3 is approximately twice that on Track Class 4, and six times the release probability on Track Class 5.

Effect of Tank Car Positions

Previous studies have suggested that the tank car derailment probability is affected by the car's position in the train. Therefore, change of tank car position was identified as a potential risk reduction strategy (*37*). In the example above, the worst case scenario was considered. In this subsection, a severity analysis is presented for two additional tank car placement scenarios. The "best" scenario is one in which all the hazmat tank cars are placed in positions with the lowest probabilities of car derailment in an accident. In the "random" scenario, all hazmat cars are placed randomly throughout the train. Similar to the analysis above, higher track classes have lower release probabilities for given tank car placements. On the same track class, release probability per train-mile varies by tank car position (Figure 7).

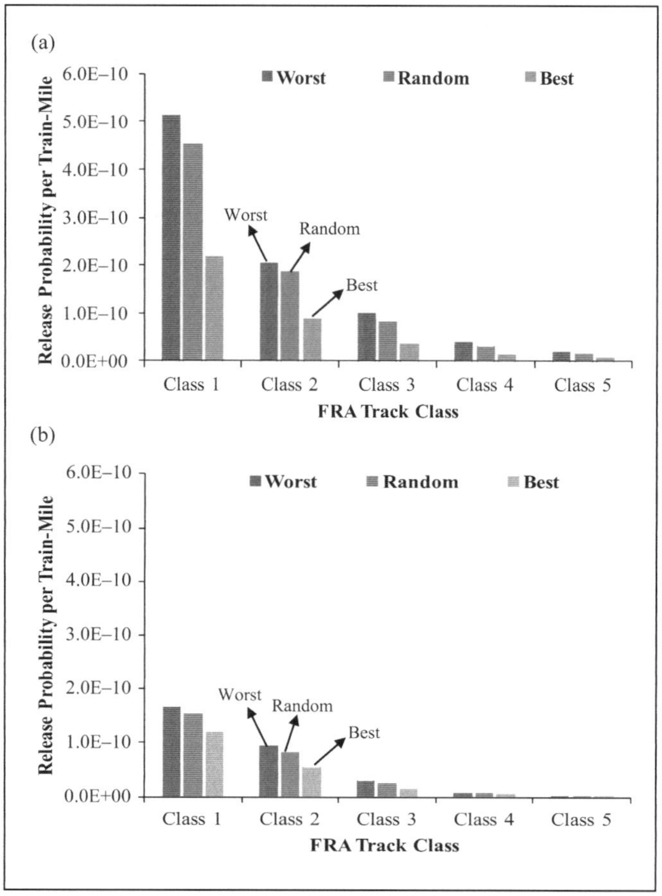

Figure 7. Hazmat release probability per train-mile by accident cause for (*a*) broken rails and (*b*) track geometry defects.

Route Transportation Risk Calculation

This section discusses how to use the information above to calculate route-specific transportation risk. For illustration, a simple, manual calculation example is presented below. Suppose that there are three track segments with different characteristics, including consequences such as the population in the evacuation zone (hypothetical and for illustration only):

- Segment 1: FRA Track Class 2, 10 mi long, 1,000 people within the evacuation zone;
- Segment 2: FRA Track Class 3, 15 mi long, 500 people within the evacuation zone;
- Segment 3: FRA Track Class 4, 20 mi long, 250 people within the evacuation zone.

Let it be assumed that there is a train length of 100 cars, carrying 10 tank cars located at the train positions that are the most likely to derail. It is also conservatively assumed that the accident speed is the maximum track speed for each track class. Again, these assumptions are only made to illustrate the calculation process. The goal is to calculate the risk when one train traverses this three-segment route for one shipment. Accident-cause-specific release probability information for this type of train is presented in Table 3 for broken rails and for track geometry defects.

For example, Segment 1 is a Class 2 track. According to Table 3, the probability of release because of broken rails is estimated to be 20.22E−11 per train-mile. The segment is 10 mi long. For one train shipment, the release probability is 20.22E−10 (20.22E−11 × 10). The consequence is 1,000 affected people. Therefore, the hazmat release risk per train shipment because of broken rails on this particular track segment is 20.22E−07 (20.22E−10 × 1,000). Similarly, the risks on segments 2 and 3 can be estimated at 7.53E−07 and 1.97E−07, respectively. In total, the route-specific risk per train shipment because of broken rails is 29.72E−07. Similarly, using Table 3, the route risk because of track geometry defects on the same route is 2.86E−07 per train shipment.

In this example, hazmat release risk because of broken rails is 10 times greater than the risk because of track geometry defects on this route. If 40% of broken rails on this route could be prevented, the release risk would be reduced proportionally, to 17.83E−07 (29.72E−07 − 29.72E−07 × 40%). This example demonstrates how the risk model can be used to understand the impact of accident cause prevention on hazmat transportation risk. This enables the assessment and comparison of different risk management strategies.

Discussions

Unit Hazmat Train Risk Analysis

The methodology developed in this research can be used to quantify accident-cause-specific risk for unit trains as well. The risk methodology first calculates train-position-specific derailment probability. Given a specific type of tank car at each position, the method can further calculate its derailment and release probability. For risk modeling, a unit hazmat train is a special case in which all train positions (excluding locomotives and buffer cars) contain hazardous materials. Note that this research only focuses on the releases caused by mechanical damage incurred by tank cars in train accidents, without accounting for releases resulting from thermal tearing, which is a process by which a fire impinging on the tank causes the steel to weaken. Accounting for thermal-tear-caused tank car release risk in unit-train operation requires a standalone study, and it is the next step of the research.

Research Contribution and Potential Implementation

This research develops a practical, generalized methodology to estimate accident-cause-specific hazmat transportation risk, accounting for track-related, train-related, and operational factors, for any train configuration. This research can be used to evaluate the potential impact of accident prevention on hazmat release risk, thereby aiding with the comparison and prioritization of alternative risk reduction strategies (e.g., broken rail prevention, track geometry quality improvement or tank car design improvement). Moreover, researchers can use the model to compare broken rail prevention versus tank car design enhancement in terms of the degree of risk reduction on a specific corridor or network. Given budget limitations, an optimal risk portfolio can be developed to allocate resources for alternative types of risk management strategies. Implementation of advanced risk models into practical use is pivotal for successful risk management. With this in mind, the authors are developing a prototype computer-aided decision support tool that can automate all the risk calculations described above. Using this tool, practitioners can easily change risk parameters, perform automated risk assessment in various "what if" scenarios, and thus compare and implement promising risk mitigation approaches.

Conclusion

This paper develops a generalized risk analysis methodology that can estimate accident-cause-specific hazmat transportation risk, accounting for various train, track, and operational characteristics, such as FRA track class, train length, speed, point of derailment, the number and placement of tank cars in a train, tank car safety design, and population density along rail lines. For illustration, the model is applied to estimate the risk created by broken rails and track geometry defects. The analysis shows that broken-rail-caused derailments tend to be associated with a higher release risk, because of their higher rate of occurrence and higher average severity (in terms of number of cars derailed). The risk analysis methodology has been implemented into a computer-aided decision support tool that automates risk calculations. The methods and tools developed here can support the railroad industry in the quantitative risk management of hazardous materials transported by rail, particularly with respect to accident cause prevention.

Future Research

Although this paper focused on the top two causes, the PRA methodology can be applied to analyzing the risks generated by other accident causes. Besides FRA track class, there could be other factors affecting derailment rates (*38*). For example, a curved track may have a different derailment rate compared with a tangent track, all else being equal (*39–41*). Future research is needed to collect safety and traffic information on this and other factors, thereby updating the safety statistics used in the methodology.

Acknowledgments

The first and third authors were partially funded by the Federal Railroad Administration (FRA) of U.S. DOT, at the time of writing this paper. The authors are grateful to anonymous reviewers for their constructive comments.

Author Contributions

The authors confirm contribution to the paper as follows: Study conception and design: Xiang Liu, Tejashree Turla, Zhipeng Zhang; Data collection, analysis and interpretation of results: Xiang Liu, Tejashree Turla, Zhipeng Zhang; Draft manuscript preparation: Xiang Liu, Tejashree Turla, Zhipeng Zhang. All authors reviewed the results and approved the final version of the manuscript.

References

1. Ling, L., M. Dhanasekar, and D. P. Thambiratnam. A Passive Road-Rail Crossing Design to Minimize Wheel-Rail Contact Failure Risk Under Frontal Collision of Trains onto Stuck Trucks. *Engineering Failure Analysis*, Vol. 80, 2017, pp. 403–415.
2. Britton, M.A., G. J. M. Read, and S. Asnaashari. Analysis of Train Derailment Cause and Outcome in Victoria, Australia, between 2007 and 2013: Implications for Regulation. *Journal of Transportation Safety & Security*, Vol. 9, No. 1, 2017, pp. 45–63.
3. Liu, X. Statistical Causal Analysis of Freight-Train Derailments in the United States. *Journal of Transportation. Part A: Systems*, Vol. 143, No. 2, 2017, p04016007.
4. Liu, X., M. R. Saat, and C. P. L. Barkan. Integrated Risk Reduction Framework to Improve Railway Hazardous Materials Transportation Safety. *Journal of Hazardous Materials*, Vol. 260, 2013, pp. 131–140.
5. Liu, X. Risk Comparison of Transporting Hazardous Materials in Unit Trains Versus Mixed Trains. *Transportation Research Record: Journal of the Transportation Research Board*, 2017. 2608: 134–142.
6. Liu, X., M. R. Saat, and C. P. L. Barkan. Probability Analysis of Multiple-Tank-Car Release Incidents in Railway Hazardous Materials Transportation. *Journal of Hazardous Materials*, Vol. 276, 2014, pp. 442–451.
7. Liu, X., M. R. Saat, and C. P. L. Barkan. Safety Effectiveness of Integrated Risk Reduction Strategies for Rail Transport of Hazardous Materials. *Transportation Research Record: Journal of the Transportation Research Board*, 2013. 2374: 102–110.
8. Kawprasert, A., and C. P. L. Barkan. Effect of Train Speed on Risk Analysis of Transporting Hazardous Materials by Rail. *Transportation Research Record: Journal of the Transportation Research Board*, 2010. 2159: 59–68.
9. Verma, M., and V. Verter. Railroad Transportation of Dangerous Goods: Population Exposure to Airborne Toxins. *Computers & Operations Research*, Vol. 34, No. 5, 2007, pp. 1287–1303.
10. Branscomb, L., M. Fagan, P. E. Auerswald, R. N. Ellis, and R. Barcham. *Rail Transportation of Toxic Inhalation Hazards: Policy Responses to the Safety and Security Externality*. RPP-2010-01. J.F.K. School of Government, Harvard University, 2010.
11. Liu, X., M. R. Saat, and C. P. L. Barkan. Freight-Train Derailment Rates for Railroad Safety and Risk Analysis. *Accident Analysis and Prevention*, Vol. 98, 2017, pp. 1–9.
12. Nayak, P. R., D. B. Rosenfield, and J. H. Hagopian. *Event Probabilities and Impact Zones for Hazardous Materials Accidents on Railroads*. Report DOT/FRA/ORD-83/20, FRA. U.S. Department of Transportation, Washington D.C., 1983.
13. Schlake, B. W., C. P. L. Barkan, and J. Edwards. Train Delay and Economic Impact of In-Service Failures of Railroad Rolling Stock. *Transportation Research Record: Journal of the Transportation Research Board*, 2011. 2261: 124–133.
14. Barkan, C. P. L. Improving the Design of Higher-Capacity Railway Tank Cars for Hazardous Materials Transport: Optimizing the Trade-Off Between Weight and Safety. *Journal of Hazardous Materials*, Vol. 160, No. 1, 2008, pp. 122–134.
15. Saat, M. R., and C. P. L. Barkan. Generalized Railway Tank Car Safety Design Optimization for Hazardous Materials Transport: Addressing the Trade-Off between Transportation Efficiency and Safety. *Journal of Hazardous Materials*, Vol. 189, No. 1, 2011, pp. 62–68.
16. United States Department of Transportation. *Hazardous Materials: Enhanced Tank Car Standards and Operational Controls for High Hazard Flammable Trains*. PHMSA-0082 (HM-251). Washington, D.C., 2012.
17. Bagheri, M., F. F. Saccomanno, S. Chenouri, and L. P. Fu. Modelling Hazardous Materials Risks for Different Train Make-Up Plans. *Transportation Research Part E: Logistics and Transportation Review*, Vol. 48, No.5, 2012, pp. 907–918.
18. Bagheri, M., M. Verma, and V. Verter. Transport Mode Selection for Toxic Gases: Rail or Road? *Risk Analysis*, Vol. 34, No. 1, 2014, pp. 168–186.
19. Yoon, H., C. J. Werth, C. P. L. Barkan, D. J. Schaeffer, and P. Anand. An Environmental Screening Model to Assess the Consequences to Soil and Groundwater from Railroad-Tank-Car Spills of Light Non-Aqueous Phase Liquids. *Journal of Hazardous Materials*, Vol. 165, No.1, 2009, pp. 332–344.
20. National Transportation Safety Board. *Railroad Accident Brief BNSF Railway Unit Ethanol Train Derailment*. Lesterville, South Dakota, 2017.
21. National Transportation Safety Board. *Railroad Accident Brief CSXT Petroleum Crude Oil Train Derailment and Hazardous Materials Release*, 2017.
22. Federal Railroad Administration. *Safe Placement of Train Cars: A Report*. U. S. Department of Transportation, Federal Railroad Administration, Washington, D.C., 2005.
23. Federal Railroad Administration. *Risk Evaluation Framework and Selected Metrics for Tank Cars Carrying Hazardous Materials*. Federal Railroad Administration, Washington, D.C., 2015.
24. Liu, X., M. R. Saat, and C. P. L. Barkan. Analysis of Causes of Major Train Derailment and Their Effect on Accident Rates. *Transportation Research Record: Journal of the Transportation Research Board*, 2012. 2289: 154–163.
25. Saccomonno, F. F., and S. M. El-Hage. Establishing Derailment Profile by Railcars Carrying Dangerous Commodities through Effective Marshalling Strategies. *Transportation Research Record: Journal of the Transportation Research Board*, 1989. 1245: 34–51.
26. Saccomonno, F. F., and S. M. El-Hage. Establishing Derailment Profile by Position of Corridor Shipments of Dangerous Goods. *Canadian Journal Civil Engineering*, Vol. 18, No. 1, 1991, pp. 67–75.
27. Glickman, T. S., E. Erkut, and M. S. Zschohke. The Cost and Risk Impacts of Rerouting Railroad Shipments of Hazardous Materials, *Accident Analysis Prevention*, Vol. 39, No. 5, 2007, pp. 1015–1025.
28. Liu, X. Collision Risk Analysis of Freight Trains in the United States. *Transportation Research Record: Journal of the Transportation Research Board*, 2016. 2546: 121–128.

29. Federal Railroad Administration. *Railroad Safety Statistics - Preliminary Annual Report*. U.S. Department of Transportation, 2011.
30. Federal Railroad Administration. *Office of Safety Analysis 3.01 – Accident Trends – Summary Statistics*. Federal Railroad Administration, 2015.
31. Martland, C., Y. Zhu, Y. Lahrech, and J. Sussman. Risk and Train Control: A Framework for Analysis. *Transportation Research Record: Journal of the Transportation Research Board*, 2001. 1742: 25–33.
32. Liu, X., M. R. Saat, X. Qin, and C. P. L. Barkan. Analysis of U.S. Freight-Train Derailment Severity Using Zero-Truncated Negative Binomial Regression and Quantile Regression. *Accident Analysis and Prevention*, Vol. 59, 2013, pp. 87–93.
33. Liu, X. Optimizing Rail Defect Inspection Frequency to Reduce the Risk of Hazardous Materials Transportation by Rail. *Journal of Loss Prevention in the Process Industries*, Vol. 48, 2017, pp. 151–161.
34. Liu, X., M. R. Saat, and C. P. L. Barkan. Analysis of Derailments by Accident Cause: Evaluating Railroad Track Upgrades to Reduce Transportation Risk. *Transportation Research Record: Journal of the Transportation Research Board*, 2011. 2261: 178–185.
35. Liu, X., and B. Schlake. Derailment Risk Analysis of Liquefied Natural Gas (LNG) Tenders on Mainline Freight Trains in the United States. *Transportation Research Part C –Emerging Technologies*, Vol. 72, 2016, pp. 77–92.
36. Liu, X. Risk Analysis of Transportation Crude Oil by Rail: Methodology and Decision Support System. *Transportation Research Record - Journal of the Transportation Research Record*, 2016. 2547: 57–65.
37. Bagheri, M., F. F. Saccomanno, S. Chenouri, and L. P. Fu. Reducing the Threat of In-Transit Derailments Involving Dangerous Goods through Effective Placement along the Train Consist. *Accident Analysis Prevention*, Vol. 43, No. 3, 2011, pp. 613–620.
38. Liu, X. Statistical Temporal Analysis of Freight-Train Derailment Rates in the United States: 2000 to 2012. *Transportation Research Record: Journal of the Transportation Research Board*, 2015. 2476: 119–125.
39. Xiao, X., X. Jin, Y. Deng, and Z. Zhou. Effect of Curved Track Support Failure on Vehicle Derailment. *International Journal of Vehicle Mechanics and Mobility*, Vol. 46, No. 11, 2008, pp. 1029–1059.
40. Silvia, M. I., I. R. Julia, H. César, and M. Laura. Derailment Risk and Dynamics of Railway Vehicles in Curved Tracks: Analysis of the Effect of Failed Fasteners. *Journal of Modern Transportation*, Vol. 24, No. 1, 2016, pp. 38–47.
41. Xiao, X., X. Jin, Z. Wen, M. Zhu, and W. Zhang. Effect of Tangent Track Buckle on Vehicle Derailment. *Multibody System Dynamics*, Vol. 25, No. 1, 2011, pp. 1–41.

The Standing Committee on Freight Rail Transportation (AR040) peer-reviewed this paper (18-01559).

All the views and opinions expressed here are those of the authors and do not necessarily state or reflect the views of U.S. DOT or FRA.

Article

TRR
JOURNAL OF THE TRANSPORTATION RESEARCH BOARD

Transportation Research Record
2018, Vol. 2672(10) 188–201

DOI: 10.1177/0361198118777386
journals.sagepub.com/home/trr
SAGE

Incorporating Economic Assessment into Capacity Allocation and Infrastructure Charging Policies for Vertically-Separated Railways

Aleksandr Prodan[1] and Paulo F. Teixeira[2]

Abstract

Capacity allocation mechanisms and infrastructure access charges in vertically separated railways generally consider only the points of view of the train operator and infrastructure manager. External and other societal costs are not considered in setting these policies and in evaluating their impact on each player. This work proposes a methodological framework for evaluating impacts of capacity allocation and infrastructure pricing policies on society by applying a set of project evaluation guidelines that is normally used for new infrastructure projects to a set of capacity allocation and pricing policies. These guidelines use a cost–benefit analysis to evaluate the project and then understand the impact on each industry player and on society. A case study of the ScanMed corridor is used, looking at a congested section over the Oresund Bridge. The case study evaluates proposed capacity allocation mechanisms that prioritize either passenger or freight traffic. The results of this evaluation show the total impact of a particular policy on each player, including society. External costs are also considered in this evaluation. This approach can be used by decision makers to make more informed decisions when setting infrastructure charging and capacity allocation policy.

Over the last two decades, several countries in Europe, Asia, and Africa have separated their railway systems, delegating infrastructure management and maintenance to one entity and train operation to a different set of entities. While market structure may differ by country, most vertically separated railways have a mechanism for allocating train paths and a system to charge for capacity use.

Motivation for this work originates from the need to better evaluate network performance of vertically separated railway systems, in which infrastructure managers (IMs) sell capacity to train operators (TOs), who, in turn, use this capacity to operate trains on the network. IMs and TOs represent two types of players in a multiplayer structure that also includes passengers and other end users, society, government, and infrastructure investors, among others. Each player has their own point of view, and each has their own financial and other motivations. While a large body of literature has been developed looking at the TO and IM points of view, society's point of view remains largely unexplored.

Having an efficient capacity allocation mechanism and a charging policy that transmits clear signals to TOs promotes efficient use of capacity. This is especially important for capacity-constrained railway corridors with different users, in which capacity management is the last option for maximizing line throughput before having to perform large-scale infrastructure investment.

Passenger long-distance and commuter trains, as well as slower freight trains, often operate in capacity-constrained corridors. When TOs submit conflicting capacity requests to the IM, it must use some mechanism to resolve this conflict. In the United States, the Northeast Corridor is a good example of having different types of operators, with commuter, long-distance, and freight trains all sharing the same right of way. While not a fully vertically separated corridor, different market segments compete for capacity, even if operated by the same operator. In Europe, congested corridors in Germany, France, Denmark, and Sweden are also faced with this problem.

[1]CPCS Transcom International Limited, Lisbon, Portugal
[2]CERIS, Instituto Superior Técnico, Universidade de Lisboa, Lisbon, Portugal

Corresponding Author:
Address correspondence to Aleksandr Prodan: aprodan@cpcs.ca

In most cases, IMs do not consider external costs or societal benefits when resolving these conflicts. The definition of "efficient use of capacity" depends on the point of view of the evaluating entity. If the IM is profit-driven, this may mean a capacity allocation mechanism that guarantees the highest financial return. If the goal is to maximize throughput, then a different set of criteria may be used. Neither of these mechanisms necessarily aligns with the societal point of view that considers maximization of total monetary and non-monetary benefits.

A great deal of work has been conducted on understanding the point of view of the IM and the TO, but not much work has been done on understanding the point of view of society. This paper provides a mechanism to quantify these societal benefits and to consider society's point of view in allocating capacity and in charging for infrastructure use. This mechanism considers both intramodal and intermodal competition and allows the system to function in a more equitable manner by taking into account externalities that are not being internalized today both by railways and by competing modes.

Literature Review

Railway capacity allocation has been a widely researched topic for several decades. Gibson has identified key capacity allocation mechanisms (*1*). Nilsson has provided innovations in algorithms for capacity allocation as well as capacity allocation mechanisms that take into account welfare-enhancing strategies in dealing with competing operators bidding for the same capacity (*2–4*). More recently, work has been done on analyzing capacity trade-offs in selecting a price or capacity-based allocation mechanism (*5*, *6*). Shared railway systems are especially challenging, with multiple types of users competing for capacity (*7*). Capacity allocation models tend to be either based on price (auctions) or on procedure (capacity-based). Although several algorithms have been suggested for implementing price-based systems, most systems have remained procedural (*8*). As the procedure of capacity allocation deals with multiple players, each player's perspective must be well understood and modeled by decision makers who set capacity allocation rules (*9*).

Infrastructure pricing in railways has been a topic of interest since the beginning of the railway reform in Europe. Charging systems have been set up to charge for infrastructure use and their structure and implementation has been studied ever since the first systems were put into place for both passenger and freight (*10*, *11*). Decision makers generally consider political and budgetary constraints only from the points of view of the infrastructure manager and train operator, ignoring the point of view of society (*12–14*).

External costs are not borne equitably by all modes of transport. Railways are considered to have fewer external costs than competing modes (*15*). At the same time, competing modes often do not internalize their external costs (*16*). Numerous approaches to quantifying external costs are possible (*17*). The European Union (EU) provides that infrastructure charges must be equitable between different modes, and that infrastructure charges can have an environmental component to internalize external costs (*18*). Several countries such as Sweden and Denmark have started charging trucks and providing incentives to trains to shift freight transport from road to rail. At the same time, rail may face congestion in some of the freight corridors that are being used for this shift. This presents a problem in how to deal with trade-offs in allocating capacity between passenger and freight trains (*19*). Since the beginning of the railway reform, these trade-offs were seen as a barrier to success (*20*).

Capital projects are often evaluated using some form of cost–benefit analysis (CBA), comparing total project costs over a project's expected lifetime to the derived benefits (*21*). Combined with the consideration of external and other societal costs, this approach provides a comprehensive overview of the project's impact on its major stakeholders (*22*). The EU suggests a set of guidelines for assessing railway projects (*23*).

Evaluating Societal Benefits of a Capacity Allocation Mechanism using CBA Under Railway Project Appraisal Guidelines

Capital projects often undergo rigorous evaluations to examine trade-offs and determine total project benefits. CBA is a common way to evaluate projects, taking into account the project's life cycle. More tactical decisions, such as infrastructure charging or capacity allocation policies, are rarely evaluated with such rigor. Capacity allocation policies are generally a procedural or a political decision to prioritize certain types of traffic. Furthermore, societal costs and benefits are rarely considered when setting these policies.

In this section we propose the use of a CBA framework, coupled together with Railway Project Appraisal Guidelines (RAILPAG) (*23*) to evaluate societal benefits for a given capacity allocation and infrastructure charging policy.

Industry Players

The authors are continuing the research that was started in Levy at al. (*9*). This paper considers five main players: society, the government, the regulator, the IM, and one or more TOs. Each entity has its own goals, motivation and driving factors:

- The society represents the view of the entire population.
- The government can wear numerous hats. It makes laws that govern the whole system. It may also be the investor in infrastructure. It does not necessarily represent the same views as society.

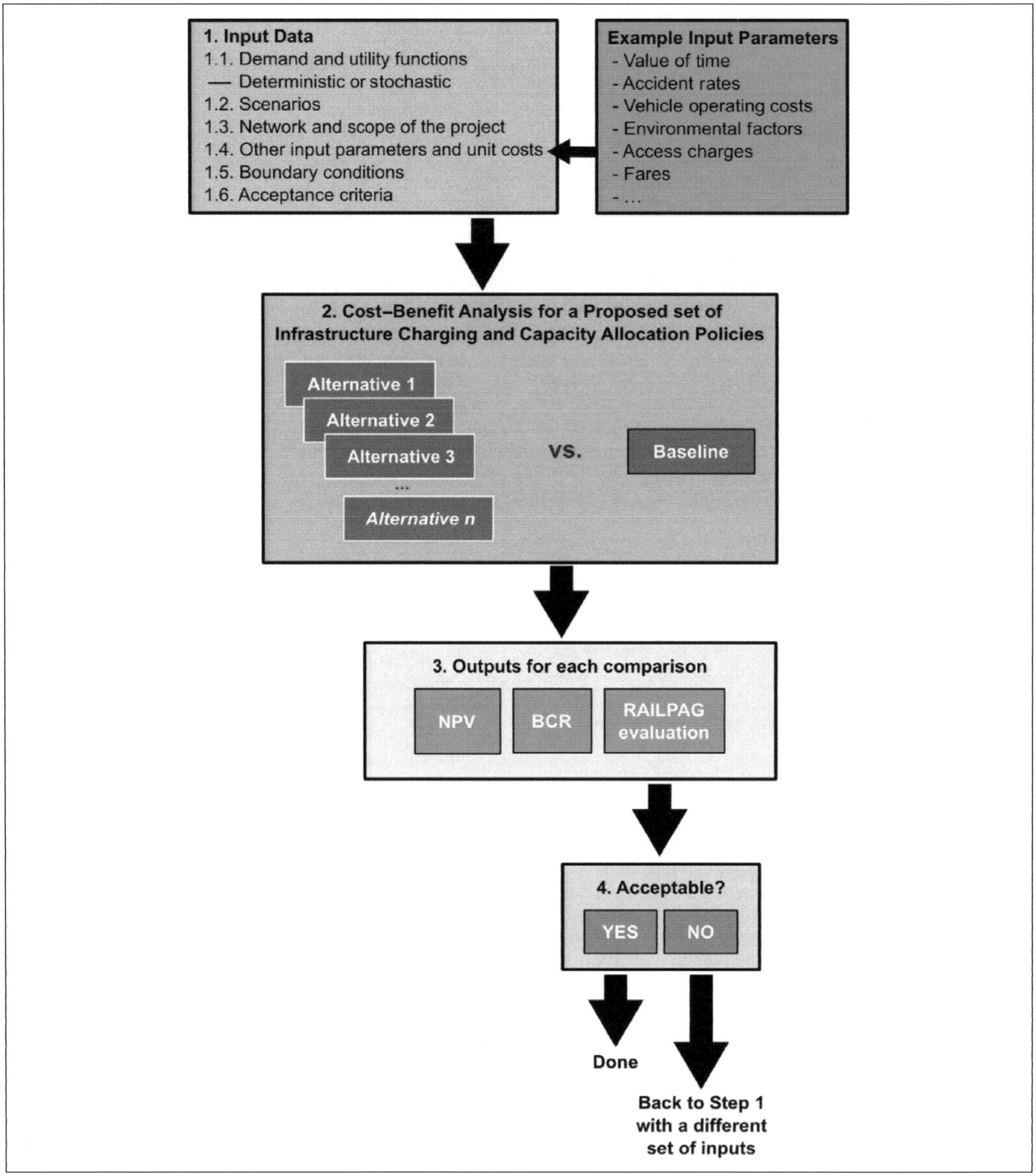

Figure 1. Proposed framework for evaluating infrastructure. BCR = benefit–cost ratio.
Source: Authors.

- The regulator is responsible for ensuring that existing laws and regulations, including those related to competition and safety, are followed.
- In a vertically separated system, IMs sell capacity, and TOs pay for using that capacity. In the EU, most countries allocate capacity using a procedural

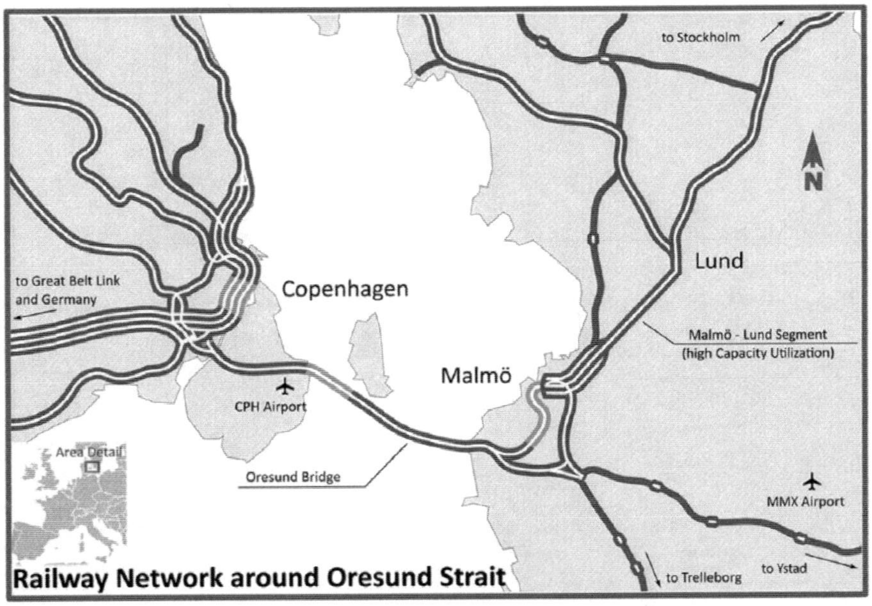

Figure 2. Oresund area railway map.
Source: Modified from Glentamara via Wikimedia Commons.

approach, though some involve capacity auctions (*9*). Capacity charges are determined by complex infrastructure charging systems that vary from one country to another (*10*).

Having already studied the points of view of the TO and IM in previous works, in this paper we focus on society's point of view.

CBA Framework Based on RAILPAG

CBA is a methodology that allows assessing the desirability of a project, in which it is important to evaluate an investment over a multiyear time period, while considering a wide range of different costs and stakeholder perspectives (*21*). The ultimate goal of a CBA is to compare different alternatives to determine which is more advantageous to implement (*22*). The assessment has been in use in some form since the early 20th century, and, perhaps, even in the 19th century (*21*).

When stating the problem to be evaluated using a CBA, it is important to consider which costs and benefits to include, how they are valued, what discount rate should be used, and the relevant constraints. In defining the project, it is important to take into account both the primary benefits and costs as well as externalities, and to remember to check if any secondary costs and benefits could be generated (*21*).

A number of local, national, and supranational guidelines have been developed on how to conduct a CBA, including those by various development banks. Since vertically separated railways are most prevalent in Europe, it makes sense to use a European set of guidelines for conducting a CBA. The European Commission handbook for performing project appraisal using CBA provides a reference approach to analyze the impact of a project in different domains using a consistent methodology (*22*). This methodology allows the user to evaluate a project during a timeline that is long enough to incorporate that project's benefits, and in a way that monetizes all costs and benefits associated with the project. The CBA methodology compares a proposed scenario with a baseline scenario, looking at changes in costs and benefits.

However, RAILPAG guidelines provide a common framework for appraisal of railway projects across the EU. Co-developed by the European Investment Bank and the European Commission, these guidelines provide an appraisal procedure that takes into account all stakeholders and all relevant effects of the project, and demonstrate how much impact the project will have on each stakeholder, and through which effects (*23*). This approach first conducts a more traditional CBA, then analyzes the same information in a stakeholders/effects (SE) matrix. This matrix lists the stakeholders on the X axis and the effects on the Y axis, and distributes the costs and benefits among all stakeholders, with each cell reflecting the net present value (NPV) of a specific effect for a specific stakeholder.

Instead of evaluating an investment project, this approach will be used here to evaluate a set of infrastructure charging and capacity allocation policies. This evaluation allows decision makers to view the impact of a proposed policy on all system players, including the point of view of society compared with the status quo.

A policy planning horizon of 10 years will be used for this assessment, which is within the European guidelines. This

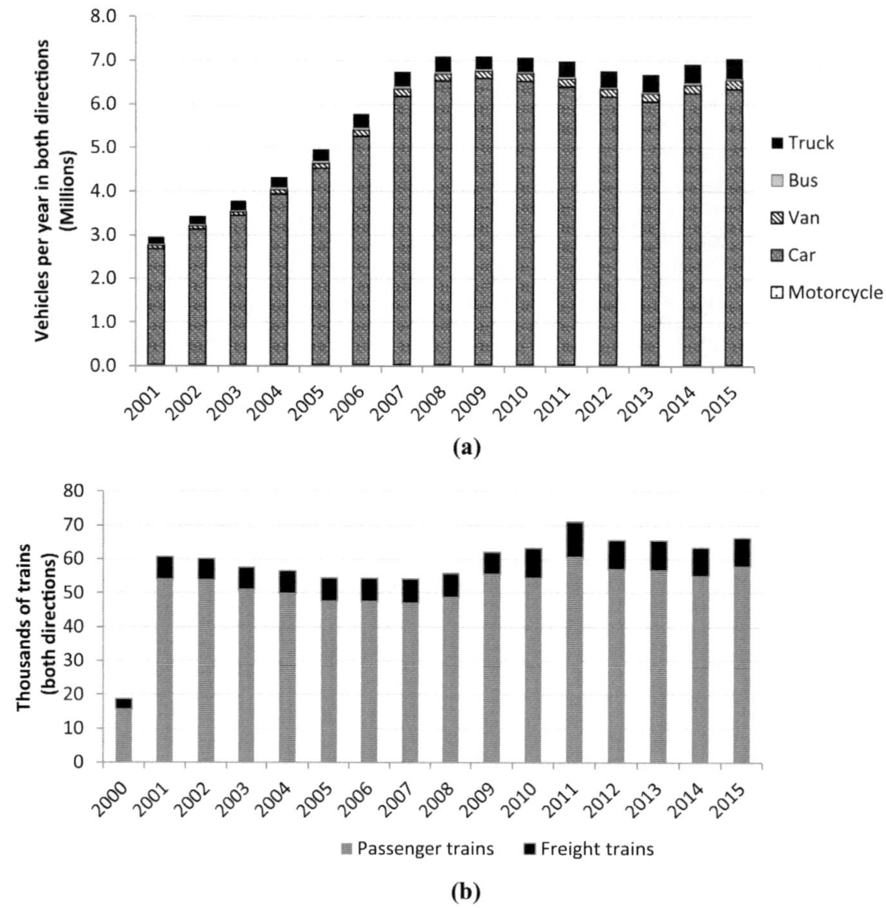

Figure 3. Traffic data for the Oresund Bridge, 2000–2015: (*a*) millions of vehicles per year; (*b*) thousands of trains per year.
Source: (*25*).

time period is stable enough to provide clear signals to TOs, and robust enough for IMs to periodically adjust their policy (every 10 years).

The model calculates the impact of traffic over the next decade for each player in the system, for IMs, TOs, final users, and non-users of the system. Competing modes are considered, looking at both road and rail. Other modes (e.g., sea, river) could also be included if they are relevant.

Evaluating Railway Infrastructure Charges and Capacity Allocation Policies

The proposed evaluation framework takes input data, processes the data using a CBA-like analysis, post-processes the results, and decides whether to refine the analysis further or if the results meet the initially set goals.

This framework consists of the following steps:

1. *Input data are gathered and scenarios are set up.* In this part of the process input data are collected and processed to be used in the analysis of alternatives.

1.1 *Demand and utility functions and elasticities are calculated and calibrated/validated.* Demand functions for all the relevant market segments need to be developed, calibrated, and validated. Depending on how demand is calculated, this approach will either be static or dynamic. Demand can either be deterministic or probabilistic. Elasticities and utility functions also need to be properly calibrated and validated to properly model demand.

1.2 *Scenarios are defined.* All test scenarios will be compared against the baseline. If there is uncertainty about a decision that is external to the project, then multiple baseline scenarios may be needed.

1.3 *Network and scope of the project is defined.* A definition is needed for the scope of the project and the extent of the network to be analyzed. A network that is inadequately small for a project will not show all network effects that may occur when testing one of the scenarios.

1.4 *Other input parameters and unit costs are defined.* Other costs that are needed to conduct a CBA need to be well defined.

Sweden			
Passenger Charge = A + B + C			
A	Passage through major cities	45.31590414	€ flat fee
B	High-traffic lines	0.68627451	€ per train-km
	Medium-traffic lines	0.250544662	€ per train-km
	Low-traffic lines	0.206971678	€ per train-km
C	Pax <20 t/axle	0.001525054	€ per tonne-km
	Pax >20 t/axle	0.00167756	€ per tonne-km
Freight Charge = A + B + C + D			
A	Oresund Bridge	324.6187364	€ flat fee
B	High-traffic lines	0.68627451	€ per train-km
	Medium-traffic lines	0.250544662	€ per train-km
	Low-traffic lines	0.206971678	€ per train-km
C	freight <22.5 t/axle	0.000675381	€ per tonne-km
	freight 22.5–25 t/axle	0.000740741	€ per tonne-km
	freight >25 t/axle	0.00080061	€ per tonne-km
D	Passage through major cities (not on the bridge)	45.31590414	€ flat fee
Denmark			
Passenger Charge = A + B			
A	Oresund Bridge	284.1420912	flat fee
B	per train-km not on the bridge	0.289544236	€ per train-km
Freight = A + B			
A	Oresund Bridge—flat fee	343.8002681	flat fee
B	per tonne-km not on the bridge	0.000509383	€ per tonne-km

Figure 4. Infrastructure charging system for Sweden and Denmark.
Sources: (26,27).

1.5 *Boundary conditions are defined.* Boundary conditions and assumptions need to be clearly defined and stated. When calculating the value of time, input passenger traffic is subdivided into business versus commute versus other type of traffic for passengers. Both the number of passengers and the number of trains or cars are estimated for the duration of the evaluation. Input freight traffic is expressed in tonnes carried, and the number of trains or trucks is calculated. Maintenance costs for the line and for the road are calculated for both scenarios. Using the value of time for passengers and freight, time savings are calculated for both categories of transport. For each parameter considered, a monetization factor is required, transforming it into a quantifiable amount of currency. This involves a conversion factor for value of time, cost per tonne of carbon emissions, cost of accidents, vehicle operating costs, and others.

1.6 *Acceptability criteria are defined.* The conditions under which a scenario should be adopted should be defined and stated.

2. *A CBA analysis is conducted.* CBA methodology and test scenarios from 1.2 are used to calculate NPV and benefit–cost ratio (BCR).

3. *Outputs are evaluated.*

3.1 *NPV and benefit–cost ratio are evaluated.* Consumer surplus includes time savings for existing traffic, time savings for diverted traffic, vehicle operating costs for diverted traffic, fare savings for the diverted traffic, and passenger benefits for generated traffic. Producer surplus includes train operating costs, track access charges, and the change in fares charged to consumers. If the NPV and benefit–cost ratio are within boundary conditions, then the scenario can be deemed acceptable, otherwise the scenario is deemed unacceptable and not evaluated further. In evaluating external costs, only the most relevant externalities will be monetized, namely accidents and avoided carbon dioxide emissions (*16*).

3.2 *The impact of the tested policies on each stakeholder is evaluated using the RAILPAG methodology.* Scenarios that were deemed acceptable in point 3.1

Scenario	Baseline	Scenario 1: Prioritize rail freight	Scenario 2: Prioritize passenger traffic
Years analyzed	Year 0: 2015, Year 1–10: 2016–2026		
Definition	• Infrastructure charges use existing charging system of each country • Capacity reductions (if needed) are proportional to existing traffic profile	• Freight traffic is prioritized • Freight TOs are not charged infrastructure charges that they pay on the bridge to IMs to incentivize shift of freight traffic to rail • Savings are passed on to shippers—freight tariffs that shippers pay to TOs are reduced by 10% • All freight growth occurs in rail and not truck traffic	• Passenger traffic is prioritized • A minimum of 10% of year 0 (2015) freight traffic is guaranteed throughout the time period
Capacity assumptions	• Malmö–Lund section limits capacity on the bridge • Assumption: All Swedish trains must be maintained, so if there is a capacity shortage, only trains on the bridge must be reduced • 10% of the freight traffic that crosses the bridge terminates in Malmö **Subscenarios** A Malmö–Lund line is two tracks through the evaluation period (400 trains/day in both directions) B Malmö–Lund line is four tracks from 2021 onwards (800 trains/day in both directions)		
Traffic growth (annual)	Rail • Intercity: 3.1% • Oresund Regional: 3.1% • Freight: 1.6% minus spillover to trucks if there are capacity shortages Road • Cars: 1.9% • Trucks: 2.1% plus spillover from rail if there are capacity shortages Source: (19)	Rail • Intercity: 3.1% • Oresund Regional: 3.1% minus spillover to cars if there are capacity shortages • **Freight: 1.6% plus all of truck freight growth = 3.9%** Road • Cars: 1.9% plus spillover from fewer available passenger trains • **Trucks: 0%** • Assumed changes in **bold**.	Same as baseline

Figure 5. Scenarios to be evaluated.
Source: Authors.

can then use the RAILPAG methodology to determine the impact of the proposed policy on each system player.

4. *Test using acceptance criteria*. If the results do not meet acceptance criteria, return to step 1, define and test other scenarios. More scenarios can be tested regardless of acceptability criteria; however, this step may allow refining of boundary conditions and scenario information.

Case Study: Capacity Allocation in the ScanMed Corridor—Oresund Bridge

Using the proposed methodology from the previous section, here we test a capacity allocation mechanism for a case study.

The ScanMed Corridor

The ScanMed corridor transits Europe from north to south, beginning in Finland and Norway, passing through Sweden, Denmark, Germany, Austria, before terminating in Italy, as well as having a sea connection to Malta. The corridor is one of Trans-European Network for Transportation's (TEN-T) corridors that aims to make freight movement in Europe more sustainable, in part by switching freight flows to more sustainable modes such as rail (*19*).

This corridor presents a key north–south link in the European transportation network, connecting the key network hubs, ports, goods, consumers, and producers. The ScanMed corridor also links to other corridors on the continent, providing even more important connections.

Table 1. Evaluation input data.

Item	Value	Unit
General information		
Currency (ECB 2015)	€1 = 7.46	DKK
Currency	€1 = 9.18	SEK
Bridge length (*28*)	20.0	km
Assumptions		
Trip purpose		
Business	15%	
Commute	70%	
Leisure	15%	
Financial items		
Discount rate	4%	
GDP growth rate	2%	
Commercial speed		
Pax trains (29)	120	km/h
Freight trains (29)	100	km/h
Car speed—average	100	km/h
Truck speed—average	80	km/h
Time penalty (waiting/last mile/...)		
Pax		
Train	20	min/pax
Car	6	min/pax
Freight		
Train	240	min/tonne
Truck	10	min/tonne
Traffic: Rail		
Demand (*19*)		
Passengers	12,273,842	pax/year
Demand NB (*19*)	2,694,000	ton/year
Demand SB (*19*)	2,992,000	ton/year
Annual growth (*19*)		
Pax trains	3.10%	
Freight trains	1.60%	
Freight		
Freight trains (*25*)	8,485	trains/year
Freight transported	7,022,420	ton/year
Average train load	828	ton
Max train load	950	ton
Passenger trains	57,757	trains/year
% intercity trains	15%	
% regional trains	85%	
Intercity		
Passengers (*25*)	1,980,590	pax/year
Trains (*25*)	8,664	trains/year
Trains	24	trains/day
Train capacity (*30*)	381	seats
Load factor	60%	
Average mass (*30*)	365	ton
Regional/Oresund		
Passengers (*25*)	10,293,252	pax/year
Trains (*25*)	49,093	trains/year
Trains	135	trains/day
Train capacity (*25*)	237	seats
% triple units	40%	
Train capacity (average)	427	seats
Load factor	49%	
Average mass (*31*)	275.4	ton
Traffic: Road		
Demand (tonnes, 2015)		
Freight—Northbound (*19*)	4,892,000	ton
Freight—Southbound (*19*)	4,448,000	ton
Passenger-car equivalents	6,902,446	cars/year
Annual growth (*19*)		
Cars	1.90%	
Truck	2.10%	
Road (2015) (*25, 32*)		
Car occupancy	1.3	pax
Trucks	448,792	vehicles
Truck capacity	26	ton
Truck load factor	80%	
Costs		
Operating costs (updated to 2015)		
Train: long distance (*33*)	€25.01	/train-km
Train: commuter (*33*)	€7.00	/train-km
Train: freight	€14.53	/train-km
Variable cost (e.g., sales) (*23*)	€0.20	per pax
Cars (*34*)	€0.44	/vkm
Trucks (*34*)	€6.92	/vkm
Maintenance costs (double track line) (*35*)		
Rail—fixed	€50,000	/km/year
Rail— variable	€1.00	/train-km
Bridge (excl. rail) (*28*)	€2,013,405	/year
Fares and tolls		
Rail tariff (SJ)		
Pax (regional)	€4.95	per trip
Pax (long distance)	€15.69	per trip
Freight	€0.05	per ton-km
Road toll (*25*)		
Average car toll (bridge)	€13.83	/vehicle
Average truck toll (bridge)	€112.50	/vehicle
Safety: Road (2015 values) (*36*)		
Fatality costs		
Denmark (€/fatality)	2,059,113	
Sweden (€/fatality)	1,887,455	
Risk (fatalities/vkm)	6.20×10^{-9}	
Underreporting factor	1.02	
Severe injury		
Denmark (€/severe injury)	254,803	
Sweden (€/severe injury)	277,010	
Risk (severe injury/vkm)	1.81×10^{-8}	
Underreporting factor	1.25	
Light injury		
Denmark (€/light injury)	19,951	
Sweden (€/light injury)	19,881	

(continued)

Table 1. (Continued)

Item	Value Unit
Risk (light injury/vkm)	1.30×10^{-7}
Underreporting factor	2.00
Safety: Rail (36)	
Passengers	€0.0006 /pax-km
Freight	€0.0002 /train-km
Emissions: CO_2 (36)	
Average car emissions	189 G-CO_2/km
Average truck emissions	715 G-CO_2/km
Cost per t-CO_2e (16)	2010: €25; 2020: €40; 2030: €55; 2040: €70; 2050: €85; 2060: €100

Value of time (2015 values) (37)			
€/pax/H or €/tonne/H	DKK	SEK	Elasticity
Business—rail, road	38.84	36.29	0.5
Commuter—rail, road	10.44	9.82	0.3
Other pax—rail, road	8.76	8.24	0.3
Freight VOT—rail	1.83	1.74	0.1
Freight VOT—road	4.47	4.23	0.1

Note: VOT = value of time; pax = passenger(s); vkm = vehicle-kilometer; ton = 1 metric tonne; t-CO2e = tonne of CO2 equivalent; DKK = Danish Krona; SEK = Swedish Krona; NB = northbound; SB = southbound; SJ = a Swedish passenger rail operator.

Demand Function: Traffic over the Oresund

Bridge traffic figures are shown in Figure 3 and will be used as the demand function for simplification purposes. In 2015, approximately 14 million tonnes of freight passed through Denmark in both directions, with traffic split evenly in the two directions. Northbound freight traffic mode share was 35.5% rail and 64.5% road, whereas southbound traffic was 40.2% rail and 59.8% road.

Road traffic over the Oresund Bridge was made up of approximately 7 million vehicles during the year of 2015, with over 90% of the vehicles being private cars, and approximately 5% being heavy duty trucks, with the remainder split between buses, vans, and motorcycles.

In 2015, 66,200 trains crossed the Oresund Bridge. Train traffic has remained relatively stable between 2001 and 2015, peaking in 2011 at just over 71,000 movements. Freight makes up 12.8% of total train traffic, with the rest of the movements being passenger (both regional and long distance). Additionally, a small number of service trains use the bridge for infrastructure maintenance activities.

Although they may present a viable alternative, ferry services between Germany and Sweden and between Denmark and Sweden were not considered in this analysis.

Regional trains operate under the brand Oresundtåg, and travel from Copenhagen and its suburbs via Copenhagen Airport, to Malmö in Sweden, continuing on to the Swedish cities of Gothenburg, Kalmar, and Karlskrona. The capacity across the bridge is severely limited by the track set-up at Copenhagen Airport Station, which does not permit passing stopped trains. There are plans to resolve this bottleneck by adding passing tracks, with no set completion date for the project. On the Swedish side, capacity is limited north of Malmö on the Malmö–Lund section, where there are only two tracks which are used by nearly 400 passenger trains and more than 60 freight trains (daily in both directions). There are plans to resolve this capacity bottleneck by doubling the number of tracks (24). West of Copenhagen, the section between Kobenhavn H and Høje Taatstrup has just over 350 passenger trains per day in both directions, operating on four tracks. Several freight trains also use that line.

The number of passengers using trains across the bridge has steadily increased from the year 2000 onwards, with nearly 12 million annual passengers using the service in 2015, and with trains having an estimated 50% load factor over the entire service day. Because this is a hybrid service for urban, suburban, and interurban passengers, several compromises in seat layout and capacity have been made in the rolling stock, which results in having standing passengers during peak periods. The percentage of intercity to regional service has remained stable since 2001, and the amount of intercity service has remained relatively small compared with the total number of seats and the total number of trains.

Before the Oresund Bridge opened, ferries operated across the strait, and ferry services are still available from Germany to Sweden, bypassing Denmark for both passengers and freight. Sufficient data are not available to compare ferry services with train services across the bridge.

Bridge Costs and Revenues, Tolls, Infrastructure Charges, and Train Fares

The Oresund Bridge is tolled for vehicles, and trains pay infrastructure charges to access the rail infrastructure. Users have several options to pay tolls, with discounts for prepayment or transponder use. Different classes of vehicles are subject to different tolls. Commuter and business account plans are occasionally available, and tolls vary based on vehicle type. In 2016, a cash toll for a car that is less than 6 meters in length was €54 one way (USD$59.40; assuming 1€ = USD$1.10 in mid-2016), with discounts given for more frequent users. Trucking companies can set up an account and receive discounts of more than 50% from a single cash toll of €145.60 for a 9–20 m truck.

Trains have to pay infrastructure access charges for using railway capacity. Because the bridge connects two countries, each country's IM is responsible for levying charges for each half of the bridge. Each country has its own infrastructure charging system. A basic per train-km or per tonne-km formula applies in each country, and the bridge passage is charged as a flat access fee in addition to, or instead of, the other charges. Passenger trains do not pay a flat access fee for crossing the Swedish part of the bridge, whereas freight

Scenario 1A versus Baseline-A													
(1A: freight traffic priority, no charges for freight trains, all growth in freight traffic goes to rail; Both: no expansion of capacity between Malmö and Lund)													
Scenario 1A compared with **Baseline-A**		**IMs**	**TOs**		**Existing users Rail**			**Diverted users Road**		**Generated users**		**Non-users**	**Effect Economic value**
			Pax	Freight	IC	Reg.	Freight	Pax	Freight	Pax	Freight		
Assets	Infra. investment												
	Infra. maintenance	0.0											**0.0**
Operating costs	Track charges	−57.6	3.3	54.3									**0.0**
	Bridge tolls	−32.2						−18.9	51.1				**0.0**
	Road operating costs							−12.1	62.9				**50.9**
	Fares		−7.4	3.2	0.0			7.4	−3.2	0.0			**0.0**
	Train operating costs		1.3	−3.6									**−2.3**
Operating benefits	Time savings				−3.4	21.7	−3.6	−7.9	17.4				**24.2**
	Generated users benefit									0.0			**0.0**
Externalities	Safety											−0.4	**−0.4**
	Climate change (CO_2)											0.1	**0.1**
Economic profitability for stakeholders		**−89.9**	**−2.8**	**54.0**	**−3.4**	**21.7**	**−3.6**	**−31.6**	**128.3**	**0.0**		**−0.4**	**72.4**

Scenario 1B versus Baseline-B													
(1B: freight traffic priority, no charges for freight trains, all growth in freight traffic goes to rail; Both: expansion of capacity between Malmö and Lund)													
Scenario 1B compared with **Baseline-B**		**IMs**	**TOs**		**Existing users Rail**			**Diverted users Road**		**Generated users**		**Non-users**	**Effect Economic value**
			Pax	Freight	IC	Reg.	Freight	Pax	Freight	Pax	Freight		
Assets	Infra. investment												
	Infra. maintenance	−0.2											**−0.2**
Operating costs	Track charges	−55.5		55.5									
	Bridge tolls	−47.3							47.3				
	Road operating costs								58.2				**58.2**
	Fares			2.4					−2.4				
	Train operating costs			−3.1									**−3.1**
Operating benefits	Time savings						−3.1		16.1				**13.0**
	Generated users benefit												
Externalities	Safety											0.2	**0.2**
	Climate change (CO_2)											0.3	**0.3**
Economic profitability for stakeholders		**−103.0**		**54.9**			**−3.1**		**119.2**			**0.5**	**68.4**

Figure 6. RAILPAG assessment of proposed scenarios.
Note: Reg. = regional; IC = intercity; Infra. = Infrastructure.

Scenario 1A-1 versus Baseline-A

(1A-1: freight traffic priority, no charges for freight trains, 50% growth in freight traffic goes to rail; Both: no expansion of capacity between Malmö and Lund)

Scenario 1C compared with **Baseline-A**		**IMs**	**TOs**		**Existing users Rail**			**Diverted users Road**		**Generated users**		**Non-users**	**Effect Economic value**
			Pax	Freight	IC	Reg.	Freight	Pax	Freight	Pax	Freight		
Assets	Infra. investment												
	Infra. maintenance	0.0											**0.0**
Operating costs	Track charges	2.9	2.0	−4.9									**0.0**
	Bridge tolls	−16.9						−11.3	28.2				**0.0**
	Road operating costs							−7.2	34.7				**27.5**
	Fares		−2.5	3.3				2.5	−3.3	0.0			**0.0**
	Train operating costs		0.5	−2.1									**−1.6**
Operating benefits	Time savings				−3.4	14.5	−2.1	−4.7	9.6				**13.8**
	Generated users benefit									0.0			**0.0**
Externalities	Safety											-0.3	**-0.3**
	Climate change (CO_2)											0.0	**0.0**
Economic profitability for stakeholders		**-14.0**	**0.0**	**-3.7**	**-3.4**	**14.5**	**-2.1**	**-20.7**	**69.2**	**0.0**		**-0.2**	**39.5**

Scenario 2A to Baseline-A

(2A: prioritizing regional passenger traffic over other traffic; Both: no expansion of capacity between Malmö and Lund)

Scenario 2A compared with **Baseline-A**		**IMs**	**TOs**		**Existing users Rail**			**Diverted users Road**		**Generated users**		**Non-users**	**Effect Economic value**
			Pax	Freight	IC	Reg.	Freight	Pax	Freight	Pax	Freight		
Assets	Infra. investment												
	Infra. maintenance												
Operating costs	Track charges	−5.0	−4.0	9.0									**0.0**
	Bridge tolls	5.7						23.8	−29.5				**0.0**
	Road operating costs							15.2	−36.3				**−21.1**
	Fares		19.8	−6.0	0.0			−19.8	6.0	0.0			**0.0**
	Train operating costs		−3.2	3.8									**0.7**
Operating benefits	Time savings				−3.4	−19.2	3.8	10.1	−10.1				**−18.7**
	Generated users benefit									0.0			**0.0**
Externalities	Safety											0.7	**0.7**
	Climate change (CO_2)											0.1	**0.1**
Economic profitability for stakeholders		**0.7**	**12.6**	**6.8**	**−3.4**	**−19.2**	**3.8**	**29.3**	**−69.9**	**0.0**		**0.8**	**−38.4**

Figure 6. continued

trains do. All trains also pay a fee per train-km and a fee per tonne-km, depending on the type of line they use, as well as for passage through areas of high railway traffic density (e.g., cities).

In Denmark the system is simpler. Trains pay either a flat fee for the bridge, or per tonne-km or per train-km not on the bridge. The Danish system also has a surcharge for using Copenhagen's main station and its approaches, in addition to a surcharge for crossing the Great Belt Link.

According to the websites of the regional public transportation authority and operator Skanetrafiken and train operator Oresundtåg, fares for using passenger trains also vary based on whether or not they are single or season tickets. The price of a single ticket for regional services varies between approximately €15 for a ticket between Copenhagen and Malmö, and €60 for a single ticket between Copenhagen and Gothenburg. A 30-day pass for the Copenhagen–Malmö service would cost approximately €270.

The Oresund Bridge outsources maintenance at a cost of €2.01million per year (*28*).

Modeling the Society's Point of View: Scenarios and Inputs

As railway capacity in this part of the corridor is constrained by the Malmö–Lund section (which in turn affects capacity on the Oresund Bridge), a decision needs to be made about which type of traffic to prioritize—passenger or freight—and how best to charge for this traffic. A scenario prioritizing freight traffic (Scenario 1) will be compared with a baseline scenario, prioritizing passenger traffic (Scenario 2). A 10-year time horizon will be used. This time horizon is important for TOs, ensures stability of infrastructure charging policies over time, and is flexible enough to allow IMs to change their infrastructure charging and capacity allocation policies.

Scenarios 1 and 2 will be tested and compared to the baseline scenario. Each of these will be tested under two conditions:

- Subscenario A: Capacity is constrained and only two tracks are present between Malmö and Lund.
- Subscenario B: Capacity on the Malmö–Lund section is expanded to four tracks by the end of the year 2020, as is foreseen in the capacity expansion plan.

This results in a total of four scenarios: 1A, 1B, 2A, and 2B. Additionally, a sensitivity analysis will be conducted for Scenario 1A; in Scenario 1A-1, only 50% of freight growth will go to rail.

Modeling the Society's Point of View: Overall Results

In all cases it is assumed that the decision of whether or not to invest in capacity on the Malmö–Lund section is an external one. Thus, scenarios considering both possibilities will be tested. Scenarios ending in "A" consider that no capacity investment will take place. Scenarios ending in "B" consider that the Malmö–Lund section will be expanded from two to four tracks, as proposed by a capacity expansion project. This expansion would take place at the start of year 6 of the evaluation. In the baseline, if a reduction in traffic because of capacity constraints is needed, traffic is reduced proportionally. In Scenario 1, all traffic reductions come from regional trains, whereas in Scenario 2, passenger trains are prioritized, and all reductions come from freight trains. Table 1 displays the results of this analysis.

Scenario 1A results in an overall NPV of €72.4 million, (benefit–cost [B/C] ratio of 1.81). This means that if implemented, the policy under scenario 1A will result in a net positive impact for all players, compared with the baseline. In Scenario 1A-1 (50% of freight growth goes to rail), the NPV is €39.5 million.

Scenario 1B results in an NPV of €68.4 million (B/C 1.66), also providing a positive impact, compared with baseline assumptions.

In Scenario 2A, the NPV is −€38.4 million, meaning that prioritizing passenger services results in a reduction of benefits to society. In Scenario 2B, the NPV is 0, because expansion in capacity allows accommodating all passenger traffic demand.

These results suggest higher total benefits for prioritizing freight traffic over passenger traffic and incentivizing it to use rail instead of trucks, regardless of capacity constraints.

Modeling the Society's Point of View: Detailed Analysis of Stakeholder Impact

This section uses the RAILPAG methodology to identify which stakeholders have the most to lose or gain from each proposed policy. An evaluation matrix is used with the names of the stakeholders on the X axis, and the effects on the Y axis, with each cell representing the impact of each effect on each stakeholder. Figure 6 provides a summary of how the benefits are redistributed between all players. It should be noted that for Scenario 2B, no redistribution takes place as capacity is not constrained.

When comparing Scenario 1A to Baseline-A, the Bridge Consortium (the entity operating the Oresund Bridge) will lose the most out of all the stakeholders because of lower revenue from bridge tolls and infrastructure charges. In this scenario, freight TOs do not pay infrastructure charges, and pass on these savings to end users (shippers) through a 10% freight tariff reduction. This leads to a reduction of the number of trucks on the road. The total toll revenue for the bridge will decline. Due to the higher number of trains, the marginal cost of rail maintenance on the bridge will increase.

Passenger train operating companies stand to lose some money because of the difference between lost fares and operating costs. The amount of time that passengers save by switching from trains to cars outweighs the accident costs. At the same time, the climate change costs of this scenario are positive, but relatively low.

It should be noted that this model is limited in its scope. It does not fully take into account all the network effects that will be caused by the tested policy, and only considers the impact on the area immediately surrounding the Oresund Bridge. The model also does not consider road capacity on bridge approaches.

For Scenario 1B, the Malmö–Lund section is expanded to four tracks. All displaced traffic is passenger. The NPV of this scenario is €68.4 million, and this scenario may also have advantages over the baseline. In this case, the bridge operator loses out on even more funds than in Scenario 1A because of a higher loss in bridge tolls, as more freight trains operate on the bridge and fewer trucks operate than in Scenario 1A. Because of the additional capacity, passenger trains do not need to be displaced to road, and passenger TOs are not affected. From a safety point of view, more passengers using the train results in a reduction in accidents and CO_2 emissions.

For Scenario 2A, the NPV is −€38.4 million. The biggest loss occurs for freight operators. Freight traffic is displaced from rail to road, meaning that freight now has to pay more in operating costs and bridge tolls. Passengers, however, save time by switching to rail from road. Overall, external cost savings are positive. Nonetheless, the overall value of time evaluation is negative.

For Scenario 2B compared with Baseline-B, there is no effect on the system, and the scenario NPV is equal to zero because the amount of available capacity is sufficient to accommodate all users.

In order to better see the impact of Scenario 1A, a sensitivity analysis of that scenario assumes that only 50% of all traffic growth will switch to rail, with the remainder using road. The same patterns that are seen in Scenario 1A can be observed in Scenario 1A-1. Both costs and benefits are proportionally reduced.

Conclusions and Future Work

In capacity allocation and infrastructure pricing, the points of view of the TO and IM have been studied in great detail, whereas impacts on society have not. This paper proposes a way to evaluate infrastructure charging and capacity allocation policies on society. The approach first uses a CBA to show the overall impact of a project, and then breaks down the impact by costs and benefits for each player. A case study of a freight corridor that passes over the Oresund Bridge is then presented. The case study tested some capacity allocation and infrastructure pricing scenarios and the impact of these scenarios on each player, including society.

Although CBA is often used to evaluate infrastructure projects, it is not often used to evaluate policy scenarios. Furthermore, to our knowledge, it has not been used to evaluate the impact of a pricing and capacity allocation policy on society. In the opinion of the authors, the proposed approach could be incorporated within a broader process for setting a capacity allocation and pricing policy, especially in cases of important network investments (approved because of positive societal impact prospects), guaranteeing policy consistency with regard to ex-ante assumptions.

The presented case study is subject to several limitations. This case study was limited to a single corridor with a single set of scenarios. It mostly tested a capacity allocation methodology (different levels of prioritization) and covered only a limited change in pricing policy. The next logical step would be to test different infrastructure charging systems to see which system provides more benefits to society. The demand function for the case study was provided by an outside corridor study. In a broader application, several deterministic and probabilistic demand functions could be tested with this approach.

The case study results show, as an example, that prioritizing freight traffic over passenger traffic over the Oresund Bridge under the tested set of scenarios and assumptions may have an overall positive impact for society. In such case, the IMs stand to lose a large amount of revenue from this policy, which would need to be compensated through government funding. However, such a scenario may often not be feasible, given usual public funding constraints that will strongly influence capacity allocation and infrastructure pricing policy, causing traffic to shift away from rail and reducing the societal benefit of the overall investment.

Of course, these results should not be interpreted beyond the specific scope of the paper, which is to illustrate the interest of using the proposed evaluation methodology within a capacity allocation and pricing policy evaluation process.

Acknowledgments

The authors would like to thank Fundação para a Ciência e Tecnologia (FCT) for funding this study (ref. SFRH/BD/51634/2011).

Author Contributions

The authors confirm contribution to the paper as follows: study conception and design: Aleksandr Prodan and Paulo F. Teixeira; data collection: Aleksandr Prodan; analysis and interpretation of results: Aleksandr Prodan and Paulo F. Teixeira; draft manuscript preparation: Aleksandr Prodan. All authors reviewed the results and approved the final version of the manuscript.

References

1. Gibson, S. Allocation of Capacity in the Rail Industry. *Utilities Policy*, Vol. 11, No. 1, 2003, pp. 39–42.
2. Nilsson, J.-E. Allocation of Track Capacity. *International Journal of Industrial Organization*, Vol. 17, No. 8, 1999, pp. 1139–1162.
3. Brannlund, U., P. O. Lindberg, A. Nou, and J.-E. Nilsson. Railway Timetabling Using Lagrangian Relaxation. *Transportation Science*, Vol. 32, No. 4, 1998, pp. 358–369.
4. Nilsson, J.-E. Towards a Welfare Enhancing Process to Manage Railway Infrastructure Access. *Transportation Research Part A: Policy and Practice*, Vol. 36, No. 5, 2002, pp. 419–436.
5. Peña-Alcaraz, M., J. M. Sussman, and D. Mort. *Analysis of Capacity Pricing and Allocation Mechanisms in Shared Railway Systems: Lessons for the Northeast Corridor*, MIT ESD Working Paper Series ESD-WP-2015-04, March 2015, http://hdl.handle.net/1721.1/97644.
6. Peña-Alcaraz, M., M. Webster, and A. Ramos. Train Timetable Design for Shared Railway Systems Using a Linear Programming Approach to Approximate Dynamic Programming. *MIT ESD Working Paper Series*, 2014, pp. 1–19.

7. Nash, A. *Best Practices in Shared-Use High-Speed Rail Systems*. Mineta Transportation Institute, San Jose State University, San Jose, CA, 2003.
8. Wong, S. K., and T. K. Ho. Intelligent Negotiation Behaviour Model for an Open Railway Access Market. *Expert Systems with Applications*, Vol. 37, No. 12, 2010, pp. 8109–8118.
9. Levy, S., M. Peña-Alcaraz, A. Prodan, and J. M. Sussman. Analyzing Financial Relationship Between Railway Industry Players in Shared Railway Systems. *Transportation Research Record: Journal of the Transportation Research Board*, 2015. 2475: 27–36.
10. Teixeira, P. F., and A. Prodan. Railway Infrastructure Pricing in Europe for High-Speed and Intercity Services: State of the Practice and Recent Evolution. *Transportation Research Record: Journal of the Transportation Research Board*, 2014. 2448: 1–10.
11. Prodan, A., and P. Teixeira. Infrastructure Charging Difference in International Passenger High-Speed and Freight Corridors: Towards an International Railway Market in Europe? Presented at 94th Annual Meeting of the Transportation Research Board, Washington, D.C., 2015.
12. Calvo, F., and J. de Oña. Opening of the European Railroad Network: A Lost Opportunity for European Unification. *Transportation Research Record: Journal of the Transportation Research Board, 2005. 1916:* 8–19.
13. Macário, R., P. F. Teixeira, and W. Rothengatter. Deploying Business Logic for Railway Infrastructure Charging. *Transportation Research Record: Journal of the Transportation Research Board*, 2001. 2159: 44–51.
14. Teixeira, P. F., A. Prodan, and A. López-Pita. *Infracharges: UIC Study on Railway Infrastructure Charges in Europe*. International Union of Railways (UIC), Paris, France, 2012.
15. Delucchi, M. External Costs of Transport in the U.S. In *Handbook of Transport Economics*, Edward Edgar Publishing, Northampton, MA, 2010, pp. 341–368.
16. Maibach, M., C. Schreyer, D. Sutter, H. P. van Essen, B. H. Boon, R. Smokers, A. Schroten, C. Doll, B. Pawlowska, and M. Bak. Handbook on Estimation of External Costs in the Transport Sector. In *Internalisation Measures and Policies for All external Cost of Transport (IMPACT)*, CE Delft, Delft, The Netherlands, 2008, p. 336.
17. Infras. *External Cost of Transport: Update Study*. Infras, Zurich, Switzerland, 2004.
18. European Commission. Directive 2012/34/EU of the European Parliament and of the Council of 21 November 2012 Establishing a Single European Railway Area (Recast) (Text with EEA Relevance). *Official Journal of the European Union*, Vol. 32, No. 343, 2012, pp. 32–77.
19. European Commission. *Scandinavian-Mediterranean Core Network Corridor Study*. Final report, 2014.
20. Nash, C., and B. Matthews. Implementing Rail Infrastructure Charging Reform: Barriers and Possible Means of Overcoming Them. *Imprint Europe Seminar*, Brussels, Belgium, 2002, pp. 1–15.
21. Prest, A. R., and R. Turvey. Cost-Benefit Analysis: A Survey. *The Economic Journal*, Vol. 75, No. 300, 1965, pp. 683–735.
22. European Commission. *Guide to Cost-Benefit Analysis of Investment Projects; Economic Appraisal Tool for Cohesion Policy 2014–2020*, 2014.
23. European Investment Bank and European Commission. *Railway Project Appraisal Guidelines*, 2005.
24. Fridlund, J., J. Wolfmaier, and H. Sylvan. *Traffic Analysis Malmö-Hamburg*, 2012.
25. Øresundsbro Konsortiet. *Traffic Statistics*, 2015. [Online]. https://www.oresundsbron.com/en/traffic-stats. Accessed July 15, 2016.
26. Trafikverket. *Statement of 2015–2018*. Trafikverket, Stockholm, Sweden, 2015.
27. Banedanmark. *Network Statement 2015*. Banedanmark, Copenhagen, Denmark, 2015.
28. Øresundsbro Konsortiet. *Oresundbron*. Annual report, 2015.
29. Engström, Lars, Rolf Sundqvist, and N. Blom Salmonsen. *Register of Infrastructure Kastrup - Lernacken*, 2013.
30. Allenbach, J.-M. *X2 Spec Sheet*, 1995. [Online]. http://documents.epfl.ch/users/a/al/allenbac/www/documents/Fich0316.pdf. Accessed August 20, 2016.
31. Oresundståg. *Fakta & siffror om Öresundstågen*. [Online]. http://www.oresundstag.se/sv/Om-oss/Oresundstagen/Fakta-Siffror/. Accessed January 21, 2015.
32. Transportation Research Board. *HCM 2010: Chapter 10*. Transportation Research Board, Washington, D.C., 2010.
33. Bundesnetzagentur - German Railway Regulator. *Railway Market Analysis 2015*. Bonn, Germany, 2015.
34. Infras. *COMPETE. Analysis of Operating Cost in the EU and the US. Annex 1*, 2006.
35. López-Pita, A., P. F. Teixeira, C. Casas, A. Bachiller, and P. A. Ferreira. Maintenance Costs of High-Speed Lines in Europe: State of the Art. *Transportation Research Record: Journal of the Transportation Research Board*, 2008. 2043: 13–19.
36. Korzhenevych, A., N. Dehnen, J. Bröcker, M. Holtkamp, H. Meier, G. Gibson, A. Varna, and V. Cox. *Update of the Handbook on External Costs of Transport*. Final Report no. 1, 2014, p. 139.
37. Bickel, P., R. Friedrich, A. Burgess, P. Fagiani, A. Hunt, G. De Jong, J. Laird, C. Lieb, G. Lindberg, P. Mackie, S. Navrud, T. Odgaard, A. Ricci, J. Shires, and L. Tavasszy. *HEATCO: Developing Harmonised European Approaches for Transport Costing and Project Assessment. Deliverable 5. Proposal for Harmonised Guidelines*. Stuttgart, Germany, 2006.

The Standing Committee on Freight Rail Transportation (AR040) peer-reviewed this paper (18-03965).

Article

TRR
JOURNAL OF THE TRANSPORTATION RESEARCH BOARD

Transportation Research Record
2018, Vol. 2672(10) 202–214

DOI: 10.1177/0361198118790325
journals.sagepub.com/home/trr
SAGE

Simulation-Based Method of Capacity Utilization Evaluation to Account for Uncertainty in Recovery Time

Yung-Cheng (Rex) Lai[1], Kuan-Ting Chen[1], Tzu-Hao Yan[1], and Ming-Hua Li[1]

Abstract

A typical objective of a railway agency is to design an efficient and reliable service plan. Although enhancing the level of capacity utilization may increase efficiency, the stability of the service plan may also decrease due to stochastic railway disruptions. Past studies used recovery time to evaluate the stability of a service plan; however, the variation of recovery time was not fully examined. Therefore, we developed a Monte Carlo simulation framework to address this issue and proposed four capacity-based indices for the evaluation of service efficiency and stability. These indices are capacity utilization efficiency, mean recovery time, dispersion of recovery time, and probability of unacceptable recovery time. The latter two new indices explored the variation of recovery time and probability of unacceptable long recovery time, which is undesired for railway operation. We conducted a real-world case study on the Taiwan Railways Administration before and after the service plan revision. Consequently, the northbound service plan needed a deliberate review and revision before implementation, whereas the southbound plan successfully increased capacity usage while not necessarily worsening stability. This result demonstrated that our study introduced uncertainty analysis into the evaluation framework and provided flexible information on the service plan. These developments can provide improved support in the decision-making of railway agencies to strike a balance between service level and asset usage.

A typical objective of a railway agency is to design an efficient and reliable service plan. The evaluation of efficiency for a service plan is often determined by the usage of its capital-intensive infrastructure, which is known as capacity utilization. Although enhancing the level of capacity utilization may increase efficiency, the stability of the service plan may decrease due to stochastic railway disruptions, such as train, signal, and power system failures. Therefore, an evaluation process should be developed to simultaneously evaluate the efficiency and stability of the service plan.

Previous studies on evaluating service stability can be divided into analytical and simulation methods. Analytical methods establish mathematical equations to compute service stability (*1–13*), whereas simulation methods enable the modeling of railway systems that is considerably close to reality (*14–23*). Previous studies commonly used delay as an index to measure service stability whether the analytical or simulation method was employed. However, Lai and Chen (*24*) indicated that using delay as an index may either over or underestimate stability. In addition, delay is indirectly connected to this relationship due to a trade-off between stability and capacity utilization. Hence, recent studies by Lai and Chen, and Lai and Ip (*24*, *25*) adopted expected recovery time to evaluate service stability. However, these studies did not consider the randomness of recovery time caused by the stochastic railway disruptions. Therefore, this research develops a simulation framework to further explore the variation of recovery time. The simulation results are used to evaluate service stability through mean, standard deviation, and probability of unacceptable failures. These additional attributes can help the railway agencies better maintain a balance between service level and asset usage.

In the following sections, we first introduce the key indices for capacity utilization evaluation and then present the simulation process for stability indices followed by case study, discussion, and conclusions.

[1]Railway Technology Research Center, Department of Civil Engineering, National Taiwan University, Taipei, Taiwan

Corresponding Author:
Address correspondence to Yung-Cheng (Rex) Lai: yclai@ntu.edu.tw

Key Attributes to Evaluate Capacity Utilization

This section presents the four key indices to evaluate the efficiency and stability of capacity utilization: capacity utilization efficiency (CUE), mean recovery time (MRT), dispersion of recovery time (DRT), and probability of unacceptable recovery time (PURT).

CUE

A desirable usage of railway systems is to minimize the waste of existing assets due to costly capital investment. The efficiency of capacity utilization can be computed as the ratio between used capacity (obtained from the service plan) to normal capacity (computed based on infrastructure and traffic characteristics), which is referred to as CUE. The higher the value, the higher the efficiency and the lesser the waste in investment.

MRT

Service stability is commonly set as the measure of a service plan to return to its normal circumstance after a disturbance. A system with stable capacity utilization can return to normal state immediately after a disruption. Lai and Chen *(24)* proposed to use recovery time to evaluate service stability. Recovery time is the required time to restore the system to normal operations after a disruption with specific downgraded capacity and disrupted trains (Figure 1).

The probability of disruptions should be incorporated in the evaluation of stability. The simulation process can estimate the actual occurrence of a disruption using the probability of disruptions. Therefore, the MRT can be computed using the mean of all recovery time in the simulation process.

The MRT represents the mean stability of capacity utilization, although a considerably high recovery time may have a substantial impact to the system. The relationship is not linear as well. Operators may also be interested in the dispersion of stability and the recovery time above the acceptable threshold. Therefore, we propose two additional attributes, namely, DRT and PURT, to evaluate the stability of capacity utilization.

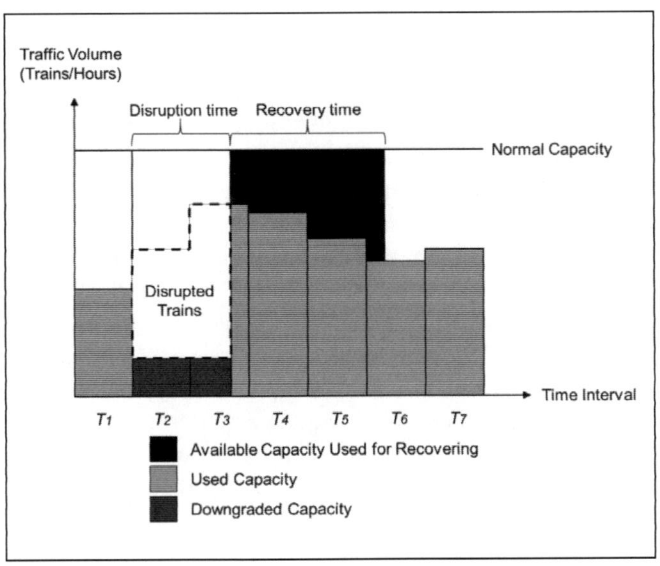

Figure 1. Illustration of the concept of recovery time from system disruption.

DRT

DRT describes the degree of dispersion for the recovery time so it can be regarded as the standard deviation of the recovery time. In the service design process, dispersion can be used to determine whether the recovery time of capacity utilization would be subject to a high degree of variation. DRT provides another insight to compare the service plan before and after revision. The revised service plan with similar MRT but high dispersion indicates that the railway service after revision is likely to experience extensive recovery time due to a high degree of variation. Caution should be exercised by the railway agency because they should not take the risk of adopting a service plan with limited stability for the possibility of facing substantially high recovery time.

PURT

Extreme condition is also of immense concern for service stability. Every operator has its own acceptable range of stability. Therefore, extreme condition is referred to as the extreme high recovery time that is beyond the acceptable standard, and the operator would substantially focus on the circumstances outside of the acceptable region. Extreme condition nearly implies a serious service failure event of capacity utilization because the resumption of operations to normal takes a long period. Accordingly, high PURT is considered undesirable because it has a high probability of becoming extremely unstable. The proposed simulation framework is capable of determining the probability of having an unaccepted recovery time. This probability reflects the probability that the operator will experience an extreme condition. Hence, the revised railway service plan ending up with high PURT could still be unsatisfactory for the operator even if the MRT remains unchanged or decreases. In this condition, we suggest to either cautiously run or deliberately modify the service plan.

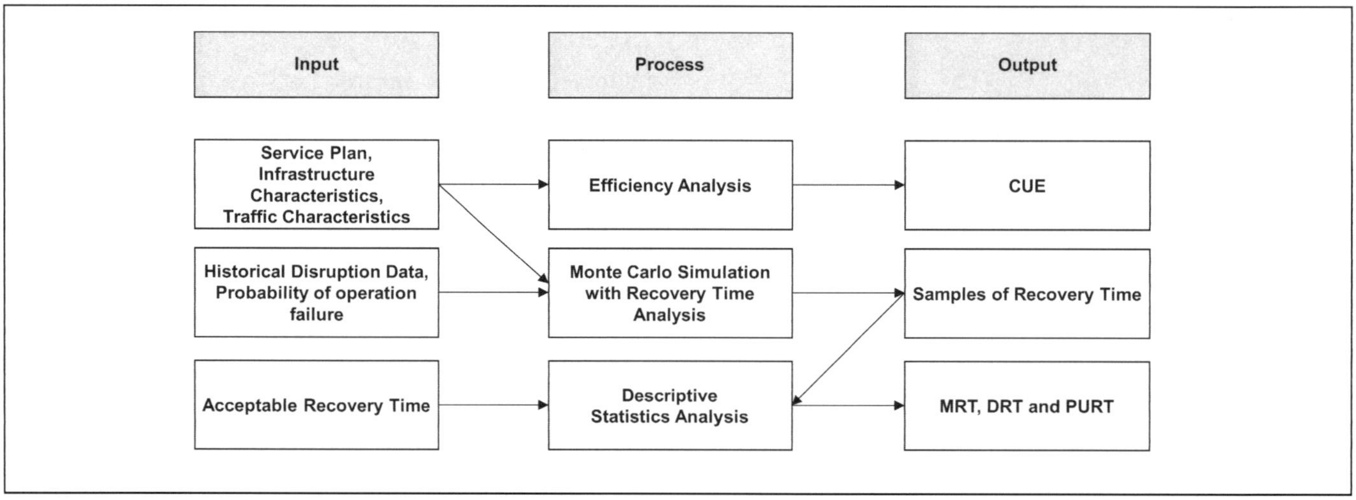

Figure 2. Overall process of capacity utilization evaluation.

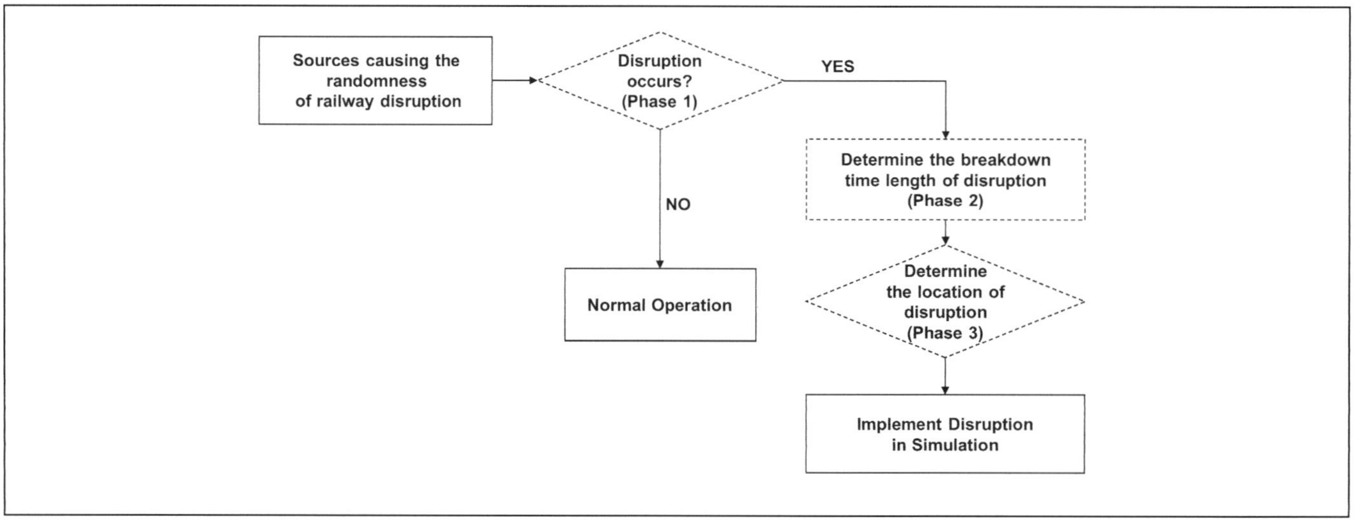

Figure 3. Source of railway disruption randomness.

Simulation-Based Methodology for Stability Indices

Figure 2 demonstrates the input, output, and required methods to compute the four key attributes of capacity utilization. The computation on efficiency was described in the previous sections. To compute the stability indices (i.e., MRT, DRT, and PURT), we developed a Monte Carlo simulation framework to simulate the recovery time based on capacity, system disruptions, and designed capacity utilization.

To account for the uncertainty, the simulation should first identify the sources of uncertainty, followed by the recovery time analysis. The uncertainty of the recovery time stems from the random disruptions induced by the usage of capacity and the probability of operation failure. Figure 3 illustrates the sources of randomness of railway disruptions surrounded by dashed borders, including whether the disruption occurs or not (phase 1), length of breakdown time (phase 2), and location of disruption (phase 3). The first phase can be determined by the probability of operation failure. The second phase is uncertain owing to different types of disruptions. The last phase considers the randomness of the occurrence of disruption location, which affects the downgraded capacity used for the calculation of recovery time.

Before engaging in the simulation process, a two-step cluster analysis was performed to categorize railway disruption types to ensure that each type is homogeneous in the length of breakdown time. In the first step, Ward's method was used to determine the number of clusters (*26*). In the second step, k-means clustering algorithm

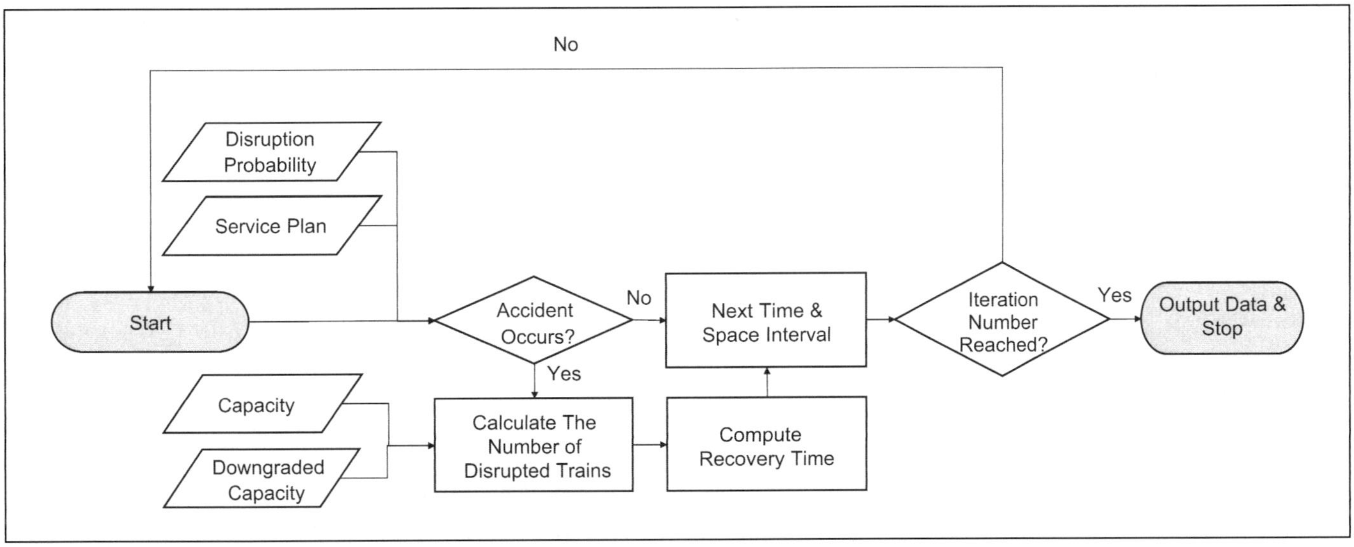

Figure 4. Simulation process for evaluating service stability.

was used to finalize the clustering process. A single probability distribution of breakdown time was not considered in the simulation process because breakdown length varies for different disruption types. Moreover, considering the influence of each clustered disruption type from the standpoint of operation would be substantially meaningful. Accordingly, stochastic variables were considered for uncertainty sources in the simulation of phases 1 and 3. The uncertainty source from phase 2 will be represented by the mean of the breakdown length according to types of disruptions in the computation process because the effect of the variance is reduced by the cluster analysis. In other words, the impact of the different approach to specific type of disruptions is embedded in the mean breakdown time.

The main idea to evaluate the stability of a given service plan is to compute its recovery time under potential disruptions. The entire simulation process is illustrated in Figure 4, in which the point is to simulate whether the disruption occurs (phase 1) and where such disruption occurs (phase 3).

The probability of disruptions is a key input to estimate the attributes related to recovery time. In this study, the phase 1 source (i.e., the occurrence of railway disruptions) is assumed to follow the Poisson process, which is a commonly used counting process (*27*). A counting process counts the total number of "events" that occur by time t. Therefore, the idea of using the Poisson process to simulate disruptions is that when a disruption is counted in a given time interval, a disruption can be seen to occur or not. The properties of the Poisson process are that its increments follow the Poisson distribution as:

$$P\{N(t+s)-N(s)=n\}=e^{-\lambda t}\frac{(\lambda t)^n}{n!}, n=0,1,\ldots \quad (1)$$

where $N(t)$ is the total number of events that occur by time t and λ is the rate of the process.

Another important property of Poisson process is that the inter-arrival time between two consecutive events is exponentially distributed with rate λ:

$$F(t<T)=1-e^{-\lambda T} \quad (2)$$

Thereafter, historical data could be used to estimate parameter λ of the exponential distribution of each type of disruption after cluster analysis.

In this study, we divided a timetable into 24 time intervals with each interval lasting for 1 h. The process of simulating the Poisson process can be completed using the inverse transform method with the following steps:

Step 1: Set time $t=0$ in a present time interval.
Step 2: Generate a uniformly distributed random number U.
Step 3: Set $t=t-\log(U)/\lambda$.
Step 4: If t is less than time unit T (one hour), then a disruption is counted; otherwise, a disruption is not counted.

For phase 3 uncertainty, this study divided the calculation of the downgraded capacity into three categories, namely, disruptions that occur at the section in between and those that occur at both end stations of the analyzed unit. Therefore, the downgraded capacity analysis for station disruptions only depends on the track layout change at the station. Moreover, we counted the historical data, that is, the ratio of the number of times of the section disruptions and station disruptions, to determine the place where the disruptions occur and the weight of the three components of the recovery time.

Thereafter, we simulate where the disruptions occur. The simulation in this research is performed by applying empirical data in Equation 3:

$$P = \begin{cases} f, \text{ section disruption} \\ \dfrac{1-f}{n}, \text{ station disruption} \end{cases} \quad (3)$$

Where f is the occurrence frequency of the section disruption and n is the possible change of the track layout of station disruptions.

Summing up the probability of all the possible outcomes changes Equation 3 into a cumulative probability form. Thereafter, we generate random numbers and see which range of the possible outcomes where the random number lies. At this moment in the simulation process, we know whether and what types of disruptions occur, along with the downgraded capacity determined from the phase 3 simulation. All this information is integrated with the concept of mean stability of capacity utilization, thereby enabling us to compute the recovery time for a single run.

After executing sufficient runs of the simulation, the next step is to analyze the result. Typical descriptive statistics are computed as evaluation indices, including mean and standard deviation of the recovery time. The PURT is calculated by determining a proper theoretical distribution, which will be explained in the case study.

Case Study

The proposed simulation framework was applied to the real-world case of Taiwan Railways Administration (TRA) before and after their annual service plan revision. A segment of the western corridor of TRA, with 22 stations from the Keelung to Hsinchu stations, is selected as the scope of this case study. This 106.4 km corridor is the busiest part of the TRA with both commuter and intercity train services, and the special program, TRA rapid transit systematization, increases the number of trains passing through this corridor after the service plan revision. As the traffic of the network further increases, the railway operator should examine the impact on stability of the new service plan.

The efficiency index of the service plan is calculated and shown in Figure 5. For the northbound and southbound directions, the overall CUE increased from 41.58% to 42.46% and from 43.20% to 44.18%. Figure 5 shows the detailed analytical results related to CUE relative to each time interval and railway section. Compared to the CUE before service revision, the CUE is generally higher, especially for rush hours. The section between Qidu and Xizhi is the most congested section in this corridor. The stability of the service plan is estimated by simulating the random events to the service plan. The current study followed Carey and Carville (*15*) and conducted simulation 1,000 times to aggregate the results. Theoretically, the more runs the simulation has, the closer the estimated mean is to the population mean. However, numerous runs will increase the computation time without significantly improving the result. Figure 6 illustrates the convergence of the MRT over 1,000 simulations. This figure also shows that the fluctuation of the mean is imperceptible after nearly 700 simulations.

We computed the mean and standard deviation of the 1,000 sample runs as the stability indices. Accordingly, the result should be sufficiently higher than the minimum iterations required for computational efficiency. For the northbound direction, the MRT increased from 69.11 min to 85.11 min and DRT is 198.73 min after revision and 172.95 min before revision. For the southbound direction, the MRT is slightly increased to 59.27 min from the original 57.58 min. The DRT is even decreased after the service plan revision. Detailed results about the MRT relative to each time interval and railway section are presented in Figure 7, which exhibits relatively high instability during rush hours and bottleneck sections (near Qidu). This type of demonstration can assist operators to identify the critical section to improve the preparation for possible system disturbances. The highest MRT occurs in the Qidu to Xizhi section. In the northbound direction, a substantial increase in MRT can be observed during peak hours. Before the revision, the morning rush hour near Qidu has already been subject to high MRT, and it becomes even higher after the service plan revision. MRT during the evening peak hours has become significantly high after the revision, due to the increases in the CUE within the continuous high-capacity utilization regions (Figure 5). The growth in MRT for the southbound direction is relatively minor compared with the case in the northbound direction because of considerably fewer continuous units with high-capacity utilization during peak hours at the bottleneck sections (see Figure 5).

The Monte Carlo simulation enables the further calculation of the probability of exceeding a selected standard of recovery time for different service plans. This probability can be considered another stability index that provides essential information to a railway agency, thereby guiding their decision in designing a service plan. However, the recovery time is a new index for evaluating service stability. For different railway agencies and railway systems, different standards should be presented to evaluate if the risk is acceptable or not. In general, the railway operator should avoid the occurrence of high recovery time because a serious disruption could perturb railway service. Therefore, for demonstration purpose and for comparison of service plan revision, a hypothetical standard of mean plus one standard deviation of

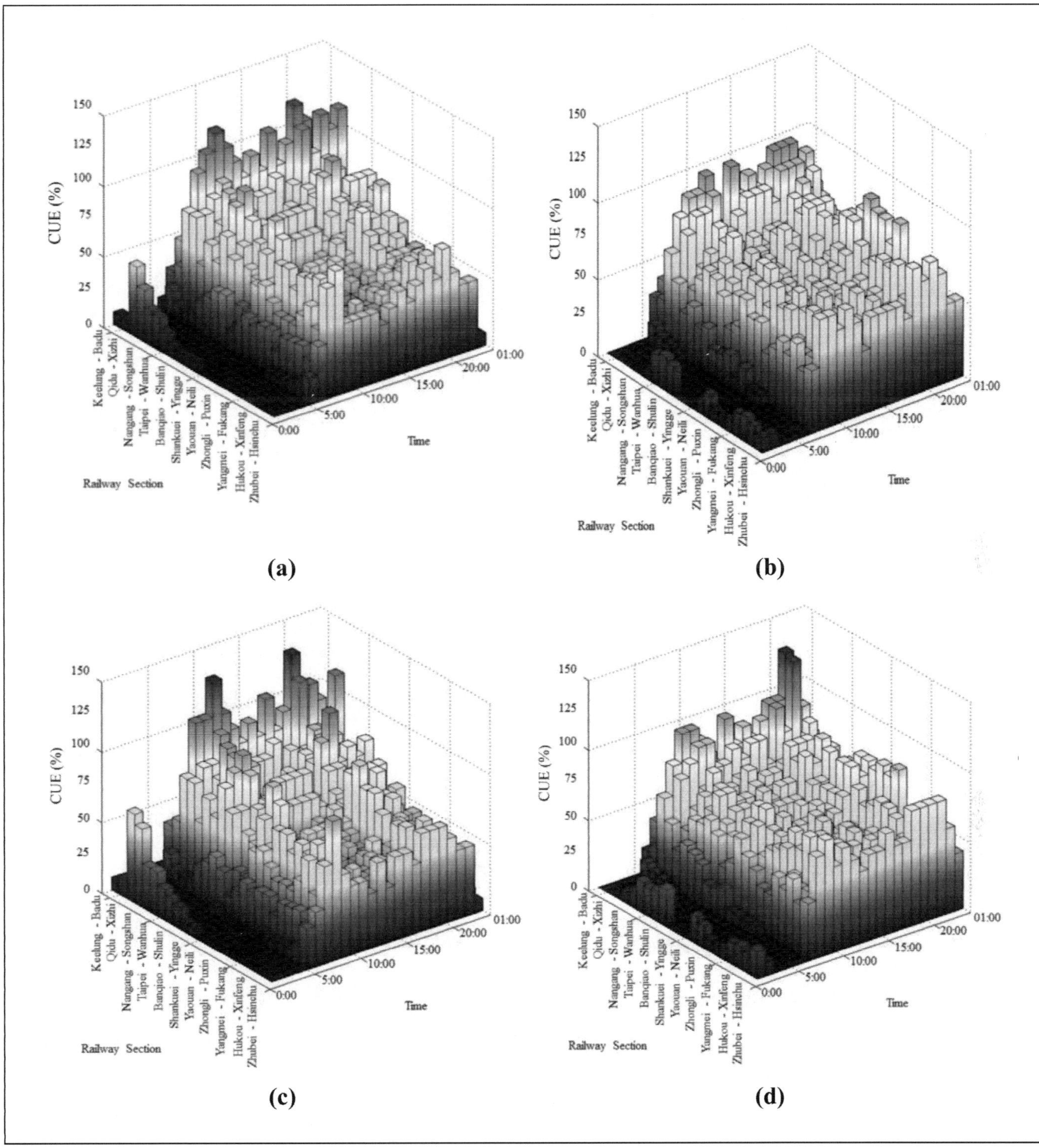

Figure 5. CUE before service plan revision, (*a*) northbound and (*b*) southbound; CUE after service plan revision, (*c*) northbound and (*d*) southbound.

recovery time of the northbound and southbound before the service plan revision (i.e., 241.96 min and 178.29 mins, respectively) were used to calculate the last index.

The simulation results enabled us to generate the relative frequency distribution for different service plans (Figure 8). A tail for recovery time distribution is seen, and extreme value theory was used to estimate the

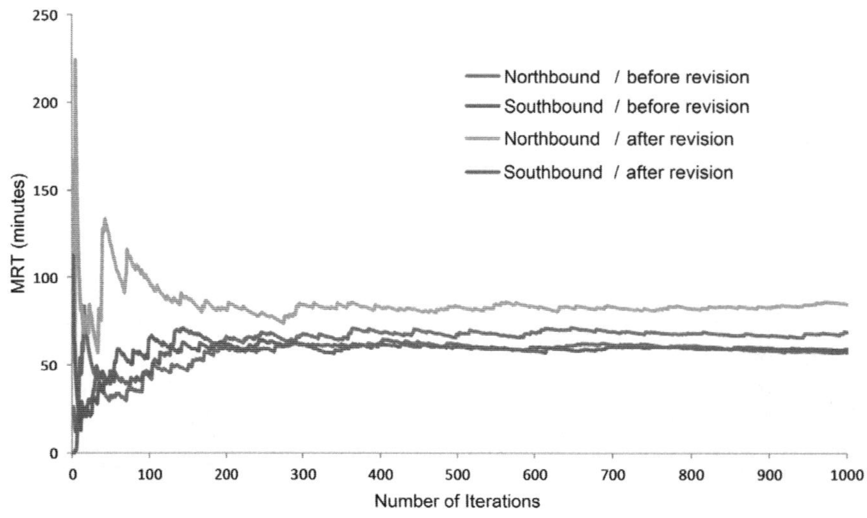

Figure 6. Number of iterations and convergence of MRT.

probability for the appearance of its long tail. We consider only the extremely high portion of the sample in the study as well.

Extreme value theory can be used to describe the tail distribution of the events without knowing population distribution. Extreme value theory can be divided into two models, namely, block maxima and peak over threshold (POT). The former mainly finds the largest value in each block that combines the usage of the generalized extreme value distribution to analyze the tail distribution. The latter uses the generalized Pareto distribution (GPD) to modify the value that is larger than specific threshold u. Coleman (*28*) explained that although an absolute reason to use either the block maxima model or POT model does not exist, we often use the former to modify seasonal or periodic samples. We use the POT model to analyze the tail distribution of recovery time because recovery time does not act periodically, and the block maxima model tends to ignore other extreme values (*29*).

The distribution function of the POT method is a conditional excess distribution function that is defined as:

$$F_u(y) = P(X - u \le y | X > u) = \frac{F(x) - F(u)}{1 - F(u)}, \ y>0 \quad (4)$$

where u is the preset threshold, X is the random variable, $X - u$ are the excesses, and $F(x)$ is the unknown distribution of recovery time. For the proper value of u, the conditional distribution (Equation 4) will be approximated using GPD, which is shown in the form:

$$G(x) = \begin{cases} 1 - \left(1 + \xi \frac{x}{\sigma}\right)^{-1/\xi}, & if\ \xi \neq 0 \\ 1 - e^{-x/\sigma}, & if\ \xi = 0 \end{cases} \quad (5)$$

where, ξ is the shape parameter and σ is the scale parameter. The plot of the sample mean excess function $e_n(u)$ to threshold u can facilitate the determination of u when the plot changes from a horizontal line to a line with a positive slope (*30*). The definition of $e_{n(u)}$ is shown in the form:

$$e_{n(u)} = \frac{\sum_{i=k}^{n} \left(x_i^n - u\right)^+}{n - k + 1}, \ k = \min\{i | x_i^n \rangle u\} \quad (6)$$

where $n - k + 1$ is the number of the data that exceeds u and the " + " sign indicates that $x_i^n - u$ should be positive.

After determining the probability distribution of the recovery time, we can calculate the existing probability of the different recovery time and further discover the existing probability of the recovery time that is extremely high to accept by the railway agency. That is, we can compute the probability of the unacceptable recovery time.

From Equations 4 and 5 we derive:

$$G(x) = \frac{F(x) - F_n(u)}{1 - F_n(u)} \quad (7)$$

which can be rearranged as:

$$F(x) = (1 - F_n(u)G(x) + F_n(u) \quad (8)$$

where $F_n(u)$ can be estimated using $(n - N_u)/n$ (Lin, 2003). Thus, Equation 8 can be rewritten as:

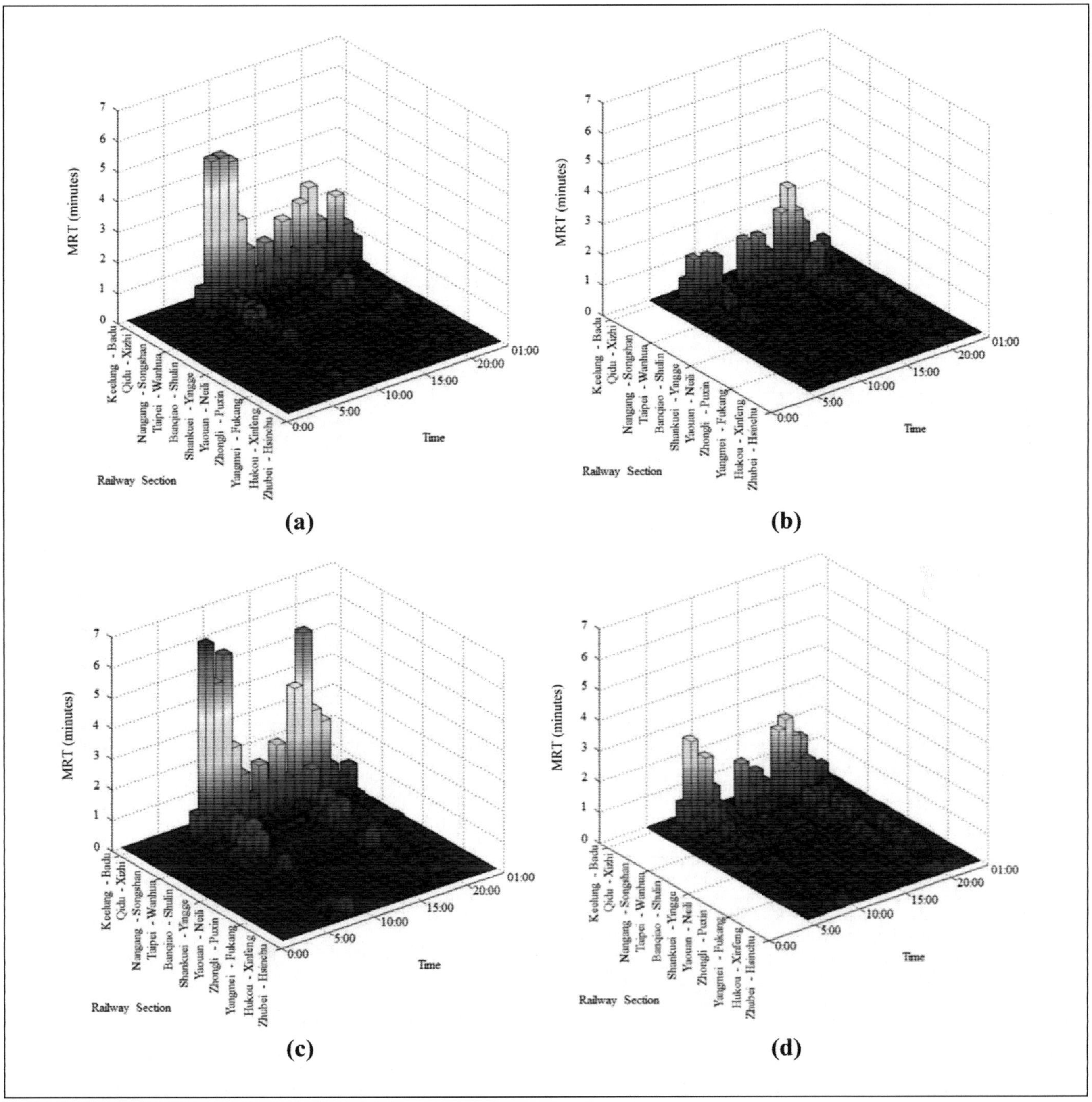

Figure 7. MRT before service plan revision, (*a*) northbound and (*b*) southbound; MRT after service plan revision, (*c*) northbound and (*d*) southbound.

$$F(x) = 1 - \frac{N_u}{n}\left(1 - \xi\frac{x-u}{\sigma}\right)^{1/\xi} \quad (9)$$

which is the estimation of the full recovery time distribution. Thus, PURT can be calculated using:

$$\text{PURT} = F(x{>}\text{UR}) = \frac{N_u}{n}\left(1 - \xi\frac{\text{UR}-u}{\sigma}\right)^{1/\xi} \quad (10)$$

In summary, the simulation results, including standard deviation and failure probability, will vary when different service plans are applied. We further conclude that having a relatively low recovery time does not mean that the standard deviation and failure probability will be low. Therefore, the stability of the service plan should be estimated using multiple factors.

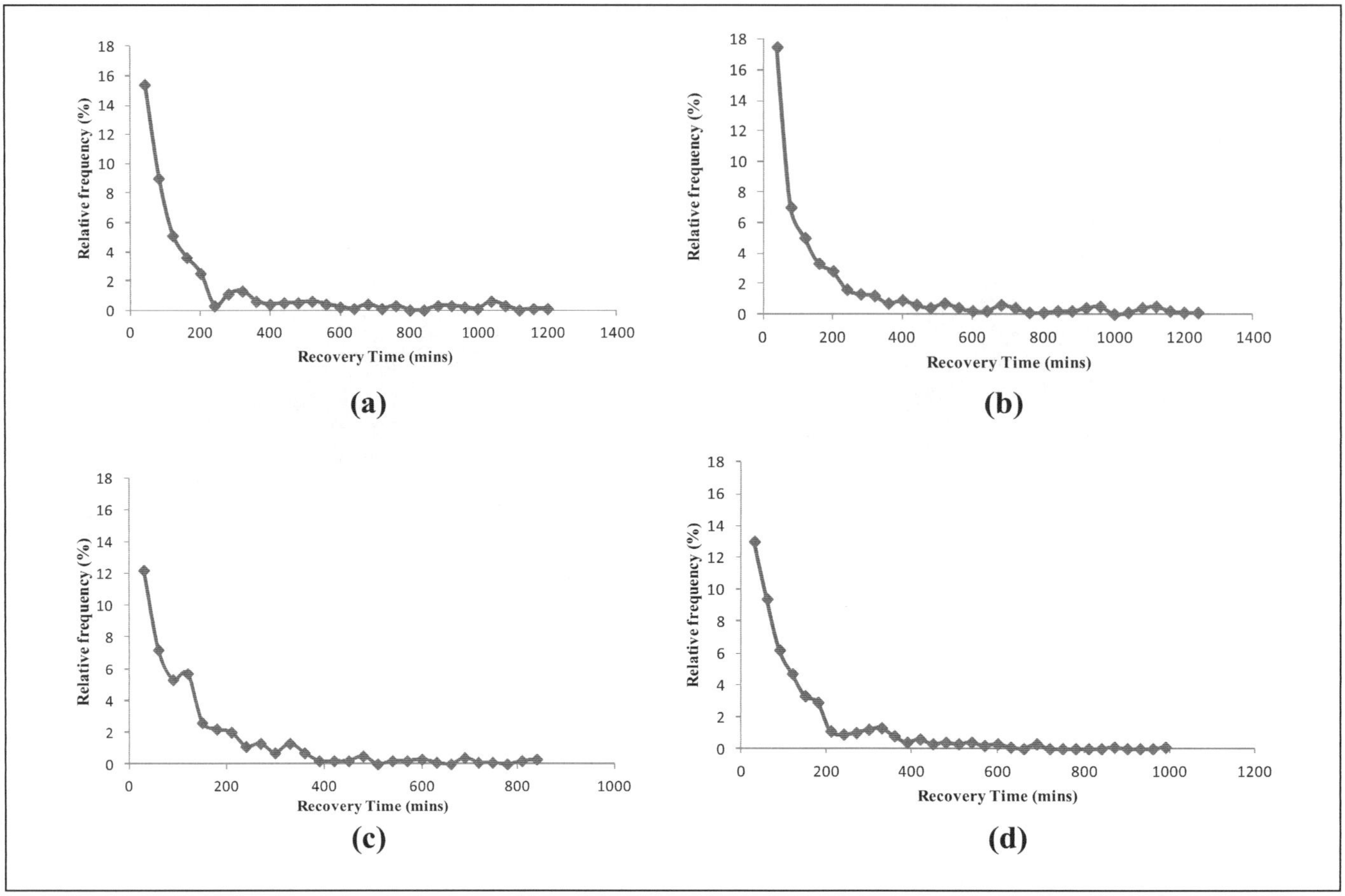

Figure 8. Relative frequency of the recovery time: (*a*) northbound, before revision; (*b*) northbound, after revision; (*c*) southbound, before revision; (*d*) southbound, after revision.

Table 1. Parameters Estimated by the Four Service Plans

	Northbound		Southbound	
Parameters	Before	After	Before	After
u	26	35	20	10
ξ	0.2331	0.0828	0.0267	0.0763
σ	159.6748	243.2664	149.6251	123.1448

The POT method determined the distribution of the data set, which is larger than a preset threshold u. Figure 9 shows the sample mean excess plot. For the northbound direction, u is set as 26 before revision and 35 after revision. For the southbound direction, u is set as 20 and 10 before and after the service plan revision, respectively.

Thereafter, the shape parameter (ξ) and scale parameter (σ) was estimated. Table 1 shows the estimated result of the four service plans. The cumulative distribution function (CDF) of excesses and fitted GPD are shown in Figure 10.

With Equation 10 and the parameter estimated from Table 1, PURT for the northbound traffic is increased from 0.065 to 0.129 (a 98.46% increase), while that for the southbound traffic is reduced from 0.126 to 0.105 (a 16.67% decrease). This attribute provides another considerably practical aspect to understand the impact of the service revision.

Discussion of Results

Table 2 summarizes the results of the four attributes before and after the service revision. More trains were scheduled to run on the corridor, resulting in the increase in CUE. Retrospective, we therefore expected a decrease in stability. Figure 1 illustrates the concept that given the same infrastructure but with increasing used capacity, limited room is left for disruption recovery. The increase in MRT manifests this trade-off relationship and is consistent with previous study (*24*), which used the analytical model.

However, if we delve into the simulation results, not all indices increased as CUE increased. For the DRT, although the value of the northbound plan increased by 14.97%, the result decreased by 3.92% for the southbound plan. This finding indicates that for the

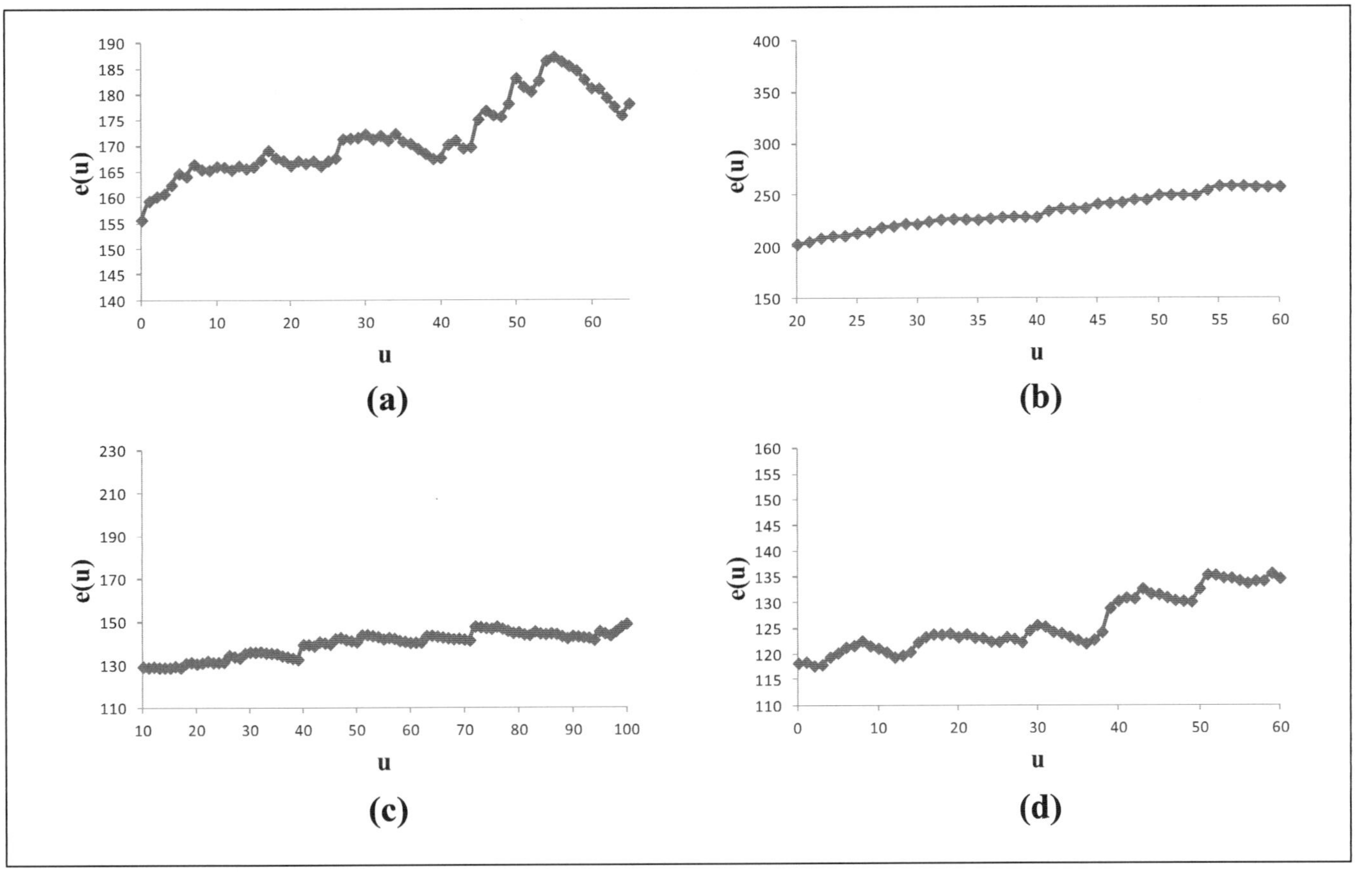

Figure 9. Sample mean excess plot: (*a*) northbound, before revision; (*b*) northbound, after revision; (*c*) southbound, before revision; (*d*) southbound, after revision.

Table 2. Overall Evaluation Indices of Exceeding the Selected Recovery Time Standard

	Northbound			Southbound		
Evaluation index	Before	After	Change	Before	After	Change
CUE (%)	41.58	42.46	+ 2.11%	43.20	44.18	+ 2.27%
MRT (min)	69.11	85.11	+ 23.14%	57.58	59.27	+ 2.94%
DRT (min)	172.85	198.73	+ 14.97%	120.71	115.98	−3.92%
PURT	0.065	0.129	+ 98.46%	0.126	0.105	−16.67%

southbound plan, recovery time has limited variance and a 90% chance would occur for recovery time to fall into a limited range after revision. This phenomenon is preferable for the operator, particularly that the MRT is only 1.69 min longer.

If we analyze PURT, the southbound plan decreased by 16.67% whereas the northbound plan significantly increased by 98.46%. Given that PURT is directly predicting the chance that the recovery time is unacceptable, the southbound plan revision instead showed improvement.

Overall, the increase in CUE does not necessarily worsen the service plan stability despite the uncertainty characteristics. For the operators, the increase in MRT may be of concern but could be acceptable while the dispersion and the PURT decreased. In our case study, the northbound service plan evidently needs a deliberate review and revision before implementation. Meanwhile, the southbound plan successfully increased capacity usage while not worsening the stability. This result shows the value of our study incorporating uncertainty analysis. Railway agencies do not need to reject their revision merely on MRT.

Interestingly, the increment of the CUE for the southbound direction is even higher compared with that of the

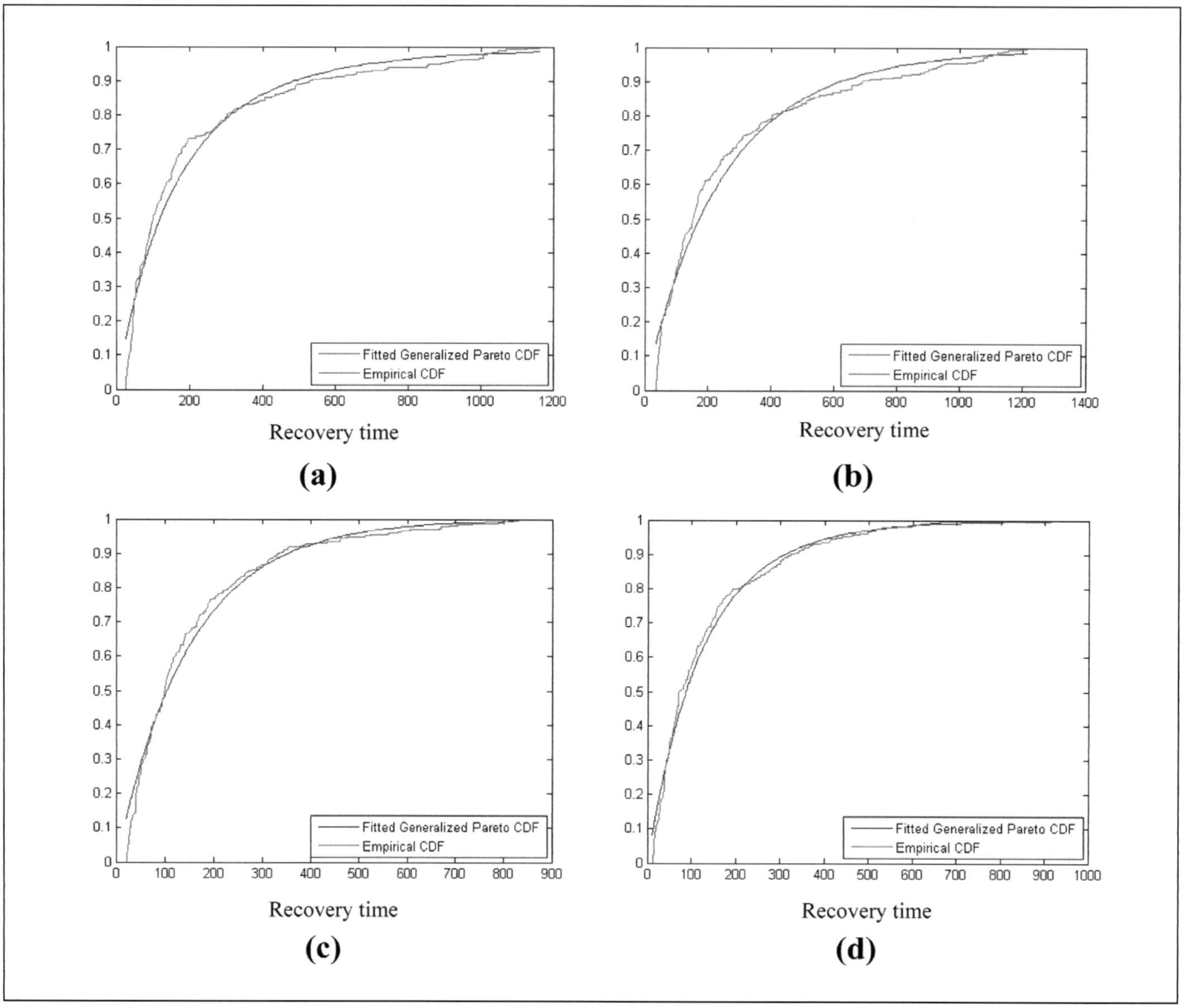

Figure 10. Fitted GPD versus CDF of excesses of the simulation result: (*a*) northbound, before revision; (*b*) northbound, after revision; (*c*) southbound, before revision; (*d*) southbound, after revision.

northbound service plan. According to Figure 5, additional trains were scheduled in the sections with less used capacity and cells with high CUE are followed by cells with low CUE (less continuous high CUE cells). This arrangement is likely to maintain considerable buffer for recovery. Further study is recommended to comprehend the relationship among CUE, MRT, DRT, and PURT, and the result could provide insights into the manner of scheduling a considerably stable service plan.

Conclusion

This research developed an evaluation framework to investigate the efficiency and stability of a service plan. We selected recovery time rather than delay to quantify service stability and further studied the variations in recovery time to capture the impact from changes in capacity utilization. Four indices were proposed, namely, CUE, MRT, DRT, and PURT. This method was completed using the proposed Monte Carlo simulation method and able to assist railway agencies to evaluate a railway service plan flexibly. The proposed process and indices (Figure 2) can easily be adopted by other railway systems with capacity constraints and potential disruptions. Evaluation results would assist railway agencies to determine whether the revision of the service plan is acceptable, and would provide an improved support to their decision-making in striking a balance between service level and asset usage.

Author Contributions

The authors confirm contribution to the paper as follows: study conception and design: Lai, Chen; data collection: Chen; analysis and interpretation of results: Lai, Chen, Yan, Li; draft manuscript preparation: Lai, Chen, Yan. All authors reviewed the results and approved the final version of the manuscript.

References

1. Carey, M., and A. Kwiecifiski. Properties of Expected Costs and Performance Measures in Stochastic Models of Scheduled Transport. *European Journal of Operational Research*, Vol. 83, No. 1, 1995, pp. 182–199.
2. Carey, M. Optimizing Scheduled Times, Allowing for Behavioural Response. *Transportation Research Part B: Methodological*, Vol. 32, No. 5, 1998, pp. 329–342.
3. Koelemeijer, G. S., R. M. P. Goverde, and R. J. van Egmond. PETER: A Performance Evaluator for Railway Timetables. *WIT Transactions on the Built Environment*, Vol. 50, 2000, pp. 405–414.
4. Hansen, I. A. Station Capacity and Stability of Train Operations. *WIT Transactions on the Built Environment*, Vol. 50, 2000, pp. 809–816.
5. Huisman, T., R. J. Boucherie, and N. M. van Dijk. A Solvable Queueing Network Model for Railway Networks and Its Validation and Applications for the Netherlands. *European Journal of Operational Research*, Vol. 142, No. 1, 2002, pp. 30–51.
6. Goverde, R. M. P. Railway Timetable Stability Analysis Using Max-Plus System Theory. *Transportation Research Part B: Methodological*, Vol. 41, No. 2, 2007, pp. 179–201.
7. Yuan, J., and I. A. Hansen. Optimizing Capacity Utilization of Stations by Estimating Knock-On Train Delays. *Transportation Research Part B: Methodological*, Vol. 41, No. 2, 2007, pp. 202–217.
8. Salido, M. A., F. Barber, and L. Ingolotti. Robustness in Railway Transportation Scheduling. *Proc., 7th World Congress on Intelligent Control and Automation*, Chongqing, China, 2008, pp. 2880–2885.
9. Cicerone, S., G. D'Angelo, G. Di Stefano, D. Frigioni, and A. Navarra. Recoverable Robust Timetabling for Single Delay: Complexity and Polynomial Algorithms for Special Cases. *Journal of Combinatorial Optimization*, Vol. 18, No. 3, 2009, pp. 229–257.
10. Goverde, R. M. P., B. Heidergott, and G. Merlet. Railway Timetable Stability Analysis Using Stochastic Max-Plus Linear Systems. *Proc., 3rd Railway Operations Modelling and Analysis Engineering and Optimization Approaches*, Zurich, 2009.
11. Andersson, E. V., A. Peterson, and J. Törnquist–Krasemann. Quantifying Railway Timetable Robustness in Critical Points. *Journal of Rail Transport Planning & Management*, Vol. 3, No. 3, 2014, pp. 95–110.
12. Cats, O., M. Yap, and N. van Oort. Exposing the Role of Exposure: Public Transport Network Risk Analysis. *Transportation Research Part A: Policy and Practice*, Vol. 88, 2016, pp. 1–14.
13. Zilko, A. A., D. Kurowicka, and R. M. P. Goverde. Modeling Railway Disruption Lengths with Copula Bayesian Networks. *Transportation Research Part C: Emerging Technologies*, Vol. 68, 2016, pp. 350–368.
14. Hooghiemstra, J. S., and M. J. G. Teunisse. The Use of Simulation in the Planning of the Dutch Railway services. *Proc., 1998 Winter Simulation Conference*, Washington, D.C., 1998, pp. 1139–1146.
15. Carey, M., and S. Carville. Testing Schedule Performance and Reliability for Train Stations. *Journal of the Operational Research Society*, Vol. 51, No. 6, 2000, pp. 666–682.
16. Middelkoop, D., and M. Bouwman. Testing the Stability of the Rail Network. *WIT Transactions on the Built Environment*, Vol. 61, 2002, pp. 995–1002.
17. Engelhardt-Funke, O., and M. Kolonko. Analysing Stability and Investments in Railway Networks Using Advanced Evolutionary Algorithms. *International Transactions in Operational Research*, Vol. 11, No. 4, 2004, pp. 381–394.
18. Vromans, J. C. M., R. Dekker, and L. G. Kroon. Reliability and Heterogeneity of Railway Services. *European Journal of Operational Research*, Vol. 172, No. 2, 2006, pp. 647–665.
19. Berbey, A., R. Galán, P. San Segundo, and J. Sanz-Bobi. Lyapunov Based Stability Analysis for Metro Lines. *Proc., 14th International Conference on Urban Transport and the Environment in the 21st Century*, Ashurst Lodge, 2008, pp. 111–119.
20. Berbey, A., R. Caballero, R. Galán, and J. Sanz-Bobi. A New Stability Index for Metro Lines. *Proc., 19th International Conference on Electronics Communications and Computers*. Piscataway, 2009, pp. 41–45.
21. Demitz, J., C. Hübschen, and C. Albrecht. Timetable Stability - Using Simulation to Ensure Quality in a Regular Interval Timetable. In I. A. Hansen (ed.), *Timetable Planning and Information Quality*, WIT Press, Southampton, UK, 2010, pp. 11–24.
22. Liu, J. R. *Development and Application of a Simulation Model on Railway Knock-on Delay*. PhD dissertation. Institute of Traffic and Transportation, National Chiao Tung University, 2011.
23. Burkhard, F., B. Seybold, T. Buker, T. Graffagnino, and H. Labermeier. OnTime - Network-Wide Analysis of Timetable Stability. *Proc., 5th International Seminar on Railway Operations Modelling and Analysis - RailCopenhagen 2013*. Copenhagen, Denmark, 2013.
24. Lai, Y. C., and C. S. Ip. An Integrated Framework for Assessing Service Efficiency and Stability of Rail Transit Systems. *Transportation Research Part C: Emerging Technologies*, Vol. 79, 2017, pp. 18–41.
25. Lai, Y. C., and K. T. Chen. Evaluating Service Risk in Railway Capacity Utilization Using Expected Recovery Time. *Journal of Transportation Engineering, Part A: Systems*, Vol. 143, No. 6, 2017, 04017016.
26. Ward, J. H. Hierarchical Grouping to Optimize an Objective Function. *Journal of the American Statistical Association*, Vol. 58, No. 301, 1963, pp. 236–244.
27. Ross, S. M. *Introduction to Probability Models*, 10th ed. Academic Press, Cambridge, MA, 2009, pp. 312–319.

28. Coleman, R. *Modelling Extremes*. International Statistical Workshop, Seoul University, Seoul, Korea, 2002.
29. Lin, X. G. Statistical Modelling of Severe Wind Gust. *International Congress on Modelling and Simulation, Townsville*, Vol. 2, 2003, pp. 620–625.
30. Kellezi, E., and M. Gilli. *Extreme Value Theory for Tail-Related Risk Measures*. International Center for Financial Asset Management and Engineering Research Paper Series, Switzerland, 2000.

The Standing Committee on Freight Rail Transportation (AR040) peer-reviewed this paper (18-01477).

Article

TRR
JOURNAL OF THE TRANSPORTATION RESEARCH BOARD

Transportation Research Record
2018, Vol. 2672(10) 215–223

DOI: 10.1177/0361198118780830
journals.sagepub.com/home/trr
SAGE

Stochastic Model of Train Running Time and Arrival Delay: A Case Study of Wuhan–Guangzhou High-Speed Rail

Javad Lessan[1], Liping Fu[1,2], Chao Wen[2,3], Ping Huang[2,3], and Chaozhe Jiang[1,2]

Abstract

Train operations are subject to stochastic variations, reducing service punctuality and thus the quality of service (QoS). Models of such variations are needed to evaluate and predict the potential impact of disturbances and to avoid service punctuality reduction in train service management and timetabling. In this paper, through a case study of the Wuhan–Guangzhou (WH–GZ) high-speed rail (HSR), we show how a wealth of train operation records can be used to model the stochastic nature of train operations at each level, section and station. Specifically, we examine different distribution models for running times of individual sections and show that the Log-logistic probability density function is the best distributional form to approximate the empirical distribution of running times on the specified line. Next, we show that the distribution of running times in each section can be used to accurately infer arrival delays. Consequently, we construct the underlying analytical model and derive the respective arrival delay distribution at the downstream stations. The results support the correctness of the model presented and show that the proposed model is suitable for constructing the distribution of arrival delays at every station of the specified line. We show that the integrated distribution models of running times and arrival delays, driven by empirical data, can also be used to evaluate the QoS at individual track sections.

The QoS has become an important performance measure for high-speed rail (HSR) train operating companies (TOC) in seeking to meet the passengers' expectations about train services. QoS includes a wide range of measures such as the accessibility, punctuality, and reliability of the transport services. Train punctuality is the percentage of trains arriving at a platform or departing from it no later than a certain time in minutes after the scheduled time (*1*). Increasing passenger expectations about QoS, and the competition between rail and other modes of transport such as airlines, has increased the importance of punctual train services (*2*, *3*). Despite the recent advances in train monitoring, communication, and control technologies, most train operations are subject to stochastic perturbations and disturbances. For instance, according to the reports from China Railway Corporation, the average punctuality for China's HSR network at the final destination stations is less than 90% (*4*). In Norway, for the best-performing routes the punctuality ratio is 94% whilst for the worst routes it drops to 80% (*5*). It should be noted that on China's conventional railway lines, just as for most European TOCs, train arrivals less than five minutes after the scheduled time are usually not considered to be delays (*6*). However, on the Norwegian State Railways, trains arriving within four minutes of the scheduled time are considered to be on time (*5*). That being said, TOCs admitted that it is impossible to eliminate the effects of unexpected delays completely. Instead, they focus on mitigating the impact through timetable design and dispatching decisions (*7*). This necessitates mining real-world train movement data to understand the characteristics of the actual operations and disturbances, and then incorporate these into the planning phase of train operations.

Recent advances, in train operation and monitoring technologies, along with data storage and processing capabilities, have enabled the building of models based on train operation records. In this regard, data-driven

[1]Department of Civil and Environmental Engineering, University of Waterloo, Waterloo, ON, Canada
[2]School of Transportation & Logistics, Southwest Jiaotong University, Chengdu Sichuan, China
[3]Railway Research Center, University of Waterloo, Waterloo, ON, Canada

Corresponding Author:
Address correspondence to Chao Wen: c9wen@uwaterloo.ca

models have gained more weight due to their potential to explain the behavior of the underlying systems realistically and with a higher level of accuracy (*8*). The statistical analysis and data mining of train operation records allow us to quantify the section- and station-specific characteristics of train operations to help improve the performance of intended services. This research aimed to conduct data-driven modeling and to study the application of heavy-tailed distribution functions to estimate the running times and arrival-delay distributions in HSR train operations. As a case study, we used real-world data from WH–GZ HSR line to find with sufficient accuracy the fits of empirical distribution for running times on each track section of the specified line. Next, we developed an analytical model that explains the relationship between the running times and arrival delays. The result of the approach we present is found to be in accordance with the results of our statistical analysis of train operation data and operational factors regarding each track section. Moreover, with the results we obtained, we showed how different track sections could be classified in terms of QoS.

This paper is structured as follows. The next section briefly reviews the relevant literature on train operation management. Section 3 provides a formal description of the terms, as well as the operational data collected from the WH–GZ HSR line. In Section 4, we show the results of statistical analysis on the train movement data, and explain the underlying model that connects the running times and arrival delays. Finally, conclusions and future research directions are discussed in Section 5.

Literature Review

In recent decades, a significant amount of research has focused on the analysis of empirical data of train operations to support rail operation management in off-line (proactive) and on-line (reactive) modes. Indeed, knowing the characteristics of train operation processes has been shown to be of practical value for improving the QoS in rail operations (*9*). Complex systems such as rail transportation networks are often exemplified by the occurrence of long-tailed probability distributions (*10*). Yuan et al. found that the location-shifted Log-normal distribution is the best distributional model to estimate both the arrival times of trains at the approach signal of the station and at the platform track, respectively. They also found that the Weibull distribution can well-approximate the non-negative arrival delays, departure delays, and free dwell times of late arriving trains (*11*). Carey and Kwiecinski assumed a shifted exponential, and a uniform distribution with a random disturbance for the running times on track sections to examine the properties of expected costs and performance measures in stochastic models (*12*).

Li et al., based on train arrival and departure times at each station, proposed a modified distribution model with four parameters to calculate the journey time of each service interval of multi-class trains (*13*). Kecman and Goverde presented predictive statistical learning models, to estimate the running times in railway traffic (*14*). Source (primary) delays have been shown to follow a negative exponential distribution at the tail (*4*, *15*) . On Dutch railways, Yuan adopted the Weibull distribution for estimating the arrival and departure delays at stations (*16*). Briggs and Beck presented a statistical model for the distribution of train delays on the British railway network, using q-exponential distribution functions (*10*). Yuan et al. proposed a method to estimate the distribution of departure delays by the Monte Carlo sampling approach based on the distributions of arrival and dwell times (*1*). Buker and Seybold evaluated the suitability of a group of existing distribution models, such as modified exponential phase-type, theta-exponential, and polynomial distributions, for the approximation of arrival delays (*17*).

Travel demand, train operation, and network elements such as track sections and stations have different static, dynamic, and stochastic characteristics, all of which are main sources of disturbances that could increase variations of travel time. Disturbances and perturbations can cause deviations in running times, arrival delays at subsequent operations, and missing transfers to connecting trains. This necessitates the consideration and integration of factors of running times into arrival delays to better represent the influence of variations in running times on arrival delays. However, most of the theoretical distribution models, such as Normal, Exponential, Weibull, and Log-normal distributions, have been used to fit statistical models to train running times (*18*). In most of the studies so far, it has been assumed that arrival delay disturbances can be approximated well with the exponential distribution. However, it has become apparent that a heavy-tailed distribution, which allows for an approximation of larger disturbances, is more appropriate than the exponential distribution (*19*). In view of this, we show how a heavy-tailed distribution can be adjusted to approximate either the distribution of running times or arrival delays. To the best of our knowledge, no work has been done to build analytically such a heavy-tailed distribution for arrival delays at a station based on the running time distributions on preceding track sections. This paper belongs in the area of statistical methods for estimating train running time distributions, and analytical models for developing arrival delay distributions from the distribution of running times.

Table 1. Descriptive Statistics Obtained from Train Operation Records

No.	Code name	Station	Arrival delay*			Departure delay*			Dwell time*		
			Avg	SD	IQR	Avg	SD	IQR	Avg	SD	IQR
1	GZN	Guangzhou North	8.48	11.87	10	8.11	12.02	9	4.35	3.82	2
2	QY	Qingyuan	4.96	8.01	4	4.93	7.97	4	1.39	1.77	2
3	YDW	Yingde West	5.18	8.49	4	5.15	8.46	4	0.61	2.51	0
4	SG	Shaoguan	4.71	8.06	3	4.66	8.07	3	0.32	1.85	0
5	LCE	Lechang East	4.85	7.82	4	4.86	7.83	4	0.40	2.07	0
6	CZW	Chenzhou West	5.03	8.18	3	4.92	8.10	3	1.49	1.94	2
7	LYW	Leiyang West	4.66	7.72	3	4.64	7.69	3	0.02	0.6	0
8	HYE	Hengyang East	4.95	7.98	4	4.94	7.96	4	0.4	1.34	0
9	HSW	Hengshan West	5.33	9.70	3	5.33	9.68	3	0.27	1.36	0
10	ZZW	Zhuzhou West	4.73	7.58	3	4.7	7.53	3	0.37	1.43	0
11	CSS	Changsha South	4.97	8.20	3	4.99	8.16	3	1.07	1.73	2
12	MLE	Miluo East	4.96	7.93	3	4.94	7.88	3	0.30	1.29	0
13	YYE	Yuyang East	6.21	10.48	4	6.35	10.52	4	1.17	2.06	2
14	CBN	Chibi North	5.08	8.17	4	5.04	8.13	4	0.78	1.71	2

Note: Avg = average; SD = standard deviation; IQR = inter quartile range.
*Values are in minutes.

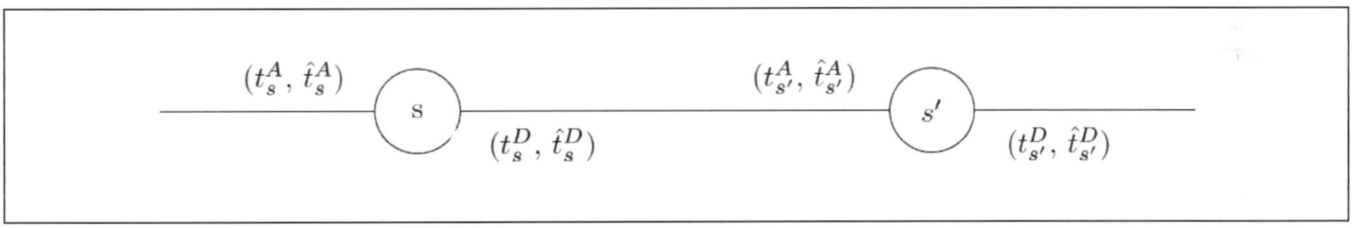

Figure 1. The general scheme of train movements at two successive stations s and $s' = s + 1$ $s' = s + 1$. Top: scheduled and actual arrivals (*A*). Bottom: scheduled and actual departures (*D*).

Description of Case Study Data

The Guangzhou Bureau provided us with train operation data, for the period from February 2015 to November 2015, from one of its HSR lines that connects Guangzhou North (GZN) and Chibi North (CBN), see Table 1. The train movement data include arrival (A) and departure (D) events between the stations on the specified line (directed from GZN to CBN), excluding early arrivals and departures. Consider a particular train passing through S number of stations according to a published timetable. The train is expected to depart at the scheduled departure time t_s^D from the station $s \in S$, and arrive at the scheduled arrival time $t_{s'}^A$ at the station $s' = s + 1 \in S$. However, due to various disturbances during its operation, the train can deviate from the schedule and have the realizations of actual departure time $\hat{t}_s^D$ and actual arrival time $\hat{t}_{s'}^A$. Figure 1 depicts the arrival and departure events at two successive stations s and s' where the parameters inside the parentheses display the scheduled and actual time of events, respectively. The (positive) differences between actual and scheduled times for the arrival $(\hat{t}_s^A - t_s^A)^+$ and departure $(\hat{t}_s^D - t_s^D)^+$ events, are defined as arrival and departure delays at station s, respectively. In the same way, the actual dwell time of the train at station s is defined as the difference between its departure and arrival times $(\hat{t}_s^A - \hat{t}_s^D)$, and the actual running time is defined, for two consecutive stations s and $s + 1$, as the time taken to traverse the track section connecting those stations $(\hat{t}_{s'}^A - \hat{t}_s^D)$. We can use the scheduled values to find the same definition for scheduled dwell time $(t_s^A - t_s^D)$ and scheduled running time$(t_{s'}^A - t_s^D)$. In practice, additional time supplements are added to the minimum required running and dwell times to reduce displacements in the timetable and recover delays from unexpected events. Based on these definitions, we have derived the summary of statistics for each station, as provided in Table 1.

In this study, we focused on the running time operations on track sections and performed statistical analysis on the train movement data obtained. The analysis showed that the majority of delays are of less than 60 minutes. In practice, severe delays are indications of disruptions and do not reflect the operation planned in the timetable. Usually, train dispatching decisions change when a train suffers from severe delay due to disruption

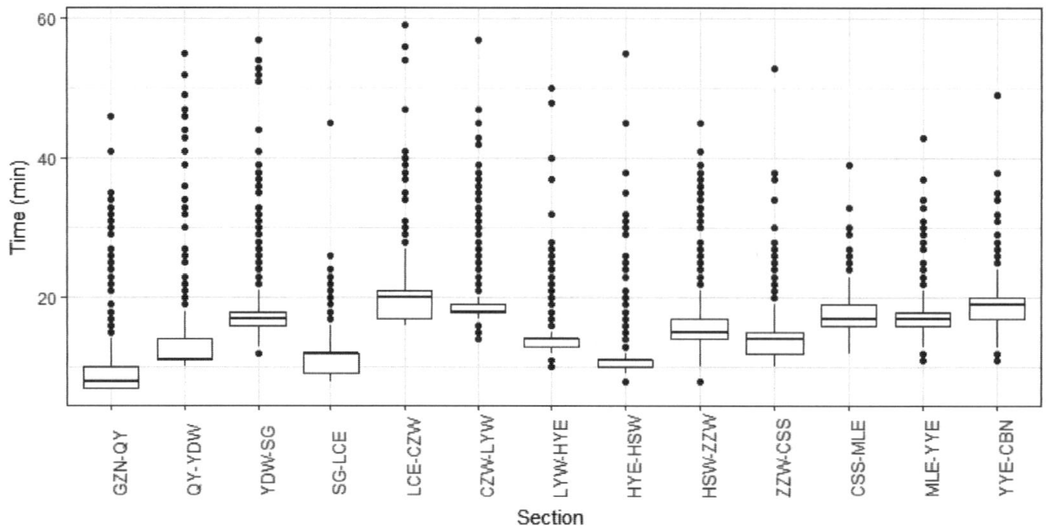

Figure 2. Section-level actual running (trip) times.

or severe disturbances. Therefore, we excluded trains with more than 60-minute delays. In the next step, the raw data is processed to extract and filter the process times of all trains that run on all of the sections on the specified line. Finally, we obtained about 60% of all the records in train movement data. These records include arrival and departure events of all the trains that pass through all the sections, from GZN to CBN and stop at all stations, successively.

The box-plots of running-time duration for each section in Figure 2 represent the approximate performance of the corresponding sections and stations. The significant variations are an indication of shifted arrival times in the timetable, which can be attributed to the variations in running times, given that we have excluded records with source delay. We can see long tails on the upper side of the running times. Therefore, one can expect that the empirical distribution of running times on track sections should follow a heavy-tailed distribution. Generally, these visualizations and statistics are limited in representing the exact features of the distributions of train operations, and they give a rough perception of the underlying processes. Next, we conduct a statistical analysis on train movement data on each track section to get a better impression of train operations on different track sections.

Statistical Analysis of Train Operations

Probability Distribution of Running Times

Fitting distributions to train operation data and choosing the right probability distribution to model each of the respective processes involves an iterative process of choosing distributions, estimating parameters, and assessing the quality of fits (*20*). Complex systems such as transport networks seem to obey heavy-tailed distributional forms. Among univariate distribution models, the Log-normal (LN), Log-logistic (LL), and Weibull (WB) are three common right-skewed, heavy-tailed distributions that occupy a central position in a wide range of process time estimations of complex systems. These distributions are selected as candidates to check further which one fits better the distribution of running times. We depicted the estimated densities of the candidate distributions and imposed them on the histogram and non-parametric density estimates with the Gaussian and Epanechnikov kernels for all of the running time records, see Figure 3. As can be seen, all the candidate distributions mimic the shape of the non-parametric distributions as well as the histogram of the running-time records. Next, we performed statistical analysis and goodness-of-fit tests for candidate distributions and report the results in Table 2. The parametric distributions were fitted to the running-time data of each section, one at a time, under the independently and identically distributed (i.i.d.) sample assumption using R-project. When fitting the candidate distributions the "Kolmogorov-Smirnov" (K-S) and the "Cramer-von Mises" (C-M) test statistics were used to evaluate and select the best distribution among those fitted (*21*). According to the results, it appears that the Log-logistic density function is the best model that fits the running time distribution for almost all of the sections, except HSW–ZZW which just the CM criterion is in favor of the Log-normal distribution. With these

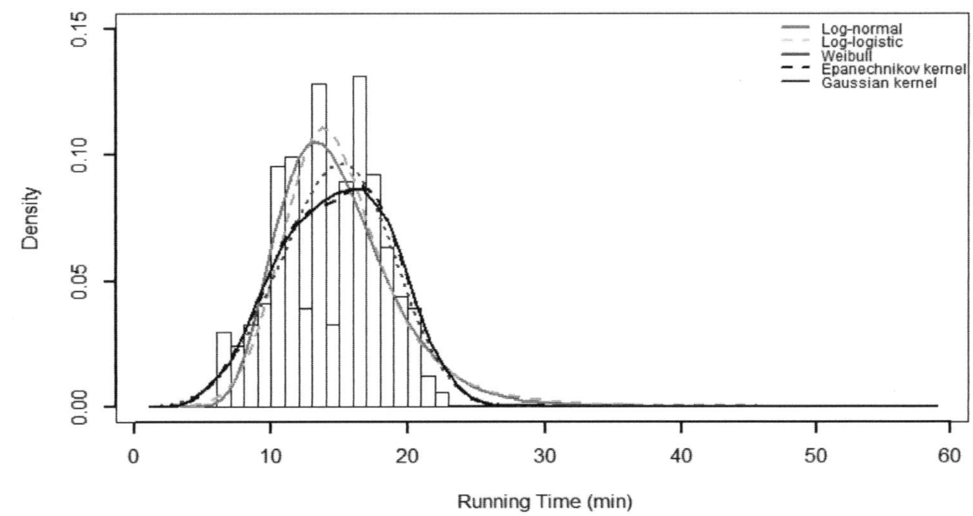

Figure 3. Posterior density estimates of LN, LL, and WB fits imposed on the histogram with the Gaussian and Epanechnikov kernels on running time records of 13 sections.

Table 2. Test Results of Density Fits on Running Time Records of 13 Sections

Section	Criteria	LN	LL	WB
GZN–QY	KS	0.23	0.22	0.27
	CM	131.30	129.08	151.81
QY–YDW	KS	0.29	0.26	0.32
	CM	218.65	199.34	294.38
YDW–SG	KS	0.30	0.27	0.28
	CM	206.37	183.39	369.93
SG–LCE	KS	0.35	0.32	0.32
	CM	343.63	313.92	316.55
LCE–CZW	KS	0.22	0.21	0.27
	CM	186.01	174.58	276.15
CZW–LYW	KS	0.31	0.29	0.32
	CM	221.39	200.10	420.71
LYW–HYE	KS	0.30	0.27	0.29
	CM	206.57	174.55	344.99
HYE–HSW	KS	0.25	0.24	0.27
	CM	157.56	142.06	339.13
HSW–ZZW	KS	0.20	0.20	0.20
	CM	106.23	111.67	120.52
ZZW–CSS	KS	0.19	0.18	0.24
	CM	68.82	67.27	150.79
CSS–MLE	KS	0.28	0.25	0.33
	CM	278.30	229.60	295.85
MLE–YYE	KS	0.26	0.23	0.26
	CM	151.23	124.76	196.30
YYE–CBN	KS	0.22	0.21	0.21
	CM	87.12	84.44	134.89

results, we have enough statistical evidence to overlook one exceptional case and postulate that the Log-logistic distribution function can approximate the stationary running times per individual track section.

Probability Distribution of Arrival Delays

Now, let us consider that $X \sim LL(\gamma, \sigma)$ denotes that the running time random variable X has Log-logistic distribution with positive shape and scale parameters γ and σ, respectively. The Log-logistic distribution model provides a two-parametric function for survival analysis. We denote $\bar{F}(x) = Pr(X \geq x)$ as its survival in the following form:

$$\bar{F}(x) = 1 - F(x) = \left[(\frac{x}{\sigma})^{\gamma} + 1\right]^{-1} \quad (1)$$

Generally, arrival delays are attributed to the disturbances that happen during train operations at upstream stations or track sections (*22*). In the latter case, we can use the idea of survival to cast the late arrivals as the running times that take longer than the scheduled (threshold) running time for the respective section, which is published in the train timetable.[1] The scheduled running time is the summation of the minimum required time, which is needed to traverse a track section between two stations, and a time supplement to recover from small disturbances during train running times. Let us define u as the threshold of scheduled running time for the section that connects stations s and $s+1$. More explicitly, if a train departs on time from station s and its actual running time to arrive at $s+1$ takes longer than u time units we call it a "delayed train". Technically, u is the

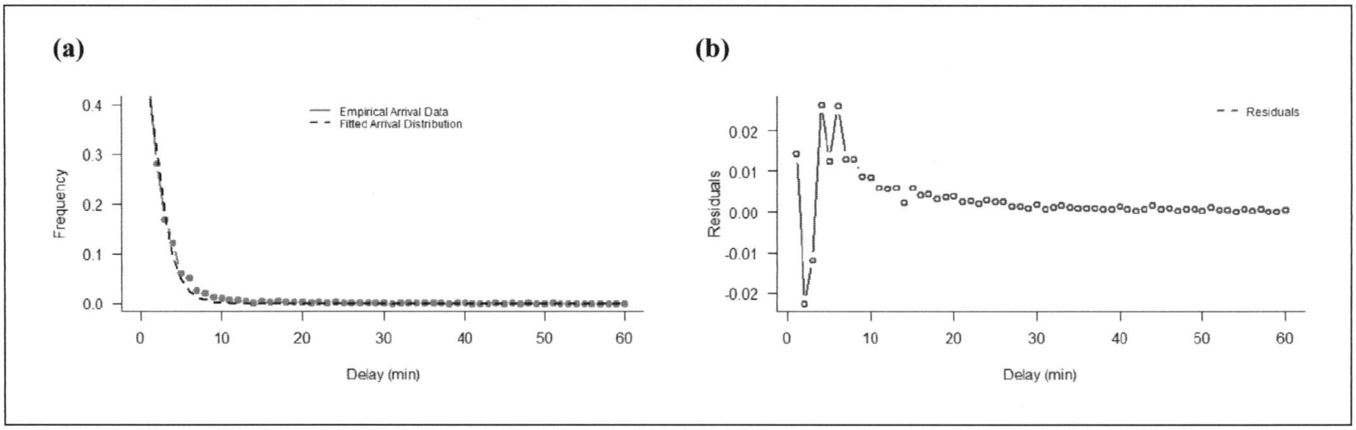

Figure 4. Empirical and fitted delay distributions with residuals for CZW: (*a*) empirical and fitted delay distribution functions and (*b*) estimation residuals.

scheduled running time that also encompasses a time supplement in addition to the minimum running time. Therefore, all the trains with $X \leq u$ are considered punctual. Theoretically, we perform a "Focus on Excesses" operation to find the excess values of X over the threshold u; a rationale for this method can be found in (*23*) and (*24*). We call these excess values "delay," which are the values of the conditional random variable X_u defined as: $X_u = X - u|x>u$. Using the Bayes' theorem we can write the conditional distribution function of the excess values as follows:

$$\bar{F}_{X_u}(x) = \Pr(X>x+u|x>u) = \frac{\bar{F}(X+u)}{\bar{F}(u)}, x>0. \quad (2)$$

Combining Equations 3 and 4, we can derive the survival function of running times as follows:

$$\bar{F}_{X_u}(x) = \frac{(\frac{u}{\sigma})^{\gamma}+1}{(\frac{x+u}{\sigma})^{\gamma}+1}, x>0. \quad (3)$$

Given that X possesses the probability density $f(x)$, then the density function of X_u is defined as: $f_u(x) = \frac{d(1-\bar{F}_{X_u}(x))}{dx}$. With this transformation, the running time random variable X_u is limited to a variable that describes arrival delays. After taking the derivative and doing some calculations to account for no delay case ($x = 0$), we can derive the following simplified density function for arrival delays:

$$f_u(x) = \begin{cases} (1-p)\frac{\gamma(\sigma^{\gamma}+u^{\gamma})(u+x)^{\gamma-1}}{(\sigma^{\gamma}+(u+x)^{\gamma})^2} & x>0 \\ p & x=0 \end{cases} \quad (4)$$

Theoretically, $p = \int_l^u f(x,\gamma,\sigma)dx$, where l is the location parameter of the distribution model, which is assumed to be zero, without loss of generality. We will use this combined probability distribution function to approximate the distribution of arrival delays at station $s' = s+1$, with the shape (γ) and scale (σ) parameters obtained from the running time distribution of the preceding section. With this model, the distribution of arrival delays at each station inherits two main factors (γ, and σ) attributed to the running times on the preceding track section.

Parameter Estimation

To estimate the parameters of the developed model for arrival delays, firstly, we estimated the optimum parameters of σ and γ for each track section by the maximum likelihood estimation (MLE) approach. Next, we fitted the arrival delay model found in (*4*) to data of each station and estimated the respective parameters for each station. All calculations were done in R-project and the results are reported in Table 3. This table reports the estimated scale($\hat{\sigma}$) and shape($\hat{\gamma}$) parameters for each section as well as the estimated threshold parameters u and p, and the residual sum-of-squares of estimations (*RES*) for each station. As the values of *RES* show, the model of the arrival delays presents a good fit to the data. As an instance, the plots of measured and fitted distribution functions of arrival delay and the respective residuals for station CZW are presented in Figure 4.

Findings

Figure 5 shows each section's position, given its scale($\hat{\sigma}$) and shape($\hat{\gamma}$) parameters. As shown in the figure, the sections can be grouped into subclasses in which their parameters of scale and shape are close to each other. For example, HYE–HSW, SG–LCE, and QY–YDW sections can be grouped in the same subclass of QoS. Moreover, given the same operational conditions, for the same value of the shape parameter, we expect a better QoS for

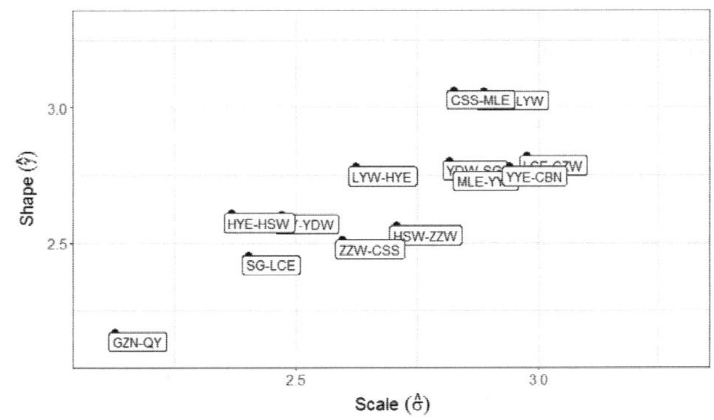

Figure 5. The estimated scale ($\hat{\sigma}$) and shape ($\hat{\gamma}$) parameters for 13 sections.

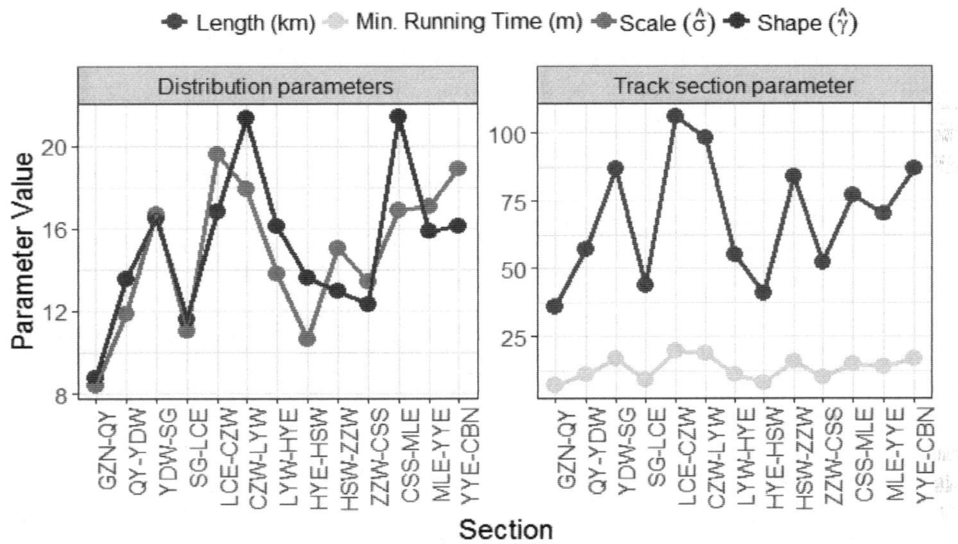

Figure 6. The behavior of the estimated scale ($\hat{\sigma}$) and shape ($\hat{\gamma}$) parameters with respect to the scheduled running time of the sections.

sections that have lower scale parameter. To see why the distributional parameters are different for each section and which operational parameters are influential on the distribution of running times, we depicted the estimated scale and shape parameters, and the lengths and scheduled running times for each section in Figure 6. We selected the minimum running time as it is a good indicator of the corresponding section's length and other factors such as geometry, the existence of junctions and other operational parameters.

The shape and scale values were subjected to the power transformation with exponent *exp.* to signify the influence of small variations. It can be seen that the scale parameter of each section follows the same pattern as the length of each section. This means that each section's operational requirements can affect significantly the scale parameter of the running time distribution. For instance, we can see that GZN–QY is a short-distance service section with a relatively low scale parameter, and LCE–CZW is the long-distance service with a relatively high scale parameter. However, the shape parameter of each section shows a different pattern considering the length of the corresponding section. We know that the variations in running times can contribute to the arrival delay, and that the variance of the Log-logistic distribution is a non-linear function of the shape parameter (*25*), i.e., $Var(x) = \sigma^2(2\pi/\gamma\ sin(2\pi/\gamma) - (\pi/\gamma\ sin(\pi/\gamma))^2)$, for $\gamma > 2$. This suggests that there is a non-linear relationship between running times and arrival delays. As a result, we expect that the long-running trains which suffered disproportionate delays at the first terminals experienced disproportionate delays at their arrivals at subsequent stations and terminals. Another observation is that CZW–LYW is the second longest section that has relatively a higher shape but a lower scale value. This can be an indication of excessive running buffer times allocated to this section. Indeed, our analysis shows that this section has a relatively higher running time supplement of about five minutes. The results show that LCE–CZW and HSW–ZZW are among the high-frequency delay occurrence sections, which is reflected in the distribution of the running times of these sections where the shape parameter falls below the scale and minimum running time parameters. One remedy to this could be allocating a larger running-time supplement to the trips on these two sections to absorb running-time variations and reduce delay propagation. It is common practice to use the statistics of train operation data in discussions about the efficiency of train operations, and decisions of time supplements allocations to design timetables or reschedule the original timetable into a new one. Our findings can be used as new metrics along with the common statistics of running times and arrival delays to help improve train operations.

Conclusion

This paper has presented the results of statistical analysis on the real-world train operation data from WH–GZ HSR in China. Based on non-parametric and parametric estimation approaches, it was shown that the Log-logistic distribution is generally the best model to approximate the running times on each section of the specified line. The relationship between running times distribution on track sections and the corresponding arrival delays at the nearby stations are then modeled analytically. The model obtained is used to estimate the arrival delay distributions at the subsequent stations. The results provided key practical implications for rail operation performance evaluation and delay management practices. This paper presented new measures to evaluate the performance of train operations on track sections. Some of the statistical results and estimated parameters may work only for the specified line. However, our analytical method, i.e., the focus on excess approach, can be generalized and applied to other train lines and networks to develop distributions of arrival delays. Developing a QoS performance measure based on the distribution of processes and operational parameters of each track section, can also be applied to other empirical models of running time distribution. Our future research will focus on improving the fits of train running and dwell times, as well as the arrival and departure delays. In the next step, we will work on developing delay prediction and time supplement allocation models based on empirical data.

Note

1. In the former case where a disturbance happens on a train operation at station s and the train departs late from the station, we can apply the same logic by transforming the departure delay to the source delay.

Acknowledgments

This work was supported by the National Key R&D Program [grant number 2017YFB1200700], National Nature Science Foundation of China [grant number 61503311] and NSERC (National Sciences and Engineering Research Council of Canada). We acknowledge the support of the Railways Technology Development Plan of China Railway Corporation [grant number 2016X008-J]. Parts of the work were supported by State Key Lab of Railway Control and Safety Open Topics Fund [grant number RCS2017K008]. We are grateful for the contributions made by our project partners.

Author Contributions

The authors confirm contribution to the paper as follows: study conception and design: Javad LESSAN, Liping FU, Chao WEN; data collection: Chao WEN, Ping HUANG, Chaozhe JIANG; analysis and interpretation of results: Javad LESSAN, Liping FU, Chao WEN; draft manuscript preparation: Javad LESSAN, Liping FU, Chao WEN. All authors reviewed the results and approved the final version of the manuscript.

References

1. Yuan, J., R. Goverde, and I. Hansen. Propagation of Train Delays in Stations. *WIT Transactions on the Built Environment*, Vol. 61, 2002, pp. 975–984.
2. Goverde, R. M. *Punctuality of Railway Operations and Timetable Stability Analysis*. PhD thesis. Delft University of Technology, TRAIL Research School, 2005.
3. Cadarso, L., V. Vaze, C. Barnhart, and Á. Marín. Integrated Airline Scheduling: Considering Competition Effects and the Entry of the High Speed Rail. *Transportation Science*, Vol. 51, No. 1, 2016, pp. 132–154.
4. China-Railway-Corporation. *Punctuality Reaches 98.8%*. Available at: http://news.gaotie.cn/yunying/2016-07-23/336706.html.
5. Harris, N. G., C. S. Mjøsund, and H. Haugland. Improving Railway Performance in Norway. *Journal of Rail Transport Planning & Management*, Vol. 3, No. 4, 2013, pp. 172–180.
6. Goverde, R. M., I. Hansen, G. Hooghiemstra, and H. Lopuhaa. Delay Distributions in Railway Stations. *Proc., 9th World Conference on Transport Research*, Seoul, South Korea, WCTRS, 2001, pp. 1–13.
7. Caimi, G., L. Kroon, and C. Liebchen. Models for Railway Timetable Optimization: Applicability and

Applications in Practice. *Journal of Rail Transport Planning & Management*, Vol. 6, No. 4, 2017, pp. 285–312.
8. Wang, R., and D. B. Work. Data Driven Approaches for Passenger Train Delay Estimation. *Proc., 2015 IEEE 18th International Conference on Intelligent Transportation Systems (ITSC)*, Las Palmas, Spain, IEEE, New York, 2015, pp. 535–540.
9. Schöbel, A. *Optimization in Public Transportation: Stop Location, Delay Management and Tariff Zone Design in a Public Transportation Network*. Springer Science & Business Media, New York, NY, 2007.
10. Briggs, K., and C. Beck. Modelling Train Delays with Q-Exponential Functions. *Physica A: Statistical Mechanics and its Applications*, Vol. 378, No. 2, 2007, pp. 498–504.
11. Yuan, J., R. M. P. Goverde, and J. A. Hansen. Evaluating Stochastic Train Process Time Distribution Models on the Basis of Empirical Detection Data. *Computers in Railways X: Computer System Design and Operation in the Railway and Other Transit Systems*, Vol. 88, 2006, pp. 631–640.
12. Carey, M., and A. Kwieciński. Properties of Expected Costs and Performance Measures in Stochastic Models of Scheduled Transport. *European Journal of Operational Research*, Vol. 83, No. 1, 1995, pp. 182–199.
13. Li, M., Z. Liu, Y. Zhang, W. Liu, and F. Shi. Distribution Analysis of Train Interval Journey Time Employing the Censored Model with Shifting Character. *Journal of Applied Statistics*, Vol. 44, No. 4, 2017, pp. 715–733.
14. Kecman, P., and R. M. Goverde. Predictive Modelling of Running and Dwell Times in Railway Traffic. *Public Transport*, Vol. 7, No. 3, 2015, pp. 295–319.
15. Schwanhäußer, W. The Status of German Railway Operations Management in Research and Practice. *Transportation Research Part A: Policy and Practice*, Vol. 28, No. 6, 1994, pp. 495–500.
16. Yuan, J. *Stochastic Modelling of Train Delays and Delay Propagation in Stations*. Eburon Uitgeverij BV, Delft, Netherlands, 2006.
17. Büker, T., and B. Seybold. Stochastic Modelling of Delay Propagation in Large Networks. *Journal of Rail Transport Planning & Management*, Vol. 2, No. 1, 2012, pp. 34–50.
18. Huisman, T., and R. J. Boucherie. Running Times on Railway Sections with Heterogeneous Train Traffic. *Transportation Research Part B: Methodological*, Vol. 35, No. 3, 2001, pp. 271–292.
19. Vekas, P., M. H. van der Vlerk, and W. K. K. Haneveld. Optimizing Existing Railway Timetables by Means of Stochastic Programming. *Proc., Stochastic Programming E-Print Series (SPEPS)*. Institut für athematik. Humboldt Universität zu Berlin, Berlin, Germany, 2012, pp. 1–7.
20. Delignette-Muller, M. L., and C. Dutang. fitdistrplus: An R Package for Fitting Distributions. *Journal of Statistical Software*, Vol. 64, No. 4, 2015, pp. 1–34.
21. D'Agostino, R. B. *Goodness-of-Fit-Techniques*. CRC Press, Boca Raton, 1986.
22. Wen, C., Z. Li, J. Lessan, L. Fu, P. Huang, and C. Jiang. Statistical Investigation on Train Primary Delay based on Real Records: Evidence from Wuhan–Guangzhou HSR. *International Journal of Rail Transportation*, Vol. 5, No. 5, 2017, pp. 170–189.
23. Leadbetter, M. R. On a Basis for 'Peaks over Threshold' Modeling. *Statistics & Probability Letters*, Vol. 12, No. 4, 1991, pp. 357–362.
24. Bercher, J.-F., and C. Vignat. A New Look at Q-Exponential Distributions Via Excess Statistics. *Physica A: Statistical Mechanics and its Applications*, Vol. 387, No. 22, 2008, pp. 5422–5432.
25. McLaughlin, M. P. *A Compendium of Common Probability Distributions*. 2006. Available at: http://www.causascientia.org/math_stat/Dists/Compendium.pdf.

The Standing Committee on Intercity Passenger Rail (AR010) peer-reviewed this paper (18-01391).

Article

TRR
JOURNAL OF THE TRANSPORTATION RESEARCH BOARD

Transportation Research Record
2018, Vol. 2672(10) 224–235

DOI: 10.1177/0361198118781665
journals.sagepub.com/home/trr
SAGE

A Bi-Level Passenger Preference-Oriented Line Planning Model for High-Speed Railway Operations

Yuxiang Yang[1,2], Jie Li[1,3], Chao Wen[1,4], Ping Huang[1], Qiyuan Peng[1], and Javad Lessan[4]

Abstract

To solve bottlenecks and capacity shortages in railway corridors, it is necessary to consider passengers' temporal-spatial preferences in conjunction with train dispatching decisions at the line planning stage. A design of the departure and stop plans which is based on the passengers' preferences would guarantee a high quality of service. This paper presents a line planning optimization model for high-speed railway (HSR) operations that addresses both the operating costs of lines and passengers' preferences through a bi-level integer programming (IP) problem. Using actual travel data from the Shenzhen–Changsha HSR in China, we investigate passengers' travel behavior during different time periods and between several origin and destination (OD) pairs to characterize their spatial-temporal preferences. An IP model is then developed that maximizes the degree of passenger departure time satisfaction (DTS) at the higher level and the operating costs of lines at the lower level. Due to the complexity of the problem, a methodology is proposed to find a conflict-free solution for the proposed model to find train frequencies in each block and their stop plans at each station between the upper- and lower-level models. It is shown that a given passenger trip demand between an origin and a destination could become more flexible by analyzing its time characteristics and the DTS. Finally, a case study is presented to show the effectiveness of the model and the solution approach.

HSR has developed significantly in China, due to its efficiency in transporting large numbers of passengers within short travel times, and since 2008 it has transformed into a complex network (*1*). As of 2017, more than 22,000 km of HSR line had been built in China, accounting for more than 65% of the total mileage of HSR networks around the world. With this large-scale network and high transport capacity, however, there are still operational issues in some sections during peak hours, because some departure times are found to be preferred by passengers. To cope with the peak time issue, train operating companies can incorporate passenger flows at different time periods between different origin–destination (OD) pairs into the line planning problem (LPP). Passenger flow can be aggregated between different OD pairs, in some stations during peak hours/seasons, when bottlenecks and capacity shortage tend to happen; see details in (*2*, *3*). This necessitates considering passengers' temporal-spatial preferences in conjunction with train dispatching decisions at the line planning stage. Especially, it is of paramount importance in HSR operations to come up with a design of the departure and stop plans (including train routes and departure times) that meets the passengers' preferences and guarantees a high quality of service. For instance, an operative combination of short-distance and long-distance trains can satisfy passengers' travel demands and their expectations regarding availability and punctuality.

In this paper, based on a detailed analysis of the train operation data from a Chinese HSR corridor, passengers' preferences between different HSR stations are characterized. Subsequently, a bi-level LPP model with a solution algorithm is put forward. The model aims to

[1]School of Transportation & Logistics, Southwest Jiaotong University, Chengdu, China
[2]Institute of Transport Science, RWTH Aachen University, Aachen, Germany
[3]Civil, Buildings and Environmental Engineering Department, SAPIENZA Università di Roma, Rome, Italy
[4]Department of Civil and Environmental Engineering, University of Waterloo, Waterloo, Canada

Corresponding Author:
Address correspondence to Jie Li: jieli20092746@126.com

achieve the maximum degree of passenger DTS and the minimum train operating time. A case study based on the actual demand data from the Guangzhou–Changsha (707 km) and Shenzhen–Guangzhou (104.6 km) HSRs is also presented to explain the proposed model and solution methodology.

The rest of this paper is organized as follows. A brief literature review of the LPP studies and research on passenger distribution is presented in the next section. The third section presents results of the analysis of passengers' spatial-temporal preferences. The fourth section gives details of the LPP model and its algorithm. A case study from the Guangzhou–Changsha and Shenzhen–Guangzhou HSRs in China is described in the fifth section.

Literature Reviews

Recently, with the development of computer science and mathematical theory, considerable research has been carried on LPP, especially in the domain of operations research and optimization theory. Moreover, different aspects of the LPP, such as passenger flow allocation, train route selection, design of train origin and terminal stations, train stop plans, rolling stock plans, and hourly train frequency design to match the travel demand, have been studied.

The first key issue for the LPP is to satisfy passenger travel demand between OD pairs. Early studies like (*4*, *5*) use the pre-allocation method, which considers a given distribution of passenger flow on tracks. Mixed integer linear programming (MILP) models with minimum operating costs, dispatching constraints, and capacity constraints are presented. Further research has looked at the allocation problem as a multi-commodity network flow problem which allows passengers to choose the travel routes in the network, see (*6*–*8*). Another issue that LPP considers is the types of objective function to optimize. Besides the studies mentioned above, several types of optimization objectives have been proposed: the cost-oriented objectives from train operating companies' view (*4*, *9*, *10*), passenger (service) oriented objectives (*11*, *12*), and multi-objective functions, which considers the views of both passengers and train operating companies (*13*, *14*). For the cost- or passenger-oriented objectives, the former takes minimizing the train frequency and the total train travel time into account while the latter considers minimizing the total passenger travel time and the passenger inconvenience. The final issue is the modeling approach and the solution algorithm. Most of these studies formulate MILP models and solve them with different methods. Branch and bound algorithm (*5*, *15*), column generation algorithm (*8*, *16*) and branch and cut algorithm (*9*, *10*) are the most common methods.

This paper contributes to the LPP literature by presenting a model that considers the operation time (or interchangeably operation cost) and passengers' spatial-temporal preferences. In addition, a parameter of flexible passenger DTS is introduced to describe the dynamic demand from train operation data and build the passenger demand matrices.

Analysis of Passenger Preferences

Due to the differences in trip length and purposes of railway passengers, they show different preferences regarding train departure times and stopping points. We looked at three high-speed railway lines, namely Guangzhou–Shenzhen HSR (105 km), Guangzhou–Changsha HSR (707 km, which is a part of Guangzhou–Wuhan HSR), and Guangzhou–Wuhan HSR (968 km), and used the OD passenger flows (calculated from the ticket data for 9 days) to observe the different passenger departure time preferences with different travel distances. The data on average passenger flows between these stations for the 9 days are shown in Figure 1, with a break = 10 min and 1 h, and the different departure time preferences of two types of passenger flows are presented in Figure 2. The real data are standardized by Equation 1, due to the confidentiality agreement with the railway operator (*2*). It should be noticed that, due to the high operating frequency of trains on HSR, it is assumed that the results of passenger flow allocation, which is based on the existing train schedules, correspond almost exactly to the actual passenger travel preference.

$$f_{standardization} = \frac{f - f_{average}}{f_{standard-deviation}} \quad (1)$$

In Figure 2, short-distance passenger flows indicate an equal distribution over the whole day while long-distance passenger flows experience distinguishable peaks at 08:00, 14:00 and 17:00. Furthermore, Figures 3 and 4*a*–*b* show the spatial-temporal propagation passenger demands between different ODs, on the Wuhan–Shenzhen HSR in China, for the normalized data. We can see that passengers departing from different stations of origin at different departure times might congregate in some blocks, causing a bottleneck in train operations. Also, the number of trains does not match the passenger demand well, which might lead to low seat occupancy rate in some blocks.

More details about the spatial-temporal distribution of passenger intentions are given by Li et al. (*2*). That paper puts forward a method to assign long-distance and short-distance trains, however, without any alternative optimized objectives. In the next section, a bi-level model is presented which aims to maximize the degree of DTS and the total train operating time.

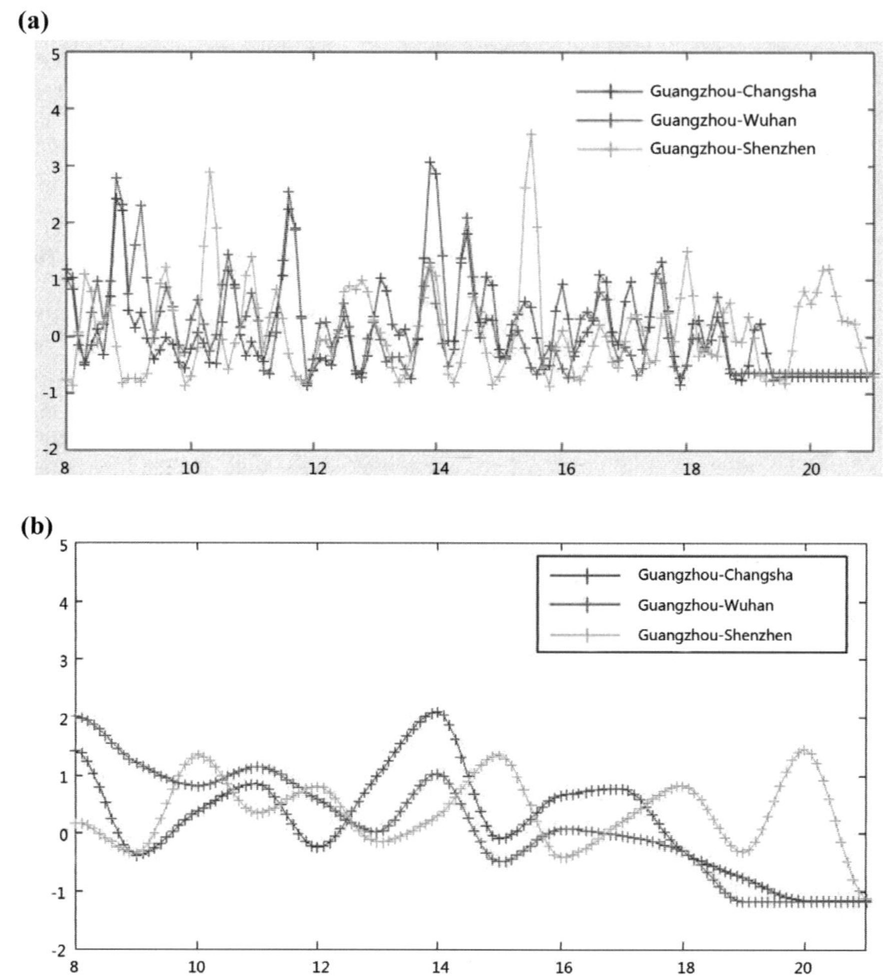

Figure 1. Average departure time data of passenger flows between stations: (*a*) break= 10 min, (*b*) break=1 h.

Passenger Preference Model

Model Assumptions

To formulate a mathematic model, we used the following assumptions:

(a) Trains have equal running times between stations, whether they stop at the next station or not. Average train operating speeds are used to calculate the operating times in blocks.

(b) Trains have equal priority. This is reasonable because, on the new HSRs in China, there are just two types of trains operating—300 km/h and 200 km/h—, but the 200-km/h trains operate at a frequency of just one or two per day in the main corridors such as the Beijing–Guangzhou HSR line. Moreover, this assumption makes our model more tractable. Due to this, overtaking will depend on the different stop plans rather than the train operation priority.

Description of Parameters and Variables

The sets, parameters, and decision variables used for modeling are as follows:

- **Sets and Parameters:**

DTS	degree of passenger departure time satisfaction,
SOR	train seat occupancy rate,
$G = (S, B)$	physical railway network,
S	set of stations, $s_i \in S, i \in [1, n]$,
n	total number of stations on the physical railway network,
B	set of blocks, $b_j \in B, j \in [1, n-1]$,
L	set of trains, $l_k \in L$,
i	index of stations,
j	index of blocks,
k	index of trains,

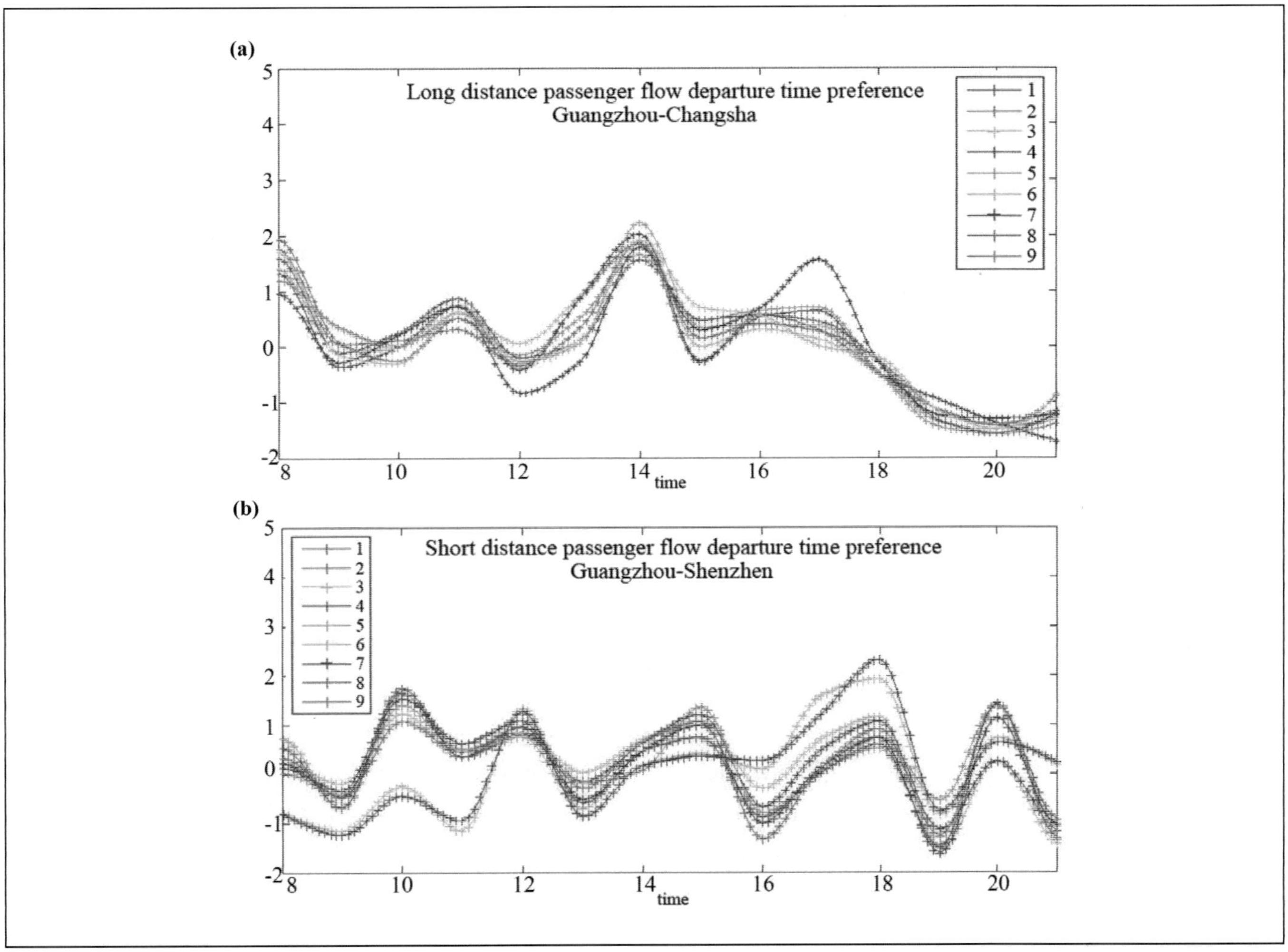

Figure 2. Departure time preferences for (*a*) long-distance passengers and (*b*) short-distance passengers.

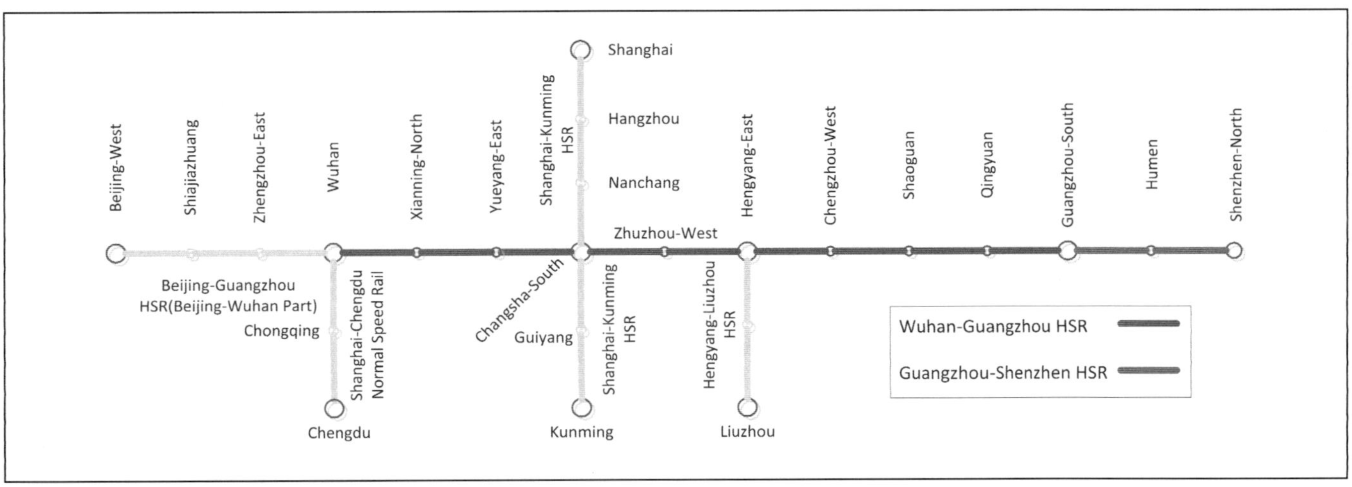

Figure 3. Physical network topology of Wuhan–Shenzhen HSR.

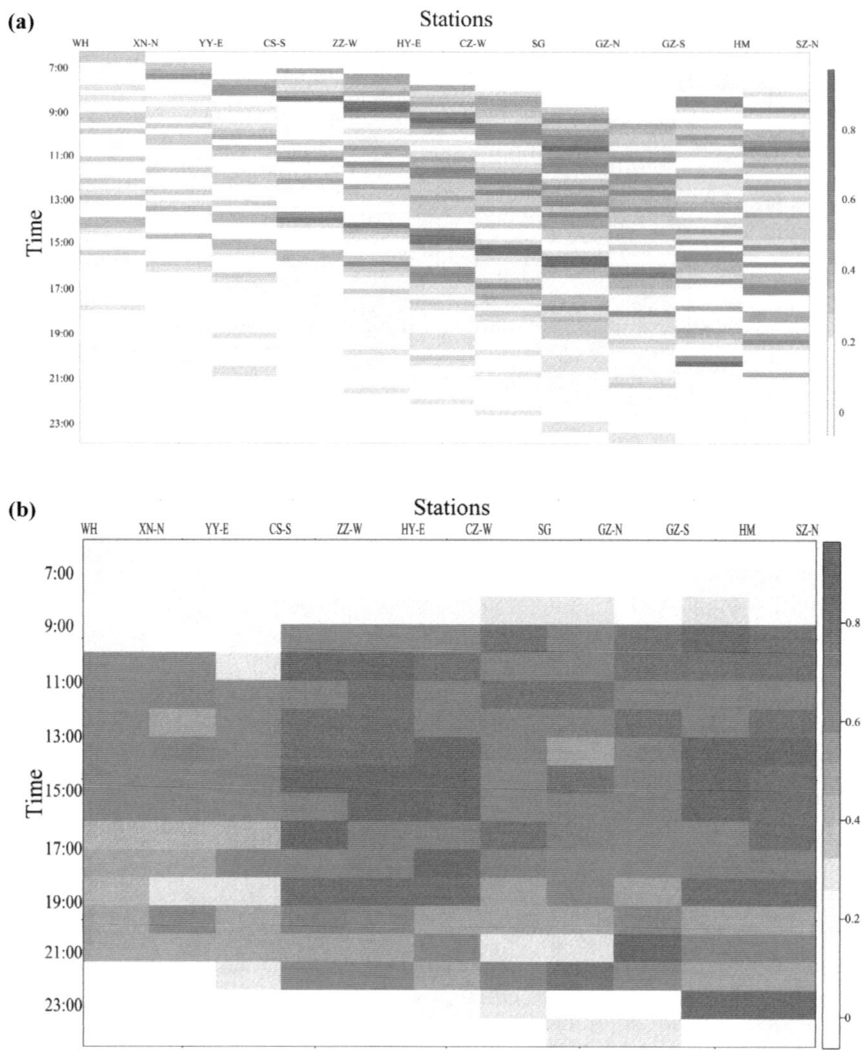

Figure 4. Propagation and contrast between passenger flows and trains on Wuhan–Shenzhen HSR: (*a*) Passenger flows in blocks; (*b*) Number of trains in blocks.

D — matrices of period passenger demands, $d_{s_i,s_{i'},\pi}$ stands for the passenger demand from station s_i to station $s_{i'}$ during time period π, $i' \geq i$, $d_{s_i,s_{i'},\pi} \in D$,

T_b — set of train operating times in different blocks,

T_d — set of train dwell times at different stations,

A — set of operating arcs in blocks and at stations, $a \in A$,

A_b — set of operating arcs in block b, $a_b \in A_b, A_b \subset A$,

A_s — set of train arcs at station s, $a_s \in A_s, A_s \subset A$,

$A_{s,dw}$ — set of train stop arcs at station s, also called dwell arcs (with dwell time) $a_{s,dw} \in A_{s,dw}, A_{s,dw} \subset A_s$,

$A_{s,op}$ — set of train operating arcs (without any operating time, in this case, train does not stop at this station) at station s, $a_{s,op} \in A_{s,op}, A_{s,op} \subset A_s$,

$R_{(s_i,s_{i'})}$ — set of the feasible train routes from station s_i to station $s_{i'}$, $r \in R$,

Or — set of origin stations; trains could only start their operating routes at these stations, $or_q \in Or$,

De — set of destination stations; trains could only end their operating routes at these stations, $de_{q'} \in De$,

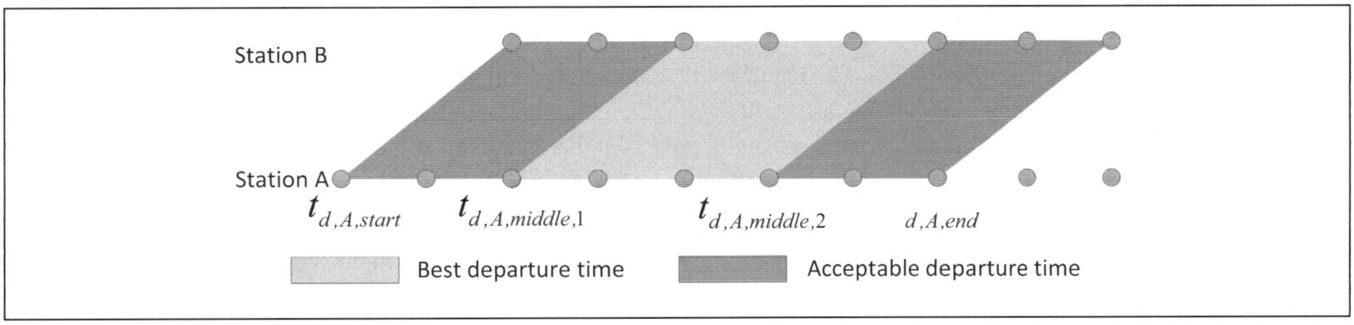

Figure 5. Method of calculation of DTS.

Π set of time divisions. In this paper, $\pi_e \epsilon \Pi$, HSR trains could only depart during the hours from 6:00 to 24:00. Since the passenger demand varies depending on the time, the departure time interval 6:00 to 24:00 should be divided into several parts. Thus, $\pi_e \in \{\pi_1, \pi_2, \ldots, \pi_{18}\}$, $\pi_1 = [6, 7)$,

L_{cap} the maximum seat capacity of a train,

η_{min} the minimum SOR, otherwise the train would be canceled if the SOR was below this minimum,

$o\left(o_{b_j}^{l_k}\right)$ the load of train l_k in block b_j,

t_a the start time of an operating arc a, $\forall t_a$, $\exists \pi_i$ that $t_a \in \pi_i$,

$t_{d,s_i,start}$, $t_{d,s_i,middle,1}$, and $t_{d,s_i,middle,2}$ $t_{d,s_i,end}$ every station is divided into two areas for OD passenger flows: the best departure time area and the acceptable departure time area. Passengers show a higher level of DTS if they are transported in the best area. If passengers are transported out of these two areas, their satisfaction degree equals m, where m is a sufficiently small number. Details are shown in Figure 5.

c_{dts} donates the passenger DTS, which is calculated by Equation 2.

$$c_{dts} = \begin{cases} c_I & t_{d,s_i,middle,1} < t_a < t_{d,s_i,middle,2}, \forall a \in A_b \\ c_{II} & t_a \in [t_{d,s_i,start}, t_{d,s_i,middle,1}] \cup [t_{d,s_i,middle,2}, t_{d,s_i,end}], \forall a \in A_b \\ m & otherwise \end{cases} \tag{2}$$

where $c_{II} < c_I$ and m is a sufficiently small number.

By using $t_{d,s_i,start}$, $t_{d,s_i,middle,1}$, $t_{d,s_i,middle,2}$, $t_{d,s_i,end}$ and c_{dts}, the given hourly OD passengers can be transported more flexibly, earlier or later. For example, passengers who prefer to depart at 09:30 could be transported between 09:00 and 10:00 with the highest DTS (c_I). However, there might be no available train departing during this period, due to the capacity limitation or low seat occupancy rate. In this case, they could accept a departure time between 08:00 and 09:00 or 10:00 and 11:00 with a lower DTS (c_{II}). It should be stated that passenger DTS is based on the period passenger demand.

- **Decision variables:**

$x_{a_s}^{t_a}$ binary variable, $x_{a_s}^{t_a} = 1$ means arc a_s starting at time t_a at station s is occupied, otherwise arc a_s is free to be taken, $\forall a_s \in A_s$,

$y_{a_b}^{t_a}$ binary variable, $y_{a_b}^{t_a} = 1$ means arc a_b starting at time t_a is occupied, otherwise arc a_b is free to be taken, $\forall a_b \in A_b$,

$\delta(l_k) = \{\delta_{s_1}^{l_k}, \delta_{s_2}^{l_k}, \ldots, \delta_{s_i}^{l_k}\}$ the set of trains arc selection configuration at all the stations, $\delta_{s_i}^{l_k} = \sum_{a_{s_i} \in A_{s_i}} x_{a_{s_i}}^{t_a}$, $i = n$, n is the total amount of stations on the network,

$\theta(l_k) = \{\theta_{s_1}^{l_k}, \theta_{s_2}^{l_k}, \ldots, \theta_{s_i}^{l_k}\}$ the set of train stop configurations at all stations, $\theta_{s_i}^{l_k} = 1$ means train l_k would stop at station s_i, $i = n$.

$\varepsilon(l_k) = \{\varepsilon_{b_1}^{l_k}, \varepsilon_{b_2}^{l_k}, \ldots, \varepsilon_{b_j}^{l_k}\}$ the set of trains arc selection configuration in blocks, $\varepsilon_{b_j}^{l_k} = \sum_{a_{b_j} \in A_b} y_{a_{b_j}}^{t_a}$ and the occupied block b_j is the one following station s_i, $j = n - 1$,

$f_{s_i,s_{i'}}^{l_k}$ passenger demand which is satisfied by train l_k from station s_i to station $s_{i'}$, $f_{s_i,s_{i'}}^{l_k} \geq 0$, $i' \geq i$, and define $f_{s_i,s_i}^{l_k} = 0$,

$\alpha_{l_k}^{or_q}$ binary variable, original station selection variable, $\alpha_{l_k}^{or_q} = 1$ means $or_q \in Or$ is the original station for train l_k,

$\alpha_{l_k}^{de_{q'}}$ binary variable, destination station selection variable, $\alpha_{l_k}^{de_{q'}} = 1$ means $de_{q'} \in De$ is the original station for train l_k,

Bi-level Programming Model

A bi-level programming model is introduced where at the upper level the objective is maximizing the passenger departure preference as much as possible. More explicitly, the objective is to maximize the passenger DTS subject to the passenger demand, block capacity, station capacity, and train capacity. For the lower-level model, the objective is to minimize the total train operating time. For the upper-level model, it is obvious that railway operating companies should operate trains as much as possible to satisfy all the passenger departure time demand. However, they also have to reduce their operating costs which is achieved by the lower-level model's objective.

Upper Level. Maximizing the passenger DTS for all OD pairs

$$\text{Max } Z_1 = \sum_{s_i, s_{i'} \in S} \sum_{l_k \in L} c_{dts} \cdot f^{l_k}_{s_i, s_{i'}} \quad (3)$$

Subject to

(a) **Travel demand**. All the demands between OD pairs are satisfied.

$$\sum_{s_{i'} \in S} \sum_{l_k \in L} f^{l_k}_{s_i, s_{i'}} = d_{s_i, s_{i'}, t}, \ \forall s_i \in S \quad (4)$$

(b) **Occupancy uniqueness**. A block arc or a station arc can only be occupied by one train at a time.

$$\sum_{l_k \in L} \varepsilon^{l_k}_{b_j} \leq 1, \ \forall b_i \in B \quad (5)$$

$$\sum_{l_k \in L} \delta^{l_k}_{s_i} \leq 1, \ \forall s_i \in S \quad (6)$$

(c) **Limited numbers of trains depart per period**. Set C_s as the maximum number of departure trains in time division π_i.

$$\sum_{t_a \in \pi_i} x^{t_a}_{a_s} \leq C_s, \forall a \in A_b \quad (7)$$

(d) **Train capacity**. Train seat occupancy in all blocks should be less than the train capacity.

$$\sum_{s_i \in S} \sum_{s_{i'} \in S, i' \geq i} f^{l_k}_{s_i, s_{i'}} \leq L_{\text{cap}}, \forall s_i, s_{i'} \in S, \forall l_k \in L \quad (8)$$

Lower Level. Minimizing total train operating time

$$\text{Min } Z_2 = \sum_{l_k \in L} \sum_{s_i \in S} \sum_{b_j \in B} \left(t_b \cdot \varepsilon^{l_k}_{b_j} + t_d \cdot \theta^{l_k}_{s_i} \right) \quad (9)$$

Subject to

(e) **Lowest seat occupancy rate (SOR)**. Set at η_{min}, theoretically the train SOR should not be less than this lowest rate. A long-distance train would break into shorter distance trains or combine with other trains if its SOR in an intermediate block is less than η_{min}, as defined in (*2*). It should be noticed that HSR trains have specified original and terminal stations, which means train routes could only start and end at specified stations. Thus this constraint might be relaxed if it conflicts with the original and terminal station constraint.

$$\sum_{s_i \in S} \sum_{s_{i'} \in S, i' \geq i} f^{l_k}_{s_i, s_{i'}} \geq \eta_{min} \cdot L_{\text{cap}}, \forall s_i, s_{i'} \in S, \forall l_k \in L \quad (10)$$

(f) **Maximum number of stops**. Set at M_{max}. To ensure the whole stop times of a HSR train, it is not allowed to stop too many times.

$$\sum_{s_i \in S} \theta^{l_k}_{s_i} \leq M_{max}, \forall l_k \in L \quad (11)$$

(g) **Original and terminal station constraint**. HSR trains could only start and end at specified stations. Generally, these stations are where the Multiple Units Operation depots are located.

$$\sum_{or_q \in Or} \alpha^{or_q}_{l_k} = \sum_{de_{q'} \in De} \alpha^{de_{q'}}_{l_k} = 1, \forall l_k \in L \quad (12)$$

$$\sum_{b_i \in B} \varepsilon^{l_k}_{b_j} = q' - q, \forall l_k \in L \quad (13)$$

$$\theta^{l_k}_{s_i} - \alpha^{or_q}_{l_k} < 0, \forall i < q, \forall l_k \in L \quad (14)$$

$$\theta^{l_k}_{s_i} - \alpha^{de_{q'}}_{l_k} \langle 0, \forall j \rangle q', \forall l_k \in L \quad (15)$$

(h) **Headway**. Set at H_{min}.

$$x^{t_{aj}}_{a_s} \cdot t_{a_j} - x^{t_{ai}}_{a_s} \cdot t_{a_i} \geq H_{min}, i < j, \forall i, \forall j, \forall s \in S \quad (16)$$

$$y^{t_{aj}}_{a_b} \cdot t_{a_j} - y^{t_{ai}}_{a_b} \cdot t_{a_i} \geq H_{min}, i < j, \forall i, \forall j, \forall b \in B \quad (17)$$

Finally, the decision variables are bounded as follows. (18)–(19) **Variables constraints**.

$$\delta^{l_k}_{s_i} = \begin{cases} 1, when \sum_{a_s \in A_s} x^{t_a}_{a_s} = 1 \\ 0, otherwise \end{cases} \quad (18)$$

$$\varepsilon^{l_k}_{b_j} = \begin{cases} 1, when \sum_{a_b \in A_b} y^{t_a}_{a_b} = 1 \\ 0, otherwise \end{cases} \quad (19)$$

Methodology

As an IP problem, the LPP has been proved to be NP-hard, see (*17*) and (*4*), so it is computationally restrictive to find the optimized solution of these problems. In this paper, the upper-level model requires more trains to operate to satisfy the passenger demand and DTS while the train operating time in the lower-level model limits the number of operating trains. Hence, our aim is to generate an optimal conflict-free solution to balance the upper- and lower-level models by the iterating train frequency and figure out the stops at intermediate stations. First of all, a candidate train path set is generated according to the passenger demands between every OD pair. Direct trains are operated to satisfy all the passenger demands with highest DTS and without considering the cost and capacity constraints. Similar to a dual problem, this initial solution is optimized, however, it is unfeasible. Then all the constraints are considered to restrict the initial solution until a feasible solution is figured out. By determining the value of t_p (see definition in Sub-method (2)) and a minimum passenger DTS_{min}, the operators could choose to satisfy passengers better or to reduce the train company's operating costs. The iterative algorithm is explained briefly as follows.

Step 1: Initialize and load the input data.
Step 2: Direct trains are operated to satisfy all the passenger demands with highest DTS. Use Sub-Method (1) to generate subset of candidate OD train path $TD^{\pi_e}_{s_i,s_{i'}}$. In this step, trains are operated as much as possible to satisfy all the passenger demands with their highest preferences without considering the dispatching constraints.
Step 3: Use Sub-method (2) to achieve train combinations. Do train combinations until there are no trains matching the combination principle and go to step 4. The combination methods are based on the algorithm which is put forward by (*18*). Trains reaching the maximum train capacity (M_{max}) would quit iteration and be put into set $TD^{\pi_e}_{s_i,s_{i'}}$.
Step 4: Update candidate train path subset $TD^{\pi_e}_{s_i,s_{i'}}$ with $TD^{\pi_e}_{s_i,s_{i'}}$ and $TD^{\pi_e}_{s_i,s_{i'}}$.
Step 5: The other case is that SOR of a train is lower than the minimum SOR in some blocks. Trains in the new candidate train path set should match the lowest SOR constraint. If $o^{l_k}_{b_j}$ in any block is less than $\eta_{min} \cdot L_{cap}$, t_p should be extended to ρ_{i+1} where $\rho_{i+1} > \rho_i$, then the combination is performed especially for these trains until t_p reaches its upper limit. As for the trains which still do not satisfy constraint (5), considering the social benefits of railway operation, constraint (5) is just relaxed for these trains.
Step 6: Calculate Z_1. A minimum passenger DTS is set as DTS_{min} to ensure that there is not a huge loss of DTS when cutting the train operating costs. If $Z_1 < DTS_{min}$ or $\sum x^{t_a}_{a_s} \leq C_s$, reduce ρ_n and go to **step 3**.
Step 7: Return the final train path set $TD^{\pi_e}_{s_i,s_{i'}}$ and the values of Z_1 and Z_2.

Sub-Method (1): Converting Passenger Flows between Each Station to Direct Train Volumes

Set another parameter $o\left(o^{l_k}_{b_j}\right)$ to stand for the load of train l_k in block b_j. It is supposed that every OD passenger demand is served by a direct train without any stops at intermediate stations. Use Equation 20 to convert OD passenger demand matrices to train demand matrices in every block.

$$TD^{\pi_e}_{s_i,s_{i'}} = \frac{d_{s_i,s_{i'},t}}{L_{cap}}, \forall i, i', i' \geq i, \forall \pi_i \in \Pi, d_{s_i,s_{i'},t} \in D \quad (20)$$

where $TD^{\pi_e}_{s_i,s_{i'}}$ denotes the matrices of the number of trains between every OD pair of stations during the period π_e, $l_k \in TD^{\pi_e}_{s_i,s_{i'}}$,

$\lceil\ \rceil$ is the ceiling function.

Sub-Method (2): Train Combination

To minimize the train operation time to reduce the cost (Z_2), it is assumed that train routes should be as long as possible and the number of trains should also be as low as possible. Thus a train combination principle is proposed to satisfy Equation 3 and to minimize Z_2. Initialize train departure time at their original stations. The arrival times for train l_k at its intermediate stations are calculated by the departure time and its operating time in different blocks.

Set w_i, a binary parameter, to identify whether train l_1 and train l_2 depart or arrive at the station s_i at the approximate period t_p (*2*). t_p is set up to ensure that the total DTS does not decrease significantly. The greater t_p is, the lower DTS and operation cost might be. Setting a suitable t_p could balance these two objectives.

$$w_i = \begin{cases} 1 & when\ t^{l_1}_a, t^{l_2}_a \in t_p \forall a \in A \\ 0 & otherwise \end{cases} \quad (21)$$

$\prod_{i=\{l_1,l_2\}}^{u} w_i = 1$ means two trains depart/operate through or arrive at the same station at an approximate time, where $u = \{\{l_1,l_2\},\{l_1,l_3\},\ldots,\{l_{k-1},l_k\}\}$ is a search set of every two trains from $TD^{\pi_e}_{s_i,s_{i'}}$.

There are three situations of train path combination:

First: Train l_1 and train l_2 are short-distance trains and the arrival time of train l_1 at station s_2 is close to the departure time of train l_2, seeing Figure 6*a*. These two trains are then combined.
Second: Train l_1 is a long-distance train and train l_2 is a short-distance train and they have approximate departure time at station s_2. In this case, train l_2 is absorbed by train l_1 and a stop would be set at station s_2 for the new train path, see Figure 6*b*. It should be noted that if train l_2 has a further path after station s_3, train l_1 would inherit it.
Third: Train l_1 and train l_2 are short-distance trains with the same OD and different approximate

departure times. These two trains are combined, see Figure 6c.

Define an operator $\oplus$ as follows (2) and (18):

$$l_{1'} = l_1 \oplus l_2 = \varepsilon(l_1 \oplus l_2) \text{ and } \delta(l_1 \oplus l_2) \text{ and } \theta(l_1 \oplus l_2) \quad (22)$$

$$\delta(l_1 \oplus l_2) = \delta(l_{1'}) = \max\left\{\delta_{s_i}^{l_1}, \delta_{s_i}^{l_2}\right\}, \forall s_i \in S \quad (23)$$

where $\varepsilon(l_1 \oplus l_2)$and $\delta(l_1 \oplus l_2)$ mean combining two train routes in blocks and stations

$\theta(l_1 \oplus l_2)$ means combining the stop plans of two trains and setting a new stop at their combining station

For example, these three situations mentioned above, there are:

First: l_1: $\delta(l_1) = (1,1,0)$, $\theta(l_1) = (1,1,0)$ and $\varepsilon(l_1) = (1,0)$; l_2: $\delta(l_2) = (0,1,1)$, $\theta(l_2) = (0,1,1)$ and $\varepsilon(l_2) = (0,1)$; l_2: $\delta(l_{1'}) = (1,1,1)$, $\theta(l_{1'}) = (1,1,1)$ and $\varepsilon(l_{1'}) = (1,1)$.
Second: l_1: $\delta(l_1) = (1,1,1)$, $\theta(l_1) = (1,1,1)$ and $\varepsilon(l_1) = (1,1)$; l_2: $\delta(l_2) = (0,1,1)$, $\theta(l_2) = (0,1,1)$ and $\varepsilon(l_2) = (0,1)$; $l_{1'} = l_1 \oplus l_2$: $\delta(l_{1'}) = (1,1,1)$, $\theta(l_{1'}) = (1,1,1)$ and $\varepsilon(l_{1'}) = (1,1)$.
Third: l_1: $\delta(l_1) = (1,1,0)$, $\theta(l_1) = (1,1,0)$ and $\varepsilon(l_1) = (1,0)$; l_2: $\delta(l_2) = (1,1,0)$, $\theta(l_2) = (1,1,0)$ and $\varepsilon(l_2) = (1,0)$; $l_{1'} = l_1 \oplus l_2$: $\delta(l_{1'}) = (1,1,0)$, $\theta(l_{1'}) = (1,1,0)$ and $\varepsilon(l_{1'}) = (1,0)$.

Combination steps are explained as follows:

1: let $t_p = \rho_i$ then **do** $l_1 \oplus l_2$
2: **if** $\exists o_{b_j}^{l_1'} > L_{\text{cap}}$ **do** delete $\{l_1, l_2\}$ from the traversal search set u, update u and go to step 1
//after the combination, if the new train is overloaded, break it back to two original trains
3: **if** $\sum_{s_i \in S} \theta_{s_i}^{l_{1'}} \leq M_{max}$ //the stop amount of new train $l_{1'}$ is no more than M_{max}
4: **if** $\sum_{s_i \in S} \theta_{s_i}^{l_{1'}} < M_{max}$
5: **do**
6: delete l_1 and l_2 from $TD_{s_i, s_{i'}}^{\pi_e}$, add $l_{1'}$ to $TD_{s_i, s_{i'}}^{\pi_e}$
7: **return** $TD_{s_i, s_{i'}}^{\pi_e}$ and update u
8: **else** $\sum_{s_i \in S} \theta_{s_i}^{l_{1'}} = M_{max}$ **do**
9: delete l_1 and l_2 from $TD_{s_i, s_{i'}}^{\pi_e}$, add $l_{1'}$ to $TD_{s_i, s_{i'}}^{\pi_e}$
10: **return** $TD_{s_i, s_{i'}}^{\pi_e}$ and $TD_{s_i, s_{i'}}^{\pi_e}$ and update u
/*if a new train reaches the maximum number of stops, put it to a new set and do not combine it any more*/
11: **else break**

Numerical Experiments: Shenzhen–Changsha HSR Corridor

To illustrate the effectiveness of the proposed methods, the Shenzhen–Changsha HSR corridor (which includes Shenzhen–Guangzhou HSR and part of Guangzhou–Wuhan HSR, as shown in Figure 3) with nine stations and eight blocks is selected as the case study in our numerical experiment. Since HSR passenger trains are usually operated in pairs, only one direction (Shenzhen to Changsha) is taken into account. For convenience of expression, S1 (Shenzhen North) to S9 (Changsha South) are used to name these stations and B1 to B8 to name the blocks between them. Trains could only start and end their operation at the original and terminal stations, which are Shenzhen North (S1), Guangzhou South (S3), Hengyang East (S7) and Changsha South (S9). In other words, Or = {S1, S3, S7}while De = {S3, S7, S9}.

Input

(a) Hourly passenger demands between each OD pair, as provided in Table 1. Experimental data are offered by Guangzhou Railway Corporation. Due to the confidentiality agreement, data are partially adjusted. For the variation of DTS according to time periods, it is supposed that:
  - If a passenger is served at the best period, set $c_{dts} = 1$,
  - If a passenger is served an hour before or after the best period, set $c_{dts} = 0.5$,
  - Otherwise, $c_{dts} = 0.1$.

For example, when $\pi_i = \pi_7$, which is 12:00–13:00, the passenger demand from S1 to S3 is 2116. The DTS would be 1*2116 if these 2116 passengers are transported during 12:00 to 13:00; DTS equals 0.5*2116 if they are served during 11:00 to 12:00 and 13:00 to 14:00, otherwise, DTS equals 0.1*2116. Certainly, these 2116 passenger demands would be served separately by different trains, the actual DTS is determined by their real departure times.

(b) Train block running times (actual average running times) are $T_b = \{16, 16, 23, 31, 30, 32, 26, 18\}$ min from B1 to B8.
(c) Train capacity is an average of 536 seats per train, $L_{\text{cap}} = 536$. It should be noticed that a 536-seat HSR train is made up of eight units. There is another type of multiple-unit train with 16 carriages. However, on the Changsha–Shenzhen HSRs, most trains are composed of eight units. For simplicity, the 16-unit trains are not considered.

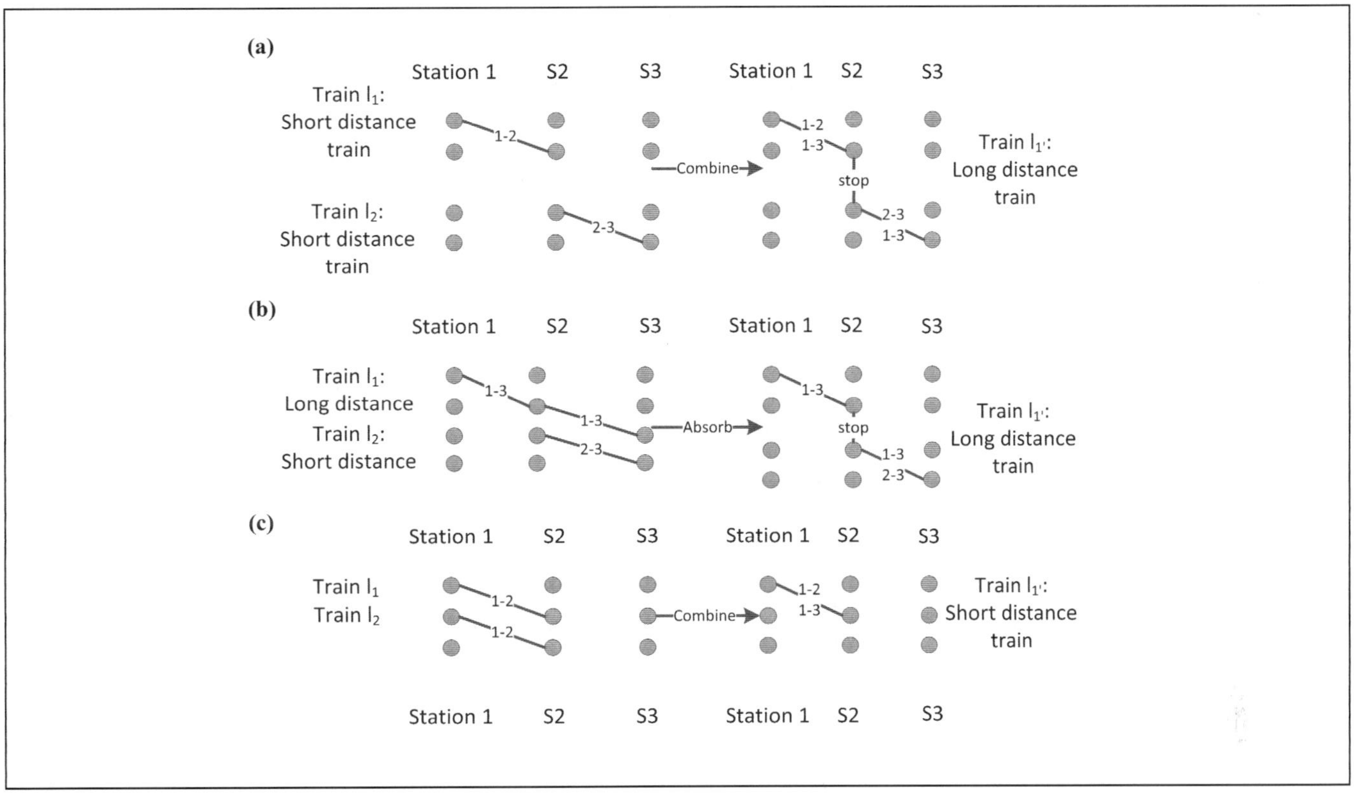

Figure 6. Examples of train combination situation: (*a*) Short-train path combination; (*b*) Long-train path absorbing; (*c*) Short-train path absorbing.

Table 1. Hourly Passenger Demand: 12:00–13:00

O/D	S1	S2	S3	S4	S5	S6	S7	S8	S9
S1	–	204	2116	0	98	206	0	0	353
S2		–	207	44	46	26	39	0	37
S3			–	58	357	178	138	204	533
S4				–	0	0	0	0	0
S5					–	17	30	5	15
S6						–	58	35	417
S7							–	0	302
S8								–	55
S9									–

(d) The departure, arrival, and operating headway is 5 min. Thus the maximum capacity for stations and blocks is 12 trains per hour, $C_s = 12$.

(e) The maximum number of stops is set at eight, including the original and terminal station stops, $M_{max} = 8$, and the minimum dwell time for trains is 3 min at intermediate stations and 5 min at original/terminal stations. Since Guangzhou South is a hub station, all trains must to stop at this station if they pass through it.

Output

The result of the stop plan and the departure time selection is presented in Figure 7, where the total train running time is 16,536 minutes compared with 20,288 minutes for the original corridor train operating plans, and the passenger DTS rate is 82.3% (DTS rate equals 100% if every passenger is served in time). A solid circle in the figure means this train stops at this station while red represents the original/terminal stations. It should be noticed that the time scale only stands for the departure time of trains at their original stations. These results support the effectiveness of the proposed model and methodology.

Conclusion

This paper has presented a bi-level model for optimizing the line planning design of HSRs that takes passengers'

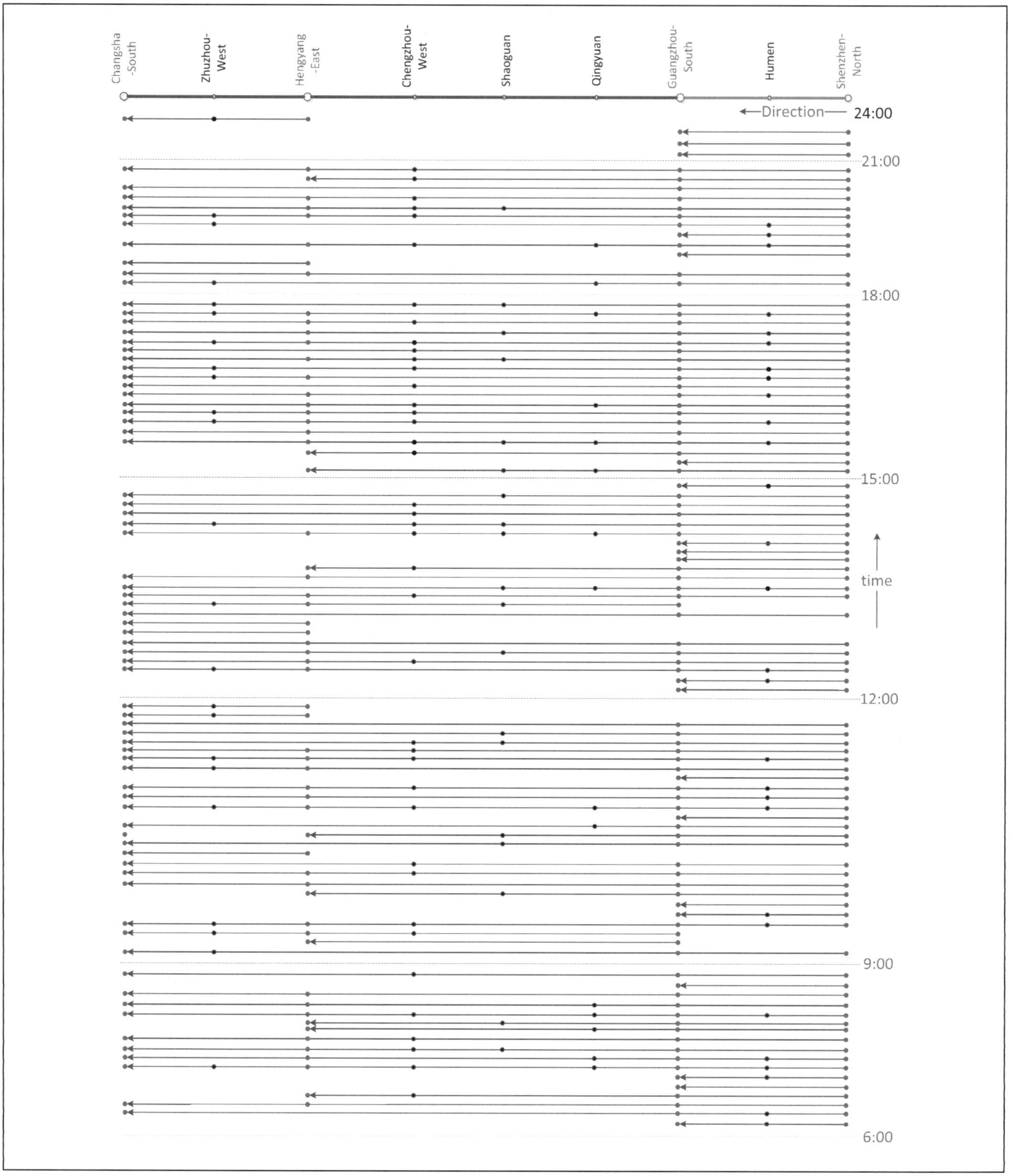

Figure 7. Numerical experiment results: daily stop plan and departure time selection on Shenzhen–Changsha HSR.

spatial-temporal travel preferences and train operating companies' concerns about the operating costs of lines into account. Based on the analysis of actual train operation data, obtained from Guangzhou Railway Corporation, an hourly OD passenger allocation is done while introducing passenger DTS. Additional parameters are proposed to produce more flexible passenger OD demands. A methodology is then a developed to cope with the complexity of the presented model. The results obtained from a case study support the finding that the proposed approach can achieve a high degree of passenger satisfaction while decreasing operational costs.

Acknowledgments

This work was supported by the National Nature Science Foundation of China [grant no. 61503311] and National Key R&D Plan [grant no. 2017YFB1200701]. We acknowledge the support of the Railways Technology Development Plan of China Railway Corporation [grant no. 2016X008-J]. Parts of the work were supported by the State Key Lab of Railway Control and Safety Open Topics Fund [grant no. RCS2017K008]. We are grateful for the contributions made by our project partners.

Author Contributions

The authors confirm contribution to the paper as follows: study conception and design: Yuxiang YANG, Jie LI; data collection: Qiyuan PENG, Ping HUANG; analysis and interpretation of results: Yuxiang YANG, Jie LI, Chao WEN; draft manuscript preparation: Yuxiang YANG, Javad LESSAN. All authors reviewed the results and approved the final version of the manuscript.

References

1. Peng, Q., J. Li, Y. Yang, and C. Wen. Influences of the High-Speed Railway Construction and Operation in Railway Traffic Organization. *Journal of Southwest Jiaotong University*, Vol. 51, No. 3, 2016, pp. 1–9.
2. Li, J., Y. Yang, Q. Peng, C. Wen, H. Yu, and P. Huang. Efficient Assignment of Long-and Short-Distance Trains Based on Spatial-Time Passenger Travel Intention. Presented at 96th Annual Meeting of the Transportation Research Board, Washington, D.C., 2017.
3. Wen, C., Z. Li, Y. Tang, H. Ping, Y. Yang and J. Lessan. Statistics of HSR Primary Delay based on Real-World Records: Case of Wuhan-Guangzhou. *Presented at RailLille 2017: 7th International Conference on Railway Operations Modelling and Analysis*, International Association of Railway Operations Research (IAROR), 1450-1469, Lille, France, 2017.
4. Claessens, M. T., N. M. van Dijk, and P. J. Zwaneveld. Cost Optimal Allocation of Rail Passenger Lines. *European Journal of Operational Research*, Vol. 110, No. 3, 1998, pp. 474 489. https://doi.org/10.1016/S0377-2217(97)00271-3.
5. Bussieck, M. *Optimal Lines in Public Rail Transport*. PhD thesis. Technische Universitaet Braunschweig, 1998.
6. Schöbel, A., and S. Scholl. Line Planning with Minimal Traveling Time. In *ATMOS 2005 – 5th Workshop on Algorithmic Methods and Models for Optimization of Railways*. Internationales Begegnungs-und Forschungszentrum für Informatik (IBFI), Schloss Dagstuhl, 2006.
7. Schoebel, A. Line Planning in Public Transportation: Models and Methods. *OR Spectrum*, Vol. 34, No. 3, 2012, pp. 1–20.
8. Borndörfer, R., M. Grötschel, and M. E. Pfetsch. A Column-Generation Approach to Line Planning in Public Transport. *Transportation Science*, Vol. 41, No. 1, 2007, pp. 123–132. https://doi.org/10.1287/trsc.1060.0161
9. Goossens, J. W., S. Van Hoesel, and L. Kroon. A Branch-and-Cut Approach for Solving Railway Line-Planning Problems. *Transportation Science*, Vol. 38, No. 3, 2004, pp. 379–393.
10. Goossens, J .W., S. Van Hoesel, and L. Kroon. On Solving Multi-Type Railway Line Planning Problems. *European Journal of Operational Research*, Vol. 168, No. 2, 2006, pp. 403–424. https://doi.org/10.1016/j.ejor.2004.04.036
11. Bussieck, M. R., P. Kreuzer, and U. T. Zimmermann. Optimal Lines for Railway Systems. *European Journal of Operational Research*, Vol. 96, No. 1, 1997, pp. 54–63. https://doi.org/10.1016/0377-2217(95)00367-3
12. Kaspi, M., and T. Raviv. Service-Oriented Line Planning and Timetabling for Passenger Trains. *Transportation Science*, Vol. 47, No. 3, 2013, pp. 295–311.
13. Fu, H., L. Nie, L. Meng, B. R. Sperry, and Z. He. A Hierarchical Line Planning Approach for a Large-Scale High Speed Rail Network: The China Case. *Transportation Research Part A: Policy and Practice*, Vol. 75, 2015, pp. 61–83. https://doi.org/10.1016/j.tra.2015.03.013
14. Brethomé, L., C. Bouvet-Agnelli, R. Chevrier, et al. Optimization of a passenger railway transportation plan considering mobility flows and service quality. *Presented at RailLille 2017: 7th International Conference on Railway Operations Modelling and Analysis*, International Association of Railway Operations Research (IAROR), 824-835, Lille, France, 2017.
15. He, B., R. Song, S. He, and Y. Xu. High-Speed Rail Train Timetabling Problem: A Time-Space Network Based Method with an Improved Branch-and-Price Algorithm. *Mathematical Problems in Engineering*, Vol. 2014, No. 1, 2014, pp. 1–15.
16. Scholl, S. *Customer-Oriented Line Planning*. PhD thesis. Technische Universität Kaiserslautern, 2005.
17. Michael, R. G., and S. J. David. *Computers and Intractability: A Guide to the Theory of NP-Completeness*. W.H. Freeman and Company, San Francisco, 1979, pp. 90–91.
18. Riejos, F. A. O., E. Barrena, J. D. C. Ortiz, and G. Laporte. Analyzing the Theoretical Capacity of Railway Networks with a Radial-Backbone Topology. *Transportation Research Part A: Policy and Practice*, Vol. 84, 2016, pp. 83–92. http://dx.doi.org/10.1016/j.tra.2015.03.018

The Standing Committee on Intercity Passenger Rail (AR01) peer-reviewed this paper (18-01367).

Article

TRR
JOURNAL OF THE TRANSPORTATION RESEARCH BOARD

Transportation Research Record
2018, Vol. 2672(10) 236–246

DOI: 10.1177/0361198118793494
journals.sagepub.com/home/trr
SAGE

Improving Intercity Passenger Rail Planning using Evidence from Passenger Survey Data

Benjamin R. Sperry[1] **and Tyler Collins**[1]

Abstract

Federal mandates require that states develop comprehensive state rail plans that clarify a vision for the role of intercity passenger rail services in the state's transportation system and a program of projects necessary to achieve that vision. In that context, this paper presents an analysis of more than 10,000 surveys obtained during the past 10 years from passengers of five U.S. state-supported intercity passenger rail routes. The purpose of the analysis is to support the development of state rail plans by gaining valuable insight into passenger behavior, the market area for passenger rail services, and the broader mobility impacts of passenger rail services. The market area analysis found that approximately three-quarters of rail passengers live within 30 mi of a rail station, affirming the use of this distance as a radius for planning purposes. The mobility analysis indicated that automobile is the primary alternative to passenger rail services among passengers of the five routes; statistical models of alternative mode preference yielded significant insight into strategies that states can implement to increase the mode shift of automobile and airplane passengers to rail services. The findings and recommendations of this paper can be used by planners and policymakers to implement a practical yet data-driven approach to the development of passenger-related components of state rail plans, providing a clear linkage between investment strategies and statewide mobility goals.

The Passenger Rail Investment and Improvement Act of 2008 (PRIIA) resulted in several new planning and policy initiatives for the U.S. intercity passenger rail community; one such initiative was the new requirement that State Departments of Transportation (formally, "State Rail Transportation Authorities" or SRTAs) develop and maintain formal state rail plans (*1*). The Federal Railroad Administration (FRA), in its September 2013 *State Rail Plan Guidance* document, outlined a standardized format, minimum content requirements, and procedures for the preparation of state rail plans to ensure compliance with the requirements set forth in Section 303 of PRIIA (*2*). Briefly, state rail plans should include a comprehensive description and assessment of the state's railroad system (i.e., inventory of facilities and initiatives), analysis of the role of rail transportation within the state's transportation system (i.e., visioning), and a list of projects necessary to achieve the vision for rail in the state (i.e., program development). Creation of a comprehensive state rail plan plays an essential role in the establishment of new intercity passenger rail routes or improvements to existing state-supported services (*3*).

The requirement that states develop a comprehensive rail plan to include both passenger and freight presents a unique and significant opportunity for the development of intercity passenger rail in the U.S. Specifically, the new requirements represent the first Federal mandate for states to adopt formal plans that clarify a vision for passenger rail service in the state and identify projects that will help achieve this vision. As SRTAs move forward with the creation of passenger elements within their state rail plans, evidence on the broader impacts of intercity passenger rail service is needed to support a practical and data-driven approach to development of rail plan content. Whereas operational data (i.e., ridership and revenue) are available for existing routes (and projected for planned routes), insight into the characteristics and preferences of rail passengers can only be obtained from personal surveys of passengers utilizing existing services.

This paper presents an analysis of more than 10,000 surveys obtained during the past 10 years from passengers of five state-supported intercity passenger rail routes

[1]Department of Civil Engineering, Russ College of Engineering and Technology, Ohio University, Athens, OH

Corresponding Author:
Address correspondence to Benjamin R. Sperry: sperry@ohio.edu

in the U.S. These surveys provide valuable insight into the market area for passenger rail services as well as the broader impacts of such services on progress toward state-level goals associated with mobility and congestion reduction. The remainder of this paper is organized as follows. Following this introduction, a brief summary of the relevant FRA requirements for the content of state rail plans and the specific questions to be analyzed in this paper is presented. Next, the details of the survey data sets analyzed in this paper are provided. Analysis of the survey data is then presented in two sections, focused on the market area for intercity passenger rail and the mobility impacts of rail service. The conclusions section presents a discussion of the research findings and recommendations for improving the development of state rail plans.

State Rail Plan Requirements

Summary of FRA Rail Plan Guidelines

In accordance with the requirements outlined in Section 303 of PRIIA, state rail plans must include at least the following elements (*2*):

- Description of the role of rail in the statewide transportation system
- Description and inventory of the state's passenger and freight rail transportation facilities
- Proposed improvements and investments for passenger and freight rail
- The state's Rail Service and Investment Program (RSIP), which describes specific projects and other elements necessary to achieve the vision for rail transportation in the state in the short term (4 years) and long term (20 years)
- Public involvement, interagency coordination, and review

As part of the description of the role of rail in statewide transportation, the state rail plan must include a conceptual analysis of rail transportation's role within the state's transportation system (49 U.S.C. § 22705[a][1]). To achieve this analysis, FRA requires states to describe mode share for intercity passenger and commuter rail operations, if such services exist. The FRA also allows states to analyze measures focused on accessibility, as follows (*2*, p. 32):

> Optionally, States can choose to describe the population with access to passenger rail service. FRA's preferred methodology to determine this population is to determine population of Census tracts within a 10- and 30-mile radius of rail stations. (FRA encourages States to investigate other measures of accessibility, including the population within a 30-minute drive of a rail station and the population within a 30-minute transit ride of a rail station).

Additionally, as part of the description and inventory portion of the state rail plan, states are required to include the following content (49 U.S.C. § 22705[a][4]):

> A general analysis of rail transportation's economic and environmental impacts in the State including, but not limited to, congestion mitigation, safety impacts including the benefit of freight rail compared to freight on public highways, trade and economic development, air quality, land use, energy use, resiliency to climate change impacts, and community impacts.

To achieve this portion of the analysis, FRA suggests a wide range of measures, including an analysis of the "Vehicle Miles of Travel (VMT) avoided" by the continued provision of passenger railroad services in the state (*2*, p. 34). States must also present a complete analysis of the RSIP's projected effect on the transportation system, the environment, and the economic effects on both the public and private sectors. Proposed passenger rail investments must include a project-level analysis of the following (*2*; p. 24):

> Projected ridership, passenger miles traveled, modal diversion from highway and air travel, revenue, and operating expenses associated with existing, 4 and 20-year passenger rail services in aggregate and broken down by commuter, intercity and high-speed rail projects.

For both inventory and RSIP evaluation of state-level passenger rail investments, the extent to which such investments currently or are anticipated to reduce congestion on other modes (particularly highway and airplane modes) contributes significantly to the assessment of the broader benefits of such investments.

Research Questions

In the broader context of the FRA state rail plan guidelines discussed in the previous section, the analysis of passenger survey data in this paper examines two key questions. First, the FRA allows states to characterize the existing or proposed role of passenger rail in the state by assessing the population with access to rail service, defined as the population within a distance of 10 and 30 mi of rail stations. The underlying assumption with this guidance is that the population within 10 and 30 mi of a rail station encompasses a meaningful amount of potential ridership such that informed planning decisions can be made. However, the extent to which the FRA-preferred radii (commonly referred to as the "catchment" or "market" area of a station) actually represent the actual market area of stations is not known. Second, the FRA requires that states consider the potential for congestion reduction resulting from passenger rail investments through an analysis of mode diversion from highway and air travel, with an estimate of the VMT avoided also being helpful to assess the impacts of rail investments on the transportation system. However, the extent that intercity passenger rail systems divert traffic from other modes of travel is not generally provided at the planning level, except for the most complex high-speed rail projects (e.g., *4–7*). Consequently, evidence on mode diversion from existing rail services, as well as the passenger and route-level characteristics that influence mode diversion, can fill the data gap and would be useful for planning and policy development activities.

Rail Passenger Survey Data Sources

Description of Routes

The analysis presented in this paper utilizes data from more than 10,000 surveys of rail passengers using five state-supported Amtrak intercity passenger rail routes in the Midwest and South Central regions of the U.S. Details of the survey data sources and a brief preliminary analysis are presented in this section.

Hiawatha Service. The Hiawatha Service route runs between Milwaukee, Wisconsin, and Chicago, Illinois. The route is 86 mi in length and the service frequency is seven round-trips per day (six on Sundays). The Hiawatha Service is supported by the Wisconsin and Illinois Departments of Transportation, which jointly manage and finance the route's operations (*8*). Surveys of Hiawatha Service passengers were conducted on all trains on a typical Thursday and Saturday during the spring of 2011 and 2016. Additional details of the survey data collection and analysis can be found in the original technical report for the respective survey wave (*9–10*).

Michigan Routes. The State of Michigan's state-supported Amtrak route network consists of three routes connecting Chicago, Illinois, with destinations in Michigan, as follows (*8*):

- Wolverine, a 304-mi route serving Kalamazoo, Battle Creek, Ann Arbor, Dearborn, Detroit, Pontiac, and intermediate stations
- Blue Water, a 319-mi route serving Kalamazoo, Battle Creek, East Lansing, Flint, Port Huron, and intermediate stations
- Pere Marquette, a 176-mi route serving Grand Rapids and intermediate stations

Service frequency on the Wolverine route is three round-trips per day whereas the other two Michigan routes have a single daily round-trip each (*8*). Surveys were conducted on all three routes in spring 2007 (*11*) and again in spring 2011 (*12*); additional details of the survey data collection and analysis can be found in the technical reports for each survey.

Heartland Flyer. The Heartland Flyer route operates along a 206-mi stretch between Oklahoma City, Oklahoma, and Fort Worth, Texas. Operations of the single daily round-trip along this route are financed by the Oklahoma and Texas Departments of Transportation (*8*). Surveys of Heartland Flyer passengers were conducted in spring and summer of 2009; additional details of the data collection and analysis can be found in the technical report of the survey project (*13*).

Passenger Characteristics Summary

The nine passenger survey administrations described in the previous section have each been thoroughly analyzed individually in their respective technical reports (*9–13*) as well as other publications offering additional insight into unique aspects such as economic development (*14, 15*) or air–rail transfer (*16*). The passenger surveys contained many similar question topics, including travel behavior (e.g., trip purpose, frequency of travel, alternative travel mode) and personal information (e.g., home ZIP code, socioeconomic characteristics). Following a thorough quality control review to ensure consistency in the response measurement across surveys, the common data elements of each data set were combined into a single master database. The assembled database, which contained a total of 10,661 data points, represents one of the most extensive collections of data describing the characteristics and preferences of U.S. rail passengers.

A summary of the passenger survey data sets examined in this analysis is provided in Table 1. Details provided in Table 1 include information about the survey sample size and response rate as well as the distribution of key trip-related and passenger demographic profile characteristics. From the data presented in Table 1, the following trends are noted:

- The most common purposes of travel among rail passengers are "visiting family or friends" and "leisure/entertainment/vacation" although some business and work commute travel is present on these routes.
- Availability of vehicles within the passenger households is strong, with less than 15% of passengers being from zero-vehicle households.
- The age distribution is fairly even across the nine surveys, with the exception of the Heartland Flyer route which has a noticeably higher share of elderly passengers.

Table 1. Summary of Rail Passenger Survey Data Sets and Selected Passenger Characteristics

Data attribute	Hiawatha Service		Wolverine		Blue Water		Pere Marquette		Heartland Flyer
Survey information									
Survey year	2016	2011	2011	2007	2011	2007	2011	2007	2009
Sample size	2,464	2,298	1,354	1,626	512	541	497	346	1,023
Response rate (%)	59.0	57.6	42.7	NA	30.1	NA	47.2	NA	50.2
Survey reference	(*10*)	(*9*)	(*12*)	(*11*)	(*12*)	(*11*)	(*12*)	(*11*)	(*13*)
Passenger trip purpose									
Work commute (%)	26	20	3	3	2	2	3	3	N/A
Business trip (%)	16	16	12	3	13	5	7	5	3
Visiting family or friends (%)	29	30	47	37	30	43	33	45	39
Leisure/entertainment/ vacation (%)	21	25	25	41	43	30	53	31	50
Education (%)	4	4	6	11	6	13	1	11	1
Other personal business (%)	3	5	6	5	6	7	3	5	6
Household vehicles									
Average household vehicles	1.79	1.90	1.77	1.86	1.95	1.71	2.04	1.68	2.07
Zero-vehicle households (%)	13	9	11	11	6	18	4	14	4
Age group									
18 to 34 years (%)	37	40	44	52	46	52	44	51	24
35 to 64 years (%)	53	53	46	40	44	40	51	43	56
65 years and older (%)	10	6	10	8	9	8	5	7	21
Annual household income									
Less than $20,000 (%)	8	10	15	14	18	23	8	18	13
$20,000 to $50,000 (%)	15	19	27	21	21	23	23	29	28
$50,000 to $100,000 (%)	27	31	30	34	35	33	42	29	37
$100,000 to $200,000 (%)	32	27	28	31	26	21	27	23	22
$200,000 or more (%)	18	14	NA	NA	NA	NA	NA	NA	NA
Home distance to rail line									
Median distance (mi)	6.7	7.9	10.4	8.9	14.5	7.4	20.0	8.1	18.2
Less than 10 mi (%)	63	59	48	52	42	58	35	54	35
10 to 30 mi (%)	21	20	32	30	21	16	24	24	27
Greater than 30 mi (%)	16	20	20	18	37	26	40	22	38

Note: NA = item not included on passenger survey or data not available. Columns may not total 100% because of rounding.

- The distribution of household income is also fairly even across the nine surveys, with between 10% and 20% of passengers being from low-income households.
- The median distance traveled by passengers to access the rail service is shorter for the routes with multiple daily frequencies whereas the single daily round-trip services attract passengers from relatively longer distances away from the rail line. This result is most likely attributed to the routes with multiple daily frequencies serving higher-population urban centers where more residents live closer to station locations.

Market Area Analysis

Analysis Methods

This section presents an analysis of the market area or catchment area for intercity passenger rail services based on the data from the passenger surveys. For the purposes of this analysis, the market area for intercity passenger rail services is defined as the spatial distribution of passengers relative to stations along the route. From each passenger survey, three geographic locations and the corresponding latitude and longitude were identified: 1) the boarding station, 2) the alighting station, and 3) the passenger's home ZIP code. Given this information, geographic information system (GIS) tools (ArcGIS 10.4) were used to calculate a "straight line" distance between each of the two stations used by the passenger and the centroid of the passenger's home zip code tabulation area (ZCTA, a U.S. Census Bureau geography definition that corresponds to ZIP codes used by the U.S. Postal Service). The smaller of the two distances was then assumed to be the minimum distance traveled by the passenger to access the passenger rail services. Most surveys of this type do not ascertain the exact address of the passenger's home and the exact location of the passenger's origin and destination before the train trip are also not available with certainty. Therefore, defining the access distance in terms of the distance between the home ZCTA and the closest station used on the day of the survey provides sufficient detail for this analysis while also balancing privacy concerns in survey implementation.

Using the latitude and longitude of the passenger's home ZCTA centroid and the closest rail station used on the day of the survey, three metrics defining the market area were estimated:

1) Straight line access distance, in miles, measured in a straight line between the ZCTA centroid and the closest station without regard for the actual travel paths available
2) Network access distance, in miles, measured as the distance between the ZCTA centroid and the closest station along the existing roadway network
3) Network access travel time, in minutes, measured as the travel time between the ZCTA centroid and the closest station along the existing roadway network

The straight line access distance was obtained as described in the previous paragraph. The network distance and travel time information were obtained using a Google Maps batch-processing tool known as Friendly Batch Routing (FBR), which allows for the distance and travel time between two locations for many thousands of records to be extracted in an efficient manner (*17*). The FBR tool uses the "best guess" travel time conditions in Google Maps, representing a historical average travel time for each part of the route. Using data from a commonly used mapping software such as Google Maps also benefits the analysis by using data on network conditions that are actually used by travelers to make decisions.

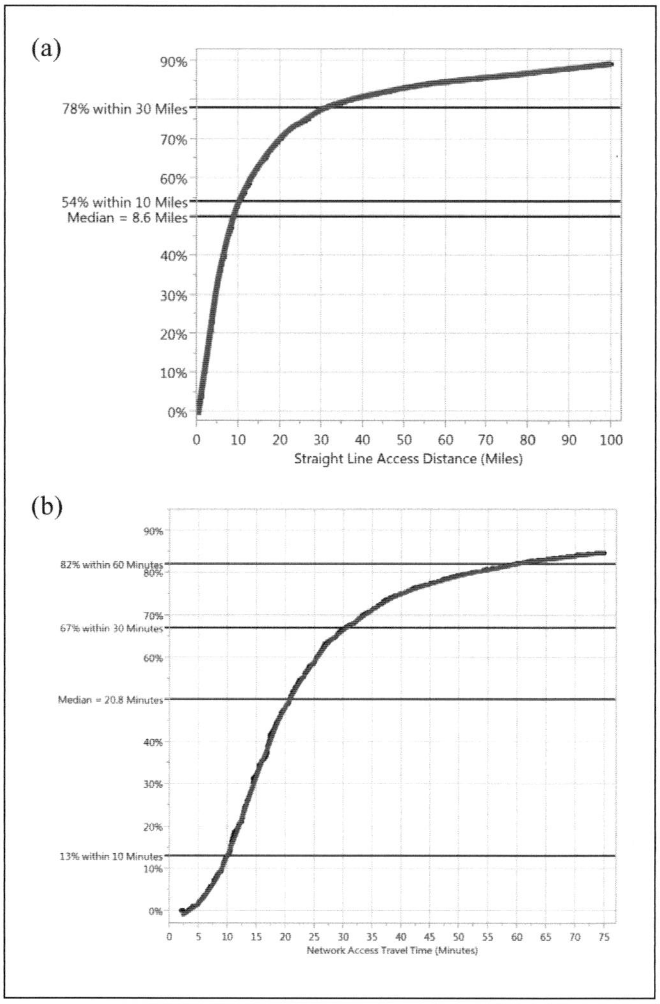

Figure 1. Distribution of rail passenger access distance and travel time, showing (*a*) straight line access distance cumulative distribution (n = 9,336 < 100 mi) and (*b*) network access travel time cumulative distribution (n = 8,892 < 75 min).

Passenger Access Distance and Travel Time Distribution

Using the process described in the previous section, the three market area metrics were calculated for each passenger survey with valid data on boarding station, alighting station, and home ZIP code. Figure 1*a* shows the cumulative distribution of straight line access distance and Figure 1*b* the network access travel time for passengers. For the sake of clarity, the horizontal axis of Figure 1*a* is constrained to surveys with a straight line access distance of less than 100 mi; even with this constraint, data from more than 9,300 passenger survey records are used. Similarly, Figure 1*b* is constrained to surveys with a network access travel time of less than 75 min, with data from more than 8,800 surveys used.

A closer examination of the cumulative distribution curves in Figure 1 provides details on the size of the market area for intercity passenger rail stations. As shown in Figure 1*a*, the median straight line access distance is approximately 8.6 mi, with approximately 54% of passengers living within 10 mi of the nearest station. A market area radius of 30 mi encompasses approximately 78% of rail passengers. The travel time distribution (Figure 1*b*) shows that only 13% of passengers live within a 10-min drive of the nearest station whereas approximately two-thirds of passengers are within a 30-min drive. The median travel time to access intercity passenger rail services was approximately 21 min.

Implications of Straight Line versus Network Distance

Figure 2 presents a scatter plot comparing the straight line distance and the network access distance for more than 9,300 passenger surveys with a straight line distance

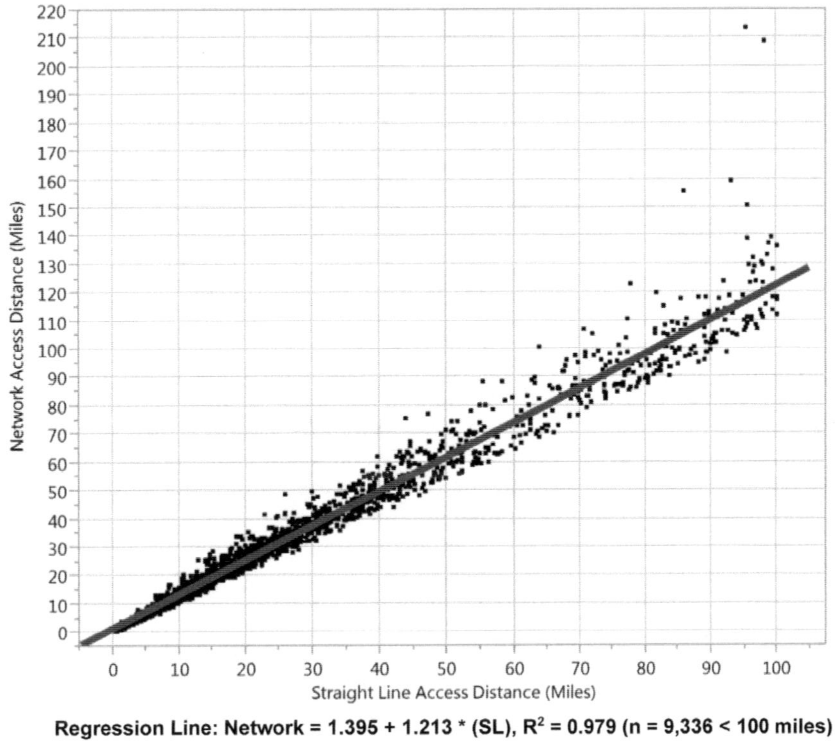

Figure 2. Comparison of straight line and network access distance.

of less than 100 mi. A linear regression line predicting network distance as a function of straight line distance is also provided below the scatter plot in Figure 2.

The relationship between the straight line access distance and the access distance based on the actual roadway network has important implications for intercity passenger rail planning. Although it is obvious that the actual distance traveled along the roadway network will be greater than the straight line distance, the trend line displayed in Figure 2 suggests that, on average, the roadway network access distance is approximately 21% greater than the straight line access distance. On average, passengers who live in a location that is approximately 30 mi from a passenger rail station will need to travel approximately 38 mi to access the rail services.

Mobility Analysis

Passenger Alternatives to Rail

Most intercity rail passenger surveys, including the five surveys analyzed in this paper, include a question about the mode of travel that might be used by the passenger if the rail service was not available. Typically, the survey response options will include several alternative travel modes available along the intercity corridor, plus the option of "would not make trip" if the rail service was not available. Responses to this question provide insight into the role of passenger rail in supporting personal mobility in intercity corridors, and address the broader policy question of the role of passenger rail services in statewide transportation networks.

Table 2 presents a summary of the distribution of passenger alternative travel modes for each of the five routes analyzed in this paper. Four alternatives to passenger rail are considered in this analysis: personal vehicle (including rental car), intercity bus/local transit, commercial airline, and the "would not make trip" option. For this paper, intercity bus and local transit alternatives were combined because of local transit only being an option for certain types of trips within intercity corridors. The 2007 Michigan routes survey data are not included in Table 2 because of inconsistency in the way the question was asked in this survey relative to the other studies.

The data presented in Table 2 show that a vast majority of rail passengers (60% to 70%) report that they would utilize a personal vehicle for travel if the rail service was not available, indicating that intercity passenger rail plays a meaningful role in providing real alternatives to driving for travelers. Intercity bus or local transit services are a common alternative to rail service in the higher-density corridors with greater service frequency

Table 2. Summary of Rail Passenger Alternative Travel Mode Responses

Route (year)	Personal vehicle (%)	Intercity bus/local transit (%)	Airplane (%)	Would not make trip (%)
Hiawatha Service (2016)	70	13	3	15
Hiawatha Service (2011)	69	14	4	14
Wolverine (2011)	60	12	20	8
Blue Water (2011)	69	10	14	8
Pere Marquette (2011)	63	6	21	10
Heartland Flyer (2009)	63	3	6	28

Note: Data from 2007 Michigan routes surveys not shown. Rows may not total 100% because of rounding.

whereas airplane is a frequent alternative for longer-distance routes. With the exception of the Heartland Flyer route, the "would not make trip" percentage is relatively low (8% to 15%), suggesting that the amount of discretionary travel being made on these routes is also very low.

Influences on Alternative Mode Preference

The summary of rail passenger alternative travel modes presented in Table 2 provides interesting insight into the general patterns of alternative mode distribution across the five routes studied for this analysis. However, looking exclusively at the distribution data in Table 2 does not provide a complete picture of how intercity passenger rail supports personal mobility. In particular, such high-level data does not distinguish the various influences on the passengers' preference for alternative travel modes. Previous analyses of these survey data have identified trends in alternative travel mode preference based on factors such as trip purpose, age, household vehicles, and household income on a route-specific basis (*9–13*). However, these analyses have been limited in that they have only examined the relevant variables individually without regard for the correlation between variables. One option to examine the influences on alternative mode preference is to construct statistical models relating the use of alternative travel modes to the characteristics of the passenger and the route being used.

For this analysis, logistic regression models were constructed to estimate the probability of a rail passenger using one of four alternative modes (personal vehicle, intercity bus/local transit, airplane, or would not make trip) as a function of several personal and route characteristics. Four separate models were specified, one for each alternative travel mode, with the independent variable data taking a value of "Yes" if the passenger indicated that they would use that mode if the rail service was not available and "No" otherwise. For this analysis, the 2007 Michigan routes passenger survey data set was included. For those surveys, passengers that responded that they would be "very likely" to use a particular alternative travel mode were considered to be "yes" for this analysis and "no" otherwise. The following dependent variables were examined in the logistic regression model development:

- Trip purpose (commute, business, or personal)
- Trip frequency (rail trips per year)
- Household vehicles
- Passenger age
- Annual household income ($1,000s)
- Access distance (miles to nearest station)
- Distance of travel on rail (miles between board and alight station)
- Frequency of rail service on the route (round-trips per day)
- Total length of the route (end-to-end miles)

For the three demographic variables (household vehicles, age, and income), three binary variables were constructed to indicate if a passenger was part of a subpopulation (zero-vehicle households, elderly passengers over age 65, and low-income passengers with annual income less than $20,000) to provide greater insight into how passenger rail service provides mobility support for these groups. The statistical analysis software package JMP 11.0 was used to estimate the logistic regression models. Variable selection utilized the "backwards elimination" technique with the goal of minimizing the Akaike information criterion (AIC) for each model (*18*). All of the resulting models were highly significant ($p < 0.0001$ for all models) and all variables retained in the final models were significant at 95% confidence level or better.

Table 3 reports the logistic regression model parameters for each of the four models of alternative travel mode preference among intercity rail passengers. Also presented in Table 3 is the odds ratio corresponding to the model parameter for selected variables of interest (because of the large sample size, certain statistically significant odds ratios were not practically significant).

Table 3. Logistic Regression Models of Alternative Mode Preference

Model variable	Personal vehicle model	Intercity bus/transit model	Airplane model	Would not make trip model
Intercept	0.655	–2.470	–3.464	–1.014
Purpose = commute	NS	NS	NS	NS
Purpose = business	0.194 (1.474)	NS	0.233 (1.592)	–0.567 (0.322)
Purpose = personal	NS	0.167 (1.397)	NS	NS
Rail trip frequency	–0.004	0.003	–0.013	0.005
Household vehicles	0.219 (1.245)	NS	NS	–0.12 (0.887)
Zero-vehicle household = yes	NS	0.556 (3.038)	NS	NS
Passenger age	NS	–0.007	NS	0.012
Age over 65 = yes	–0.113 (0.798)	NS	0.185 (1.448)	NS
Household income	0.004	NS	NS	–0.003
Income < $20,000 = yes	NS	0.318 (1.889)	NS	NS
Access distance	–0.001	NS	0.002	NS
Rail trip distance	–0.003	–0.002	0.01	NS
Route service frequency	NS	0.171 (1.187)	-0.079	–0.161 (0.852)
Route length	NS	0.003	NS	–0.004

Note: The odds ratio for each variable of interest is reported in parenthesis below the corresponding model parameter; not all odds ratios are shown. NS indicates that the variable was not significant in the model at the level of α = 0.05 and was removed from the estimate.

The odds ratios presented in Table 3 provide the greatest insight into the contribution of each variable toward the probability of choosing the respective alternative travel mode, when controlling for all the different factors examined in the model specification. Business travelers were approximately 50% more likely to use personal vehicle or airplane as an alternative to rail service and about one-third as likely to forgo the trip. Individuals traveling for personal reasons were about 40% more likely to use intercity bus or local transit as compared with other travelers. The availability of an automobile contributed significantly to three of the four models reported in Table 3. An increase in the number of vehicles in the passenger's household increased the likelihood of using automobile as an alternative to rail service by about 25% with a corresponding decrease in the odds of forgoing the trip entirely. Passengers from zero-vehicle households were three times as likely to use intercity bus or local transit as compared with other passengers, the strongest relationship noted in Table 3. The significance of the variables associated with automobile availability is not surprising given that automobile is a common alternative for a majority of rail passengers. Elderly passengers were less likely to drive and more likely to use airplane if the rail service was not available. Low-income passengers were nearly twice as likely to use intercity bus or local transit services in the absence of the rail service. The distance traveled by the passenger to access the rail station as well as the distance traveled on the rail service itself did not contribute to the probability of choosing one of the four alternatives examined in a meaningful way. However, an increase in the frequency of rail service in a corridor resulted in an 18% increase in the probability of passengers using intercity bus or local transit as an alternative to rail while decreasing the probability of not making the trip if the rail service was not available. This result is not surprising as an increase in service frequency allows for more trips (particularly those of a non-discretionary nature) to be accommodated on the passenger rail service.

Conclusion

Summary of Findings

This paper presented an analysis of the characteristics and preferences of U.S. intercity rail passengers using data from more than 10,000 surveys obtained over the last 10 years from passengers on five Amtrak routes in the Midwest and South Central U.S. Analysis of the survey data provides unique insight into the behavioral characteristics of rail passengers. As noted in Table 1, a majority of passengers were traveling for personal reasons, such as visiting family or friends or leisure/entertainment/vacation; however, some business and work commute travelers were present among passengers of the five routes analyzed. Analysis of the passengers' socioeconomic distribution revealed that rail services accommodate a diverse group of travelers, with a wide range of ages and household income levels present in the data set.

The analysis in this paper focused on two distinct dimensions of passenger behavior that are relevant for the creation of passenger elements within state rail plans. First, an analysis of the market area or catchment area for passenger rail stations was presented based on data related to the passenger home ZIP code and station usage on the day of the survey. This analysis found that approximately half of the passengers on these five routes live within 8 to 10 mi of the rail station, whereas approximately three-quarters of passengers live within 30 mi (Figure 1*a*). Approximately two-thirds of rail passengers live within a 30-min drive of the rail station (Figure 1*b*). On average, the distance traveled by passengers on the highway network to access rail stations is 21% greater than the corresponding straight line distance between the passenger's home and the station (Figure 2). The second aspect of the analysis focused on the personal mobility supported by intercity passenger rail services, based on data from survey questions on passenger alternative mode of travel if the rail service was not available. A majority of passengers on the five routes analyzed would utilize automobile as an alternative to rail service, indicating that the provision of rail service in an intercity corridor plays a significant role in reducing highway congestion. Analysis of the personal and route characteristics that influence the likelihood of using different alternative travel modes was also presented using logistic regression models (Table 3). Traveler characteristics such as the business trip purpose, availability of household vehicles, age, and household income all contributed significantly to the probability of using various alternative travel modes.

Recommendations and Future Research

In accordance with Section 303 of PRIIA, SRTAs must develop comprehensive state rail plans that incorporate both freight and passenger elements. Such plans require states to establish a vision for intercity passenger rail service as part of the state's transportation system and to identify a list of projects necessary to achieve that vision. With decision-makers and their constituents (i.e., taxpayers and the traveling public) demanding a more transparent and data-driven approach to policy development, rail planners have limited evidence on how passenger rail investments contribute to broader policy goals. The analysis presented in this paper, based on responses from more than 10,000 surveys of passengers on five U.S. intercity rail corridors, attempts to bridge this data gap and provide the evidence needed to inform the development of robust and goal-oriented state passenger rail programs.

Based on the results of this research, the following recommendations for the development of passenger elements of state rail plans are provided:

- The market area analysis presented in this paper indicated that approximately half of passengers live within 10 mi of rail stations and approximately three-quarters live within 30 mi. Given these results, as well as the shape of the distribution curve above 30 mi presented in Figure 1*a*, there is no evidence to suggest that the FRA's preferred methodology for rail service accessibility (the population within a 10- and 30-mi radius of rail stations) is inadequate for state rail planning activities.
- However, the market area analysis also indicated that the actual distance traveled by passengers along the roadway network to access rail services was approximately 21% longer than the corresponding straight line distance. Although a market area analysis using a simple buffer around a station location is easier to implement than a network-based analysis, this finding could have significant implications for intercity rail planning. Given these results, if a SRTA wishes to account for the actual distance traveled along the highway network to access rail services, the straight line market area radius used for planning purposes should be reduced by approximately 20%.
- The mobility analysis presented in this paper indicated that between 60% and 70% of passengers would utilize an automobile for travel if the rail service was not available. Based on these results, it is recommended that rail planners assume that 60% of passengers would use automobile as an alternative to rail for the purposes of calculating estimates of VMT avoided by the provision of passenger rail services.
- The mobility analysis indicated that passengers traveling for business purposes were more likely to use automobile or airplane as an alternative to rail services. Consequently, the extent to which rail services can attract new business-oriented travelers plays a substantial role in the success of rail services in diverting automobile and air trips. Based on these results, it is recommended that state rail plans consider projects within the passenger RSIP that will increase the attractiveness of rail services

to business travelers; such projects may include improving service reliability, enhanced station and on-board services, and marketing initiatives focused on business travelers.
- The mobility analysis also indicated that intercity passenger rail services provide personal mobility for certain population groups (specifically, zero-vehicle households, elderly individuals, and low-income individuals). Based on these results, it is recommended that state rail plans consider projects within the passenger RSIP that improve rail service access for these individuals, including the location of station facilities and the first mile/last mile connectivity to stations.
- Evidence from the mobility analysis indicates that increasing service frequency results in a lower probability of "would not make trip" if the rail service was not available. The practical result of this finding is that increasing service frequency provides greater opportunities for a wider range of non-discretionary trips to be accommodated on passenger rail services. Based on these results, if SRTAs desire to increase the broader appeal and impact of rail services, it is recommended that state rail plans consider projects within the passenger RSIP that will contribute toward an increase in service frequency; such projects may include route-level infrastructure, new rolling stock, or additional corridor-specific service development planning activities.

One obvious direction for future research is to grow the rail passenger survey database analyzed in this paper by adding data from new routes or new survey waves on the routes analyzed here. SRTAs should consider conducting passenger surveys on routes under their jurisdiction, allowing for local or corridor-specific data to be used in the state rail plan. Although any new data would enhance this analysis, additional surveys from routes with higher service frequencies or other geographic regions would provide greater evidence on the broader impacts of rail services. A similar analysis conducted on the higher-density U.S. rail corridors (such as the Northeast Corridor or the California Amtrak routes network) would provide further insight from these critical regions. A more in-depth theoretical analysis could examine the relationship between rail passenger behavior patterns and underlying variables such as population density, other population-level characteristics (e.g., vehicle availability, household income), regional transit accessibility, or urban/suburban/rural area type. Examining these relationships could add new dimensions to the analysis presented in Figure 2, allowing for greater refinement of the market area for intercity passenger rail service that recognizes network conditions as well as context and demographics. At a minimum, the results presented in this paper can be used as a starting point for those SRTAs that do not currently sponsor service or do not have the resources to conduct surveys. As the practice of state rail planning matures across the U.S., incorporating data-supported evidence of the relationship between statewide mobility goals, investments in passenger rail systems, and transportation system outcomes will undoubtedly result in improved resource allocation, greater transparency, and informed decision making for all modes of travel.

Acknowledgments

The authors gratefully acknowledge the State Departments of Transportation of Illinois, Michigan, Oklahoma, Texas, and Wisconsin for supporting the rail passenger survey research studies from which the data analyzed in this paper were drawn. Financial support provided by the Department of Civil Engineering and the Russ College of Engineering and Technology at Ohio University is also gratefully acknowledged. The authors also wish to acknowledge the five anonymous reviewers who provided helpful feedback on the initial draft of this paper.

Author Contributions

The authors confirm contribution to the paper as follows: study conception and design: Benjamin Sperry, Tyler Collins; data collection: Benjamin Sperry, Tyler Collins; analysis and interpretation of results: Benjamin Sperry, Tyler Collins; draft manuscript preparation: Benjamin Sperry, Tyler Collins. All authors reviewed the results and approved the final version of the manuscript.

References

1. Passenger Rail Investment and Improvement Act of 2008. U.S. Public Law 110-432, Division B: Amtrak. October 16, 2008.
2. U.S. Federal Railroad Administration. *State Rail Plan Guidance.* U.S. Department of Transportation Federal Railroad Administration, September 2013. https://www.fra.dot.gov/Elib/Document/3382. Accessed August 25, 2016.
3. Morgan, C. A., J. E. Warner, E. S. Horowitz, D. P. Simpson, B. Sperry, and W. E. Zullig. *Guidebook for Intercity Passenger Rail Service and Development.* Report No. 6. National Cooperative Rail Research Program, Transportation Research Board, National Research Council, Washington, D.C., 2016.
4. Parsons Brinckerhoff. *California High-Speed Train Ridership and Revenue Model: Development, Application, and Project-Level EIR/EIS Forecasts.* California High-Speed Rail Authority, Sacramento, CA, 2011.
5. Louis Berger Group, Inc. *All Aboard Florida Ridership and Revenue Study.* Florida East Coast Industries, Coral Gables, FL, 2015.

6. AECOM. *Southeast High-Speed Rail Ridership Study*. Southeast High-Speed Rail (Raleigh to Richmond) Tier 2 Study, 2013. https://connect.ncdot.gov/resources/Rail-Division-Resources/Documents/2015%20SEHSR%20FEIS%20-%20Appendix%20C.pdf. Accessed May 12, 2016.
7. Steer Davies Gleave. *HSIPR Best Practices: Ridership and Revenue Forecasting*. Office of Inspector General, U.S. Department of Transportation, Washington, D.C., 2011.
8. *Amtrak System Timetable*. National Railroad Passenger Corporation, Washington, D.C., 2017. https://www.amtrak.com/ccurl/887/420/System-Timetable-020117.pdf. Accessed February 1, 2017.
9. Sperry, B. R., and C. A. Morgan. *Intercity Passenger Rail: Implications for Urban, Regional, and National Mobility*. Report No. UTCM 11-10-75. University Transportation Center for Mobility, Texas A&M Transportation Institute, College Station, TX, 2011.
10. Sperry, B. R., and T. Collins. *Analysis of the Spring 2016 Hiawatha Service Passenger Surveys*. Prepared for the Wisconsin Department of Transportation, Rails and Harbors Section. Ohio University, Athens, OH, 2016.
11. Grengs, J., C. Hu, and M. Weitz. *Intercity Bus and Passenger Rail Study*. Prepared for the Michigan Department of Transportation, College of Architecture and Urban Planning, University of Michigan, Ann Arbor, MI, 2009.
12. Sperry, B. R., and C. A. Morgan. *Analysis of the 2011 Michigan DOT Intercity Rail and Bus Passenger Surveys*. Prepared for the Michigan Department of Transportation, Texas A&M Transportation Institute, College Station, TX, 2012.
13. Sperry, B. R., and C. A. Morgan. *Measuring the Benefits of Intercity Passenger Rail: A Study of the Heartland Flyer Corridor*. Report No. SWUTC/10/169116-1. Southwest University Transportation Center, Texas A&M Transportation Institute, College Station, TX, 2010.
14. Sperry, B. R., and C. A. Morgan. Economic Benefits of Intercity Passenger Rail Service: Evidence from Passenger Surveys. *Transportation Research Record: Journal of the Transportation Research Board*, 2011. 2261: 25–30.
15. Sperry, B. R., J. C. Taylor, and J. L. Roach. Economic Impacts of Amtrak Intercity Passenger Rail Service in Michigan: Community-Level Analysis. *Transportation Research Record: Journal of the Transportation Research Board*, 2013. 2374: 17–25.
16. Sperry, B. R., S. Larson, D. Leucinger, S. Janowiak, and C. A. Morgan. Intercity Passenger Rail Access to Airports: Case Study in Milwaukee, Wisconsin. *Transportation Research Record: Journal of the Transportation Research Board*, 2012. 2300: 22–30.
17. Médard de Chardon, C., and G. Caruso. *Friendly Batch Routing with Google Maps API*. http://geow.uni.lu/apps/fbr/. Accessed June 7, 2017.
18. Agresti, A. *Categorical Data Analysis*, 2nd ed. John Wiley and Sons, Hoboken, NJ, 2003.

The Standing Committee on Intercity Passenger Rail (AR010) peer-reviewed this paper (18-00925).

The views expressed in this paper are solely the opinions of the authors and do not represent the official views of Ohio University or any other organization. Any errors, inaccuracies, or omissions are the sole responsibility of the authors.

Article

TRR
JOURNAL OF THE TRANSPORTATION RESEARCH BOARD

Transportation Research Record
2018, Vol. 2672(10) 247–259

DOI: 10.1177/0361198118786809
journals.sagepub.com/home/trr
SAGE

Economic Impact Analysis on Regional Industries by High-Speed Rail Investments: Application of an Input-Output Model

Changju Lee[1], Jiaqi Ma[2], and Kwan Kyo Oh[3]

Abstract

High-speed rail (HSR) systems have advantages over road and air transport, but they require significant financial investment. The economic viability of HSR systems has been analyzed through various methods. The approaches used cannot consider all intangible variables, or cannot satisfy the needs of all stakeholders. Particularly, in situations when policy makers are considering the construction of a new HSR system to stimulate the regional economy, detailed economic effects need to be measured. Hence, this research focuses on: (a) formulating a methodological framework to utilize an input-output (I-O) model in a given region where the regional I-O table is not provided, and (b) quantifying actual economic effects on regional industries by new HSR investment as a case study. Specifically, a four-step process with an I-O model is proposed, and a three-point effect (output, employment, and income) is assessed in Sejong, Republic of Korea. The effects of building a new HSR station in Sejong might generate over $100 million in output, over $21 million in income, and almost 1,000 jobs. Construction of railroad tracks would have more effects on the regional economy with over $1,671 million output, over $37 million income and over 145,000 jobs. In this study, only publicly available data sources are considered to propose the methodology. This approach, therefore, can be repeated for policy analyses in other locations where the data is limited. Further, regardless of the type of transportation investment, this approach can be used by policy makers seeking secure decisions to quantify economic effects on regional industries.

High-speed rail (HSR) has been regarded as one of the most technologically advanced passenger transportation modes since the Second World War. There is robust evidence for globally increasing interest; for example, as of 2011, approximately 10,513 miles of HSR lines were in operation, 4,827 miles were under construction, and annual HSR ridership reached 928 million across 13 countries (*1*). Although HSR systems provide reduced travel time, increased rail capacity, and more comfortable travel experiences (*2*), policy makers from many agencies are hesitant to invest in them because of the high costs resulting from complicated infrastructure, stations, maintenance and operations of HSR systems (*3*).

Nonetheless, the prospect of extending HSR systems in many countries has recently received a great deal of attention as an effective way to redevelop cities and vitalize regional economies (*2*, *4*). For example, a new HSR station at Kakegawa in Japan brought increases of 12% to 20% in employment and 38% to 63% in production (*5*). With a new HSR system in Tianjin, China, sales at 10 franchise restaurants increased by over 20% (*6*). As another recent example, the Korean government has considered a HSR system in a new Multifunctional Administrative City (MAC) within Sejong for the purpose of economic vitalization of regional industries and for its socioeconomic advantages.

While a few researchers have measured the impacts of HSR systems from different perspectives, most of them have focused on quantifying (in)direct benefits from changed demands and travel times for the benefit-cost analysis (e.g., *7–10*). In other cases, empirical studies have been conducted after an HSR system has been constructed (e.g., *5*, *6*, *11*). This is because verifying the economic effects of HSR investments might be difficult with previous approaches, and is even harder for newly

[1]Virginia Transportation Research Council, Charlottesville, VA
[2]University of Cincinnati, Cincinnati, OH
[3]Korea Transportation Safety Authority, Gimcheon-si, Gyeongsangbuk-do, South Korea

Corresponding Author:
Address correspondence to Changju Lee: cl8ax@virginia.edu

planned cities owing to insufficient specific information. Given that developing countries could enjoy economic ripple effects from transportation investments (*12*), it is noteworthy for policy makers to observe a HSR system's economic effects on regional industries in a new planned city within a developing country. Ultimately, these economic effects can be used by transportation policy makers as one of the primary elements to support them in making rational decisions.

Purpose and Scope

This research sheds light on finding and verifying a reliable methodology that can be used for transportation policy analysis by capturing regional economic effects from HSR investments. To this end, this research pursues four objectives:

1. Investigate a proper approach to assess the economic effects of HSR investments on regional industries.
2. Identify a suitable region considering a new HSR system for a case study.
3. Develop a methodology to utilize a selected approach in a given region.
4. Quantify actual economic effects on regional industries through a proposed methodology in a given region.

Because new planned cities in developing countries generally have deficient detailed data for analysis, this research only uses publicly accessible data sources to develop a methodology.

Literature Review

Possible Methods for Evaluating Policies for Transportation Investments

Benefit-cost analysis (BCA) is the most widely used method for ex ante evaluation of transportation investments including HSR systems (*13*). In principle, BCA compares total expected costs (e.g., infrastructure, maintenance, and operations) with total expected benefits in order to evaluate the feasibility of transportation investments. Improved air quality, less noise, lower energy consumption, and economic competitiveness have recently been considered as benefits in addition to traditional benefits such as travel-time savings, reduced traffic congestion, and fewer motor vehicle accidents (*2*, *7*, *9*). Because all the factors are expressed as monetary values in the BCA, it is possible to prevent expensive mistakes in large transportation investment costs (*13*). On the other hand, it is almost impossible to monetize all the invisible benefits. Also, compared with costs, calculating benefits is subjective, depending on current conversion units which are based on certain locations and times.

Multi-criteria analysis (MCA) is another popular method used to evaluate transportation investments. MCA is used to help policy makers find the best alternative among a set (*13*). Unlike BCA, MCA does not require monetary units for effects but can express effects in physical units or qualitative terms (*14*). Further, MCA uses various criteria with weights that can reflect the objectives of the policy maker, meaning that decision makers can apply their concerns to specific issues. Therefore, MCA can be utilized in any field, including economics, finance, business, and engineering, in which there is a need to make an optimal choice among many options (*15*). Yet because the above characteristics of MCA are based on weighted indicators among alternatives, MCA is not congruent with quantifying economic effects in a sole project; for example, the construction of HSR in a certain area.

The input-output (I-O) model has been a valuable method for national and regional economic analysis since Wassily Leontief first introduced it in 1936. Thanks to its advantages for identifying the structural relationships among the different industries of an economy, the I-O model has been used both in planning processes and in policy design, as well as being applied more narrowly in the transportation field (*16*). As examples, the I-O model was used in estimating the inter-regional economic impacts of earthquakes on intermodal transportation hubs (*17*); analyzing the relationship between transportation and communications as industrial inputs (*16*); and assessing the regional economic impacts of charging heavy goods vehicles in Spain (*18*). In the U.S., a few HSR-related I-O model studies exist focusing on estimated effects on employment (e.g., California HSR project [*19*]). Similarly, although there are other studies using the I-O model for transportation expenditure (e.g., public transportation or highway), they mostly adopt the Regional Input-Output Modeling system (RIMS) of the U.S. Department of Commerce's Bureau of Economic Analysis or the model of Regional Economic Models Incorporated (REMI) to estimate job creation (*20*, *21*).

Some researchers have proposed other methods for evaluating the effects of transportation investments, yet these have generally focused on predicting changes in travel time, pattern, and volume, and tried to measure the effects of these changes. For example, conventional gravity-based models were used to evaluate how accessibility was changed by and affected the economic performance of transportation investments (*22*). Sasaki et al. (*11*) constructed a supply-oriented regional econometric model and simulated hypothetical scenarios to evaluate the spatial dispersion of economic activities and population resulting from HSR systems in Japan. The changes

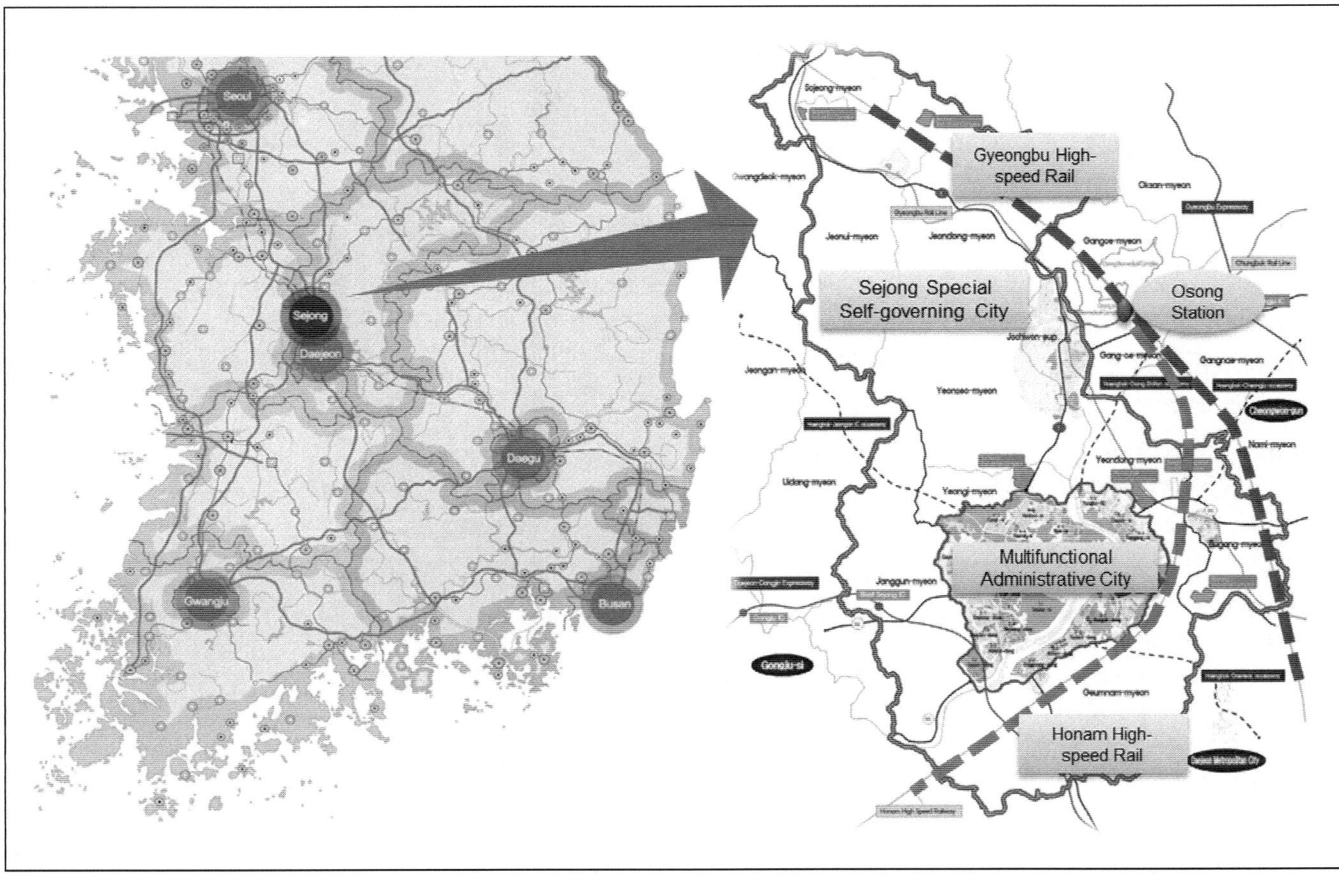

Figure 1. Location of Sejong and (right) high-speed rail in the region (*29*).

in the spatial pattern of the working population caused by transportation investments were predicted by an economic geography, inter-industry model (*23*).

Many transportation projects consider several alternatives, and previous approaches (e.g., BCA, MCA) are relatively effective at evaluating transportation policies for investments in this context. However, if the policy goal of constructing a HSR system is to stimulate a regional economy (industries), the economic effects of the HSR need to be quantified to make a decision, which is considered one of the limitations of previous methods (*24*). The I-O model is regarded as one of the most useful tools for identifying the inter-industry effects of a unit of investment, by using the I-O table (also known as I-O coefficients) (*25*). Further, the I-O model has comparative advantages of simplicity, transparency, and costs in terms of procedures. In that sense, this research develops a new regional I-O model to attain the objectives, rather than simply adopting the provided I-O models.

Brief Background of Target City

Seoul, the capital of Korea, is notorious for its high-density development and overflowing population, which is almost half of the national population. Political, economic, and social concentration in Seoul continuously generates various negative externalities. The Korean government has attempted to suppress the expansion by relocating public organizations outside of Seoul. Finally, the Korean government constructed the new MAC in Sejong (see Figure 1), 100-miles distant from Seoul, in order to relocate 36 Korean government ministries and 16 national research institutes there (*26*).

According to the master plan of Sejong (*27*), in 2030 the population of the city will be 0.8 million, including the MAC's planned population of 0.5 million. In addition to government ministries and national research institutes, the MAC will be filled with universities, diverse related businesses, and many state-of-the-art knowledge-based industries. The first residents moved there in late 2011 and the first stage of relocation of government organizations began in late 2012. In 2013, Sejong's population was just around 0.12 million and several industries dominated city's economy (*28*). Sejong has therefore proposed a new HSR system to the Korean government to stimulate the regional economy, in accordance with the plan to construct a HSR system from Seoul to Mokpo (located at the southwestern tip of Korea). With this, Sejong expects more convenient accessibility to other cities and economic effects on regional industries. In this regard, policy makers need to see the possibility that a new HSR system can create actual economic effects on the regional economy, not only (in)direct benefits. Because Sejong is a new planned city per se, as well as including a new sub-city (MAC), Sejong is an ideal region to study for the purpose of this research.

Methodology

Framework for Assessing the Economic Effects on Regional Industries

From the concept of the I-O model, a customized regional I-O table—a matrix explaining the interrelationship between input and demand for goods and services among industries during a certain period—is basically required to quantify the economic effects on regional industries. In developing countries, including Korea, this regional I-O table is rarely provided by governments. Accordingly, this research formulates a methodological framework to make a new regional I-O table by indirect methods (*30*). Indirect methods are based on the fact that the regional I-O table is consistent in principle with the national I-O table, except for its target area (*31*). For the reliability of the developed regional I-O table, this research uses a combination of two indirect methods: Regional Weights (RW) and Location Quotient (LQ), both known as relatively high accuracy methods (*31–35*). A stepwise approach of four steps is proposed, as shown in Table 1.

Contrary to the national I-O table, in a regional I-O table, the regional technical coefficient is conceptually different from the regional input coefficient which concerns only intraregional inputs for outputs in a specific region (*31*). This feature is embraced in Step 2. Note that, unlike previous studies concerning job creation alone (e.g., *19–21*), this research identifies regional output, employment, and income effects that are generally of interest to policy makers. The four-step process is elaborated with actual data from Sejong in the following section, which can also be regarded as the calibration process to make a reliable regional I-O table and multipliers. In the Application part, the proposed regional I-O table and multipliers are tested with actual budgets of the HSR station and railroad tracks in Sejong, Korea.

Details of Four-Step Process

Identification of the Regional Technical Coefficient (Step 1). In Step 1, the regional technical coefficients are identified from the national I-O table provided by the Bank of Korea (*36*). To use the national I-O table, industries in Sejong are aggregated according to the industrial category based on the Korean Standard Industrial

Table 1. Methodological Framework with Four-Step Process and Application

Task	Detailed process[a]	Main source	Method
Step 1 Regional technical coefficient	(1) Categorization of regional industry (2) Calculation of regional outturn from each industry (3) Modification of national technical coefficient (4) Calculation of regional intermediate transactions between industries (5) Identification of regional technical coefficient	National I-O table, Type of regional industry Regional outturn of each industry	Regional weights
Step 2 Regional input coefficient	(1) Computation of location quotient coefficient (2) Identification of regional input coefficient	Regional/national economic statistic (e.g., employment)	Local quotient
Step 3 Regional I-O table	(1) Recreation of regional intermediate transaction table between industries (2) Calculation of total inputs, outputs, added values	Regional outturn of each industry	
Step 4 Regional economic effect multiplier	(1) Calculation of Leontief Inverse Matrix (2) Identification of economic effect multipliers (i.e., output, employment, and income)	Regional outturn, employment, income	
Application Economic effect analysis	(1) Categorization of expenditure or budget by each industry (2) Analysis of output, employment, and income effects	Budgets of high-speed rail systems	

[a]Numbers in parenthesis indicate the order of process.

Classification which has 28 major class levels and 78 middle class levels as of 2010 (*36*). Consequently, 15 industries in Sejong are classified, excluding small manufacturing industries (under 2% of total employment) of low influence to the region.

With 15 industries categorized, the national I-O table is modified and new national technical coefficients are calculated using Equation 1.

$$a_{ij}^{N} = \frac{X_{ij}^{N}}{X_{j}^{N}} \quad (1)$$

where

a_{ij}^{N} = national technical coefficients of 15 industries (i, j = 1, 2,..., 15)

X_{ij}^{N} = national intermediate transactions of 15 industries (i, j = 1, 2,..., 15)

X_{j}^{N} = national total inputs (or outputs) of 15 industries (j = 1, 2,..., 15)

Regional technical coefficients are acquired by the RW method (Equation 2). The RW method uses the differences between national and regional industrial structures identified by the regional outturn of each industry (*35*, *37*). Specifically, the RW method uses modified national technical coefficients and regional outturn of 15 industries in order to calculate regional intermediate transactions between industries ($a_{ij}^{N} \cdot P_{j}^{R}$). Regional technical coefficients are eventually drawn with the same principle of national technical coefficients (see Table 3).

$$a_{ij}^{Rt} = \frac{a_{ij}^{N} \cdot P_{j}^{R}}{X_{j}^{R}} \quad (2)$$

where

a_{ij}^{Rt} = regional technical coefficients of 15 industries (i, j = 1,2,...,15)

P_{j}^{R} = regional outturn of 15 industries (j = 1,2,...,15)

X_{j}^{R} = regional total inputs (or outputs) of 15 industries (j = 1,2,...,15)

Identification of the Regional Input Coefficient (Step 2). In the analysis of regional economic effects, the regional input coefficient is conceptually more important than the regional technical coefficient (*30*). The LQ method is employed to produce regional input coefficients from regional technical coefficients in this step. The LQ method quantifies how concentrated a particular industry is in a region compared with the nation. By doing so, the LQ method can reveal which industry makes a particular region unique in comparison to the national average. The LQ method is computed as an industry's share of a regional total for some economic statistic (income, outputs, employment, etc.) divided by the industry's share of the national total for the same statistic (Equation 3). In this research, the number of employees is used as a statistic because of its relative availability and reliability.

$$LQ_{j}^{R} = \frac{e_{j}^{R}/e^{R}}{e_{j}^{N}/e^{N}} \quad (3)$$

where

LQ_{j}^{R} = regional LQ of j industry (j = 1,2,...,15)

e_{j}^{R} = regional employment of j industry (j = 1,2,...,15)

e^{R} = regional employment of all industries

e_{j}^{N} = national employment of j industry (j = 1,2,...,15)

e^{N} = national employment of all industries

Table 2 shows the numbers of employees and the result of the LQ of 15 industries in Sejong. If LQ_{j}^{R} is less than 1, local employment of j industry is less than expected for a given industry. In other words, j industry does not satisfy the local demand for a given good or service, which has to be imported from other regions. On the contrary, if LQ_{j}^{R} is equal to or greater than 1, then j industry satisfies local demand, assuming that the outputs of j industry are not exported. Additional modification of regional technical coefficients is not necessary in those industries.

Based on LQ and regional technical coefficients, regional input coefficients are finally calculated by Equation 4 (see Table 4).

$$a_{ij}^{Ri} = LQ_{j}^{Ra} \cdot a_{ij}^{Rt} \quad (4)$$

where,

a_{ij}^{Ri} = regional input coefficients of 15 industries (i, j = 1,2,...,15)

a_{ij}^{Rt} = regional technical coefficients of 15 industries (i, j = 1,2,...,15)

LQ_{j}^{Ra} = regional applied LQ of j industry (j = 1,2,...,15)

Completion of the Regional I-O Table (Step 3). To create the regional I-O table in Sejong, a regional intermediate transaction table between 15 industries is computed by regional outturn and regional input coefficients. Because the national I-O table provides added values for each industry, regional added values are calculated in tandem with intermediate transactions. Further, final demands are calculated by intermediate demands subtracted from total outputs because the regional I-O model does not consider external trading with other regions.

$$T_{ij}^{R} = a_{ij}^{Ri} \cdot P_{j}^{R} \quad (5)$$

where

T_{ij}^{R} = regional intermediate transactions of 15 industries (i, j = 1,2,...,15)

Table 2. Employment and Location Quotient in 15 Industries[a]

Industrial category	Employment (person)		Location quotient	
	Sejong	Nation	Computed LQ	Applied LQ
1. Agriculture, Forestry, and Fishing	389	26,069	1.46	1
2. Mining and Quarrying	275	14,552	1.85	1
3. Manufacturing	56,064	3,246,547	1.69	1
4. Electricity, Gas, Steam, and Water Supply	635	54,471	1.14	1
5. Construction	6,674	1,134,996	0.58	0.58
6. Wholesale and Retail Trade	14,802	1,869,873	0.78	0.78
7. Accommodation and Food Service Activities	12,298	1,232,889	0.98	0.98
8. Transportation	5,030	675,440	0.73	0.73
9. Information and Communications	842	412,407	0.20	0.20
10. Financial and Insurance Activities	2,567	703,613	0.36	0.36
11. Real Estate Activities and Business Services	4,965	355,277	1.37	1
12. Public Administration and Defense	3,664	619,007	0.58	0.58
13. Education, Human Health, and Social Work Activities	16,493	1,813,635	0.89	0.89
14. Arts, Sports, and Recreation Related Services	2,082	209,835	0.97	0.97
15. Other Services	5,263	575,447	0.90	0.90
Total	132,045	12,944,058		

[a]This table was created by the authors based on data from Statistics Korea (38).

a_{ij}^{Ri} = regional input coefficients of 15 industries (i, j = 1,2,...,15)

P_j^R = regional outturn of 15 industries (j = 1,2,...,15)

As a result, total output is \$28,564 million, manufacturing occupies 73% (\$20,748 million) and is followed by construction (5.7%, \$1,630 million). In terms of intermediate inputs, manufacturing and construction take 81% (\$15,809 million) and 4.8% (\$931 million), respectively (see Table 5).

Identification of the Regional Economic Effect Multipliers (Step 4). Final demand (purchases) for goods and services affects not only certain industries associated with that demand but also whole industries, from the production process to use. In a regional I-O table, to investigate total effects, including direct effects and cumulative indirect effects, the multiplier matrix (known as the Leontief inverse matrix) is necessary (*39*). The Leontief inverse matrix is formulated as shown in Equation 6.

$$\begin{aligned} X &= AX + Y \\ X - AX &= Y \\ X(I - A) &= Y \\ X &= (I - A)^{-1}Y \end{aligned} \tag{6}$$

where

A = regional input coefficient matrix (or direct requirements matrix)

X = regional output matrix

Y = regional final demand matrix

I = identify matrix

$(I - A)^{-1}$ = Leontief inverse matrix

By the Leontief inverse matrix, economic effects (expressed as regional outputs of each industry) can be obtained according to the variation of regional final demands. Generally, economic effects are assessed with respect to changes in output, gross regional product, personal income, and employment (*40*). Recall that the effects on output, employment, and income of regional industries are analyzed in terms of the investment in a new HSR system. Each multiplier can be computed by Equations 7 to 9. A summary of multipliers is provided in Table 6.

$$O_j^R = \sum_i \sum_j \lambda_{ij}^R \tag{7}$$

$$E_j^R = \sum_i \sum_j (\lambda_{ij}^R \cdot \frac{e_j^R}{P_j^R}) \tag{8}$$

$$I_j^R = \sum_i \sum_j (\lambda_{ij}^R \cdot \frac{i_j^R}{P_j^R}) \tag{9}$$

where

O_j^R = regional output multiplier of 15 industries (j = 1,2,...,15)

λ_{ij}^R = regional Leontief coefficients of 15 industries (i, j = 1,2,...,15)

E_j^R = regional employment multiplier of 15 industries (j = 1,2,...,15)

I_j^R = regional income multiplier of 15 industries (j = 1,2,...,15)

P_j^R = regional outturn of 15 industries (j = 1,2,...,15)

Table 3. Regional Technical Coefficients (Step 1)

		Output														
		1	2	3	4	5	6	7	8	9	10	11	12	13	14	15
Input	1	0.068189	0.000664	0.028641	0.000001	0.001967	0.000000	0.070203	0.000000	0.000000	0.000000	0.000065	0.002262	0.002818	0.000392	0.021244
	2	0.000034	0.000000	0.017949	0.395972	0.002456	0.000000	0.000272	0.000000	0.000000	0.000000	0.000003	0.000075	0.000031	0.000003	0.000306
	3	0.274794	0.088309	0.573368	0.036607	0.404493	0.033440	0.350382	0.093894	0.054100	0.005862	0.037784	0.049771	0.102777	0.137635	0.194519
	4	0.007377	0.025589	0.016198	0.177208	0.002782	0.020944	0.031619	0.012230	0.016858	0.008192	0.019722	0.013812	0.027114	0.028034	0.002770
	5	0.000697	0.001182	0.000348	0.008395	0.000169	0.001410	0.002482	0.000520	0.002631	0.000508	0.024552	0.027336	0.003089	0.002528	0.000000
	6	0.035887	0.009569	0.034907	0.005776	0.037677	0.028531	0.064614	0.014744	0.106777	0.003160	0.005857	0.010281	0.016882	0.021403	0.047456
	7	0.000000	0.000000	0.000000	0.000000	0.000000	0.000000	0.000000	0.000000	0.000000	0.000000	0.000000	0.000000	0.000000	0.000000	0.594393
	8	0.010597	0.107364	0.017679	0.002925	0.009098	0.076986	0.004779	0.252015	0.009867	0.012957	0.010093	0.017944	0.007700	0.011719	0.002888
	9	0.003962	0.002689	0.002492	0.001979	0.002874	0.054693	0.004953	0.006721	0.146333	0.026649	0.028112	0.010042	0.006950	0.015791	0.015433
	10	0.014576	0.022289	0.010579	0.006365	0.016035	0.032514	0.013602	0.036036	0.018415	0.244741	0.053729	0.029317	0.026801	0.025031	0.001517
	11	0.010477	0.028742	0.048218	0.036682	0.073293	0.120535	0.042731	0.087505	0.146947	0.092319	0.077442	0.051319	0.043273	0.116132	0.003017
	12	0.001830	0.000000	0.000000	0.000000	0.000000	0.000000	0.000000	0.011330	0.000000	0.000000	0.000000	0.000000	0.000000	0.000000	0.008786
	13	0.005463	0.002064	0.002584	0.002333	0.006427	0.006251	0.009250	0.004470	0.003794	0.002965	0.005106	0.006430	0.019457	0.005282	0.000600
	14	0.001017	0.002786	0.001139	0.000371	0.004424	0.004811	0.001627	0.008119	0.024159	0.003231	0.013915	0.005181	0.006602	0.044160	0.107071
	15	0.015527	0.039798	0.007832	0.003844	0.009292	0.038358	0.005080	0.015003	0.020597	0.022125	0.041258	0.057844	0.046769	0.066507	0.000001

Note: The codes of industrial category used in Table 3 are the same as those used in Table 2.

Table 4. Regional Input Coefficients (Step 2)

		Output														
		1	2	3	4	5	6	7	8	9	10	11	12	13	14	15
Input	1	0.068189	0.000664	0.028641	0.000001	0.001134	0.000000	0.068646	0.000000	0.000000	0.000000	0.000065	0.001313	0.002512	0.000381	0.019046
	2	0.000034	0.000000	0.017949	0.395972	0.001416	0.000000	0.000266	0.000000	0.000000	0.000000	0.000003	0.000043	0.000027	0.000003	0.000275
	3	0.274794	0.088309	0.573368	0.036607	0.233155	0.025950	0.342607	0.068546	0.010825	0.002097	0.037784	0.028883	0.091624	0.133863	0.174393
	4	0.007377	0.025589	0.016198	0.177208	0.001604	0.016253	0.030917	0.008929	0.003373	0.002930	0.019722	0.008015	0.024172	0.027266	0.002484
	5	0.000697	0.001182	0.000348	0.008395	0.000097	0.001094	0.002427	0.000380	0.000526	0.000182	0.024552	0.015864	0.002754	0.002459	0.000000
	6	0.035887	0.009569	0.034907	0.005776	0.021717	0.022140	0.063180	0.010764	0.021366	0.001130	0.005857	0.005966	0.015050	0.020816	0.042546
	7	0.000000	0.000000	0.000000	0.000000	0.000000	0.000000	0.000000	0.000000	0.000000	0.000000	0.000000	0.000000	0.000000	0.000000	0.532894
	8	0.010597	0.107364	0.017679	0.002925	0.005244	0.059742	0.004673	0.183981	0.001974	0.004634	0.010093	0.010413	0.006864	0.011398	0.002589
	9	0.003962	0.002689	0.002492	0.001979	0.001657	0.042443	0.004843	0.004906	0.029281	0.009532	0.028112	0.005828	0.006196	0.015358	0.013836
	10	0.014576	0.022289	0.010579	0.006365	0.009243	0.025231	0.013300	0.026308	0.003685	0.087539	0.053729	0.017013	0.023892	0.024345	0.001360
	11	0.010477	0.028742	0.048218	0.036682	0.042247	0.093537	0.041783	0.063882	0.029404	0.033021	0.077442	0.029781	0.038577	0.112950	0.002705
	12	0.001830	0.000000	0.000000	0.000000	0.000000	0.000000	0.000000	0.008271	0.000000	0.000000	0.000000	0.000000	0.000000	0.000000	0.007877
	13	0.005463	0.002064	0.002584	0.002333	0.003705	0.004851	0.009044	0.003263	0.000759	0.001061	0.005106	0.003731	0.017345	0.005138	0.000538
	14	0.001017	0.002786	0.001139	0.000371	0.002550	0.003733	0.001591	0.005927	0.004834	0.001155	0.013915	0.003007	0.005886	0.042950	0.095993
	15	0.015527	0.039798	0.007832	0.003844	0.005356	0.029766	0.004967	0.010953	0.004122	0.007914	0.041258	0.033568	0.041694	0.064685	0.000001

Note: The codes of industrial category used in Table 4 are the same as those used in Table 2.

Table 5. Regional I-O Table in Sejong (Step 3)

		Output (million USD, year: 2010)																	
		1	2	3	4	5	6	7	8	9	10	11	12	13	14	15	Intermediate demand	Final demand	Total output
Input	1	78	0	594	0	3	0	33	0	0	0	0	2	3	0	8	721	426	1,147
	2	0	0	372	86	4	0	0	0	0	0	0	0	0	0	0	463	−367	96
	3	315	8	11,896	8	660	16	163	45	3	1	32	35	97	31	73	13,384	7,365	20,748
	4	8	2	336	38	5	10	15	6	1	1	17	10	26	6	1	482	−265	217
	5	1	0	7	2	0	1	1	0	0	0	21	19	3	1	0	56	1,574	1,630
	6	41	1	724	1	61	14	30	7	5	0	5	7	16	5	18	937	−446	491
	7	0	0	0	0	0	0	0	0	0	0	0	0	0	0	223	223	242	465
	8	12	10	367	1	15	38	2	120	0	2	9	13	7	3	1	599	−124	474
	9	5	0	52	0	5	27	2	3	7	4	24	7	7	4	6	152	−102	51
	10	17	2	219	1	26	16	6	17	1	33	46	20	25	6	1	438	−301	137
	11	12	3	1,000	8	120	59	20	42	7	13	67	36	41	26	1	1,454	−594	860
	12	2	0	0	0	0	0	0	5	0	0	0	0	0	0	3	11	688	699
	13	6	0	54	1	10	3	4	2	0	0	4	4	18	1	0	110	837	946
	14	1	0	24	0	7	2	1	4	1	0	12	4	6	10	40	113	114	227
	15	18	4	162	1	15	19	2	7	1	3	35	40	44	15	0	368	8	375
	Intermediate Input	516	32	15,809	147	931	205	280	257	28	58	273	197	294	108	375	19,511	426	28,564
	Added Value	630	64	4,940	70	699	285	185	217	23	79	587	502	653	119	0	9,054		
	Total Input	1,147	96	20,748	217	1,630	491	465	474	51	137	860	699	946	227	375	28,564		

Note: The codes of industrial category used in Table 5 are the same as those used in Table 2.

Table 6. Summary of Multipliers for Economic Effect Analysis[a]

Industrial category	Multiplier		
	Output (USD)	Employment (jobs)	Income (USD)
1. Agriculture, Forestry, and Fishing	2.22	6.38	0.245
2. Mining and Quarrying	1.77	7.85	0.23
3. Manufacturing	3.16	12.43	0.22
4. Electricity, Gas, Steam, and Water Supply	2.35	8.95	0.27
5. Construction	1.90	6.78	0.18
6. Wholesale and Retail Trade	1.61	27.84	0.24
7. Accommodation and Food Service Activities	2.55	33.59	0.22
8. Transportation	1.79	12.81	0.23
9. Information and Communications	1.19	4.82	0.12
10. Financial and Insurance Activities	1.23	8.28	0.22
11. Real Estate Activities and Business Services	1.64	10.03	0.27
12. Public Administration and Defense	1.37	5.68	0.17
13. Education, Human Health, and Social Work Activities	1.67	19.98	0.29
14. Arts, Sports, and Recreation Related Services	2.09	16.07	0.27
15. Other Services	3.27	35.71	0.36
Total	29.82	217.14	3.57

[a]Multipliers were computed through steps 1 to 4 based on data from multiple sources (28, 36, 38).

Table 7. Planned Budgets for Three High-Speed Rail Stations[a]

High-speed rail station		Gongju	Jeongeup	Songjeong
Budget (Million USD)	Industrial Category			
	4. Electricity, Gas, Steam, and Water Supply	4.32 (11.7%)	7.43 (15.1%)	5.84 (10.5%)
	5. Construction	31.51 (85.5%)	39.96 (81.2%)	48.31 (86.7%)
	9. Information and Communications	1.02 (2.8%)	1.85 (3.8%)	1.54 (2.8%)
	Total Budget	36.85 (100.0%)	49.24 (100.0%)	55.69 (100.0%)
Population of municipality (persons)		127,260	51,812	362,097

[a]This table was created by the authors based on data from Korail for 2013 (42).

e_j^R = regional employment of j industry (j = 1,2,...,15)
i_j^R = regional income of j industry (j = 1,2,...,15)

Applications and Economic Effect Analysis

In this section, the actual economic effects (output, employment, and income) on regional industries in Sejong associated with the investment in a new HSR system are evaluated. The calibrated regional I-O table and multipliers based on the proposed process can be also tested through this application. Note that, because of the base year of the national I-O table, it is presumed that actual effects become effective from this base year but there is no fixed time period for the total effects computed, which is the nature of an I-O model. Generally, HSR systems impact the regional economy from three perspectives: the manufacture of HSR vehicles and the construction of stations and of railroad tracks (*41*). Because HSR vehicles are not manufactured in the target area, the effect of this perspective is negligible although there may be indirect effects from the manufacturing areas. Hence, only station and railroad track investments are considered as factors in determining the economic effects in Sejong.

To compute the regional economic effects of station and railroad track investments, specific construction expenditures (or budgets) are necessary. However, Sejong and the Korea Railroad Corporation (Korail) have not yet drawn up specific budgets for a new HSR station; thus, recently established budgets for HSR stations in other Korean cities were reviewed to determine the most appropriate substitute for this analysis (see Table 7).

Among the three candidates presented in Table 7, Songjeong was chosen for this analysis because it had a population size similar to that of Sejong (363,447 persons) in 2013. According to the three budgets, the HSR station requires investments in three types of industries which would produce direct economic effects on those industries in a region.

Table 8. Planned Budgets of New Railroad Tracks in Sejong[a]

Industrial category	Budgets (2006)	
	Million USD/km	Million USD
3. Manufacturing	3.77 (9.7%)	68.09 (9.7%)
4. Electricity, Gas, Steam, and Water Supply	2.17 (5.6%)	39.15 (5.6%)
5. Construction	25.85 (66.7%)	466.42 (66.7%)
9. Information and Communications	1.51 (3.9%)	27.24 (3.9%)
11. Real Estate Activities and Business Services	2.64 (6.8%)	47.66 (6.8%)
15. Other Services	2.83 (7.3%)	51.07 (7.3%)
Total Budget	38.77 (100.0%)	699.63 (100.0%)

[a]This table was created by authors based on data from Korail (43).

Secondly, Korail plans to construct new HSR tracks (18.044 km) in Sejong. Table 8 presents details of the budgets for these new railroad tracks as of 2006. Unlike a HSR station, new railroad tracks need investments in the "manufacturing," "real estate activities and business services," and "other services" industries in addition to "electricity, gas, steam, and water supply," "construction," and "information and communications." Hence, these industries are expected to be directly affected by the construction of new railroad tracks.

While the base year of the proposed regional I-O table is 2010, the two planned budgets analyzed are based on different years. For the sake of reliability, these budgets are converted to those of 2010 by the Consumer Price Index (CPI) (*38*). The CPI in 2006 (budget year for HSR tracks), 2010 (base year) and 2013 (budget year for HSR station) was 88.07, 100, and 106.8, respectively.

With the derived multipliers (Table 6), the economic effects on 15 regional industries in Sejong are analyzed from a three-point perspective. Because three multipliers are generic for analyzing economic effects, multiplying HSR-related investments results in HSR-specific regional economic effects in Sejong. Tables 9 and 10 present the total inductive effects of output, employment, and income of investment in the HSR station and tracks, respectively.

In sum, output effects worth $100,635,308 and income effects worth $21,145,402 are generated, as well as the creation of 988 jobs, by building a new HSR station in Sejong. In the output effects, "construction" and "manufacturing" generate $45,439,007 (45.15%) outputs and 308 (31.17%) jobs, and $28,776,485 (28.59%) outputs and 358 (36.21%) jobs, respectively. From the income perspective, "construction" contributes $8,297,956 (39.24%) in income effects and "manufacturing" $6,286,731 (29.73%). Noticeably, the construction of a new HSR station does not greatly affect other industries, except "construction" and "manufacturing." Only the "electricity, gas, steam, and water supply," "wholesale and retail trade," and "real estate activities and business services" industries are affected relatively more than other industries. The construction of new HSR tracks in Sejong is predicted to yield a total of $1,671,616,127 in output effects, 145,059 jobs, and $37,575,509 income effects. Unlike building a new station, the "wholesale and retail trade," "accommodation and food service activities," "real estate activities and business services," and "other services" industries are all affected by constructing new HSR tracks, in addition to "manufacturing" and "construction." Interestingly, 34,304 jobs, $76,416,742 of outputs and $4,439,234 of income are generated in "other services." Compared with the station, construction of railroad tracks has more effect on the regional economy, including a 146.8-times greater employment effect.

Concluding Remarks

This research is the most recent endeavor that responds to the need for verifying economic effects on regional industries. As the review of the literature showed, widely used methods (e.g., BCA, MCA) have a weakness when it comes to quantifying regional economic effects. Comparing previous methods like BCA or MCA with the proposed methodology is, however, beyond the scope of this research. Rather, this research focuses on providing a methodological framework using an I-O model and testing it on an actual example.

In this context, several contributions are provided from this research for policy analysis and development in transportation fields. First, the proposed methodology can be repeated in other cities where a regional I-O table does not exist, but only a national I-O table is publicly provided. Second, the suggested methodology can be used for the impact analysis regarding any type of transportation investment. Third, the applicability of the proposed methodology was proven through the actual case study. Fourth, the regional I-O table and three multipliers that were computed in Sejong can be employed to evaluate any other future policy analysis, and even

Table 9. Total Inductive Effects of a New High-Speed Rail Station

Industrial category	Inductive effects		
	Output (million USD)	Employment (no. of jobs)	Income (million USD)
1. Agriculture, Forestry, and Fishing	1.00 (1.00%)	6 (0.64%)	0.25 (1.16%)
2. Mining and Quarrying	3.61 (3.59%)	28 (2.87%)	0.85 (4.01%)
3. Manufacturing	28.78 (28.59%)	358 (36.21%)	6.29 (29.73%)
4. Electricity, Gas, Steam, and Water Supply	7.66 (7.61%)	69 (6.94%)	2.07 (9.78%)
5. Construction	45.44 (45.15%)	308 (31.17%)	8.30 (39.24%)
6. Wholesale and Retail Trade	2.33 (2.32%)	65 (6.58%)	0.57 (2.68%)
7. Accommodation and Food Service Activities	0.53 (0.53%)	18 (1.80%)	0.12 (0.56%)
8. Transportation	1.69 (1.67%)	22 (2.19%)	0.39 (1.86%)
9. Information and Communications	1.95 (1.93%)	9 (0.95%)	0.23 (1.10%)
10. Financial and Insurance Activities	1.37 (1.36%)	11 (1.15%)	0.31 (1.46%)
11. Real Estate Activities and Business Services	4.56 (4.54%)	46 (4.64%)	1.22 (5.76%)
12. Public Administration and Defense	0.02 (0.02%)	0 (0.00%)	0.00 (0.00%)
13. Education, Human Health, and Social Work Activities	0.33 (0.33%)	7 (0.67%)	0.10 (0.46%)
14. Arts, Sports, and Recreation Related Services	0.37 (0.37%)	6 (0.60%)	0.10 (0.47%)
15. Other Services	0.99 (0.99%)	35 (3.59%)	0.36 (1.71%)
Total	100.64 (100.00%)	988 (100.00%)	21.15 (100.00%)

Table 10. Total Inductive Effects of New High-Speed Railroad Tracks

Industrial category	Inductive effects		
	Output (million USD)	Employment (no. of jobs)	Income (million USD)
1. Agriculture, Forestry, and Fishing	23.47 (1.40%)	50 (0.03%)	0.81 (2.14%)
2. Mining and Quarrying	41.17 (2.46%)	929 (0.64%)	1.47 (3.91%)
3. Manufacturing	592.18 (35.43%)	19,889 (13.71%)	5.89 (15.66%)
4. Electricity, Gas, Steam, and Water Supply	75.12 (4.49%)	1,966 (1.35%)	2.11 (5.62%)
5. Construction	534.02 (31.95%)	8,539 (5.89%)	10.42 (27.73%)
6. Wholesale and Retail Trade	43.07 (2.58%)	28,060 (19.34%)	1.73 (4.59%)
7. Accommodation and Food Service Activities	40.72 (2.44%)	35,354 (24.37%)	0.78 (2.07%)
8. Transportation	27.98 (1.67%)	2,776 (1.91%)	0.90 (2.40%)
9. Information and Communications	42.25 (2.53%)	678 (0.47%)	0.49 (1.29%)
10. Financial and Insurance Activities	25.27 (1.51%)	1,405 (0.97%)	1.07 (2.85%)
11. Real Estate Activities and Business Services	131.02 (7.84%)	7,591 (5.23%)	6.59 (17.53%)
12. Public Administration and Defense	0.88 (0.05%)	15 (0.01%)	0.02 (0.06%)
13. Education, Human Health, and Social Work Activities	5.50 (0.33%)	1,706 (1.18%)	0.36 (0.95%)
14. Arts, Sports, and Recreation Related Services	12.56 (0.75%)	1,797 (1.24%)	0.51 (1.37%)
15. Other Services	76.42 (4.57%)	34,304 (23.65%)	4.44 (11.81%)
Total	1,671.62 (100.00%)	145,059 (100.00%)	37.58 (100.00%)

special events (e.g., natural disasters). Finally, the result can provide a good reference for policy makers when determining transportation policies or plans for the new HSR system in new planned cities, such as Sejong.

Nevertheless, some limitations remain. First, because of a lack of usable data for Sejong, alternative budgets for the new HSR station were used. Although the alternative budgets were chosen based on a location with a similar population to Sejong, assuming that might lead to similar demands, there might still be different economic effects of an actual new HSR station. Second, indirect methods were incorporated in a four-step process to develop the regional I-O table. This approach was the only reasonable way verified through previous related studies, but the indirect method could be less accurate in nature than the direct method. To acquire more precise results, a direct method should be applied. Third, several assumptions were made; for example, that industrial structures would be consistent and relatively unchanged within a certain period, thereby using a 2010 national I-O table to find the regional I-O table. Although CPI is applied to convert the prices of 2006 and 2013 to those of 2010 to overcome this limitation, the most recent national I-O table (if applicable) would provide more reliable results. For future research, further investigation at the national level needs to be conducted as investing a new HSR system could have economic effects on both the nation and the individual city. Developing new HSR-related multipliers (e.g., reduced travel time or increased accessibility) based on the I-O model is also worthy of exploration.

Shifting the viewpoint of policy analysis for transportation systems, sustained financial support is one of the instrumental components to keep transportation systems well operated and sustainable. Worldwide, many transportation projects have failed because of hasty policy judgements based on incorrect initial assumptions for estimated costs, construction periods, maintenance needs, and future demands. Naturally, this results in financial issues which interfere with the sustainability of the system. In part, economic effects captured by the proposed methodology can provide another rational source of evidence to pursue sustainable transportation investments. Although the proposed methodology cannot prevent mistakes that are made based on assumptions, at least it can support more mature policy decisions that consider economic effects on regional industries.

Acknowledgments

The authors wish to express their sincere gratitude to John S. Miller for introducing the principle of the input-output model and his invaluable feedback, and to Hyung Wook Choi for helping collect data sets essential to completing this work.

Author Contributions

The authors confirm contribution to the paper as follows: study conception and design: Changju Lee; data collection: Changju Lee and Kwan Kyo Oh; analysis and interpretation of results: Changju Lee and Jiaqi Ma; draft manuscript preparation: Changju Lee, Jiaqi Ma, and Kwan Kyo Oh. All authors reviewed the results and approved the final version of the manuscript.

References

1. Todorovich, P., D. Schned, and R. Lane. *High-Speed Rail: International Lessons for U.S. Policy Makers.* Lincoln Institute of Land Policy, Mass., 2011.
2. Peterman, D. R., J. Frittelli, and W. J. Mallett. *High Speed Rail (HSR) in the United States.* CRS Report of Congress, Congressional Research Service, Washington, D.C., 2009.
3. Rutzen, B., and C. M. Walton. *High Speed Rail: A Study of International Best Practices and Identification of Opportunities in the U.S.* Center for Transportation Research. The University of Texas, Austin, 2011.
4. Feigenbaum, B. *High-Speed Rail in Europe and Asia: Lessons for the United States.* Policy Study 418, Reason Foundation, Washington, D.C., 2013.
5. Okada, H. Features and Economic and Social Effects of the Shinkansen. *Japan Railway & Transport Review*, Special Feature, October Issue, 1994, pp. 9–16.
6. Sun, X., and L. Xu. Economic Effect of Chinese High-Speed Rail. *SERI China Review*, 2009, pp. 10–18.
7. Brand, D., M. R. Kiefer, T. E. Parody, and S. R. Mehndiratta. Application of Benefit-Cost Analysis to the Proposed California High-Speed Rail System. *Transportation Research Record: Journal of the Transportation Research Board*, 2001. 1742: 9–16.
8. De Rus, G. *The Economic Effects of High Speed Rail Investment.* Discussion Paper 2008-16 revised. Joint Transport Research Centre, International Transport Forum, Paris, France, 2012.
9. Taniguchi, M. *High Speed Rail in Japan: A Review and Evaluation of the Shinkansen Train.* The University of California Transportation Center, University of California, Berkeley, 1992.
10. Vople National Transportation Systems Center. *Evaluation of High-Speed Rail Options in the Macon-Atlanta-Greenville-Charlotte Rail Corridor.* Georgia Department of Transportation, Atlanta, Georgia, 2008.
11. Sasaki, K., T. Ohashi, and A. Ando. High-Speed Rail Transit Impact on Regional Systems: Does the Shinkansen Contribute to Dispersion? *The Annals of Regional Science*, Vol. 31, 1997, pp. 77–98.
12. Banister, D., and J. Berechman. *Transport Investment and Economic Development.* UCL Press, London, UK, 2000.
13. Bruinsma, F., E. Pels, H. Priemus, P. Rietveld, and B. van Wee. *Railway Development: Impacts on Urban Dynamics.* Physica-Verlag, Heidelberg, Germany, 2008.
14. De Brucker, K., C. Macharis, and A. Verbeke. Multi-Criteria Analysis in Transport Project Evaluation: An Institutional Approach. *European Transport*, No. 47, 2011, pp. 3–24.
15. Velasquez, M., and P. T. Hester. An Analysis of Multi-Criteria Decision Making Methods. *International Journal of Operations Research*, Vol. 10, No. 2, 2013, pp. 56–66.
16. Lee, T., and P. L. Mokhtarian. *An Input-Output Analysis of the Relationships between Communications and Travel for Industry.* Research Report UCD-ITS-RR-04-46. Institute of Transportation Studies, University of California, Davis, 2004.
17. Ham, H., T. J. Kim, and D. Boyce. Assessment of Economic Impacts from Unexpected Events with an Interregional Commodity Flow and Multimodal Transportation Network Model. *Transportation Research Part A: Policy and Practice*, Vol. 39, No. 10, 2005, pp. 849–860.
18. Guzman, A. F., and J. M. Vassallo. A Methodology for Assessing Regional Economic Impacts of Charging HGVs in Spain: An Integrated Approach through a Random Utility Based Multiregional Input-Output and a Road Transportation Network Model. *Transportation Research Record: Journal of the Transportation Research Board*, 2013. 2378: 129–139.
19. Parsons Brinckerhoff. *California High-Speed Rail Project: Economic Impact Analysis Report.* California High Speed Rail Authority, Sacramento, California, 2012.
20. Economic Development Research Group, Inc. *Economic Impact of Public Transportation investment: 2014 Update.* American Public Transportation Association, Washington, D.C., 2014.
21. Levine, L. *Job Loss and Infrastructure Job Creation during the Recession.* CRS Report for Congress, 7-5700 (R40080), Congressional Research Service, Washington, D.C., 2009.

22. Barrón, I., J. Campos, P. Gagnepain, C. Nash, A. Ulied, and R. Vickerman. *Economic Analysis of High Speed Rail in Europe*. Fundación BBVA, Bilbao, Spain, 2009.
23. Elhorst, J. P., and J. Oosterhaven. Forecasting the Impact of Transport Improvements on Commuting and Residential Choice. *Journal of Geographical Systems*, Vol. 8, 2006, pp. 39–59.
24. De Rus, G. *Economic Evaluation of the High Speed Rail*. The Expert Group for Environmental Studies, Elanders Sverige AB, Stockholm, Sweden, 2012.
25. Choi, S. G. Economic Impact Analysis of Regional Event. *Journal of Economic Studies*, Vol. 24, No. 4, 2006, pp. 19–40.
26. Multifunctional Administrative City Construction Agency. *Development Plan of Multifunctional Administrative City*. Sejong, Korea, 2006.
27. Sejong Metropolitan Autonomous City. *A Master Plan of Sejong City*. Sejong, Korea, 2013.
28. Sejong Metropolitan Autonomous City. *2013 Present Situation of Manufacturers by Industry and Scale*. Sejong, Korea, 2013.
29. National Agency for Administrative City Construction. *Discover NAACC – Mapping the City*. Sejong, Korea, 2017.
30. Choi, S. K. A Study on Estimation Method of Regional Input-Output Model. *The Korean Economic Association*, Vol. 58, No. 2, 2010, pp. 91–119.
31. Korea Research Institute for Human Settlements. *Regional Input-Output Analysis (III)*. Gyeonggi, Korea, 2003.
32. Daegu Gyeongbuk Development Institute. *An Application of the Regional Input-Output Modelling Technique to Daegu Region*. Daegu, Korea, 1994.
33. Morrison, W. I., and P. Smith. Non-Survey Input-Output Techniques at the Small Area Level: An Evaluation. *Journal of Regional Science*, Vol. 14, No. 1, 1974, pp. 1–14.
34. Schaffer, W. A., and K. Chu. Nonsurvey Techniques for Constructing Regional Interindustry Model. *Papers of the Regional Science Association*, Vol. 23, 1969, pp. 83–101.
35. Shen, T. Y. An Input-Output Table with Regional Weight. *Papers of the Regional Science Association*, Vol. 6, No. 1, 1960, pp. 113–119.
36. The Bank of Korea. *2010 Input-Output Tables*. Seoul, Korea, 2012.
37. Yun, Y. S. Development and Application of Regional Input-Output Analysis. Korean Research Institute For Human Settlements, Planning and Policy, Vol. 12, No. 1, 1994, 47–54.
38. Statistics Korea. *Population/Employment Census*. Seoul, Korea, 2013.
39. Lee, C. K. *Regional Input-Output Analysis*. Hakmun Publishing Co., Seoul, Korea, 2006.
40. McConnell, C. R., and S. L. Brue. *Economics: Principles, Problems and Policies*, 14th ed. McGraw-Hill, Boston, 1999.
41. Litman, T. *The First Casualty of a Non-Existent War*. Victoria Transport Policy Institute, British Columbia, Canada, 2011.
42. The Korea Railroad Corporation. *2013 Final Adjusted Budgets of Honam Korea Train Express (KTX) Exclusive Stations*. Daejeon, Korea, 2013.
43. The Korea Railroad Corporation. *2013 Total Budgets of Honam Korea Train Express (KTX)*. Daejeon, Korea, 2013.

The Standing Committee on Intercity Passenger Rail (AR010) peer-reviewed this paper (18-00677).

The views expressed in this research are those of the authors and do not necessarily reflect those of the authors' institute, university, and government. The institue, university, and government do not endorse any of the views stated, nor any claims or representations made in this research. The information and expression of opinion contained in this research are not intended to be a comprehensive study, nor to provide actuarial advice or advice of any nature and should not be treated as a substitute for specific advice concerning individual situations.

TRR
JOURNAL OF THE TRANSPORTATION RESEARCH BOARD

Article

Transportation Research Record
2018, Vol. 2672(10) 260–268

DOI: 10.1177/0361198118796359
journals.sagepub.com/home/trr
SAGE

Impact of Train Drivers' Cognitive Responses on Rail Accidents

Bahareh Hani Tabai[1], Morteza Bagheri[1], Vahid Sadeghi-Firoozabadi[2], Vahideh Shahidi[3], and Hadi Mirasadi[4]

Abstract

Despite the innovations in automatic train control (ATC) systems to reduce the risk of driver error, many rail accidents still occur due to defects in these systems, emphasizing the essential role of the driver in preventing rail accidents and proper control of the train. This paper studies the influence of drivers' cognitive performance, including attention and visual perception, on the occurrence of rail accidents. The research is conducted using so-called Ex-Post facto method on a random sample of 56 train drivers with a minimum of three years of experience. The research instruments included drivers' cognition test system including WAFV (perception and attention function) sustained attention test, COG (cognitrone) selective attention test, LVT (visual pursuit) visual perception test, demographic questionnaire, and drivers' safety history. Results of this research showed that there is no significant relationship between age and education level of train drivers, and rate of occurrence of rail accidents. A comparison on train drivers' cognitive characteristics, between drivers with accident record(s) and those without, showed that the drivers who had experienced rail accident(s) had lower levels of sustained attention. However, no significant difference was found between the two groups in selective attention and visual perception. Investigating the association of drivers' ages with their levels of sustained attention, the drivers with high levels of sustained attention were found to be significantly older than other drivers. According to practical implications of these findings, cognitive rehabilitation courses are recommended for train drivers to attenuate the risk of rail accidents.

In February 2012, in Ontario, Canada, a train driver's lack of attention to red signal and consequent failure to reduce the train speed properly ended up with derailment of locomotive and five cars, three deaths, 45 injuries, and leak of 4,300 l of diesel (*1*). In April 2014, in Bakran station in Iran, a train driver just ignored the yellow signal and failed to reduce the train speed, causing derailment of eight cars which ended up with five injured passengers. Similarly, in June 2014, a Tehran–Mashhad train driver did not comply with the general rules of mobility, passed red signal and hit the freight stopped at the station; accordingly, six cars derailed and one passenger passed away. Train driver's inattention was reported as the main cause of the rail accident (*2*). Relatively frequent occurrence of accidents caused by train driver's inattention raises the question, How could it be possible that some train drivers do not pay attention to the signals or keep moving despite having seen the red signal?

According to Islamic Republic of Iran Railways (RAI), during 2009–2014, more than 1,900 rail accidents occurred in Iran. Among others, human factors were recognized as the cause of the largest share of these accidents (36%). Among all human factors, train driver's fault has been found as responsible for the largest fraction of rail accidents (47%). Also, inattention to red signal or speed control signals has caused 35% of the accidents (*3*). This shows the importance of driver's attention to red and speed reduction signals.

Studies show that the occurrence of red signal passing is a result of one of three situations: first, inappropriate train driver's conditions for seeing the signal, abnormality of signal operation, or lack of sufficient knowledge to understand the signal; second, the train driver's inattention (or inadequate attention); and third, wrongly showing the signal on the route or failure to provide sufficient information by the controller (*4*).

[1]School of Railway Engineering, Iran University of Science and Technology, Tehran, Iran
[2]Department of Psychology, Shahid Beheshti University, Tehran, Iran
[3]Department of Psychology, Wilfird Luier University, Waterloo, ON, Canada
[4]Tehran University, Tehran, Iran

Corresponding Author:
Address correspondence to Morteza Bagheri: Morteza_Bagheri@iust.ac.ir

Working conditions of a train driver increase the chance of a rail accident. These conditions include job complexity, high levels of stress, exhaustion caused by monotonous nature of the job, unfavorable cabin conditions, long working shifts, and need for continuous attention to signals (*5*). Tolerating these conditions while maintaining safety is possible only with a proper level of sustained attention (*6*).

Research on train drivers' attention began in the 1970s. The importance of train drivers' vigilance and attention have been emphasized in many studies (e.g., *4*, *6–8*). Michaut and McGaughey concluded that, the highly undesirable conditions in which a train driver works (e.g., unfavorable nature of locomotive cabin in terms of temperature, humidity, and different types of noise) largely distort the driver's attention (*9*). Similarly, Hilderbrandt, Rohmert, and Rutenfranz introduced repetitive duties and the monotonous nature of the train driving job as the main causes behind decreased vigilance and attention in fatigue times (*10*). Haga reviewed red-signal-passing accidents and argued that lack of sustained attention is the main cause of the drivers' distraction and inattention (*7*). Presenting a discussion on unfavorable conditions of locomotive cabins, Smiley also suggested lack of sustained attention as the main cause of rail accidents (*8*). In a supplementary study, Edkins and Pollock reviewed rail accidents in Australia during 1990–1994 and found lack of the train driver's sustained attention (especially in the form of inattention to signals) as the most frequent human error (*6*). They stipulated that the unfavorable conditions and repetitive nature of the train-driving job necessitate the driver's sustained attention, and found no significant relationship between train driver's age and deal of experience and the rate of occurrence of rail accidents (*6*). The ability to maintain vigilance and sustained attention under monotonous conditions is a requirement for any train driver (*11*). Philips and Sagberg introduced inattention or inappropriate attention by train driver as the most common factor among the set of the occurrence factors of passing-red-signal accident (*4*).

Some studies have discussed distraction of train driver in relation to job conditions. Naweed introduced train driver's inattention and distraction as a cause of passing-red-signal accidents (*12*). By using focus groups in eight railway regions across New Zealand and Australia and simulating a train driver's job conditions, he found the drivers' distraction as an implication resulted from drivers' job conditions and recognized the accidents caused by train drivers' distraction as being a result of a chain of factors including excessive working hours, control room inconsistencies, and limited visibility at intersection station. Many studies have been performed on the role of train drivers' job conditions in drivers' distraction and inattention and occurrence of rail accidents (*6*, *7*, *13*).

In addition, many studies have considered the role of train driver's fatigue in the occurrence of rail accidents. Recently, Filtness and Naweed considered passing red signal as the main cause of rail accidents (*14*). They found inappropriate and erratic working shifts of train drivers as the main cause of driver fatigue and inefficiency in fulfilling their tasks. Many similar studies found driver fatigue as the main cause of rail accidents (*15–18*). On the other hand, Dunn and Williamson concluded that a little increase in train driver's cognitive level could considerably reduce negative effects of the monotonous nature of driver's job and driver fatigue (*11*). Other studies highlighted the importance of the role of selective attention in road driving, identifying selective attention as an important predictor of road accidents (*19–21*).

Luke et al. considered the train driver's job as a visual job in the first place (*22*), because the driver has to monitor inside and outside of the locomotive cabin every now and then. Poor visual accuracy of the train driver can cause red-signal-passing accidents (*23*). Numerous studies have been conducted to check train drivers' visual behavior during their job (*22*, *24–26*). Recently, Horiguchi et al. surveyed drivers' visual behavior or visual pattern during driving and emphasized the important role played by the driver's visual accuracy in this job and the occurrence of rail accidents (*25*).

A review of previous studies shows that many studies have ended up introducing train drivers' job conditions as resulting in fatigue and inattention, affecting the drivers' performance; also, numerous studies have been performed to identify the cognitive factors that are affected by these conditions, emphasizing the role of driver (in terms of sustained attention) in accident management (*6–10*). On the other hand, to the best of the authors' knowledge, no study has been reported on the evaluation of the level of sustained and selective attention of train driver, even though numerous studies have been performed within the scope of road transportation to evaluate selective and sustained attention of car drivers and its relation to driving style, crime rates, traffic violations, and accidents, with some studies introducing the level of selective attention as an important predictor for driving accidents (*19*, *20*, *27*). The role of visual perception of the train driver has also been studied, but the studies in the scope of railway transportation have mostly surveyed drivers' behavior when driving (*22*, *24–26*) with no study conducted to examine the role of drivers' visual perception in the occurrence of rail accidents. All in all, no study has been done to evaluate the role of cognitive characteristics in the occurrence of rail accidents, and this is the main motivation behind performing this research.

The main purpose of this research is to study the role of train drivers' selective and sustained attention and visual perception in the occurrence of rail accidents. For this purpose, train drivers' attention and visual perception were evaluated using cognitive Vienna Test System (WAFV computerized sustained attention test, COG computerized selective attention test, and LVT computerized visual perception test). Demographic data was collected through a questionnaire. In the next step, records of previous accidents were supplied by the Department of Railway Safety and Network Monitoring of RAI. To study the relationship between cognition characteristics and the occurrence of accidents, statistical analysis of test results and the number of accidents committed by train drivers were used.

Methodology

This section provides definitions and concepts of cognition characteristics, gives a thorough discussion on cognition tests and the procedure followed in this research, and presents demographic characteristics and work experience of train drivers taking part in this research. Intended definition of experience is the years of service as a locomotive engineer.

Definitions and Concepts

Attention is of the most important aspects of human cognition which is also known as awareness (*28*). Attention is a cognitive process in which a person focuses selectively on a particular subject while neglecting other disturbing subjects and stimulus. Here is an example of attention: listening to a person's words accurately in the presence of different competitive environmental stimulus like noise or pictures; this disturbing stimulus can be either physical or mental, that is, apart from environment or human. In psychology, attention is known as consciousness and reaction to it; attention is also interpreted as a person's ability to focus on and continue an activity (*29*). In 1890, William James said, "Everyone knows what attention is; it is the taking possession by the mind, in clear and vivid form, of one out of what seems several simultaneously possible objects or trains of thought. Focalization, concentration of consciousness is of its essence. It implies withdrawal from some things in order to deal better with others." (*30*)

Sohlbert and Mateer presented a hieratical model to evaluate post-coma attention in patients with brain damages. They classified attention into four levels according to its function: sustained attention, selective attention, alternative attention, and divided attention (*31*). Sustained attention is the most fundamental level of attention which provides a basis for all other levels of attention; a defect at this level may lead to defects at other levels of attention (*31*). Requiring drivers' sustained attention, the repetitive and monotonous nature of train-driving job increases probability of making mistakes (*32*). Selective attention refers to the capacity of maintaining attention in the presence of disturbing stimuli. At this level of attention, the person must stay focused on an activity or subject in the presence of disturbing stimuli, so that the stimuli may not interrupt them (*31*).

Participants

A total of 203 train drivers live within Tehran, Iran, of which 56 (obtained from Cochran Formula) drivers were sampled randomly at a precision of 90%. Average age of the sampled train drivers was 39.18 years (range: 28–63; standard deviation (SD): 6.921). 28% of the sampled drivers had college certificates, 14% of them had associate diploma, 41% held bachelor degrees, and 16% had master degrees. In order to realize the main research objective, the train drivers were divided into two groups: those who had experienced at least one accident (hereinafter referred to as "with accident") and those had no accident so far (hereinafter referred to as "without accident"), according to their safety histories.

- **Train drivers without accident.** This group was composed of 34 train drivers who had not been involved in any rail accident according to their safety history. Average age of this group was 38.94 years (range: 28–63; SD: 7.75), average work experience was 15.62 years (range: 3–41; SD: 8.62), and 20%, 11%, 47%, and 20% of them had college certificates, associate diploma, bachelor degrees, and master degrees, respectively.
- **Train drivers with accident.** This group was made up of 22 train drivers who had one or more accidents in their safety history. Average age of this group was 39.55 years (range: 31–55; SD: 5.54), average work experience was 16.68 years (range: 5–32; SD: 6.20), and 40%, 18%, 31%, and 9% of them had college certificates, associate diploma, bachelor degrees, and master degrees, respectively.

Method of Research

This research was performed by School of Railway Engineering at Iran University of Science and Technology in cooperation with Psychological Department of Shahid Beheshti University. Prior to undertaking the tests, training sessions were held separately for every subject in the sample, with the subjects filling questionnaire on demographic data. The participants were then provided with a suitable environment for test execution; psychological standards were respected before the test execution, and the project observer was present in the test environment. Because of software access limitations, the tests were performed separately for each subject, so that the entire process took a total of 70 min to execute (inclusive of break times between successive tests).

Once the train drivers completed the test, their safety history containing their records of accidents, if any, was received from the Department of Safety and Network Monitoring of RAI, according to which the drivers were divided into two groups. For the purpose of this research, the term accident may refer to an event which occurred solely because of a train driver's error, that is, signal passed at danger (SPAD).

According to the third European railway package (European Railway Commission, 2007), only drivers with driving certificate are allowed to drive trains across the European rail network. According to this certificate, train drivers have some basic requirements in terms of age, education level, mental health, sufficient knowledge, train-driving skills, and cognitional characteristics. This standard places an emphasis on drivers of passenger trains. In 2013, Schuhfried GmbH presented the cognitive Vienna Test System according to the third European railway package for traffic via three transportation modes (road, railway, and aviation). In the package published for train drivers, three tests, namely WAFV, COG, and LVT, were emphasized in terms of attention and visual perception. Train drivers' standard by the Railway Safety and Standard Board in 2013 states that train driver applicants must be cognitively evaluated, and that execution of WAFV computer test is mandatory for evaluation of vigilance and sustained attention of train drivers (*33*).

Perception and Attention Function (WAFV; Sturm [34]). WAFV test is a credible computerized test to evaluate people's level of sustained attention in vigilance state (*34*). Reliability of the test was confirmed by a Cronbach's alpha of 0.96, and construct validity was calculated as 0.42 by means of factor analysis. This test took 32 min to perform. In this, a grey square appears and disappears reciprocally on a white screen. When the stimulus color changes to black, the examinee must press a green button on the keyboard. Total number of stimuli is 900, of which 50 (5%) are black and must be recognized by the examinee during the test. Main variables measured in this test are omission error (failure to respond to target stimulus within 1,500 ms) and average reaction time (average time before a true reaction is taken to the target stimulus). Omission error measures possible disorder in a person's sustained attention, and average reaction time estimates processing rate of brain in vigilance state. According to this test, the subjects were categorized as follows: the subjects with scores less than 15% of average score of the entire sample were classified as being in warning state in terms of sustained attention, those who attained scores of 16 to 84% of average score of the entire sample were classified as normal, those with scores above 85% of average score of the entire sample were classified as being in good state.

Reduction of a train driver's sustained attention occurs due to monotonous condition of the cabin and routine job flow. Given weak stimuli and monotonous environment of train cabin, WAFV test simulates cabin conditions in the course of driving and measures the driver's sustained attention in constant conditions. In WAFV, the respondent is given visual and auditory stimuli that occasionally diminish in intensity. The respondent's task is to respond to these occasional cases; sustained attention is measured at 25% of the stimuli, while vigilance is measured at 5% of the stimuli. In train driving, given the absence of stimuli, train driver's vigilance is measured.

Cognitrone (COG; Wagner and Karner [35]). COG test is used for assessment of attention and concentration through comparison of figures with regard to their congruence. COG test is a computerized test that evaluates a person's selective attention in the process of comparing shapes and their proportionality (*35*). The reliability of this test was confirmed by a Cronbach's alpha of 0.95 and the validity coefficient was calculated as 0.68 by using an artificial neural network. The execution time of this test was 15 min. The examinee was trained before the test. This test lets the participants compare a particular shape with four other ones and decide on their proportionality. In case the aforementioned shape is the same as one of the other four shapes, the examinee shall press a green button on keyboard, otherwise, a red button must be pressed. In this test, examinee's attention is evaluated by precision and reaction time. The main variable evaluated in this test is average time to pass wrong stimulus, indicating disorders in driver's selective attention and energy level to maintain a certain level of precision. Selective attention level is deemed as acceptable provided the participant succeeds in correctly answering at least 85% of the items. Any score below the 85% may raise concerns about appropriate job fulfilment.

Visual Perception (LVT; Biehl [36]). LVT test is a computerized test that measures the subject's visual perception accuracy (*36*). The reliability of this test was confirmed by a Cronbach's alpha of 0.96. This test evaluates subject's ability to navigate visually and ability to get a general view of the route. The test begins with a combined instruction and practice phase. Provided eight training items are worked out with fewer than three errors, the respondent moves on to the actual test phase items. The

Table 1. Comparison of Demographic Characteristics between the Two Groups by *t*-Test and Chi-Square Test

Demographic characteristics	Group		DF	*t*	Chi-square	Sig
	With accident	Without accident				
Age	39.55 (5.54)	38.94 (7.75)	54	0.31	–	0.75
Experience	16.68 (6.20)	15.62 (8.62)	54	0.5	–	0.61
Education level	–	–	3	*	4.17	0.24

Note: *Significant at the 0.05 Level (2-tailed).

Table 2. Comparison of Cognitive Characteristics between the Two Groups by *t*-Test

Measures	Group		Mean differences	*t*	Sig
	With accident	Without accident			
Mean time correct rejection at COG (min)	3.06 (1.008)	2.96 (0.96)	–0.1	0.43	0.66
Number of missed reactions at WAFV	3.59 (2.63)	2.12 (1.62)	–1.47	2.35	0.025*
Mean reaction time at WAFV(min)	520.41 (77.75)	486.97 (59.23)	–33.43	1.82	0.07
Score at LVT	54.68 (18.47)	61.79 (15.62)	7.11	1.54	0.12

Note: *Significant at the 0.05 Level (2-tailed).

respondent is provided with an array of lines and must find the end of a specified line as quickly as possible. The respondent can work at their own speed. They must find and dial the path among nine meandrous paths. After this stage and some practice, the execution stage is launched wherein the respondent is provided with 80 pictures and must press red and green buttons together to see each picture; subsequently, as soon as they find the end of the specified line, they must release the buttons and dial the target number. The main research variables examined in this test are the number of true paths and reaction time, which are measured by accuracy, pace, and number of correctly identified items by the respondent and shows visual perception accuracy.

Statistical Analysis

In this research, three groups of analysis were performed. The first group compared the two groups of train drivers (with and without accident) in terms of demographic data. For this purpose, independent *t*-test, Chi-square test and ANOVA at the significance level of 0.05 were used. The second group of analysis compared the two groups of train drivers in terms of cognitive characteristics using independent *t*-test at the significance level of 0.05. The final group of the analysis was performed to show the correlation between demographic variables and cognition using Pearson's correlation test at the significance level of 0.05. All the aforementioned analyses were performed using the SPSS software package (version 24.0, SPSS an IBM Company, Chicago, IL).

Results

Demographic Characteristics Analysis in Both Groups

Table 1 shows the results of a comparison on average age and job experience between the two groups of train drivers using independent *t*-test, and further compares the level of education in either group using independent Chi-square test.

As seen in Table 1, no significant difference was found between the two groups of train drivers as far as demographic data was concerned.

Cognitional Characteristics Analysis in Both Groups

Table 2 shows average and standard deviation of each statistic in either of the groups and the difference in the average value of each statistic between the two groups using independent *t*-test at a significance level of 0.05.

According to Table 2, it can be seen that there is a significant difference in average of omission error between

Table 3. Model Summary

Step	−2 Log likelihood	Cox & Snell R square	Nagelkerke R square
1	66.599[a]	0.140	0.190

[a]Estimation terminated at iteration number 4 because parameter estimates changed by less than .001.

the two groups of train drivers; accordingly, the drivers without accident had significantly less omission error (T: 2.35, $p < 0.05$). In other words, there is a significant difference in the sustained attention in vigilance conditions between the two groups of train drivers. It shows that train drivers without accident have higher sustained attention level compared with the other group.

Correlation Analysis

In this section, correlations between test variables and the relation between test variables and train drivers' demographic data are studied using Pearson's correlation coefficient at the significance level of 0.05.

Correlation analysis shows that average time of passing wrong items in COG has a significant relationship with the job experience ($r = 0.3$, $p < 0.05$), but no significant relationship was found between this variable and others. In other words, distraction of selective attention of train drivers has a significant, negative relationship with their job experience.

A significant, positive relationship was discovered between omission error in WAFV and train drivers' number of accidents ($r = 0.3$, $p < 0.05$). In addition, a significant, positive relationship was observed between omission error in WAFV test and average reaction time ($r = 0.67$, $p < 0.01$), meaning that lack of sustained attention is directly related to a driver's brain performance in vigilance conditions, so that the drivers who achieved higher sustained attention level showed higher brain performance in vigilance conditions.

Drivers' visual perception had a negative, significant relationship ($r = -0.3$, $p < 0.05$) with their age, meaning that an increase in the driver's age tends to decrease their visual cognitive accuracy. Visual perception was also observed to have a negative, significant relationship with the driver's job experience ($r = -0.26$, $p < 0.05$).

Regression Analysis

Logistic regression analysis was performed to determine the predictor variables. For this purpose, the two groups of train drivers were taken as dependent variables, and sustained attention, selective attention, and visual perception were considered as predictor (independent) variables. The total of 56 train drivers were tested and the model analysis was significant (Chi-square = 8.44, df = 3, $p < 0.05$). As can be seen in Table 3, this model measures the variance of accident between 14 and 19%. Table 4 presents coefficients, Wald statistic, degrees of freedom (df), and probability values for each predictor variable. The results show that only sustained attention was able to significantly predict the membership in each of the two groups of train drivers. With increasing sustained attention, likelihood of train drivers falling in the group without accident increased. The results also show that selective attention and visual perception of train drivers cannot significantly predict the possession of a subject to either group (with accident and without accident).

Discussion and Conclusion

As discussed earlier, WAFV sustained attention test, COG selective attention test, and LVT visual perception test were used to evaluate attention and visual perception of train drivers.

Results showed no significant difference in age, job experience, and level of education between the two groups of train drivers. The result is in agreement with road studies, indicating independence of the occurrence of accidents on drivers' demographic characteristics including age and experience (*37*). This suggests that, as far as improving safety in railways and decreasing human errors are concerned, researchers should put aside demographic characteristics and rather look for causes of rail accidents in cognitive characteristics of train drivers; this shows the importance of the present research in studying the role of sustained and selective attention of train driver in rail accidents.

A significant relation was found between sustained attention of train drivers in either group and average level of sustained attention in both groups. It shows that sustained attention of the train drivers without accident is significantly higher than that of those with accident, that is, the train drivers who had accident(s) in their safety history exhibited weak performance in this test. Regarding previous papers which emphasized the monotonous and repetitive nature of train drivers' job and the importance of sustained attention for train drivers (*6–8*), results of this paper further confirmed this hypothesis and clearly proved the role of sustained attention in the occurrence of rail accidents. These results show that enhancement of sustained attention of train drivers could reduce the occurrence of accidents by them. Driver's selective attention has been emphasized in road studies, and many researchers have introduced selective attention as an important predictor of accidents (*19–21*). This

Table 4. Variables in the Equation

		B	S.E.	Wald	df	Sig
Step 1[a]	*COG S1 MTRN Mean time correct rejection	0.185	0.399	0.214	1	0.644
	*LVT S1 S Score {based on viewing times	–0.020	0.020	1.057	1	0.304
	*WAFV S2 ASN Number of missed reactions	0.344	0.151	5.224	1	0.022*
	Constant	–0.755	2.146	0.124	1	0.725

Note: *Significant at the 0.05 Level (2-tailed).

Table 5. Analyze Variance of Age and Work Experience at Three Levels of Sustained Attention

Measure	Statistical indicators	Sum of squares	df	Mean square	*F*	Sig
Age	Between groups	1313.08	2	656.54	26.339	0.000*
	Within groups	1321.13	53	24.92		
Experience	Between groups	1398.46	2	699.23	19.73	0.000*
	Within groups	1877.46	53	35.42		

Note: *Significant at the 0.01 Level (2-tailed).

Table 6. Scheffe Post Hoc Test for Age and Experience at Various Levels of Sustained Attention

		Age		Experience	
Level of sustained attention		Mean differences	Sig	Mean differences	Sig
Low	Normal	3.01	0.26	2.56	0.5
	Good	−18.44*	0.000	−19.66*	0.000
Normal	Low	3.01	0.26	–2.56	0.5
	Good	–21.45*	0.000	–22.23*	0.000
Good	Low	18.44*	0.000	19.66*	0.000
	Normal	21.45*	0.000	22.23*	0.000

Note: *Significant at the 0.01 Level (2-tailed).

result can be justified by mentioning the different nature of driving a train rather than a car. For a train driver, because of the low number of stimuli and monotonous nature of the job, the job requires sustained attention, while on a road, selective attention is necessary because of the large number of distracting stimuli (*38*). Driving on a road also needs an appropriate level of selective attention as there are lots of distracting stimuli in the driving environment that must be seen, understood, and responded to by the driver appropriately (*39*).

The relationship between demographic variables and cognitive variables of train drivers showed that selective attention is reduced with increasing job experience, while Amado et al. showed, using COG test, that there is no significant relationship between driver's age, level of education, and driving experience and driver's selective attention (*20*). The result also revealed that, visual perception decreases with increasing age and job experience.

Data resulting from ANOVA among the classification of people in this test's statistics and their age and job experience was performed, and results are as follows. Table 5 shows ANOVA analysis of different levels of train drivers' sustained attention in terms of age and job experience.

Regarding Table 5, it is observed that there is a significant difference between groups in terms of age and job experience and this difference is significant at the level of $p < 0.001$. Regarding Scheffe post hoc test in table 6, there is a significant difference between high and normal sustained attention groups and the average of high and low sustained attention groups.

It can be seen in Table 6, there is a meaningful difference between high-sustained attention groups and groups with low-sustained attention who are normal in terms of age and job experience. These results show that train drivers who have a high level of sustained attention are significantly older than other train drivers and have higher experience. In road studies also the relation between age and drivers' attention has been estimated to be positive but the relation between job experience and drivers' attention has been denied (*35, 37*). This shows that the job of a train driver needs more skill and profession and a train driver improves their sustained attention level by the increase in their job experience because of the job repetitiveness and need for sustained attention. The results of regression analysis also showed that only the level of a train driver's sustained attention can predict the occurrence of accident. It should be noted that restricted sampling represents the main limitation of this study.

Accident can be caused by equipment failure, errors by crew member(s), et cetera. Results of this research showed that a train driver's sustained attention plays a determinant role in the occurrence of rail accidents. In addition, the results showed that the occurrence of a rail accident is independent of a train driver's age, job experience, and level of education, while it is closely related to the sustained attention of the driver, so that improving sustained attention of train drivers can decrease rail accidents and enhance railway safety. Given that sustained attention is affected by emotions, job conditions and family relationships play important roles in train drivers' sustained attention (*40*). Future studies may examine the effect of counseling sessions on the improvement of sustained attention level and transportation safety. In this study, all locomotive drivers were men (according to Iran's laws), so the role of gender was not investigated. Focus of this research was on studying the role of sustained attention of train drivers. For future studies, it is suggested to study the role of another level of attention (such as divided attention) in the occurrence of rail accidents.

Acknowledgments

The research here was partially funded by the Islamic Republic of Iran Railways project, Investigating the Effect of Train Drivers' Cognitive Characteristics on Rail Accidents and Promoting the Occupation Personality of the Train Drivers. We thank all those involved with this project especially the Fleet Deputy of the Islamic Republic of Iran Railways for providing us with case study data.

Author Contributions

The authors confirm contribution to the paper as follows: study conception and design: M.Bagheri, V.Sadeghi-Firoozabadi; data collection: B. Hani Tabai; analysis and interpretation of results: M.Bagheri, V.Sadeghi-Firoozabadi, B. Hani Tabai, V.Shahidi, H.Mirasadi; draft manuscript preparation: B. Hani Tabai. All authors reviewed the results and approved the final version of the manuscript.

References

1. Transportation Safety Board of Canada. *Railway Investigation Report R12T0038*, Burlington, 2016.
2. Iran Ministry of Roads and Urban Development. *Rail Accidents Caused by the Train Driver*, 2016.
3. Iran Bureau of Rail Mobility. *Statistical Data of Iran Railway Accident (2009–2014)*, 2015.
4. Phillips, R. O., and F. Sagberg. *Helping Train Drivers Pass Signals Safely: Lessons from Ten Case Studies*. Transportøkonomisk institutt, 2010.
5. Railway Safety Promotion Unit. *Job Requirement of Train Driver*, 2015.
6. Edkins, G. D., and C. M. Pollock. The Influence of Sustained Attention on Railway Accidents. *Accident Analysis and Prevention*, Vol. 29, 1997, pp. 533–539.
7. Haga, S. An Experimental Study of Signal Vigilance Errors in Train Driving. *Ergonomics*, Vol. 27, No. 7, 1984, pp. 755–765.
8. Smiley, A. M. The Hinton Train Disaster. *Accident Analysis and Prevention*, Vol. 22, No. 5, 1990, pp. 443–455.
9. Michaut, G. M. E., and T. P. McGaughey. *Work Conditions and Equipment Design in Diesel Locomotives: Feasibility Study and Recommendations*. Canadian Institute of Guided Ground Transport, Queen's University, Canada, 1972.
10. Hildebrandt, G., W. Rohmert, and J. Rutenfranz. 12 and 24 h rhythms in error frequency of locomotive drivers and the influence of tiredness. *International Journal of Chronobiology*, Vol. 2, No. 2, 1974, pp. 175–180.
11. Dunn, N., and A. Williamson. Driving Monotonous Routes in a Train Simulator: The Effect of Task Demand on Driving Performance and Subjective Experience. *Ergonomics*, Vol. 55, No. 9, 2012, pp. 997–1008.
12. Naweed, A. Psychological Factors for Driver Distraction and Inattention in the Australian and New Zealand Rail Industry. *Accident Analysis and Prevention*, Vol. 60, 2013, pp. 193–204.
13. Mazloumi, A., Z. Kazemi, G. N. Saraji, and S. Barideh. Quality of Working Life Assessment among Train Drivers in Keshesh Section of Iran Railway. *International Journal of Occupational Hygiene*, Vol. 6, No. 2, 2014, pp. 50–55.
14. Filtness, A. J., and A. Naweed. Causes, Consequences and Countermeasures to Driver Fatigue in the Rail Industry: The Train Driver Perspective. *Applied Ergonomics*, Vol. 60, 2017, pp. 12–21.
15. Buck, L., and F. Lamonde. Critical Accidents and Fatigue Among Locomotive Engineers. *Safety Science*, Vol. 16, No. 1, 1993, pp. 1–18.
16. Wilde, G. J. S., and J. F. Stinson. The Monitoring of Vigilance in Locomotive Engineers. *Accident Analysis and Prevention*, Vol. 15, No. 2, 1983, pp. 87–93.
17. Austin, A., and P. D. Drummond. Work Problems Associated with Suburban Train Driving. *Applied Ergonomics*, Vol. 17, No. 2, 1986, pp. 111–116.
18. Rajaratnam, S. M. W., and C. B. Jones. Lessons About Sleepiness and Driving from the Selby Rail Disaster Case: R v Gary Neil Hart. *Chronobiology International*, Vol. 21, No. 6, 2004, pp. 1073–1077.
19. Arthur, W., and D. Doverspike. Locus of Control and Auditory Selective Attention as Predictors of Driving Accident Involvement: A Comparative Longitudinal Investigation. *Journal of Safety Research*, Vol. 23, No. 2, 1992, pp. 73–80.

20. Amado, S., M. Koyuncu, and G. Kaçaroğlu. Evaluation of Factors Affecting Safe Driving: Demografic Factors, Experience, Personality and Psycho-Technical Evaluation. *Türk Psikoloji Dergisi*, Vol. 19, No. 53, 2004, pp. 45–47.
21. Ferreira, I. S., M. R. Simões, and J. Marôco. Cognitive and Psychomotor Tests as Predictors of On-road Driving Ability in Older Primary Care Patients. *Transportation Research Part F: Traffic Psychology and Behaviour*, Vol. 21, 2013, pp. 146–158.
22. Luke, T., N. Brook-Carter, A. M. Parkes, E. Grimes, and A. Mills. An Investigation of Train Driver Visual Strategies. *Cognition, Technology & Work*, Vol. 8, No. 1, 2006, pp. 15–29.
23. Groeger, J. A., M. F. Bradshaw, J. Everatt, N. Merat, and D. Field. *Pilot Study of Train Drivers' Eye-movements.* University of Survey Technical Report for Rail Safety and Standards Board, London, 2003.
24. Park, K. S., and T. Ohkubo. A Study on Visual Characteristics of Train Drivers for the Differences of Display Layout. *Japanese Journal of Ergonomics*, Vol. 31, No. 1, 1995, pp. 31–38.
25. Horiguchi, Y., T. Suzuki, T. Sawaragi, H. Nakanishi, and T. Takimoto. Extracting Train Driver's Eye-Gaze Patterns Using Graph Clustering. *IFAC-PapersOnLine*, Vol. 49, No. 19, 2016, pp. 621–626.
26. Mizuta, J. Visual Behavior of Motive Power Car Drivers: Ergonomic Study of Driving Information. *Bulletin of Railway. Labour Science Research Institute*, Vol. 28, 1975, pp. 129–142.
27. Ferreira, I. S., M. R. Simões, and J. Marôco. Cognitive and Psychomotor Tests as Predictors of On-road Driving Ability in Older Primary Care Patients. *Transportation Research Part F: Traffic Psychology and Behaviour*, Vol. 21, 2013, pp. 146–158.
28. Sternberg, R. J. Testing and Cognitive Psychology. *American Psychologist*, Vol. 36, No. 10, 1981, p. 1181.
29. Anderson, J. R. *Cognitive Psychology and Its Implications.* WH Freeman/Times Books/Henry Holt & Co, 1990.
30. James, W. The Principles of Psychology, New York: Dover, 1950
31. Sohlberg, M. M., and C. A. Mateer. *Introduction to Cognitive Rehabilitation: Theory and Practice.* Guilford Press, 1989.
32. Sümer, N., T. Lajunen, and T. Özkan. Big Five Personality Traits as the Distal Predictors of Road Accident. *Traffic and Transport Psychology: Theory and Application*, 2005, p. 215.
33. Rail Safety and Standards Board. *Rail Industry Standard for Train Driver Selection.* London, 2013.
34. Sturm, W. *Manual Perception and Attention Function (WAFV)*, 2011.
35. Karner, T., and M. Wagner. *Manual Cognitrone (COG).* Schuhfried GmbH, Modling, 2001.
36. Biehl, B. *Manual Visual Perception (LVT)*, 1996.
37. Reimer, B., B. Mehler, Y. Wang, and J. F. Coughlin. A Field Study on the Impact of Variations in Short-Term Memory Demands on Drivers' Visual Attention and Driving Performance Across Three Age Groups. *Human Factors Journal of Human Factors and Ergonomics Society*, Vol. 54, No. 3, 2012, pp. 454–468.
38. Zare, H., F. Valiyollah, A. Alipour, and M. Nazer. Comparison Types of Attention Between Drivers With and Without Accident. *Journal of Transportation*, 2012.
39. Trick, L., and J. T. Enns. *Driving and Selective Attention: A Conceptual Framework for Understanding the Role of Selective Attention in Driving.* Brown ID, Haslegrave C. Taylor SP Vis. Veh. X., Amsterdam, North-holl., 2004.
40. Sanjani, M. S. *The Impact of Using Transcranial Direct-Current Stimulation (TDCS) on Executive Functions of Brain*, 2015.

The Standing Committee on Railroad Operational Safety (AR070) peer-reviewed this paper (18-03139).

TRR
JOURNAL OF THE TRANSPORTATION RESEARCH BOARD

Article

Transportation Research Record
2018, Vol. 2672(10) 269–277

DOI: 10.1177/0361198118792751
journals.sagepub.com/home/trr
SAGE

Video Analytics for Railroad Safety Research: An Artificial Intelligence Approach

Asim Zaman[1], Xiang Liu[1], and Zhipeng Zhang[1]

Abstract

The volume of video data in the railroad industry has increased significantly in recent years. Surveillance cameras are situated on nearly every part of the railroad system, such as inside the cab, along the track, at grade crossings, and in stations. These camera systems are manually monitored, either live or subsequently reviewed in an archive, which requires an immense amount of human resources. To make the video analysis much less labor-intensive, this paper develops a framework for utilizing artificial intelligence (AI) technologies for the extraction of useful information from these big video datasets. This framework has been implemented based on the video data from one grade crossing in New Jersey. The AI algorithm can automatically detect unsafe trespassing of railroad tracks (called near-miss events in this paper). To date, the AI algorithm has analyzed hours of video data and correctly detected all near-misses. This pilot study indicates the promise of using AI for automated analysis of railroad video big data, thereby supporting data-driven railroad safety research. For practical use, our AI algorithm has been packaged into a computer-aided decision support tool (named AI-Grade) that outputs near-miss video clips based on user-provided raw video data. This paper and its sequent studies aim to provide the railroad industry with next-generation big data analysis methods and tools for quickly and reliably processing large volumes of video data in order to better understand human factors in railroad safety research.

The availability of video data in the railroad industry is increasing every year. The cameras are sited on nearly every part of the railroad system, such as inside the cab, along the track, at grade crossings, and in stations. The Fixing America's Surface Transportation (FAST) Act requires all passenger railroads to install inward-facing cameras to better monitor train crews and assist in accident investigations, and outward-facing cameras to better monitor track conditions (*1*). The Los Angeles Metro Transit Authority in California began utilizing video cameras for law enforcement at grade crossings (*2*). In the New York area, Metro-North and the Long Island Rail Road received $5 million from the Federal Railroad Administration (FRA) for grade crossing improvements. Approximately 40% of those funds were committed to installing a closed-circuit television (CCTV) system on high-risk grade crossings (*3*). While video big data have been collected, analyzing these data quickly and reliably remains a challenge. In many cases, these camera systems are manually monitored by railroad staff, either live or subsequently reviewed in an archive.

There exist many scenarios in the rail industry where "near-misses"—or dangerous situations without actual incidents—occur. Because no actual harm occurs, these near-misses are typically not recorded in FRA safety databases. For example, if a pedestrian trespasses on a railroad track when the red signal is on, but this action does not result in an accident, we call it a near-miss. Although near-misses do not cause actual damage, they indicate certain characteristics that may ultimately cause severe consequences if they occur repeatedly. Learning from near-miss data is an important research topic in proactive risk management (*4*).

The pervasive presence of surveillance cameras provides a big data platform for collecting and analyzing near-miss data in support of railroad safety and risk management. Despite its value, video data analysis can be extremely laborious, usually taking hours or days to process and analyze. To address this technological challenge, this paper describes an artificial intelligence (AI)

[1]Department of Civil and Environmental Engineering, Rutgers, The State University of New Jersey, Rutgers, NJ

Corresponding Author:
Address correspondence to Xiang Liu: xiang.liu@rutgers.edu

technology to let the computer program "watch," "identify," and "understand" near-miss clips automatically and efficiently utilizing an existing video infrastructure. Once this technology is practice-ready, it can be adapted to various applications in which video big data are used to support railroad safety decisions.

Objectives of Research

This paper aspires to develop an AI framework to gather useful information from video footage in support of railroad safety research. Specifically, this research aims to produce the following deliverables:

- Development of a general AI methodological framework for railroad video big data analytics.
- Application of the technology to a particular use-case, which is grade crossing near-miss detection.
- Implementation of the AI algorithm into a computer-aided decision-support tool that automatically processes video big data and outputs near-miss video clips.

Literature Review

A literature review was conducted to understand the state of the art and practice in two major categories: (1) how video big data are utilized in the railroad industry for safety research; and (2) how AI is used for video analytics in railroad and other relevant domains.

Video Data for Railroad Safety Research

In the railroad industry, the extraction of useful information from video data has largely been based on manual reviewing of the gathered footage. For example, Ngamdung and daSilva (*5*) conducted a study to understand illegal trespassing on railroad property in Pittsford, New York. The video analysis required a large amount of human resources to accomplish (*6*). In addition, there have been studies on the effectiveness of humans watching CCTV cameras; they show that after 20–40 min of active monitoring, operators often suffer from "video blindness," which reduces their ability to effectively complete their task (*7*). Currently, there is minimal prior work on how AI can assist us in analyzing video big data, which is a principal knowledge gap that this research aims to fill. Effort has been made to quantify the frequency and severity of highway–rail grade crossing incidents. Previous studies (*8, 9*) employed the U.S. Department of Transportation (U.S. DOT) Accident Prediction Model to estimate the number of collisions occurring at grade crossings. An understanding of driver behavior and human factors can contribute to grade crossing safety improvement (*10*). A comprehensive overview of grade crossing research was summarized by Chadwick et al. (*11*). Since grade crossing incidents account for a large portion of casualties on U.S. railroads (*11, 12*), it is important to better understand this type of risk so as to develop proper risk-mitigation strategies.

AI Technologies for Video Analytics

AI has the potential to tremendously reduce the laborious effort required to process video data. Similar sectors, such as roadways and airports, have begun to implement these techniques for video big data analysis. Selected AI techniques include background subtraction, region of interest, and Kalman filtering (*13–16*). The first and most fundamental tool in video analytics is background subtraction. When attempting to isolate moving objects in a frame, the removal of the landscape against which they are moving can improve processing time and accuracy. Originally, cameras at airports were used to provide visual confirmation of a plane's identity, and infrared cameras were used to ensure security from trespassers. In recent years, a network called the Autoscope Solo Wide Area Video Vehicle Detection System has been deployed in European airports. This system utilizes background subtraction in its AI to identify moving objects within the field of view (*13*). Other techniques of video big data analysis, region of interest (ROI) and line of interest (LOI), were implemented in a study using a stationary CCTV camera to count pedestrians and cyclists crossing an intersection. A user can define a line or polygon of pixels in the frame which an AI can use as a reference. In that study, pedestrians and cyclists were tracked in the frame and only counted as "crossing" if they passed through the ROI (*16*). Another AI technique is the Kalman filter, which is a set of mathematical equations to estimate the state of a process (*14*). This technique has been used to track vehicles within a camera view for highway applications (*15*).

While AI has the potential to provide useful data analysis capabilities, there are privacy concerns that may occur when collecting personally identifiable information (*17, 18*). For example, a survey showed that 88% of Americans "do not wish to have someone watch or listen to them without their permission" (*19*). A total of 63% of respondents "feel it is important to be able to go around in public without always being identified" (*19*). This opinion has fueled legal and technological changes to preserve the privacy of individuals. For example, in 1974 the United States Congress enacted the Federal Privacy Act, which regulated governmental databases in how they could store and publish information on U.S. citizens (*20*). Therefore, it is important to recognize and manage these privacy concerns. In 2009 the Federal Trade Commission (FTC) published a general set of principles for the collection of information, including awareness, consent, access, security, and enforcement (*21*). To maintain these principles and still extract useful information, specialized video-processing techniques have been developed to preserve privacy. Google's Street View's anonymization techniques are among the examples of how these concerns are technologically considered. The anonymization techniques involved an intricate neural network approach that first identifies faces and then performs a post-processing obfuscation resulting in a final anonymized image (*22*). In a full-scale implantation of video analysis on grade crossings, a similar anonymization algorithm could be implemented to preserve privacy.

Knowledge Gaps

Currently, AI-driven video big data analytics are still at an early stage in railroad safety research. Video analysis occurs largely on a manual basis. A customized AI algorithm would significantly expedite the video analysis process.

Intended Contributions of this Paper

This paper intends to develop a unique, AI-aided methodological framework for video analytics that can be adapted to different application scenarios in which railroads need to analyze video big data in support of their safety decisions. Using an illustrative application in grade crossing near-miss detection using surveillance camera videos, we provide a step-by-step analytical procedure showing how AI can be developed and used to generate near-miss video clips. The methodology can be adapted to other scenarios toward automated, real-time, video monitoring and analysis. Near-miss data, which supplements accident data, provides additional useful information for understanding risky behaviors.

AI-Aided Railroad Video Analytics

There are a variety of resolutions, frame rates, opacities, and brightness levels in railroad video data. Each of these presents a challenge when training an AI to process and extract information from these data. There are several performance requirements for the AI in analyzing video data. First, it must accurately identify vehicles, trains, artifacts, shadows, and other objects. Second, the algorithm needs to be robust in diverse environmental conditions. This includes inclement weather (e.g., rain, fog, snow) and varying light conditions (twilight, nighttime, daytime). During the night those opacity levels change,

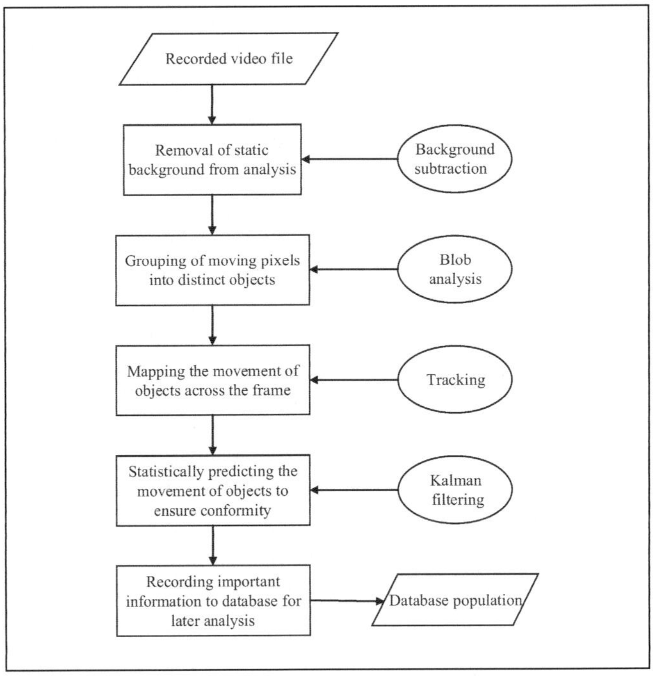

Figure 1. General AI framework for railroad video data analytics.

and when vehicles drive by headlights may cause a false detection. New opacity levels and extra checking techniques need be implemented to alleviate this issue.

To address the above-mentioned challenges, we introduce general AI approaches for video analytics, including background subtraction (*13, 23–25*), blob analysis (*26*), and Kalman filtering (*14, 16, 27, 28*) for potential application to railroad video analysis (Figure 1). These techniques isolate the moving objects and track their movement. Background subtraction is particularly useful because most cameras are static (e.g., those in stations, at grade crossings, or on bridges). The removal of the background allows for the isolation of the moving objects (humans or vehicles) in the frame. Each pixel is derived in color scale and averaged over several frames as appropriate to the application. This is important as the environment causes light and vegetation to shift slightly, and an average value with inbuilt tolerances allows for a more dynamic background. The subtraction occurs on a frame-by-frame basis as well, where each color-scaled pixel is subtracted from the learned background, resulting in a binary mask.

In another approach, an AI algorithm establishes pixel ranges known as LOI or ROI, which aid in the counting and recording of objects' behavior as they traverse the frame. By isolating part of the frame, fewer pixel-to-pixel calculations are required, which is particularly useful in high-resolution footage where there are many pixels. Finally, Kalman filtering can predict the movement of objects. This can also aid in the classification of specific types of objects that are tracked. With the values of objects' sizes and acceleration obtained and/or predicted, the differentiation between vehicle and pedestrian or vehicle and train can be ascertained (*14*). These techniques—removing the stationary background, identifying the moving objects, determining if they are traversing an area of interest, and removing the non-conforming objects—establish a framework for AI-aided railroad video data analytics.

The developed AI-based techniques should be trained to test and verify its robustness. A training program for an AI application for railroads would require the development of an initial algorithm with established environmental parameters. This draft algorithm analyzes a training set of data, comparing the algorithm's results to the knowns. A successful verification would require the algorithm to correctly "see" images of trains and pedestrians independently from the background, using techniques such as background subtraction (*13*). The AI can then be retested with various weather conditions and diverse daylight conditions, such as dawn, day, dusk, and dark. After undergoing this training, an AI application is able to explicitly capture the images and moving paths of trains and highway users, such as cars, pedestrians, and cyclists, under a wide array of external conditions. Then the AI tool is able to record critical video information automatically, which is compiled into a database for future study.

Application to Grade Crossing Near-Miss Analysis

Grade crossing trespassing accounts for many incidents and fatalities annually (*29*). An AI algorithm was developed and implemented with the data based on one grade crossing in New Jersey. The CCTV video footage of this grade crossing was obtained, and a customized AI algorithm was developed to detect near-misses. A near-miss event occurs when a pedestrian or vehicle traverses the crossing while the red signal is on. Almost all prior studies in the field of grade crossing safety have focused on using accident data (*9, 30*), without accounting for a larger number of near-misses that share similar behavioral characteristics but (fortunately) did not cause any harm. The following section details the process of using AI to automatically detect near-misses from grade crossing video data. The general methodology can be adapted to other use cases in the future.

Algorithm Flow Chart

This AI reads the video file looking for a red signal, processes the image (details will be presented later), and evaluates whether a near-miss has occurred. Detailed analytical steps are presented below.

Step 1: Reading Video Frames Sequentially. The first step of the algorithm is to start reading the video file frame by frame. During this reading, the prime objective is to determine whether the active signalized crossing light has been triggered. To increase processing speed, a frame-skip segment is included, which advances the reading in 10-s intervals and stops when a red light is detected; this is practical in this application because the duration of a stop signal is greater than 10 s for this grade crossing. Frame-skip algorithms also allow for adaptability to high frame-rate video and reduce analysis time.

Step 2: Detection of Stop Signal. After a frame has been isolated, the stop signal (red signal) is recognized in that frame. A check of the red pixel values in the small area

Figure 2. Stop signal under day and night conditions.

Table 1. Results for AI-Aided Detection of Near-Misses

Date	From	To	Duration (hours)	Red signals	Near-misses
Day 1	08:00	15:00	07:00	21	0
Day 2	00:19	09:00	08:41	20	2
Day 3	12:00	21:00	09:00	26	0
Total			24:41	67	2

Figure 3. Computer-recognized background using training data.

of the frame where the signal lies determines its status (Figure 2). The user can configure the location and the opacity threshold for this application. If a stop signal is detected, the algorithm performs a frame-by-frame check backwards to determine the beginning of the stop signal. Then, the subroutine of near-miss detection is activated.

Step 3: Background Template Learning. The near-miss detection subroutine follows several steps. The first is to learn and subtract the background template at the beginning of the stop signal. Non-moving objects are captured in the field of view at this time. For each stop signal that is encountered in the video, a new background is learned. This overcomes the challenge of the gradual changing of light levels throughout the day. Other environmental conditions such as passing rainstorms, parked cars in the background, and others are also captured in the background template learning (Figure 3).

Step 4: Objective Tracking. Moving objects are detected in the foreground with the background subtraction technique (*13, 23–25*). With background subtraction, the total number of moving pixels can be tracked and recorded from frame to frame; this detection continues until the red signal turns off.

Step 5: Identifying Near-Misses. The algorithm identifies a near-miss event based on the total number of moving pixels. One main challenge here is to recognize and remove the "noise" from moving pixels of a train. It was noted that the number of pixels that a train occupies in the foreground during a crossing is much larger than that of highway users (e.g., a pedestrian or a vehicle). Therefore, a proper threshold can be established to separate near-miss objectives from trains. If a near-miss is detected, all frames of the red signal are extracted to a video file for further review. After stop signal processing concludes, the algorithm skips 5 min and continues the analysis from Step 1. This 5-min skip further reduces processing time and does not compromise the accuracy of the analysis since no stop signals re-occur within this short interval in this case study. These parameters can be easily changed for different applications.

Results

The goal of our algorithm is to complete the analysis much faster and with equal or greater accuracy than manual reviewing. In this case study, the processing of the video took roughly 2% of the total video duration to complete. This duration is highly dependent on the number of stop signals encountered. Two near-miss events were detected on a 25-h video dataset, covering three different days. The processing time for this video was less than 40 min. A detailed summary is listed in Table 1.

The algorithm's output showed two near-miss events occurring within a single stop signal in the morning of one day. In the first near-miss, before the train arrived, two pedestrians entered the grade crossings while the stop signal was active (Figure 4*a*). Five seconds after the two pedestrians crossed the track, the train arrived. The second near-miss occurred when a cyclist, who had stopped at the deployment of the arm gates and stop signal, crossed after seeing that the train was gone, without waiting for the signal to be deactivated (Figure 4*b*).

The results of this study epitomize two different types of highway users and two typical non-compliance behaviors. The two pedestrians perceived the timing of train arrival from their judgment and were confident in their ability to cross the track before the train arrived. The second case illustrates the assumption that no second train would cross, despite the presence of multiple

Figure 4. Two near-miss incidents detected by the AI algorithm.

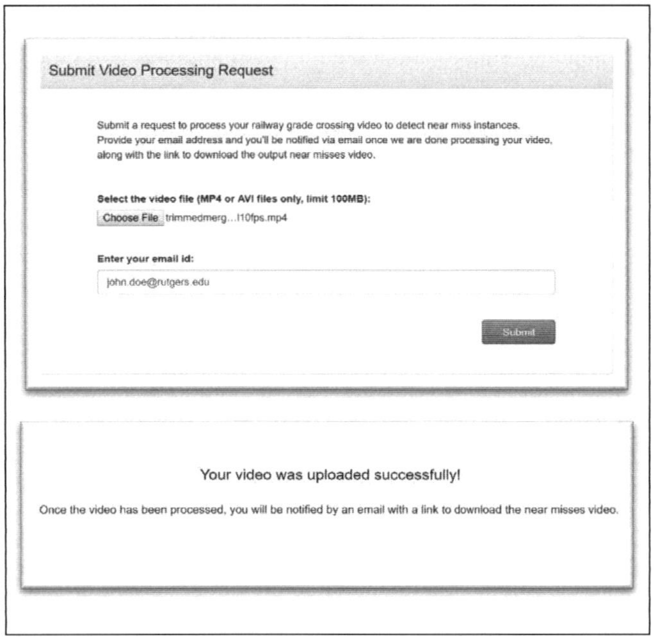

Figure 5. AI-Grade decision-support tool user interface.

tracks and the continuing signal. Both near-misses represent risky behaviors with potentially catastrophic consequences, which have been seen in past accident data (*12, 30*).

Wed-Based Decision-Support Tool (AI-Grade)

The AI algorithm described above has been implemented into a web-based decision-support tool called "AI-Grade" (Figure 5). The web-based AI-Grade streamlines the automatic processing of railroad grade crossing data through the following steps:

- Step 1: Log in to the application website.
- Step 2: Select the video file that needs to be analyzed and enter the user's email address.
- Step 3: Click "Submit" and the processing will begin.
- Step 4: Once processing is complete, users will receive an email that provides the cropped near-miss video, if any.

Tool Validation

To ensure the usefulness of this AI tool, results must be accurate and achieved faster than via manual processing. A validation of this criteria was completed using the collected video data. There are four possible results: (1) an illegal trespass occurs, and a detection is recorded (correct); (2) no illegal trespass occurs, but a detection is recorded (false positive); (3) an illegal trespass occurs, but there is no detection (false negative); and (4) there is no illegal trespass and there is no resulting detection (correct).

For comparison, several students manually reviewed all the footage and compared their results to the output of AI-Grade. To date, AI-Grade is 100% accurate without any false positives or negatives (Table 2). In addition, the AI program completed processing the 25-hour video within 40 min, totaling 2% of the video time. We are further developing and training this algorithm using more video data (e.g., 1-year data) from our industry partners. Ultimately, we hope to design a tool for real-time analysis of video data in support of railroad safety decision-making.

Contributions to Research and Practice

Contribution to Academic Research

This paper describes an AI technological framework for automatically detecting near-misses at grade crossings. Before the advent of AI technology, it was not practical to collect diverse information (e.g., the time, type, and environmental conditions surrounding illegal

Table 2. Tool Validation Outcomes for Near-Miss Detection

	Trespassing	No trespassing
Detection	100%	0%
No detection	0%	100%

trespassing) from video big data because of an inordinate amount of human resources required for the acquisition of such information. The expected contribution of this research to railroad safety parallels what the FHWA-sponsored study on Naturalistic Driving did for highway traffic safety, which used sensors to collect vehicle movement and driver attention data and used this information for highway safety analyses (*31*). Similarly, we aim to empower AI to analyze a large amount of railroad video data for better understanding of human factors in various application scenarios.

Contribution to Practice

The practical contribution of the AI framework is its applicability to this and other scenarios in the rail industry (e.g., inside cabs, at stations, rail yards, and on platforms). This information will help railroad agencies make decisions about the allocation of limited safety budgets. AI can be trained to recognize a variety of environmental factors (e.g., weather, track geometry, the population surrounding the rail facility), as well as risk-prone human behaviors (e.g., illegal trespassing, operator fatigue). Further, AI can be developed to quantitatively measure the association between risky behaviors and their influencing factors. These results enable development of proactive strategies to prevent or reduce near-misses or incidents in railroad systems, thereby improving their safety. Additionally, the implementation of this framework has a low cost. It utilizes an already existing video-recording infrastructure and has no additional hardware costs.

Conclusion

This paper proposes the use of a customized AI algorithm for automatically analyzing railroad video data to solicit useful information for understanding human behavioral characteristics. An example implementation and decision-support tool are developed based on grade crossing surveillance video data. In the study period, our AI algorithm correctly detected all the near-miss events associated with unsafe trespassing of the studied grade crossing. The near-miss data can be used to develop safety strategies and to prevent the occurrence of risk-prone behaviors and resultant accidents. This research indicates the promising applications of AI to other research areas in the railroad industry in the future, such as in-cab video analysis for distraction detection or security surveillance in railway stations.

Future Work

To take this research further we are increasing the volume of the training set to include more environmental conditions and possibly more near-misses. Once the AI algorithm is trained via a very large and diverse amount of video data, it can be used to "recognize" and "understand" a wide array of scenarios in real-time settings. Real-time video analytics in other locations and applications within the railroad industry will be developed, validated, and implemented. Another area of future research would be the analysis of video from the cameras installed in locomotives based on an adaptation of the AI algorithm described in this paper.

There are several major considerations when implementing a real-time system, some of which are as follows:

- Ethical: maintaining privacy of individuals in analysis and protection against sensitive data breaches;
- Economical: balancing costs and benefits of the technology;
- Accuracy: continually improving accuracy with a growing database;
- Demand: adding data types and metrics as per stakeholder request;
- Support: responding to system failures and correcting errors;
- Adaptability: ensuring the ability to perform under unforeseen or untested scenarios;
- Availability: maintaining access for stakeholders;

Additionally, a potential future step is to use the developed database for railroad safety risk analysis. As mentioned above, most previous studies were based on accidents instead of near-misses. If near-miss data can be collected, additional insights (particularly behavioral characteristics) could be drawn to further support railroad safety research (*9*). This would be combined with potential cost–benefit analyses to understand the practical value of AI implementation in the rail industry.

Acknowledgments

The first author has been supported by the New Jersey Department of Transportation (NJDOT). The second and third authors are partially funded by the Federal Railroad Administration (FRA) of U.S. DOT at the time of writing this paper.

Author Contributions

The authors confirm contribution to the paper as follows: study conception and design: AZ, XL, ZZ; data collection, analysis, and interpretation of results: AZ, XL, ZZ; draft manuscript preparation: AZ, XL, ZZ. All authors reviewed the results and approved the final version of the manuscript.

References

1. *The Fixing America's Surface Transportation Act (P.L. 114-94)*. Federal Railroad Administration, U.S. Department of Transportation, Washington, D.C., 2015.
2. *Grade Crossing Safety Improvement Program*. Federal Transit Administration, U.S. Department of Transportation, Washington, D.C., 2016.
3. *FRA Awards $5 Million to Improve Metro-North and Long Island Rail Road Crossings*. Metro Transit. http://www.mta.info/news-metro-north-grade-crossings-fra/2016/09/12/fra-awarded-about-5m-improve-metro-north-and-long. Accessed March 12, 2017.
4. Phimister, J. R., U. Oktem, P. R. Kleindorfer, and H. Kunruther. Near-Miss Incident Management in the Chemical Process Industry. *Risk Analysis*, Vol. 23, No. 3, 2003, pp. 445–459.
5. Ngamdung, T., and M. daSilva. Driver Behavior Analysis at Highway–Rail Grade Crossings using Field Operational Test Data: Light Vehicles. DOT/FRA/ORD-13/28. Federal Transit Administration, U.S. Department of Transportation, Washington, D.C., 2013.
6. daSilva, M. P., W. Baron, and A. A. Carroll. *Highway Rail–Grade Crossing Safety Research: Railroad Infrastructure Trespassing Detection Systems Research in Pittsford, New York*. DOT-VNTSC-FRA-05-07. John A. Volpe National Transportation Systems Center, Cambridge, Mass., 2012.
7. Dee, H. M., and S. A. Velastin. How Close Are We to Solving the Problem of Automated Visual Surveillance? *Machine Vision and Applications*, Vol. 19, Nos. 5–6, 2008, pp. 329–343.
8. Faghri, A., and M. J. Demetsky. *Evaluation of Methods for Predicting Rail Highway Crossing Hazards*. No. VHTRC86-R32. Virginia Transportation Research Council, 1986.
9. *Railroad–Highway Grade Crossing Handbook*, Revised 2nd ed. Federal Railroad Administration, U.S. Department of Transportation, Washington, D.C., 2007.
10. Caird, J. K. *A Human Factors Analysis of Highway–Railway Grade Crossing Accidents in Canada*. Transportation Development Centre, Montreal, Canada, 2002.
11. Chadwick, S. G., N. Zhou, and M. R. Saat. Highway–Rail Grade Crossing Safety Challenges for Shared Operations of High-Speed Passenger and Heavy Freight Rail in the US. *Safety Science*, Vol. 68, 2014, pp. 128–137.
12. *Highway–Rail Grade Crossings Overview*. Federal Railroad Administration, Washington, D.C. https://www.fra.dot.gov/Page/P0156. Accessed February 1, 2017.
13. Dimitropoulos, K., N. Grammalidis, D. Simitopoulos, N. Pavlidou, and M. Strintzis. Aircraft Detection and Tracking Using Intelligent Cameras. *Proc., IEEE International Conference on Image Processing 2005*, Genova, Italy, IEEE, New York, 2005.
14. Patel, H. A., and D. G. Thakore. Moving Object Tracking Using Kalman Filter. *International Journal of Computer Science and Mobile Computing*, Vol. 2, No. 4, 2013, pp. 326–332.
15. Ismail, K. A. *Application of Computer Vision Techniques for Automated Road Safety Analysis and Traffic Data Collection*. University of British Columbia, Vancouver, Canada, 2010.
16. Kocamaz, M. K., J. Gong, and B. R. Pires. Vision-Based Counting of Pedestrians and Cyclists. *Proc., 2016 IEEE Winter Conference on Applications of Computer Vision (WACV)*, Lake Placid, New York, IEEE, New York, 2016.
17. Tene, O., and J. Polonetsky. Big Data for All: Privacy and User Control in the Age of Analytics. *Northwest Journal of Technology & Intellectual Property*, Vol. 11, 2013, p. 239.
18. Metcalf, J., E. F. Keller, and D. Boyd. *Perspectives on Big Data, Ethics, and Society*. Council for Big Data, Ethics, and Society, 2017. https://bdes.datasociety.net/council-output/perspectives-on-big-data-ethics-and-society/. Accessed July 1, 2017.
19. Madden, M., and L. Rainie. Americans' Attitudes about Privacy, Security and Surveillance. *Pew Research Center: Internet, Science & Tech*, 20 May 2015, www.pewinternet.org/2015/05/20/americans-attitudes-about-privacy-security-and-surveillance. Accessed November 11, 2017.
20. Raul, A. C. *The Privacy, Data Protection and Cybersecurity Law Review*. Law Business Research Ltd, London, United Kingdom, 2016.
21. *Fair Information Practice Principles*, Federal Trade Commission. https://web.archive.org/web/20090331134113/http://www.ftc.gov/reports/privacy3/fairinfo.shtm. Accessed November 11, 2017.
22. Frome, A., G. Cheung, A. Abdulkader, M. Zennaro, B. Wu, A. Bissacco, H. Adam, H. Neven, and L. Vincent. Large-Scale Privacy Protection in Google Street View. *Proc., 2009 IEEE 12th International Conference on Computer Vision*, Kyoto, Japan, 2009, pp. 2373–2380.
23. Shah, M., O. Javed, and K. Shafique. Automated Visual Surveillance in Realistic Scenarios. *IEEE Computer Society*, Vol. 14, No. 1, 2007, pp. 30–39.
24. Sheikh, Y. A. Visual Monitoring of Railroad Grade Crossing. *Defense and Security*. In Sensors, and Command, Control, Communications, and Intelligence (C3I) Technologies for Homeland Security and Homeland Defense III, International Society for Optics and Photonics. Vol. 5403, 2004, pp. 654–661.
25. Fakhfakh, N., L. Khoudour, E. M. El-Koursi, J. Jacot, and A. A. Dufaux. A Video-Based Object Detection System for Improving Safety at Level Crossings. *Open Transportation Journal*, Vol. 5, 2010, pp. 45–59.
26. Cucchiara, R., C. Grana, M. Piccardi, and A. Prati. Detecting Moving Objects, Ghosts, and Shadows in Video Streams. *IEEE Transactions on Pattern Analysis and Machine Intelligence*, Vol. 25, No. 10, 2003, pp. 1337–1342.
27. Salmane, H., L. Khoudour, and Y. Ruichek, A Video-Analysis-Based Railway–Road Safety System for Detecting

Hazard Situations at Level Crossings. *IEEE Transactions on Intelligent Transportation Systems*, Vol. 16, No. 2, 2015, pp. 596–609.
28. Chen, C. H. *Handbook of Pattern Recognition and Computer Vision*. World Scientific, River Edge, NJ, 2015.
29. Table 2-39: Railroad and Grade Crossing Fatalities by Victim Class. National Transportation Statistics. https://www.rita.dot.gov/bts/sites/rita.dot.gov.bts/files/publications/national_transportation_statistics/html/table_02_39.html. Accessed April 16, 2017.
30. Saccomanno, F., L. P. Fu, and L. Miranda-Moreno. Risk-Based Model for Identifying Highway–Rail Grade Crossing Blackspots. *Transportation Research Record: Journal of the Transportation Research Board*, 2004. 1862: 127–135.
31. Valera, M., and S. A. Velastin. Intelligent Distributed Surveillance Systems: A Review. *IEE Proceedings: Vision, Image and Signal Processing*, Vol. 152, No. 2, 2005, pp. 192–204.

The Standing Committee on Railroad Operational Safety (AR070) peer-reviewed this paper (18-04595).

Article

TRR
JOURNAL OF THE TRANSPORTATION RESEARCH BOARD

Transportation Research Record
2018, Vol. 2672(10) 278–288

DOI: 10.1177/0361198118776508
journals.sagepub.com/home/trr
SAGE

Feedforward Tactical Optimization for Energy-Efficient Operation of Freight Trains: The Swiss Case

Valerio De Martinis[1], Ambra Toletti[1], Francesco Corman[1], Ulrich A. Weidmann[1], and Andrew Nash[2]

Abstract

The optimization of rail operation for improving energy efficiency plays an important role for the current and future market of rail freight services and helps rail compete with other transport modes. This paper presents a feedforward simulation-based model that performs speed profile optimization together with minor rescheduling actions. The model's purpose is to provide railway operators and infrastructure managers with energy-efficient solutions that are tailored especially for freight trains. This work starts from the assumption that freight train characteristics are completely defined only a few hours before actual departure; therefore, small specific feedforward adjustments that do not affect the surrounding operation can still be considered. The model was tested in a numerical example. The example clearly shows how the optimized solutions can be evaluated with reference to energy saved and robustness within the rail traffic. The evaluation is based on real data from the North–South corridor crossing Switzerland from Germany to Italy.

Increasing railway energy efficiency is an important objective in making railway transport more attractive in the face of competition with other transport modes. Consequently, energy efficiency is currently a major topic of research. This research ranges from evaluating new zero emission technologies for traction units, to increasing energy savings for secondary services (heating, air conditioning, communication, and entertainment, etc.), to new operating methods; essentially railways are looking for all types of ideas that can help reduce energy use and help minimize operating costs. This paper describes research carried out in Switzerland to develop methods for a more efficient operation of freight trains in mixed traffic networks with the goal of reducing energy consumption.

Switzerland operates one of the densest railway schedules in the world, and freight trains account for a non-negligible part of this traffic. Thus, it is important to optimize freight train operations both to make rail freight more attractive and to maintain capacity for passenger traffic. Optimizing freight and passenger rail services shares certain requirements; above all any optimized solution for a specific train operation must be carefully evaluated in terms of impacts on other rail services. However, there are also several key differences; for example, rail freight has more freedom in operation, including more flexible departure and arrival times as well as wagons composition. Yet freight trains generally have lower performance and lower priority than passenger services. These differences are especially relevant in mixed traffic contexts, such as in Switzerland.

This paper assesses the potential energy savings for freight trains operating in mixed traffic networks that can be achieved by deploying small feedforward actions which can be specified a few hours before departure, that is, before train characteristics and departure time are completely defined.

These possible actions include early departures and stop avoidance, which are acceptable for freight trains, and speed profile optimization. These actions are more appropriate for freight traffic than most state-of-the-art traffic management strategies, which are mostly designed for passenger traffic. Moreover, current traffic management methodologies are conceived for the operational level and, therefore, are based on conflict detection and

[1]ETH Zurich, Zurich, Switzerland
[2]Emch + Berger AG Bern, Bern, Switzerland

Corresponding Author:
Address correspondence to Valerio De Martinis:
valerio.demartinis@ivt.baug.ethz.ch

complex information and communications technology-(ICT) intensive feedback loops from disrupted operation. In contrast, the proposed model is based on feedforward planning only, providing a much simpler and operations-ready setup. A feedforward control determines beforehand a new plan to be followed and to be transmitted to the drivers, that is, a disposition timetable. A disposition timetable is here intended as a tactical adjustment of the original timetable before real operation. This concept comes from a common practice, which consists of slightly modifying the planned trajectory of a specific train according to new information about its characteristics or infrastructure availability on its itinerary. This practice is mainly based on traffic manager's knowledge on usual traffic conditions of a given area, that is, without applying a shared methodology.

This paper first presents a brief literature review on railway operation and Swiss rail freight practice. It then describes main assumptions on train motion models and the optimization framework. Following this, it describes the case study and optimization results. Finally, the paper presents conclusions and perspectives for model enhancement and further investigation.

Review on Energy-Efficient Rail Operation

Current Literature on Planning and Real-Time Control

There is a wide literature on energy-efficient strategies for train operation, but few works consider specific traffic management measures in detail. Most energy-efficient train operation strategies described in existing research are based on the a priori generation of particular speed profiles that are then transformed into instructions for drivers (*1*). In the last 20 years, several models have been developed following different approaches, from analytical to heuristics. Many solutions proposed in literature are based on the optimal control problem approach, which has been continuously enhanced for simple to more complex conditions, for example, different track characteristics (*2*). Recent theoretical studies proved the convexity of the local energy function. Thus, for each holding speed and for each steep section of track, the optimal switching points for coasting regime are uniquely defined (*3*). Many approaches have been based on metaheuristics to improve computing speed and keep the quality of results (*4*). Several optimal solutions have been evaluated using different optimization routines, and within these some very interesting results have been found by combining adaptive processes and parallel computing. Genetics algorithms (GA) is one of the most used optimization algorithm types in the field as it generally gives good results and it is not time consuming (*5*). Unfortunately, most of the proposed traffic management solutions are founded on hypotheses based on passenger train configurations that do not easily fit freight trains' operation. Typically, wagons' composition, loads, and traction unit characteristics significantly affect models' applicability.

The other extreme is real-time control, that is, with an explicit feedback loop from realized operations, reporting either the trajectories of a single train or multiple trains. This latter case, which is now the subject of several studies, is based on the combination of rescheduling and speed control procedures. Main approaches consider either a sequential combination, or a double feedback loops between the train trajectory control module and the traffic control module (*6–9*). In real-time control applications, a fundamental aspect to be considered is information transfer to the drivers. Although planning approaches can deliver a precomputed plan of operations, real-time instructions need to consider the human ability to follow multiple instructions and adhere to them (*10*). The process of humans driving inevitably decreases the potential energy savings that are estimated in the algorithms. In addition, frequent stops for overtaking and the fact that freight train performance is usually lower than that of passenger trains make it difficult to implement energy-efficient freight train operations in practice (*11*).

In terms of computation tools, simulation has mostly been used for validation rather than in the methodology to produce solutions. However, increasing computing performance and the opportunity for comprehensive modeling are making simulation more attractive both for planning and for real-time applications. Many applications of simulation-based optimization routines that comprehensively consider the railway operation have been recently proposed (*12–14*).

In summary, there is a clear trend both in academic and industrial research to enhance simulation methods. Nonetheless, there is still a knowledge gap, particularly with respect to practical experience on freight trains, together with the development of more sophisticated tools.

Current Practice on Freight Rail Services and Energy Optimization in Swiss Networks

In the light of current trends, freight transport volumes are expected to increase dramatically. Moreover, because of the European Union's efforts to increase rail market share, railways will face huge challenges in the near future. Recent studies estimate the expected growth and outline the measures needed to handle it (*15*). The Swiss railway network, which is already among the most heavily used in the world, will approach its stability limit.

In Switzerland, periodic timetables are used to facilitate passenger transfers at hub stations. Periodic

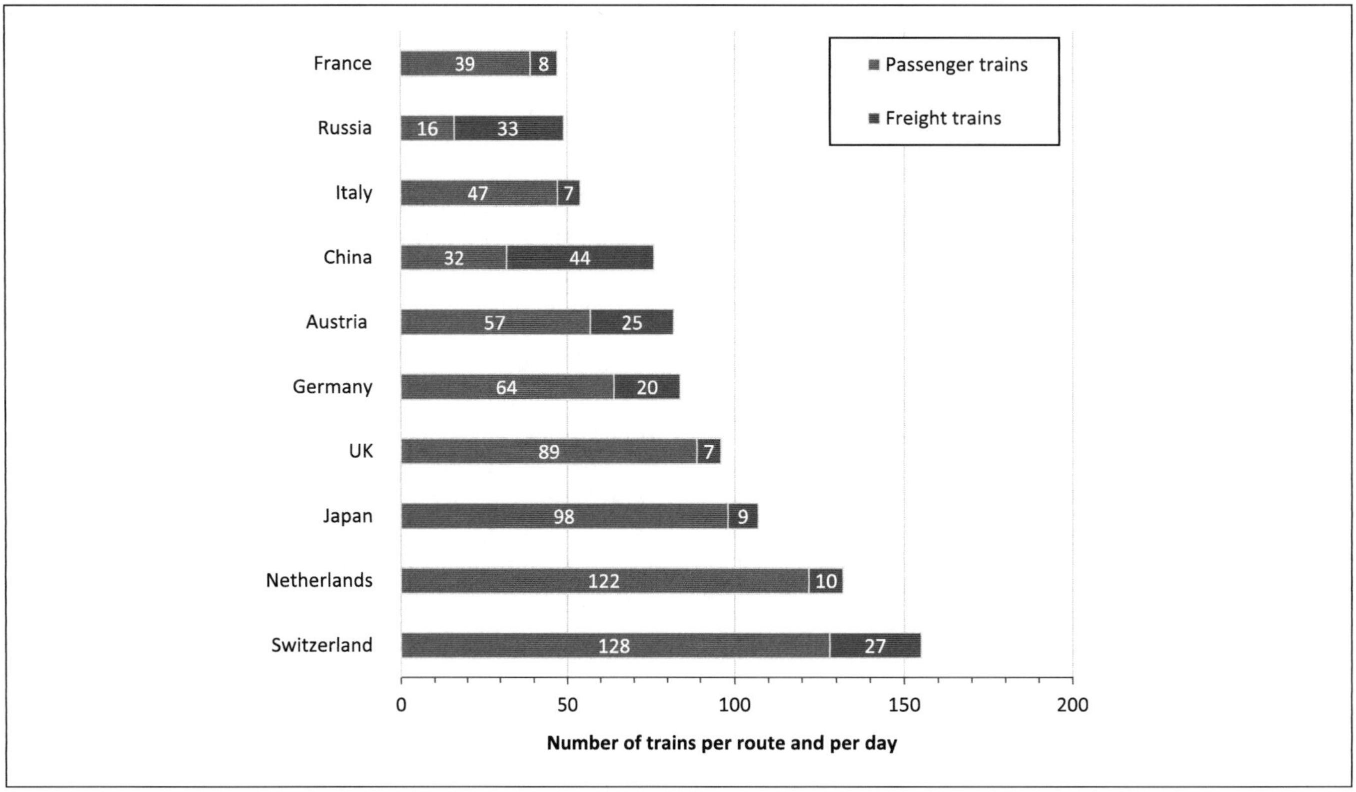

Figure 1. Number of trains per route and per day spent on rail network in 2014. (Source: UIC statistics, 2014.)

timetables include specific time slots for freight trains, which are sold to various operators. The time slots are designed using minimum performance requirements, fixed time supplements, and fixed buffer times. Figure 1 compares track utilization in 2014 and shows how Switzerland is characterized by mixed traffic.

The growing amount of freight traffic means railways need to significantly improve infrastructure utilization by optimizing capacity. This entails using a set of strategic, tactical, and real-time traffic management tools, the main aims of which are to improve both the use of available capacity (reducing stops) and system performances (fewer delays, less energy consumption). To this end, several driving assistance tools have been developed and successfully adopted, such as the ADL system owned by SBB or the Leader system Adaptive Lenkung (ADL), Schweizerische Bundesbahnen (SBB) (*16,17*). These tools compute energy-efficient speed recommendations for locomotive drivers.

We instead address a complementary task that focuses on defining the best tactical plan for a freight train when detailed information about its characteristics are known, i.e., a few hours before departure, and on minor changes to the actual freight train's schedule to enhance its operation without affecting the passenger traffic. We further fill the knowledge gap with respect to practical experience on freight trains together with the development of more sophisticated tools.

Modeling Freight Train Operation

Train Motion Modeling and Specification to Freight Trains

The literature describes several approaches for train motion modeling. The present work considers the simplest and most studied one, which assumes the entire train as a mass point and Newton equation for train motion dynamics. Although the assumptions in this approach are far from a precise representation of reality, it has several advantages, such as fast computation and wide experience. For the full description, please refer to reference (*18*).

The speed profile of a train is composed of a series of motion phases. Previous research argues that the best representation consists of four phases: acceleration, cruising, coasting, and braking. Although there is general agreement on this assumption, it is worth highlighting that this specific sequence is quite hard to implement in real life due to constraints on arrival times and speed limit variation along the track. These factors are especially true when considering long journeys and mixed traffic networks.

The acceleration and cruising phases are the most energy consuming phases because of tractive efforts applied. Acceleration is highly expensive with reference to energy and, therefore, unplanned stops should be avoided. Coasting has no tractive effort term as the motion depends on the kinetic energy acquired and the resistances.

The braking phase is one of the most studied phases and, up until now, there is no comprehensive (shared) representation of all the phenomena that contribute to the braking curve definition. Therefore, constant braking values within speed intervals were used in the proposed model. Train motion models can be easily specified for freight trains when all characteristics are known. In normal practice, the exact configuration of a freight train is known only a few hours before its actual departure, which can slightly differ from the planned departure considered in the timetable. Therefore, when the configuration is known, an optimization on the operation of the specific freight train can be performed. Given freight trains' degrees of freedom, such as possible early departures and possible stop avoidance, an optimal departure time and a minimum set of stops, coupled with optimal speed profiles, can help save energy without affecting rail traffic.

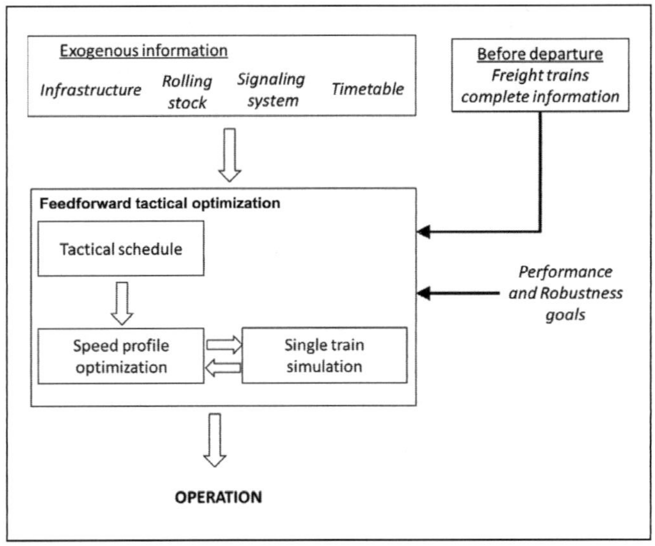

Figure 2. The simulation framework with fundamental relationships. Bold text represents the framework's main modules; italic text highlights required data and information.

Optimization of Freight Train Operation

A disposition timetable is here defined as a set of tactical adjustments on the original timetable for considering specific characteristics of the trains, which cannot be known at the time of yearly timetable definition (such as in the case of freight trains). Such a timetable can be put into operation directly. In this section, the framework for generating the disposition timetable is described. Apart from the framework's scope, note that the traffic monitoring process will continuously check the overall traffic conditions. Potential delays and possible conflicts, caused by imperfect adherence to planning conditions during operation, will be the input for other decision processes, for example, traffic management.

The scheme of the framework is illustrated in Figure 2. It is based on the feedforward use of two optimization models (Tactical Schedule and Speed Profile Optimization), whereas a simulation tool helps evaluate the ability to introduce new train trajectories (that respect current rail traffic conditions given by the timetable) and the feasibility of small deviations from the original schedule. The optimization refers to a single freight train operation and it aims at both saving energy and avoiding conflicts with other rail traffic.

The tactical schedule model aims at minimizing the number of planned stops; it assumes a planned schedule composed of m stops (M is the set of stops for that specific train). For each planned stop i, the arrival time is modeled as T_{Ai}, and the departure time as T_{Di}. For each planned stop i we also introduce the binary variable X_{Di}. It is equal to 1 when $T_{Ai} \neq T_{Di}$, (i.e. when the train must stop) otherwise the value is 0. The planned schedule has all X_{Di} equal to 1.

In the present formulation, travel times TT between two successive stops are fixed. The objective function shown in Equation 1 can be used to find the optimal set of values $[T_{Ai}, T_{Di}]$ for all the i-stops in the specific train's planned schedule:

$$(T_{D1}, \ldots, T_{Ai}, T_{Di}, \ldots, T_{Am})^{opt} = arg\ \min \sum_{i \in M} X_{Di}(T_{Ai}, T_{Di}) \tag{1}$$

Constraints are on: time arrival at final destination, for example, scheduled $\hat{T}_{Am}$ (Equation 2); conflict avoidance (Equation 3); time consistency at stops (Equation 4). Fixed travel times are expressed by Equation 5.

$$T_{Am} = \hat{T}_{Am} \tag{2}$$

$$T_{Di} \in \bigcup_j \left[\hat{\tau}^j_{Di} + MH_i(j,f), \hat{\tau}^{j+1}_{Di} - MH_i(f,j+1)\right] \quad \forall i \in M \tag{3}$$

$$T_{Ai} \leq T_{Di} \forall i \in M \tag{4}$$

$$T_{Ai} - T_{Di-1} = TT_{(i-1,i)} \forall i \in M \tag{5}$$

Equation 3 takes inspiration from the interval scheduling defined in (*19*) for avoiding conflicts at stations, with some simplification given from considering fixed travel times. It ensures conflict-free solutions as it allows choosing only those times T_{Di} for the freight train f, which are within the (possibly many) time intervals related to trains j and $j+1$ stopping at or passing through the station under investigation. In the formulation, the intervals are defined as being before or after an arrival, departure, or passage of a train by a headway MH. The headways are considered for both positions (j, f) and $(f, j + 1)$, that is, assuming that the key capacity bottleneck can be identified at those positions (typically, stations or passing loops).

The speed profile optimization model finds the optimal values of speed profile parameters (SP) (here cruising speeds as SP have been considered) constrained by maximum train performances and speed limits on infrastructures, travel time available, and distance to cover. These constraints are represented with Equations 7, 8, and 9. The last two constraints are defined by the previous optimization through the set of values $[T_{Ai}, T_{Di}]$. The model is specifically described as follows:

$$SP^{opt} = arg \min E(Sim(SP)) \tag{6}$$

subject to the following constraints:

$$SP \leq SP^{\max} \tag{7}$$

$$T(Sim(\mathrm{SP})) \leq \hat{\mathrm{T}} \tag{8}$$

$$S(Sim(\mathrm{SP})) = \hat{\mathrm{S}} \tag{9}$$

where:
SP is the vector of speed profile parameters chosen for optimization;
SP^{opt} is the vector of their optimal values; and
SP^{max} is the vector of maximum allowed values of SP parameters.

These values are defined as the minimum values between rolling stock maximum performance and the maximum performance allowed by the infrastructure/service. $E(.)$ is the total tractive energy used, estimated with the function Sim, which represents the simulation process. Equations 8 and 9 ensure the respect of time $\hat{T}$ and space $\hat{S}$ constraints given by the timetable and the tactical schedules. The simulation process Sim updates train motion data (time, position, acceleration, speed) and train traction data (tractive efforts, resistances, power, energy consumed) for each time step, to estimate the optimization output (i.e. speed profile parameters) at each iteration.

This model takes inspiration from the simulation-based optimization approach described by De Martinis and Weidmann (*14*). The additional contributions are as follows:

- In (*14*) the influence of traffic is considered only as a feasibility check. In the current framework, it is integrated in the tactical schedule.
- Given the previous point, we adopted a single train simulator instead of a rail traffic simulator, which also results in a significant reduction of the overall computing efforts and time.
- The stop pattern is no more fixed. It is an optimization variable specific for freight train operation.

The proposed model assumes that possible enhancements can be developed because of a better specification of the rolling stock a few hours before departure. Regarding the assumptions proposed here, fixed *TT*s are considered because variable *TT*s would require an iterative loop between speed profile generation for travel time definition and a new scheduling process for timetable generation; this would be beyond the meaning of a tactical optimization. Clearly, a suitable choice of the set M is important, depending on the network characteristics and the characteristics of the specific freight train (e.g. length). With the aim to further enhance the timetable from realized operation, the realized set of stops M has been considered as the optimal set of stops for the specific freight train.

Case Study Application

Case Study Description

The case study considers the rail path between Basel and Domodossola via Bern (see Figure 3). Data refers to the rail traffic between 9:45 p.m. and 1:33 a.m. The train performs a single wagonload service with 17 wagons with an approximate weight of 1,170 tons. The locomotive is a Bombardier TRAXX F140 AC unit, known as BLS Re485, configured for ETCS/ERTMS (European Train Control System/European Rail Traffic Management System) operations.

The route is approximately 240 km and includes several tunnels that generate additional resistance to the train motion. The two longest tunnels are the Loetscheberg tunnel (34.6 km) and the Simplon tunnel (19.8 km).

Existing Rail Traffic: Planned versus Observed

BLS, a Swiss train operator, provided data from the train's onboard monitoring equipment that tracks the train position and its main performance indicators (e.g. journey time, track speed, energy consumption, and energy recovered from braking with 1-min time-steps).

Data on the planned traffic situation was extrapolated from the official timetable (*20*). The rail traffic mainly consists of passenger trains until the city of Thun and other freight trains from Thun until the final destination, Domodossola. This is due to the late arrival time in

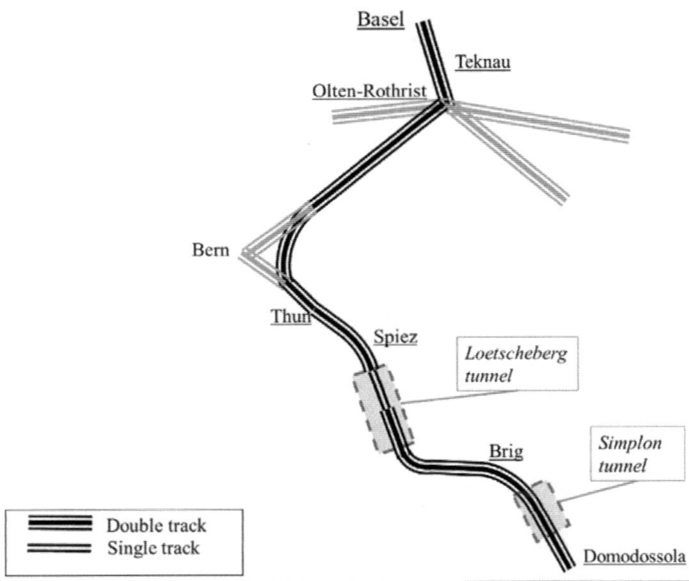

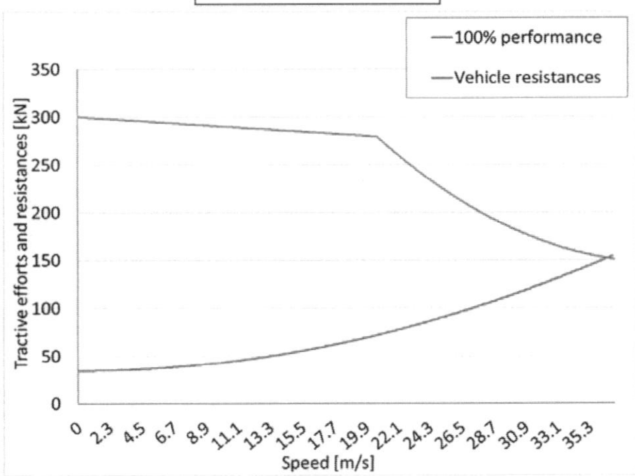

BLS Re485	Characteristics
2002–2003	Year
84 tons	Weight
87 mph (140km/h)	Max speed
5.6 MW	Power class
18.9 m	Length
300 kN	Tractive effort
240 kN	Braking effort

Figure 3. Top center: the considered train path (in black); bottom left: the traction speed diagram of the train used in the analyses; bottom right: main data on the train under analysis.

Thun, which is close to the end of regular passenger service operations.

During the journey the analyzed train crosses several critical junctions: Olten (merging lines to Bern both from Basel and from Zurich), Bern bypass (merging lines from Biel and from Bern station), Spiez (diverging lines to Loetscheberg tunnel and Interlaken) and Brig (diverging lines to Simplon tunnel and to the east side of Switzerland).

The train trajectory provided by BLS includes several stops: Teknau (5 min), Olten-Rothrist (3 min), Spiez (21 min), Brig (4 min). To consider the real behavior of the specific train, these stops will compose the set M of possible stops for the generation of the optimized solution. The stop in Teknau is planned to allow a train overtaking; the stop in Olten-Rothrist is planned for train crossing. The long stop in Spiez is because of the rail traffic of the Loetscheberg tunnel, which is partly on single track. In Brig, the train had a 4-min stop because of a crossing train before it entered the Simplon tunnel. The total travel time was about 230 min.

A planned stop in the Bern area, toward Spiez, was avoided. This is a common procedure when freight trains arrive earlier than scheduled and are able to continue without affecting the successive train.

Optimized Rail Traffic Solution

Figure 4 illustrates the planned, realized, and optimal solutions. The optimized solution considers the Basel SBB marshaling yard (BRBD) as the starting point and Domodossola (Dom.) as the final station. Dotted lines before BRBD represent the approaching trajectories

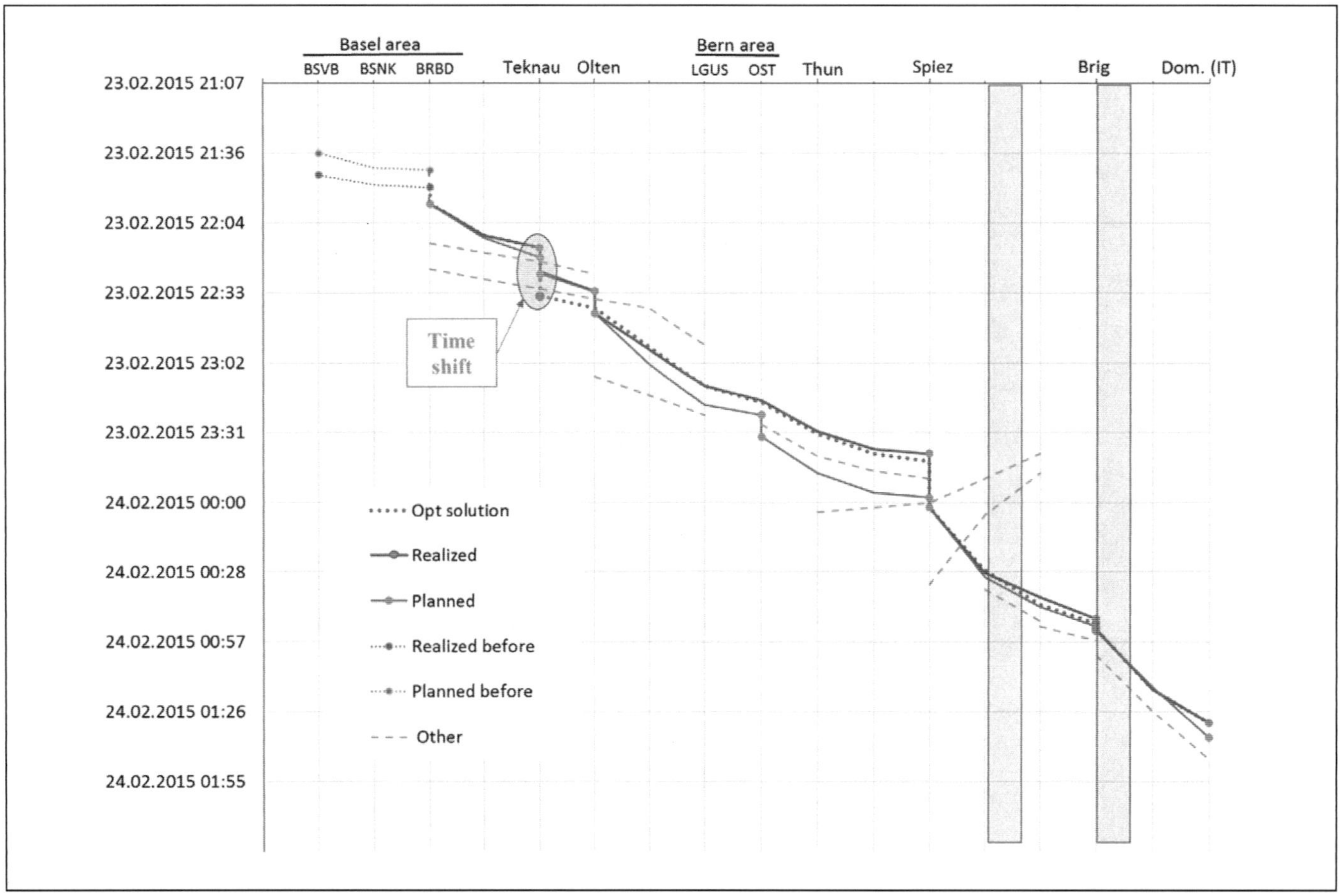

Figure 4. Case study trip time distance diagram showing selected other traffic (shown as "Other"); the departure time shift in Teknau is highlighted.

(Planned/Realized) to Basel: these were not considered in the case study modeling.

The most significant change between the optimized and realized solutions is the time shift of 10 min in Teknau. This allows the freight train to avoid stopping in Olten. New trajectories based on speed profiles optimized to reduce energy consumption have been generated. An important consideration when generating optimized speed profiles is the ability of drivers to absorb and react to instructions. Therefore, it is necessary to minimize the number of instructions. For example, one cruising speed should be assigned to several adjoining track sections even if they have different characteristics.

This contrasts with the microscopic simulation environment that is used in both rescheduling and train operation optimization procedures. Indeed, the choice of aggregating data at a high level versus using more detailed output increases the difference between model output and the optimal solution.

The existing rail traffic is known. The model considers time supplements and buffer times following Union Internationale des Chemins de fer (UIC) guidelines (*21*). The solution was constrained based on a maximum delay of 3 min, that is, the amount of time a train can be delayed and still considered punctual, although some specific freight trains could be considered punctual even with a 10-min delay. The optimized solution is based on new scheduled departure times and new cruising speed targets. In the case study example, the optimized solution was able to obtain an energy saving of circa 12%.

Table 1 presents the most relevant data for comparing actual with optimized scenarios: travel times and energy consumption. It is worth also noting that a train following the proposed energy saving solution would arrive on time (see last value of Table 1: the optimized solution arrives within 1 min of actual arrival time).

In Table 1, values in brackets of the optimal solution refer to the actual stop-to-stop trips to compare the optimized results with the measured values.

The computing time for developing the optimized solution was 6 min. Both the optimization problems and the simulation model have been implemented in Matlab on an Intel I7 2.1 GHz laptop. The computation time depends on the track length and number of cruising

Table 1. Stop-to-Stop Travel Times and Actual Energy Consumption versus Energy Optimized Solution

	Basel–Teknau		Teknau–Olten		Olten–Spiez		Spiez–Brig		Brig–Dom. (IT)	
	(kW-h)	(min.s)	(kW-h)	(min.s)	(kW-h)	(min.s)	(kW-h)	(min.s)	(kW-h)	(min.s)
Realized	900	17.42	1412	8.12	4195	55.24	6014	46.21	6892	40.18
Optimal	848	18.15	(1724)	(7.39)	3501	(58.23) 69.02	5174	47.14	6085	41.16

speeds along the examined section. In the case study, manual driving has been assumed, which leads to several related assumptions:

- The presence of a driving assistance system. A system like the ADL used by the SBB has been assumed. The ADL system gives recommendations on cruising speeds only. This means there is no indication with regard to other motion phases (e.g. coasting). Therefore, coasting phases have not been considered.
- The number of instructions. In conflict-free conditions, every change of infrastructure characteristic along the journey could be associated with an optimal speed value. This leads to a large list of instructions for drivers that is difficult to follow. In the case study, the number of energy saving cruising speeds was set equal to the number of different speed limits.
- The values of energy saving cruising speeds. The suggested speed values should be easily understood by drivers. Therefore, speeds are presented in whole numbers (no decimals) and there are only small speed variations (less than 2 km/h) for successive sections.
- The simulation tool. The optimization process is based on energy consumption estimated via simulation. For the related computation, train builder data on traction unit and default values of Davis formula for resistances have been used.

Speed profiles and related energy consumption profiles are shown in Figure 5. Simulation tools are not able to replicate real driving behavior and the speed variations from a change of gradients, curves, and so forth. This is due to the entity of power used; whereas there is some reserve of power (e.g. in cruising regimes), the simulation tool covers a possible increase in resistance immediately by providing an increase of power. As an aside, a possible delay in response would give results that are more realistic. Furthermore, it seems that simulation tools overestimate the energy consumption; this is quite evident, especially during the acceleration phase, although the same acceleration rate has been used. In the authors' opinion, the issue concerns calibration of resistance parameters. Although Davis formulation is undoubtedly valid, its parameters refer to a different technology; therefore, an "actualization" is required.

Even with these assumptions, the proposed model can be used to develop optimized solutions for freight train operation.

The feedforward nature of the model proposed here is ensured by a robustness analysis that can be performed offline. The analysis has been conducted on the Teknau–Spiez track, which is the one subject to departure time shift. By varying the departure time as an initial delay, a set of 1,000 stochastic simulations has been performed. The initial delay values have been extracted from their distribution according to SBB statistics (*21*). Stochastic simulations represent the different ability of drivers to follow instructions; speed values have been extracted uniformly within the interval [–5 km/h, + 5 km/h] defined by each optimal speed target of the proposed solution, with respect to speed limits.

Figure 6 shows the trend of arrival delays given by the simulations set. When the initial delay is around 5.5 min, the arrival delay increases by almost 9 min from the previous value. In fact, with the above-mentioned initial delay the train has to stop in Bern area (OST) and let the successive train pass as planned (i.e. it must follow the orange trajectory in Figure 4 from OST to Spiez). However, delays equal or higher than the above-mentioned value have a very low occurrence probability, as shown in the heat-chart (data from International Union of Railways (*22*)).

Discussion of Results

The case study results show that it is possible to generate a disposition timetable that reduces energy consumption by modifying speed profiles and making small shifts in departure times, while still respecting the surrounding traffic conditions as planned. Given the feedforward nature of the procedure, the proposed solution must be shown to be robust, that is, the solution should be able to keep the surrounding traffic conditions unaltered.

It is clear that possible variation of the surrounding traffic may affect the applicability of the solution; however, the only critical point is where the shift of the

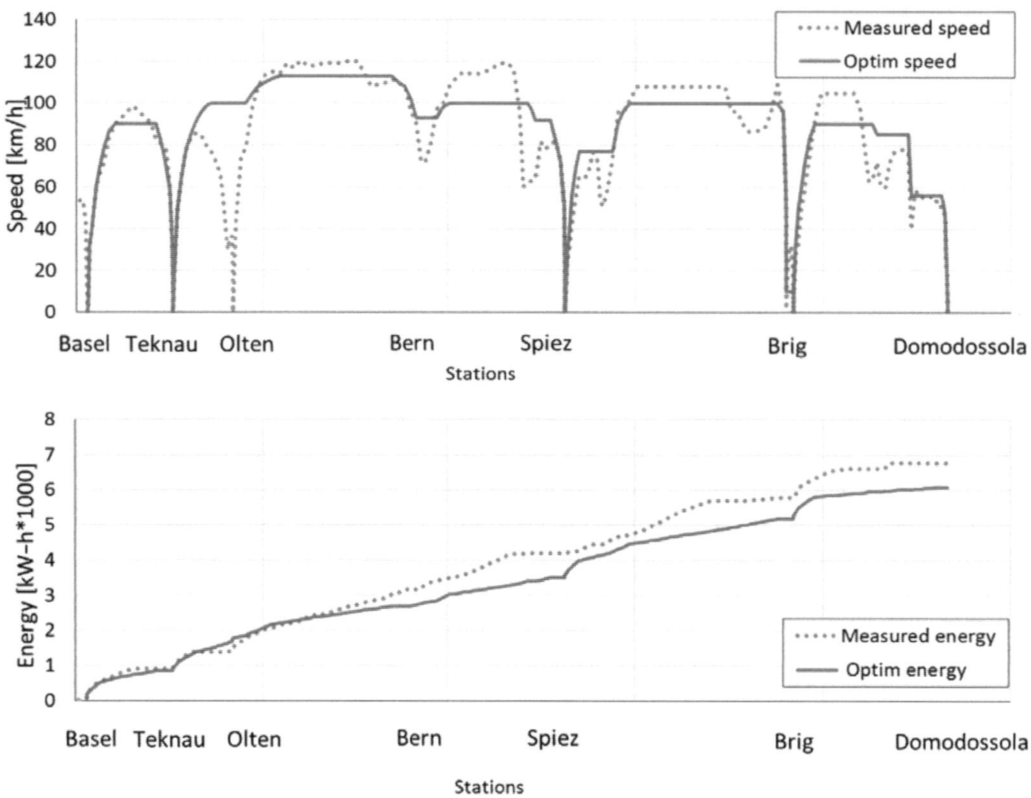

Figure 5. Comparison between real measurements and the optimized solution in terms of speed and energy consumption.

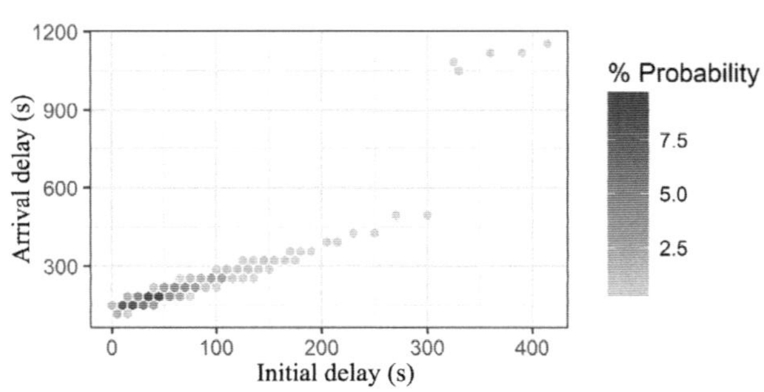

Figure 6. Robustness evaluation of the solution considering the distribution of initial delay in Switzerland. Spots represents the set of simulations; the tones of green represent the probability distribution according to SBB statistics (*22*).

departure time occurs (Teknau). Here, if the overtaking train is delayed, there would also be a delay for the freight train being optimized. In this case, it should be possible to use real-time rescheduling to generate a new optimized speed profile taking into consideration the delayed train.

The energy optimal solution's robustness can be evaluated through a set of simulations, in which the random variation of performance is generated by varying cruising speed values. In this way, it is possible to consider the variation of drivers' capability to follow the prescribed speeds. However, the relevance of this aspect requires a more specific evaluation of robustness and therefore it will be the subject of further research.

Furthermore, the proposed optimization model is based on the idea of including simulation processes within the optimization procedure. The research shows this approach has several advantages. Above all, simulation allows all possible aspects (and their variability) of real operation to be considered in developing the optimized solution. Although the processes for data retrieval and scenario construction are still relatively slow, the increasing use of common data exchange formats is improving the ability to import entire railway networks, even at a microscopic level, into simulation environments.

Conclusions and Further Developments

The feedforward tactical optimization model proposed in this paper enables railways to consider specific measures for increasing freight train energy efficiency: stop avoidance, itinerary changes, and early departures. These measures are often neglected by rescheduling approaches focused on passenger traffic. In a case study, the model was able to reduce estimated energy consumption by up to 12% based on real operating conditions and specific strategies. Nevertheless, it is important to highlight that this level of energy savings represents ideal driving conditions, which are difficult to approach without onboard driving assistance systems.

By using energy consumption data measured on board, and train trajectories, it will be possible to improve the model's calibration for both train motion and energy consumption. Earlier research is based on ideal conditions estimated through simulation-based approaches and some specific data (e.g. train position on the track, track occupancy). In contrast, further research will consider the detailed energy consumption related to the track position. This will lead to the development of new methodologies that will better specify the train motion modeling and, at the same time, will enhance current energy-efficient strategies.

This research is being extended by examining methods for speeding up the computing time for real-time applications and by the integration of rescheduling procedures with speed profile optimization. The main idea for the latter activity is to define an appropriate bi-level solution with reference to departure time and optimal speed profile. From the rail operation perspective, the introduction of stop priorities will be considered for critical points of the network, such as between two areas managed by different traffic control centers.

Acknowledgments

This research project was financially supported by the Swiss Innovation Agency Innosuisse and is part of the Swiss Competence Center for Energy Research SCCER Mobility. The authors are grateful to BLS Cargo AG for their help in data collection.

Author Contributions

The authors confirm contribution to the paper as follows. Study conception and design: De Martinis, Toletti, Corman; data collection: De Martinis, Weidmann; analysis and interpretation of results: De Martinis, Toletti, Corman, Weidmann, Nash; draft manuscript preparation: De Martinis, Toletti, Corman, Weidmann, Nash. All authors reviewed the results and approved the final version of the manuscript.

References

1. Hoffrichter, A., J. Silmon, S. Iwnicki, S. Hillmansen, and C. Roberts. Rail Freight in 2035 –Traction Energy Analysis for High-Performance Freight Trains. *Proceedings of the Institution of Mechanical Engineers, Part F: Journal of Rail and Rapid Transit*, Vol. 226, 2012, pp. 568–574.
2. Liu, R., and I. M. Golovitcher. Energy-Efficient Operation of Rail Vehicles. *Transportation Research Part A: Policy and Practice*, Vol. 37, 2003, pp. 917–932.
3. Albrecht, A. R., P. G. Howlett, P. J. Pudney, and X. Vu. Energy-Efficient Train Control: From Local Convexity to Global Optimization and Uniqueness. *Automatica*, Vol. 49, 2013, pp. 3072–3078.
4. Wang, Y., B. De Schutter, T. J. J. van den Boom, and B. Ning. Optimal Trajectory Planning for Trains – A Pseudospectral Method and a Mixed Integer Linear Programming Approach. *Transportation Research Part C: Emerging Technologies*, Vol. 29, 2013, pp. 97–114.
5. Bocharnikov, Y. V., A. M. Tobias, C. Roberts, S. Hillmansen, and C. J. Goodman. Optimal Driving Strategy for Traction Energy Saving on DC Suburban Railways. *IET Electric Power Applications*, Vol. 1, 2007, pp. 675–682.
6. D'Ariano, A., and T. Albrecht. Running Time Re-Optimization During Real-Time Timetable Perturbations. *WIT Transactions on the Built Environment*, Vol. 88, 2006, pp. 531–540.
7. Albrecht, T., C. Gassel, A. Binder, and J. Van Luipen. Dealing with Operational Constraints in Energy Efficient Driving. *IET Seminar Digest*, Vol. 2010, 2010, pp. 22–28.
8. Mazzarello, M., and E. Ottaviani. A Traffic Management System for Real-Time Traffic Optimisation in Railways. *Transportation Research Part B: Methodological*, Vol. 41, No. 2, 2007, pp. 246–274.
9. Luthi, M. *Improving the Efficiency of Heavily Used Railway Networks through Integrated Real-time Rescheduling*. PhD thesis, ETH Zurich No. 18615, 2009.
10. Smith, P., M. Kyriakidis, A. Majumdar, and W. Ochieng. Impact of European Railway Traffic Management System on Human Performance in Railway Operations. *Transportation Research Record: Journal of the Transportation Research Board*, 2013. 2374: 83–92.
11. Sicre, C., A. P. Cucala, A. Fernández, and P. Lukaszewicz. Modeling and Optimizing Energy-Efficient Manual Driving on High-Speed Lines. *IEEJ Transactions on Electrical and Electronic Engineering*, Vol. 7, No. 6, 2012, pp. 633–640.
12. Lukaszewicz, P. Energy Saving Driving Methods for Freight Trains. *Advances in Transport*, Vol. 15, 2004, pp. 901–909.
13. Zhao, N., C. Roberts, S. Hillmansen, and G. Nicholson. A Multiple Train Trajectory Optimization to Minimize Energy Consumption and Delay. *IEEE Transactions on Intelligent Transportation Systems*, Vol. 16, No. 5, 2015, pp. 2363–2372.
14. De Martinis, V., and U. A. Weidmann. Definition of Energy-Efficient Speed Profiles Within Rail Traffic by Means of Supply Design Models. *Research in Transportation Economics*, Vol. 54C, 2015, pp. 41–50.
15. Islam, D. M. Z., S. Ricci, and B.-L. Nelldal. How to Make Modal Shift from Road to Rail Possible in the European Transport Market, As Aspired to in the EU Transport White Paper 2011. *European Transport Research Review*, Vol. 8, No. 3, 2016, p. 18.

16. Voelker, M. *Adaptive Lenkung bei den SBB.* 5th IT15 Rail Conference, June 11-13, Zurich, Switzerland.
17. New York Air Brake, *LEADER* http://www.nyabproducts.com/wp-content/uploads/2015/08/LEADER_InfoCard.pdf
18. Hansen, A., and J. Pachl. Conclusions. In *Railway Timetabling and Operations: Analysis, Modelling, Optimisation, Simulation, Performance, Evaluation* (I. A. Hansen, and J. Pachl, eds.), Eurail Press, Hamburg, 2014, pp. 307–310.
19. Lamorgese, L., and C. Mannino. An Exact Decomposition Approach for the Real-Time Train Dispatching Problem. *Operations Research*, Vol. 63, No. 1, 2015, pp. 48–64.
20. Federal Office of Transport. *Official Timetable Switzerland.* http://www.fahrplanfelder.ch/en/archives/graphic-timetables.html.
21. UIC Code 451-1. *Timetable Recovery Margins to Guarantee Timekeeping – Recovery Margins.* International Union of Railways, Paris, France, 2000.
22. Ullius, M. Delay Analysis of Rail 2005 1st Phase Using OpenTimeTable. *Proc., 1st International Seminar on Railroad Operations Modelling and Analysis*, Delft, 2005.

The Standing Committee on Railroad Operating Technologies (AR030) peer-reviewed this paper (18-01196).

Sage Publications USA
sagepub.com

ISBN 978-1-5108-8179-2
9 781510 881792